Beutel, Friedrich, Ge, Yang
Insect Morphology and Phylogeny
De Gruyter Graduate

Rolf G. Beutel, Frank Friedrich, Si-Qin Ge,
Xing-Ke Yang

Insect Morphology and Phylogeny

A textbook for students of entomology

DE GRUYTER

Authors

Rolf G. Beutel
Friedrich-Schiller-Univ., Institut f.
Spezielle Zoologie u. Evolutionsbiologie
Erbertstr. 1
07743 Jena
Germany

Si-Qin Ge
Chinese Academy of Sciences
Institute of Zoology
1 Beichen West Road, Chaoyang
100101 Beijing
China

Frank Friedrich
Universität Hamburg
Biozentrum Grindel u. Zoologisches Museum
Martin-Luther-King-Platz 3
20146 Hamburg
Germany

Xing-Ke Yang
Chinese Academy of Sciences
Institute of Zoology
1 Beichen West Road, Chaoyang
100101 Beijing
China

ISBN 978-3-11-026263-6
e-ISBN 978-3-11- 026404-3

Library of Congress Cataloging-in-Publication data
A CIP catalog record for this book has been applied for at the Library of Congress.

Bibliographic information published by the Deutsche Nationalbibliothek
The Deutsche Nationalbibliothek lists this publication in the Deutsche Nationalbibliografie; detailed bibliographic data are available in the Internet at http://dnb.dnb.de.

Typesetting: Werksatz Schmidt & Schulz GmbH, Gräfenhainichen
Printing and binding: Hubert & Co. GmbH & Co. KG, Göttingen
Cover image: *Mengenilla moldrzyki* Pohl et al., 2012 (Strepsiptera, Mengenillidae)
Nikon D 90, 25 mm Zeiss Luminar macro lense, adjustable extension bellows, flashlight.
Combined stack of several partially focused images using Helicon Focus. Courtesy H. Pohl
♾ Printed on acid-free paper
Printed in Germany

www.degruyter.com

The carpet makers of Isfahan deliberately knot tiny flaws into their rugs, because perfection is an attribute reserved for God.

We dedicate this book to Niels-Peder Kristensen who has set a shining example in insect morphology and phylogeny.

Foreword

This book emerged from a close cooperation between scientists from the Institute of Zoology of the Chinese Academy of Sciences (CAS) and two German institutions, the Institut für Spezielle Zoologie und Evolutionsbiologie mit Phyletischem Museum of the Friedrich-Schiller-Universität Jena and the Biozentrum Grindel & Zoologisches Museum of the University of Hamburg. Between these institutions, joint research projects have focused on insect anatomy and innovative morphological techniques and on the phylogeny and evolutionary history of different hexapod lineages. Our progress and interest in these topics are reflected in the contents of this work.

The tremendous importance of Hexapoda was highlighted in numerous contributions and will not be treated in detail here. However, the most outstanding feature of this clade is its unparalleled diversity. With approximately 1,000,000 described species, they comprise more than half of the known total species diversity on this planet. However, what is presently known is apparently only the tip of the iceberg. Estimates of the real diversity range between 2 million species and a staggering number of 30 million. Hexapod species often occur in extremely dense local populations and can form an immense biomass. Up to 100,000 springtails in only one m^3 of forest topsoil or millions of mosquitos forming gigantic swarms are only two examples of such incredible population density, among many others. Hexapods are largely and primarily missing in marine habitats, but they play a crucial role in nearly all terrestrial ecosystems and occur in a broad variety of limnic habitats. They have a huge impact on human health as vectors of many diseases (e.g., malaria, sleeping sickness), and many species are important plant pests or pests of stored products. Positive aspects of hexapods include their role as predators or parasitoids of pest species (mostly pest insects) and as pollinators of plants including important crops. Insects are an important food source for numerous animal species and traditionally also for humans in many parts of the world. Last but not least, the production of silk and honey have been important economic factors going back several thousand years. The combination of unusually complex morphology, fascinating biology, remarkable species attractiveness and charisma, economic and medical impact, and various other aspects have made hexapods a highly attractive group for researchers and dedicated amateurs for centuries. Moreover, the grave threat posed by an unparalleled biodiversity crisis to the seemingly inexhaustible hexapod diversity presents one more very serious reason to intensify the study and detailed documentation of this fascinating group of organisms.

Insect morphology was a flourishing discipline in the first two thirds of the 20th century, with outstanding researchers such as J. Chaudonneret and H. Weber in European countries, but also excellent entomologists in other parts of the world. Morphology based systematic entomology arguably reached a peak with the publication of Willi Hennig's groundbreaking work "Die Stammesgeschichte der Insekten" in 1969. Towards the end of the 20th century, the detailed anatomical study of insects became less and less popular, a development apparently linked to the rise of molecular systematics. However, in the last ten years innovative techniques and new theoretical concepts (e.g., "evolutionary morphology") have led to a remarkable renaissance of the investigation of structures and functions of Hexapoda.

Molecular systematics has "evolved" with breathtaking momentum in the last ten years (see e.g., 1KITE.org). Robust "molecular phylogenies" will likely be available for Hexapoda and other groups of organisms in the very near future. Nevertheless, morphology will continue to play a vital role for different reasons. First of all, it provides an independent source of information for critically evaluating molecular trees (and vice versa), a procedure referred to as the "model of reciprocal enlightenment" by W. Hennig. Organisms cope with their environment using their morphological structures, which are the main target of natural selection. Body functions cannot be understood without solid morphological data, and detailed and meaningful evolutionary scenarios cannot be developed without knowing the changes on the phenotypic level. Another obvious reason is that morphology is the only source of information regarding fossils. To reconstruct the evolution of Hexapoda in its historical dimension is only possible using morphological data for the placement of extinct taxa.

The primary purpose of this book is to provide a comprehensive overview of hexapod morphology, mainly, but not exclusively, for investigations in an evolutionary context. On one hand an overwhelming richness of available data is made easily accessible here, including also extensive and highly valuable sources in non-anglophone languages (see below). On the other hand, extensive results of our own morphological investigations are integrated in this volume, including comprehensive tables of muscles with recently introduced nomenclatures, high quality SEM micrographs, and computer-based 3D reconstructions. The second main aim is to outline the state of the art in hexapod phylogenetics. The almost unprecedented progress in hexapod systematics in the last years, arguably comparable to Hennig's "Stammesgeschichte der Insekten", provides an almost ideal background. Long disputed questions, such as the position of Strepsiptera ("the Strepsiptera problem"), are now settled, and it is probably not overoptimistic to assume that a more or less completely resolved hexapod phylogeny (on the interordinal level) will be available in the very near future. In this context it should be emphasized that this is not only owed to the immense progress in molecular systematics, but also to several coordinated morphology-based projects, including phylogenetic studies of Polyneoptera and Holometabola.

The first main part of this book covers general hexapod morphology (**1. Morphology**) which is followed by a concise treatment of the development and immature stages (**2. Reproduction, development and immature stages**) and an extensive glossary (**3. Glossary of hexapod morphology**). A broad spectrum of traditional and innovative morphological techniques is described briefly in the next chapter (**4. Traditional and modern morphological techniques**) followed by a brief introduction into morphology-based phylogenetics (**5. Phylogenetic reconstruction based on morphology**). The second main part (**6. The hexapod orders**) covers all currently recognized hexapod orders and their systematic relationships. The main focus in the ordinal chapters is on the morphology, but these chapters also contain shorter sections on the distribution and diversity, taxonomy, biology, reproduction, fossil record, and economic importance of the different orders.

The information presented in this volume is based on numerous sources (see **7. Literature**). Works extensively used are Snodgrass' classical "Principles of Insect Morphology", the German "Handbook of Zoology" series (De Gruyter), the "Traité de Zoologie" (edited by P.P. Grassé), some textbooks in German language (e.g., "Entomologisches Praktikum", G. Seifert), and last but not least "Evolution of the Insects" by D. Grimaldi and M. Engel. It should be emphasized that numerous specialists have made valuable contributions to this volume by carefully reviewing chapters (see Acknowledgements). Few chapters were written by invited specialists Assoc. Prof. Dr. M. Bai (Chinese Academy of Sciences), Dr. B. Wipfler, and Dipl. Biol. K. Schneeberg (Institut für Spezielle Zoologie und Evolutionsbiologie, University Jena).

This book addresses students of entomology, especially those interested in morphology, phylogeny and evolution, but also researches dealing with hexapod systematics or other aspects of entomology. A slightly modified Chinese version of this book is presently in preparation. We hope that this contribution will not only promote the study and investigation of insect morphology and evolution but also stimulate international exchange and joint research projects in systematic entomology and related disciplines.

Rolf G. Beutel
Frank Friedrich
Si-Qin Ge
Xing-Ke Yang

Acknowledgements

To finish this book project would not have been possible without the help of many persons. First of all we would like to acknowledge the invaluable contribution made by Hans Pohl (FSU Jena) by providing numerous SEM micrographs and macro light images of superb quality, including the one used for the cover photo, and also extremely helpful comments on many chapters. Numerous drawings of outstanding quality were provided by Wenzhu Li, which is also greatly appreciated. Additional valuable images were made available by Romano Dallai (Universita di Siena), Ryuichiro Machida, Yuta Mashimo, Apisit Thipaksorn (Sugadaira Montane Research Center), Reinhard Predel (Universität zu Köln), Rico Spangenberg, Lars Möckel, Katrin Friedemann and Frank Hünefeld (FSU Jena). The chapters of this book were carefully reviewed by colleagues, almost all of them externally. This was crucial for the success of the project. The external reviewers include Horst Aspöck (Universität Wien) (Neuropterida), Ulrike Aspöck (Naturhistorisches Museum Wien) (Neuropterida), Matt Bertone (North Carolina State University) (Diptera), Alexander Blanke (Universität Bonn) (Odonata), Sven Bradler (Universität Göttingen) (Phasmatodea), Daniel Burckhardt (Coleorrhyncha, Sternorrhyncha), Fabian Haas (Dermaptera), Bruce Heming (University of Alberta) (development, Thysanoptera), Wieland Hertel (FSU Jena) (morphology), Frank Hünefeld (FSU Jena) (Lepidoptera), Karl Kjer (Rutgers University) (Trichoptera), K.-D. Klass (Senckenberg Naturhistorische Sammlungen Dresden) (techniques, Dictyoptera, Mantophasmatodea), Markus Koch (University Hamburg) (entognathous orders), Günther Köhler (FSU Jena) (Orthoptera), Lars Krogmann (Naturkundemuseum Stuttgart) (Hymenoptera), Robert E. Lewis (Iowa State University) (Siphonaptera), Ryuichiro Machida (University of Tsukuba) (development, Archaeognatha, Zygentoma), Rudolf Meier (National University of Singapore) (claclistics), Christine Mißbach (MPI for Chemical Ecology, Jena) (central nervous system), Carsten Müller (Universität Greifswald) (eyes), Michael Ohl (Naturkundemuseum Berlin) (Neuroptera), Thomas Simonsen (The Natural History Museum, London) (Lepidoptera), Adam Ślipiński (CSIRO Canberra) (Coleoptera), Hans Strümpel (Universität Hamburg) (Sternorrhyncha, Auchenorrhyncha), Claudia Szumik (CONICET, Argentina) (Embioptera), Gert Tröster (Universität Göttingen) (Phthiraptera), Christiane Weirauch (University of California, Riverside) (Hemiptera), Brian Wiegmann (North Carolina State University) (Diptera), Kazunori Yoshizawa (University of Hokkaido) (Embioptera, Zoraptera, Psocoptera), and Peter Zwick (MPI of Limnology, Plön) (Plecoptera). We are very grateful for the immense support by these highly competent reviewers, which helped greatly to improve the quality of the book. Jasmin Tannert did the lettering of all drawings. Renate Walter helped with the preparation of SEM samples. Janin Naumann and Rommy Petersohn provided excellent microtome sections, which were important for different parts of the book. Kathrin Streitberger helped proofreading the manuscript. This is also gratefully acknowledged, as is the multiple support with illustrations provided by Gunnar Brehm. Finally our great thanks go to the Chinese Academy of Sciences (Beijing) and the Friedrich-Schiller-Universität Jena for various kinds of supports for this project.

Contents

1 Morphology

1.1 Integument

1.1.1 Cuticle and epidermis

Like other euarthropods (Chelicerata, Myriapoda, crustaceans [probably paraphyletic]) Hexapoda (=insects in the widest sense)[1] are characterized by a differentiated exoskeleton formed by the external cuticle. It is composed of **sclerites, membranes**[2] and semimembranous areas. The cuticle is a biological composite material containing chitin, proteins, lipids and catecholamines (e.g., N-acetyl-dopamine). Catecholamines cross-link proteins and chitin filaments, which results in specific mechanical properties. The exoskeleton is usually robust in most areas and results in an improved mechanical protection of the body, but it also provides differentiated attachment areas for a complex muscular system. It is a precondition for the formation of a complex locomotor apparatus with true articulations and complex appendages (arthropodia), which was a key evolutionary innovation of Euarthropoda. Protection against desiccation is another function in most terrestrial arthropods, usually linked with the presence of an external wax layer (see below).

The cuticle does not only cover the surface. Endoskeletal structures are formed as ingrowths, referred to as **apodemes** if they are solid and as **apophyses** (or entapophyses) if they are hollow. They play an important role in most hexapods, especially as muscle attachment areas, but also increase the mechanical stability of certain body parts, such as the **tentorium** or **postoccipital ridge** in the head, or the **furcae** and **pleural ridges** in the thoracic segments. Internal organs such as the tracheae and fore- and hindgut are also covered by a very thin cuticle, the **intima**.

The cuticle is secreted by the single-layered epidermis, which is also referred to as **hypodermis** due to its position below the cuticle (Fig. 1.1.1.1). It is mainly formed by cubic or more or less strongly flattened cells with a basal lamina (0.2–0.5 µm), but contains also different types of gland cells (see **1.1.6 Integumental gland cells**), cells forming setae (**tormogen** and **trichogen cells**, see **1.1.4 Cuticular sensilla**), sensorial cells and **oenocytes**. The basal lamina is formed by epidermal cells but also by **plasmatocytes**. Its main components are collagen, glycoproteins and glycosaminoglycans (Chapman 1998). In contrast to most other cells of the hypodermis, the oenocytes have no contact with the cuticle. They are often large (more than 100 µm

---

1 Insects is the commonly used name for the entire Hexapoda (see title). In the text of the book we use consistently Hexapoda/hexapods for all insects including the entognathous orders (Collembola, Protura, Diplura) and Insecta/insects for Ectognatha, i.e. Archaeognatha, Zygentoma and Pterygota.

2 Terms in bold face in the parts 1 and 2 (Morphology and Development) are covered in the glossary (part 3).

in diameter) and characterized by a large nucleus, an extensive endoplasmatic reticulum, a low number of mitochondria, and crystalline inclusions. Oenocytes synthesize hydrocarbons that contribute to the epicuticle (Chapman 1998).

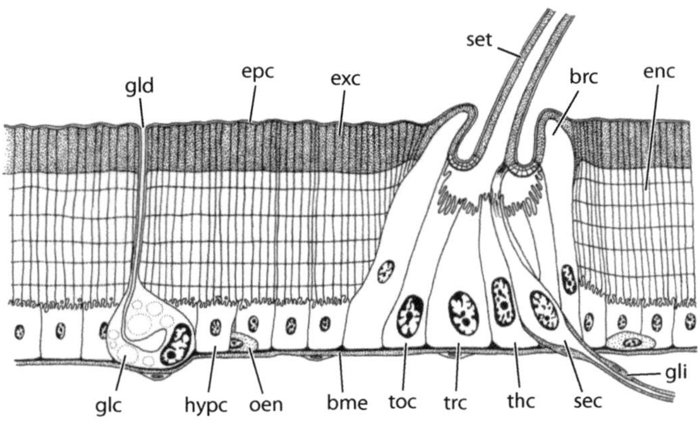

Fig. 1.1.1.1: Integument. Abbr.: bme – basement membrane, brc – basal ring cell, enc – endocuticle, epc – epicuticle, exc – exocuticle, glc – gland cell, gli – glia cell, hypc – hypodermal cell, oen – oenocyte, set – seta, sec – sensory cell, thc – thecogen cell, toc – tormogen cell, trc – trichogen cell. Redrawn from Seifert (1995).

The unmodified epidermal cells are held together by zonulae adherens near their apical regions and connected by septate junctions more basally. Desmosomes, hemidesmosomes and septate junctions also occur (Seifert 1995; Chapman 1998). The apical membrane forms a series of short ridges or projections (resembling microvilli) which are flattened apically. These plasma membrane plaques are the sites of the secretion of the **epicuticle** and chitin fibers (Chapman 1998). All epidermal cells have a glandular function as they secrete cuticle and also enzymes involved in its production and digestion.

The cuticle is composed of three layers, the external **epicuticle**, the **exocuticle**, and the internal **endocuticle** (Fig. 1.1.1.1). The two inner layers are initially secreted as a soft and more or less homogenous **procuticle** by the epidermis. It contains chains of alpha-chitin (poly-N-acetylglucosamine: $[C_8H_{13}NO_5]x$) connected by hydrogen bonds as larger units with a parallel arrangement, the micelles or microfibrils (2.5–3 nm). These are embedded in a matrix of silk-like and globular proteins. The micelles lie parallel to each other in each plane, but the arrangement differs in successive layers of the cuticle. A regular helicoidal arrangement in a series of lamellae is a typical pattern.

The poly-N-acetylglucosamine molecules form the main component of the procuticle. The matrix protein of the procuticle and endocuticle is the flexible and water-soluble **arthropodin**. In the outer layer a hardening process takes place involving dehydration and tanning mediated by phenoloxidases. This transforms the arthropodin into the rigid, brownish and water-resistant **sclerotin** of the exocuticle. A specialized matrix material is the rubber-like, highly elastic protein **resilin**. It occurs in sockets of true hairs (**setae**), in wings, and in mechanically highly active areas such as for instance the wing articulations.

The thickness of the epicuticle, which forms a multilayered external barrier, varies strongly (ca. 30 nm in culicid larvae, maximum ca. 4,000 nm). It is always free of chitin. In pterygote insects it is covered by a wax layer secreted by oenocytes. It is composed of paraffins and esters, which reduce water loss via evaporation. Its thickness varies between 10 nm and 1,000 nm and different surface modifications can occur (e.g., as whitish dust in Aleyrodoidea [white flies] and Coniopterygidae [dustywings]). An additional external cement layer occurs in some groups of insects (e.g., Blattaria), in some cases as an open meshwork. The very thin intermediate lamina of the epicuticle (ca. 15 nm) is mainly formed by the hardened protein **cuticulin**, which is similar to the sclerotin of the exocuticle. The homogenous inner layer is called the **dense lamina**. It is highly robust mechanically.

The exocuticle is strongly developed in the sclerites of the exoskeleton and can be half as thick as the entire cuticle in some cases. It is very strong under compressive forces, but comparatively weak under tension. It is very thin in the membranous areas, which are mainly formed by the endocuticle, which is flexible and able to resist tensile forces. Membranous areas occur at articulations but also on other body regions in most groups, notably between the segments (intersegmental membranes), in the pleurotergal (thorax) or pleural regions (abdomen), and on the ventral sides of the thoracic segments. The flexible parts have a higher proportion of chitin.

The hardening process transforming arthropodin into sclerotin takes place in several steps. **Prosclerotin** is an intermediate product. A cuticle where the tanning process terminates at an intermediate stage is referred to as **mesocuticle**. It is hardened but not fully pigmented, and can be stained with acid fuchsine (Chapman 1998). This type occurs in transition areas between sclerites and membranes (semimembranous areas).

1.1.2 Canals and pores

The endo-, meso- and exocuticle are perforated by **pore canals**. Very thin cellular processes of the hypodermis are involved in their formation during the secretion of the cuticle. They are usually withdrawn after the process is complete. The shape of the canals is often helical, following the arrangement of the chitin micelles in different layers. Single epidermal cells can form numerous pore canals, up to 200 in *Periplaneta americana* (Seifert 1995), which is equivalent to more than a million per mm^2. The lumen can vary between ca. 15 nm (*Periplaneta americana*) and 100 nm. The hypodermal cytoplasmic processes are maintained in the proximal parts of the channels in some groups, but usually very fine chitin filaments are formed in the lumen. In the exocuticle of sclerites they are often filled with chitin-free sclerotin, which increases the mechanical stability in the vertical direction (Seifert 1995). At the base of the epicuticle each pore canal divides into several branches, which are filled with wax and perforate the dense lamina and the cuticulin layer (Seifert 1995).

1.1.3 Surface structures, microtrichia and setae

The cuticle is usually characterized by fine surface patterns reflecting the arrangement of subtending hypodermal cells. A multitude of surface structures and modifications occur in different lineages. Minute spines or tubercles can be formed by the epicuticle alone, but the exocuticle is usually involved in the formation of surface structures. Simple surface modifications are the solid **microtrichia** (=**trichomes**), which are not articulated and not in contact with the hypodermis after their formation. They are never associated with sensory cells and are referred to as **acanthae** when they are formed by a single cell. Specialized microtrichia, usually with a soft, flexible cuticle and a spatulate apical part, form attachment devices of the hairy type (e.g., hairy soles of **tarsomeres**; see **1.3.5 Attachment structures**).

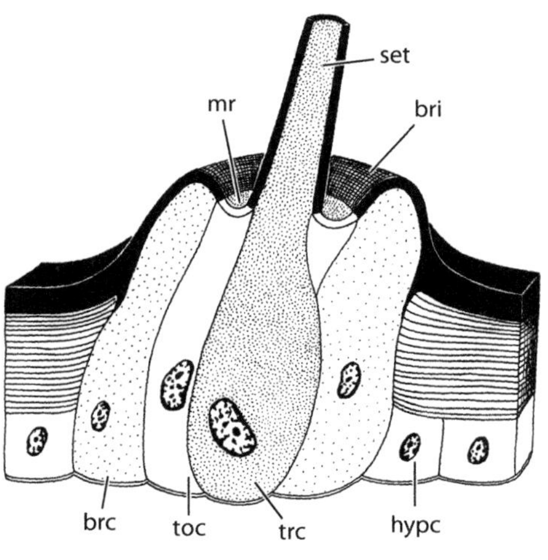

Fig. 1.1.3.1: Seta and associated cells, schematized. Abbr.: brc – basal ring cell, bri – basal ring, hypc – hypodermal cell, mr – membrane ring, set – seta, toc – tormogen cell, trc – trichogen cell. Redrawn from Seifert (1995).

Setae or true hairs articulate in a flexible socket or diaphragm permitting movement (Figs 1.1.1.1, 1.1.3.1). They are formed by a specific hypodermal cell, the **trichogene cell**, and are always in contact with the hypodermis. The socket is formed by the **tormogene cell** and has usually three layers, the external joint membrane, a ring of suspensory fibers, and the thin and fibrous socket septum. The joint membrane is enclosed by the cuticular basal ring. Setae are primarily hair-like but can be modified in many different ways, especially in the context of sensorial functions (see **1.1.4 Cuticular sensilla**). Cuticular scales occur in Collembola, Diplura, Archaeognatha and Zygentoma, and are possibly an apomorphic groundplan feature of Hexapoda. The scales of Lepidoptera and Archostemata (Coleoptera) are neoformations derived from hair-like setae. Club-shaped setae occur in 1st instar nymphs of Orthoptera and spear-shaped defensive hairs are present in larvae of Dermestidae (Coleoptera). Long and flexible swimming hairs occur in different lineages of aquatic beetles (e.g., Dytiscidae, Hydrophilidae) and also in aquatic groups of Heteroptera (e.g., Corixidae, Notonectidae).

1.1.4 Cuticular sensilla

Sensilla are the basal functional and structural units of cuticular mechanoreceptors and chemoreceptors. They include the cuticular component (e.g., the cuticle of a seta), the sensory neuron (or neurons), the associated sheath cells with the cavities they enclose and the structures they produce (Figs 1.1.1.1, 1.1.4.1) (Chapman 1998).

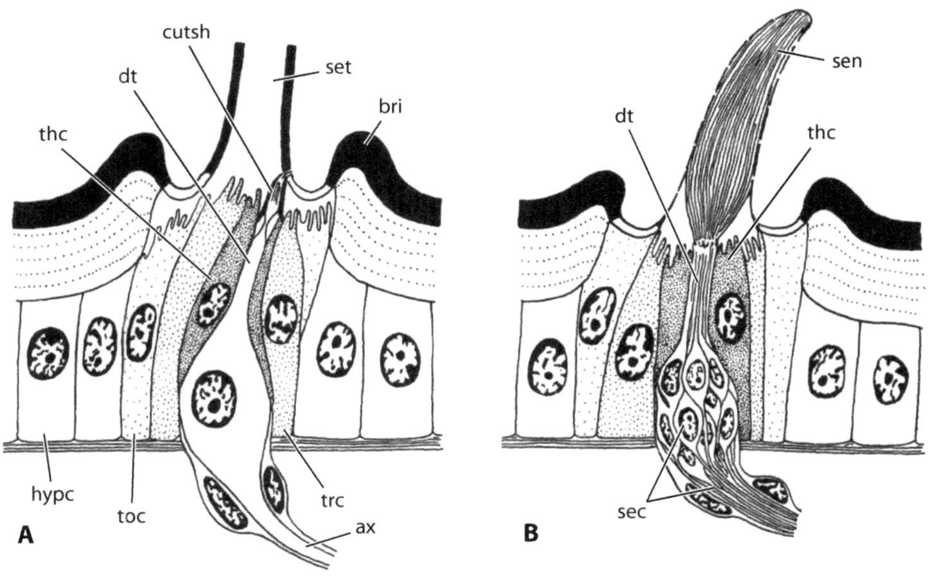

Fig. 1.1.4.1: Cuticular sensilla, schematized. A, mechanoreceptive seta; B, chemoreceptive sensillum. Abbr.: ax – axon, bri – basal ring, cutsh – cuticular sheath, dt – denrite, hypc – hypodermal cell, sec – sensory cell, sen – chemoreceptive modified seta, set – mechanoreceptive seta, thc – thecogen cell, toc – tormogen cell, trc – trichogen cell. Redrawn from Seifert (1995).

The least modified type is the **sensillum trichodeum** (hair sensillum). It is composed of a primary sensory neuron and a hair-like seta with a wall consisting of exocuticle and epicuticle (Figs 1.1.1.1, 1.1.4.1). The typical mechanoreceptive seta tapers from the base towards its apex. The displacement of the hair in the socket results in the neural stimulus (Chapman 1998). **Sensilla chaetica** are shorter and thick-walled hair-like mechanoreceptors. Very long and thin sensilla not tapering apically are sometimes referred to as **trichobothria**. They are not homologous with the true trichobothria occurring in Arachnida.

The cells involved in the formation of a sensillum are derived from the same hypodermal cell (sense organ precursor cell, sense organ mother cell). In addition to the **tormogen** and **trichogen cells**, a small **thecogen cell** is present between the latter and the neuron. It secretes a cuticular layer around the distal dendrite, the dendrite sheath, which usually ends at the base of the hair.

Modified types of cuticular sensilla are the apically rounded **sensillum basiconicum**, and the **sensillum campaniformium**, which scarcely protrudes beyond the basal ring enclosing it (Fig. 1.1.4.2). The **sensillum placodeum** is entirely flat and the **sensilla coeloconica** and **ampullacea** are sunk below the external cuticular surface to different degrees. The shape of the outer cuticular element of the sensillum does not necessarily indicate its function. Larger hair-like sensilla are in most cases mechanoreceptors, but can also function as contact chemoreceptors. A function as olfactory chemoreceptor is indicated by fine pores in the wall of the sensillum. Pores (apart from the molting pore) are always absent from sensilla solely functioning as mechanoreceptors (e.g., Chapman 1998). Therefore they are also referred to as aporous sensilla.

The sensorial neurons associated with sensilla trichodea are usually of the bipolar type (Figs 1.1.1.1, 1.1.4.1). Basally they are enclosed by the **neurilemma** (glial sheath). The distal dendrite is ensheathed by the thecogen cell and is strongly narrowed where it enters the base of the seta. Its distal part is characterized by one or two concentric rings of microtubule doublets ($9 \times 2 + 0$ pattern) and is consequently called the sensorial cilium. The receptive element is the membrane of its apical part, the tubular body, which comprises a bundle of numerous microtubules (between 30 and up to 1,000) connected by an electron-dense substance. The apical part of the tubular body is firmly embedded in the cuticle of the base of the sensillum trichodeum. The sensorial cilium (including the tubular body) is enclosed in the cuticular dendritic sheath, which is secreted by the thecogen cell. In contrast to mechanoreceptors, the distal dendrites of chemoreceptive cells do not insert into the cuticle at the base of the sensillum by way of a tubular body. Chemoreceptors are often equipped with several sensorial cells of different modality. Each individual dendrite can be enclosed in a tubular sheath but often all dendrites are enclosed within a multilocular dendritic sheath with one in each loculus. The dendrites usually branch after entering thin-walled olfactory sensilla, which are characterized by pores arranged in hexagonal groups. The diameter usually ranges between 15 and 20 nm. Each pore widens within the cuticle and forms a pocket.

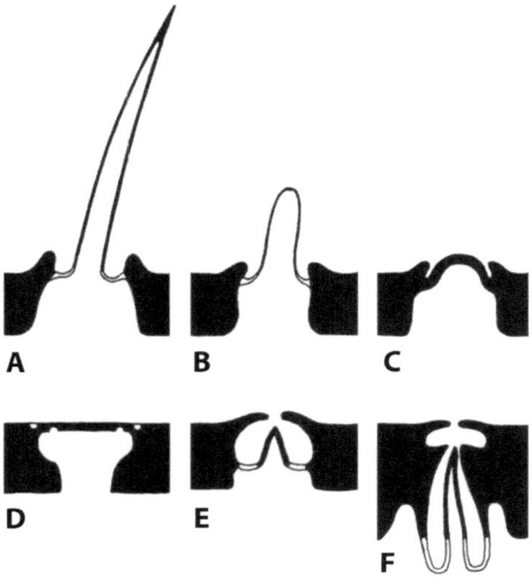

Fig. 1.1.4.2: Different types of cuticular sensilla. A, sensillum trichodeum; B, s. basiconicum; C, s. campaniformium; D, s. placodeum; E, s. coeloconicum; F, s. ampullaceum. Redrawn from Seifert (1995).

Extremely fine pore tubules (5–6 nm) originate from the pockets and end in the liquor of the sensillum adjacent to the dendritic membrane (Seifert 1995).

1.1.5 Scolopidia

Scolopidia are specialized internal (subcuticular) mechanoreceptors (Fig. 1.1.5.1) and probably derived from sensilla trichodea. They consist of the scolopale cell which is homologous to the thecogen cell, the scolopale cap cell which is equivalent to the trichogen cell (enclosed by a tormogen cell), and one or several bipolar sensory neurons. Scolopidia often function as proprioreceptors or are sensitive to vibrations of air or substrates.

An external hair-like element is not present. The cell body of the sensorial cell is sunk below the hypodermis. It is connected with it by the external **scolopale cap cell (=attachment cell)**, which is placed on top of the **scolopale cell** and the dendrite, which is proximally covered by a sheath cell and apically narrows to a cilium-like process containing a peripheral ring of nine microtubule doublets with proximally extending roots. The doublets are often (or perhaps always) connected with the cell membrane near their origin at the basal body by a structure called the ciliary necklace (Chapman 1998). The cuticular scolops is secreted by the scolopale cell and rests on the apical part of the dendrite like a cap. The scolopale cell contains the **scolopale**, which consists of fibrous material containing actin

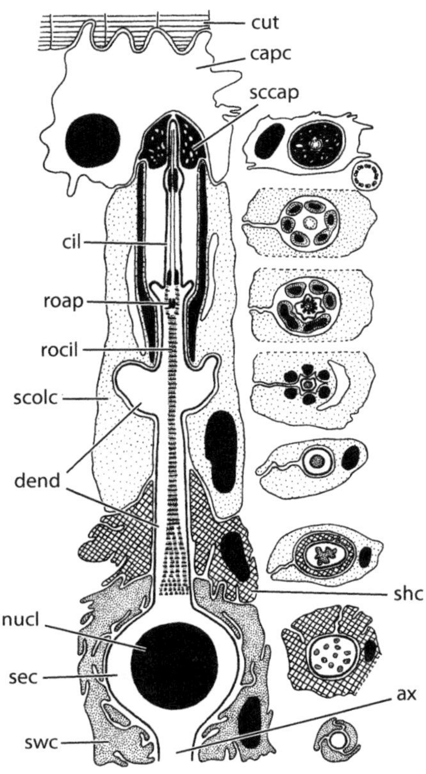

Fig. 1.1.5.1: Scolopidium. Abbr.: ax – axon, capc – cap cell, cil – cilium, cut – cuticle, dend – dendrite, nucl – nucleus, roap – root apparatus, rocil – root of cilium, sccap – scolopale cap, scolc – scolopale cell, sec – sensory neuron, shc – sheath cell, swc – Swann cell. Redrawn from Seifert (1995), after Gray (1960).

arranged in ring or a series of rods (scolopale rods) (Chapman 1998). In scolopidia of the subintegumental (=mononematic) type the distal end of the cap lies completely below the body surface. In integumental (=amphinematic) scolopidia the scolops is attached to the external cuticle by a thin cuticular terminal extending though a narrow fold of the scolopale cap cell.

Chordotonal (or scolopophorous) organs (Fig. 1.1.5.2) are formed by scoloparia, groups of scolopidia stretching between two movably connected sclerites. They are specialized mechanoreceptive organs (either proprioreceptive or exteroceptive). Examples are Johnston's organ in the second antennomere (pedicel) of Insecta, or the subgenual organs in the distal parts of the legs which perceive substrate vibrations.

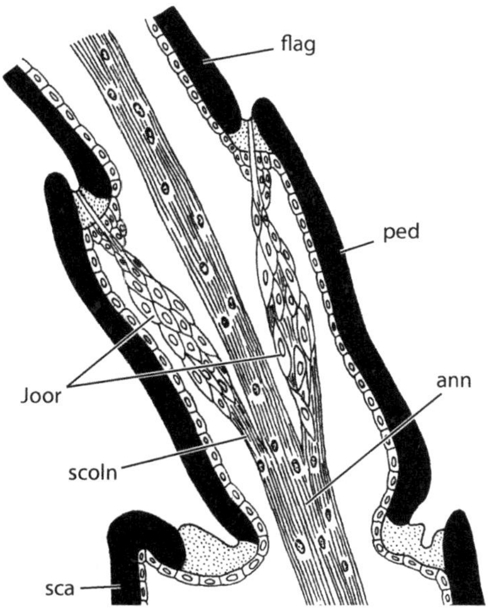

Fig. 1.1.5.2: Chordotonal organ, Johnston's organ of the pedicellus, *Melolontha vulgaris* (Coleoptera, Scarabaeidae).
Abbr.: ann – antennal nerve, flag – 1st flagellomere, Joor – Johnston's organ, ped – pedicellus, sca – scapus, scoln – scolopale nerve. Redrawn from Seifert (1995), after Snodgrass (1935).

1.1.6 Integumental gland cells

Gland cells associated with the integument are embedded in the hypodermis (Fig. 1.1.1.1: glc, see above) and much rarer than the unmodified epidermal cells surrounding them. They usually produce secretions permanently. The shape is more rounded compared to other cells of the hypodermis and they are often extended towards the body cavity. In the typical case the nucleus appears enlarged, irregularly lobate or star-shaped. Endopolyploidy is common. Three types of cells with specialized glandular functions were distinguished by Noirot & Quennedey (1974). Class 1 gland cells have the apical membrane produced as microvilli or lamellae, which are in direct contact with the cuticle secreted by themselves. They are often involved in the production of pheromones. Class 2 (only present in termites) and class 3 cells have no contact with the cuticle. Microvilli are present around vesicles structurally associated with a duct forming a connection to the exterior is absent in the former but present in the latter (Chapman 1998). The duct is an invagination of the external cuticular layer (Seifert 1995). Microvilli are absent from class 2 cells (Noirot & Quennedey 1974).

1.1.7 Ecdysis

The **ecdysis**, i.e. the molting, shedding and replacement of the cuticle, was traditionally considered as an autapomorphy of Arthropoda in the widest sense, i.e. also including Onychophora and Tardigrada ("Panarthropoda"). This interpretation was based on the Articulata-concept, with Annelida ("ringed worms") and arthropods as sistergroups. It is assumed today that it evolved earlier, as a derived groundplan feature of a clade Ecdysozoa, which includes Arthropoda and the Cycloneuralia (Nematoda, Nematomorpha, Priapulida, Kinorhyncha, Loricifera).

Due to mechanical properties of the cuticle the extensibility of the integument is very limited. Therefore, before reaching their maximum size and maturity, hexapods and other arthropods molt several times. During these intervals the integument, i.e. the hypodermis (=epidermis) + cuticle, undergoes a period of expansion, and this allows an increase in size of the body. However, ecdysis affects not only the body surface, but also endoskeletal elements (tentorium, furcae, pleural ridges etc.) and other chitinized internal invaginations such as the tracheae and also the ectodermal fore-and hindgut (see **1.9 Digestive tract**). The cuticle of these structures is also replaced during molts.

The succession of ecdyses divides the life cycle of hexapods and other arthropods into a series of stages or instars. The number of stages in the postembryonic development differs strongly between groups and depends on different factors, such as for instance, availability of food, temperature, or humidity. It is mainly taxon-specific and usually relatively constant but may even vary between individuals of the same species in some groups. Molting stops after maturity is reached in most groups of hexapods. However, this is not the case in the basal apterygote lineages, which are characterized by a large number of molts (e.g., up to 50 in Collembola). The number of ecdyses is still relatively high in basal pterygote orders (e.g., Ephemeroptera, Odonata, Plecoptera) but most insects molt only 4–6 times before reaching the adult stage. Ephemeroptera are the only insects molting as an immature winged instar, the subimago.

Molting comes at an evolutionary cost corresponding to the various benefits of a solid integument (see **1.1.1 Cuticle and epidermis**). Ecdyses are always critical intervals in the life cycle: hexapods and other arthropods lack their mechanical protection during this process and their mobility is strongly restricted. The condition of hexapods just after ecdysis is called teneral.

Molting starts with the **apolysis**, the separation of the old cuticle from the epidermis. This is induced by an increased level of ecdysteroids functioning as molting hormones. The size of the epidermal cells increases and a series of mitoses take place subsequently. Shortly before and immediately after apolysis vesicles within each epidermal cell release their electron-dense contents at the cell apex. These are principally enzymes involved in the degradation of the old cuticle or material for building the new one. The vesicles are still recognizable below the old endocuticle before they release their contents. A thin hyaline and homogenous lamina formed from the inner layers of the endocuticle is called ecdysial membrane. Due to a specific sclerotization process it is not affected by enzymes in the following stages of ecdysis. That part of the cuticle separating from the epidermis and the ecdysial membrane is referred to as the **exuvia** and the gradually expanding space below it as **ecdysial space**. The latter is filled with exuvial fluid secreted by the epidermal cells. The enzymes begin with the degradation of the old endocuticle after a short period of inactivity. The formation of the new cuticle is not affected due to the barrier provided by the ecdysial membrane. The degradation products of the old cuticle are absorbed by the epidermis and used to build the new exoskeleton in the following process (Dettner & Peters 2003).

The secretion of new cuticle starts during the degradation of the old one. This process is initiated by the formation of the epicuticular cuticulin layer on top of the projections or ridges of the apical membrane of the epidermal cells. After consolidation and hardening of this lamina the secretion of the new inner layers of the epicuticle and of the procuticle starts. The precise modalities of the hardening of the external procuticle, i.e. the formation of the exocuticle, are still disputed. Phenoloxidases (tyrosinases, laccases) apparently play a role in this process, but also in other functional contexts such as the repair of damaged cuticle or melanization (Dettner & Peters 2003). Movements of epidermal microvilli and plaques at the apices of these minute structures are probably responsible for the regular arrangement of chitin filaments.

The final stage of molting is the ecdysis in the narrow sense, the shedding of the old epi- and exocuticle. In most groups it splits open at the epicranial sutures (frontal- and coronal sutures) and the dorsomedian ecdysial line of the postcephalic tergites. Muscle contractions usually increase the haemolymph pressure in the anterior body, which results in the rupture of the old exoskeleton at the dorsal preformed zones of weakness. The teneral exoskeleton is soft, unpigmented and wrinkled. During this stage, the expansion of the body takes place involving locally increased haemolymph pressure (e.g., in the limbs) and often also air uptake, especially in larger hexapods. In the typical case the tanning process in the exocuticle results in the re-formation of a hardened and pigmented exoskeleton within a few hours after eclosion, but this process can also take days or even weeks in the members of some groups. After the normal mechanical properties of the cuticle are restored, the animal has regained its full mobility and mechanical protection.

[Snodgrass (1935); Gray (1960); Noirot & Quennedey (1974); Seifert (1995); Chapman (1998); Dettner & Peters (2003)]

1.2 Head

1.2.1 Segmentation, sutures and cephalic regions

The head (Figs 1.2.1.1, 1.2.2.1) is a compact and complex **tagma** with a concentration of a broad array of structures und functions. It is equipped with sense organs such as the highly complex **compound eyes** (Fig. 1.6.2.1) and the **antennae**, an elaborate set of mouthparts (Fig. 1.2.4.1), and a complex muscle apparatus (see **Table I** for all cephalic muscles). It also contains central elements of the nervous system, the anterior part of the digestive tract including the **cibarium** (preoral cavity) and the **salivarium**, and also **neurohaemal organs** (Fig. 1.11.2). It is likely that the hexapod head is homologous to a six-segmented **cephalon** which may represent a groundplan apomorphy of Euarthropoda (Chelicerata, Myriapoda, Pancrustacea [=Tetraconata, crustaceans and Hexapoda]). A six-segmented head was likely present in the extinct †Trilobita and other fossil lineages, and is preserved in the extant Myriapoda and Pancrustacea (fused with thoracic segments in most crustacean subgroups: cephalothorax). In contrast to †Trilobita, the primary head segmentation in hexapods cannot be traced directly by external segmental borders, but it can be deduced from the tripartite **cerebrum** (brain) and tripartite **suboesophageal ganglion** (Figs 1.5.1.1–1.5.1.3) and the appendages, i.e. the antennae, **mandibles**, **maxillae** and **labium** (the segmental affiliation of the **labrum** is disputed). Territories of the head were identified by different authors, but it was pointed out by Denis & Bitsch (1973) that they do not exactly correspond with the primary segments. Unlike in crustaceans, the 3rd head segment (=intercalary segment) of hexapods, which is associated with the **tritocerebrum**, lacks appendages. The posterior three segments, associated with the three parts of the suboesophageal ganglion and the mouthparts, are sometimes referred to as **gnathencephalon**. The "**occipital ridge**" was postulated as a border separating the maxillary and labial segments. However, this line often depicted in schematic drawings is either absent or only present as a relatively short dorsolateral furrow, which does *not* delimit a primary "occipital segment".

Different head regions separated by "**sutures**" can be distinguished in most hexapods (Fig. 1.2.1.1). However, it has to be noted that what is generally addressed as "sutures" comprises two very distinctly different structural modifications of the external cuticle (e.g., Wipfler et al. 2011):

a) Molting lines (**frontal** and **coronal sutures** [=epicranial lines or sutures]). In the following only these will be addressed as sutures.
b) Internal strengthening **ridges** (e.g., clypeofrontal "suture"). In the following they will be addressed as ridges or strengthening ridges. They are not lines of fusion or molting lines.

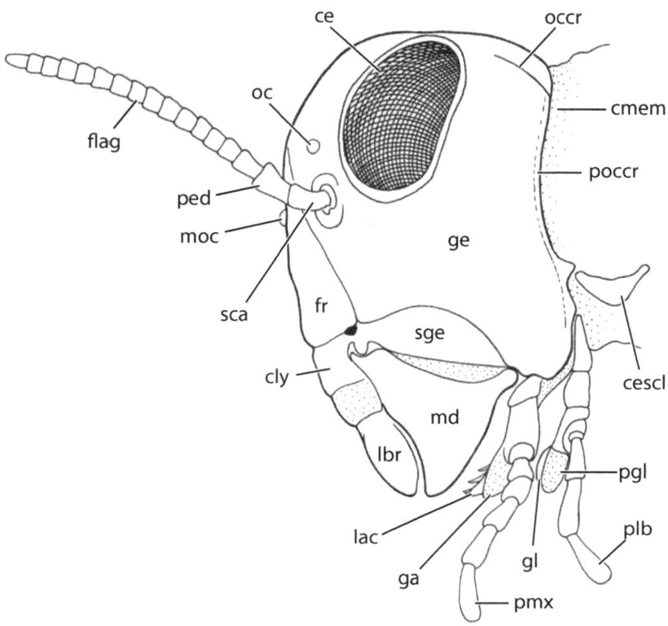

Fig. 1.2.1.1: Generalized hexapod head, lateral view. Abbr.: ce – compound eye, cescl – cervical sclerite, cly – clypeus, cmem – cervical membrane, flag – flagellum, fr – frons, ga – galea, ge – gena, gl – glossa, lac – lacinia, lbr – labrum, md – mandible, moc – median ocellus, oc – ocellus, occr – occipital ridge, ped – pedicellus, pgl – paraglossa, plb – palpus labialis, pmx – palpus maxillaris, poccr – postoccipital ridge, sca – scapus, sge – subgena. Redrawn from v. Kéler (1963).

1.2.2 Head capsule

The head capsule is almost always a rigid, well sclerotized structural unit, equipped with a defined set of appendages (Fig. 1.2.1.1). It is reinforced by strengthening ridges and the endoskeletal **tentorium** (Fig. 1.2.2.1). The shape is varying from more or less globular (e.g., Orthoptera) (Fig. 6.12.1) to strongly flattened and elongated (e.g., Raphidioptera) (Fig. 6.28.1A). It is connected with the prothorax by a more or less wide **cervical membrane**, often reinforced by articulatory **cervical sclerites**. The posterior head region can be fully exposed (e.g., Zoraptera, Hymenoptera, Diptera) or more or less strongly retracted into the prothorax (e.g., Blattodea, Coleoptera) (Fig. 6.29.1). The **foramen occipitale** can be very narrow (e.g., Diptera), moderately wide (most Coleoptera), or very wide (e.g., Orthoptera). It is usually divided by the **tentorial bridge** into a larger upper part (**alaforamen**) and a smaller lower part (**neuroforamen**), and strengthened by the more or less wide **postoccipital ridge**, which also serves as a muscle attachment area (Fig. 1.2.1.1).

In hexapods with an **orthognathous head** (e.g., Orthoptera, Zoraptera, Hymenoptera) (Figs 6.12.1, 6.13.1, 6.25.1) the mouthparts are ventrally directed, whereas they are anteriorly directed in those with a **prognathous head** (e.g., Grylloblattodea, Raphidioptera) (Fig. 6.14.2). A prognathous head is usually but not always found in predacious hexapods. Subprognathism is an intermediate condition, with a slightly to moderately inclined head. In the **hypognathous head** the mouthparts are posteriorly directed (e.g., Auchenorrhyncha, Sternorrhyncha) (Figs 6.24.1, 6.24.4). Hyperprognathism with dorsally directed mouthparts is an unusual condition occurring in larvae of Hydrophilidae (Coleoptera).

The different head regions usually bear a characteristic vestiture of setae (articulated hairs) (Figs 6.13.2, 6.25.1). The distribution pattern of setae and pores (chaetotaxy) is taxonomically informative in some groups, especially in larvae (e.g., Lepidoptera). The coloration and surface structure of the cuticle can differ strongly in different groups.

The compound eyes (see **1.6 Photoreceptor organs**) are usually well-developed and more or less strongly convex and round, oval or kidney-shaped. In some groups they are small and flat (Fig. 6.14.2) or even largely (e.g., Siphonaptera) or completely reduced (e.g., Protura, Diplura [=Nonoculata]). They are almost always placed laterally. A partial or complete subdivision occurs in few groups (e.g., males of Ephemeroptera, Gyrinidae). In Gyrinidae (Coleoptera) the upper and the lower parts are usually widely separated and differ in their fine structure. The transparent cuticle covering the **ommatidia** is thin, thus representing a zone of mechanical weakness of the head capsule. This is usually compensated for by the presence of a broad internal **circumocular ridge**. **Ocelli** (median eyes) (Fig. 1.6.2.3) are also present in most groups (e.g., Fig. 6.7.1). Three of them are usually arranged in a triangle on the frontal region. They are reduced in different groups (e.g., Dermaptera, Embioptera), often in correlation with flightlessness (e.g., Zoraptera [apterous morphs], Grylloblattodea).

The **frontal** and **coronal** (=**epicranial**) **sutures** are usually present (Figs 1.2.1.1, 1.2.2.1A). The former enclose a triangular **frons** in most groups (sometimes U-shaped). Both sutures split open during the molting process. The area laterad the coronal suture is referred to as **vertex** and the lateral head region as **gena**. However, both areas are not delimited by a suture or ridge. A lateral area above the lateral mandibular base, the **subgena**, is separated from the genal region by the **subgenal ridge**, which can be divided into an anterior **pleurostomal ridge** above the mandibular base and a posterior **hypostomal ridge**. The posterior part of the subgena, traditionally assigned as **postgena**, is usually not defined as a separate element. The antennal foramen is enclosed by a circumantennal ridge, which often bears a small articulatory process, the **antennifer**.

The **clypeus** is often trapezoid (Figs 1.2.1.1, 1.2.2.1A). Its broader posterior margin is primarily separated from the frontal region by a transverse **frontoclypeal strengthening ridge** (=frontoclypeal "suture", epistomal "suture"). The clypeus is sometimes divided into an anterior, transparent **anteclypeus** without muscle attachment, and

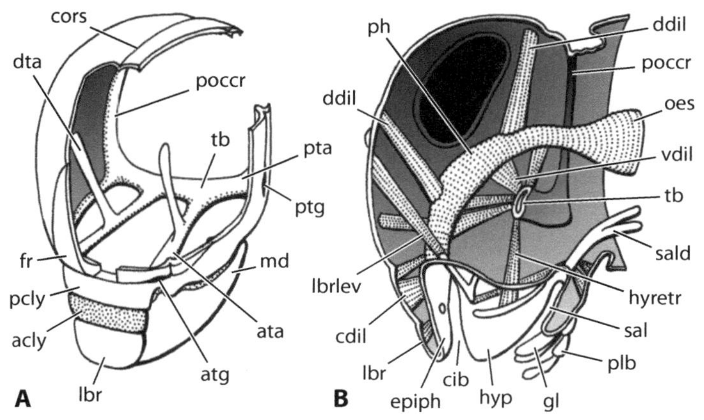

Fig. 1.2.2.1: Generalized hexapod head, internal structures. A, anterolateral view, frontal side of head capsule opened; B, sagittal section, brain and suboesophageal complex removed. Abbr.: ata – anterior tentorial arm, acly – anteclypeus, atg – anterior tentorial groove, cib – cibarium, cdil – cibarial dilator (M. clypeobuccalis), cors – coronal suture, ddil – dorsal pharyngeal dilator (M. fronto-/verticopharyngalis), dta – dorsal tentorial arm, epiph – epipharynx, fr – frons, gl – glossa, hyp – hypopharynx, hyretr – hypopharyngeal retractor, lbr – labrum, lbrlev – external labral levator (M. frontohypopharyngalis), md – mandible, oes – oesophagus, pcly – postclypeus, ph – pharynx, plb – palpus labialis, poccr – postoccipital ridge, pta – posterior tentorial arms, ptg – posterior tentorial groove, sal – salivarium, sald – salivary duct, tb – tentorial bridge, vdil – ventral pharyngeal dilator (M. tentoriopharyngalis). Courtesy of H. Pohl, redrawn from Weber & Weidner (1974), with modifications.

a posterior sclerotized **postclypeus** (Fig. 1.2.2.1A), which serves as attachment area for epipharyngeal muscles (M. clypeopalatalis[3]). The dorsolateral clypeal margin is marked by the anterior tentorial pits or grooves, which represent the areas of invagination of the anterior tentorial arms.

The foramen occipitale may be narrowed by a sclerotized, unpaired **gula**, especially in prognathous forms (e.g., Coleoptera, Megaloptera, Raphidioptera). The gula is likely formed by a sclerotized ventromedian region of the cervical membrane and is usually distinctly separated from the laterally adjacent parts of the head capsule by internal **gular ridges**. The landmark between the gular region and the posterior labial (submental or postmental) margin is marked by the **posterior tentorial pits** or grooves, the posterior invagination sites of the tentorium, which are generally present in hexapods, with very few exceptions (e.g., Strepsiptera). An alternative partial closure

3 In the following names of Wipfler et al. (2011) are used for cephalic muscles. A muscle table (Table I) with the nomenclature for head muscles is presented at the end of the chapter **1.2 Head**.

of the foramen occipitale can be formed by a **hypostomal** (or **postgenal**) **bridge**, i.e. a mesally projecting duplicature of the posterolateral head capsule (e.g., Hymenoptera partim).

Muscles not associated with movable appendages or the digestive tract are largely or completely absent in hexapods. Exceptions are muscles linking the tentorium with the external wall of the head capsule and muscles associated with the **antennal hearts** (**pulsatile organs**).

1.2.3 Cephalic endoskeleton

The head endoskeleton (Fig. 1.2.2.1) comprises invaginated sclerotized elements and also mesodermal ligamentous structures in the most basal hexapod lineages (e.g., Archaeognatha, *Tricholepidion*). Ridges play an important role as muscle attachment areas and mechanical strengthening elements. The **postoccipital ridge** around the foramen occipitale is often an attachment area for bundles of the mandibular flexor and extensor and also for extrinsic head muscles. The **circumocular ridge** strengthens the zone of weakness resulting from the thin cuticle of the compound eyes and is sometimes an area of origin of bundles of the mandibular extensor. The transverse strengthening ridge separates the clypeus from the frons. Main elements of the tentorium of Insecta are the strongly developed **posterior arms**, which usually arise immediately anterad the foramen occipitale. The invagination sites are the posterior tentorial pits or grooves. The posterior arms are usually connected with each other by the **tentorial bridge** (=**corpotentorium**) and also fused with the **anterior tentorial arms** in Pterygota (connected by muscles in apterygote hexapod lineages). The tentorial bridge of Pterygota (and Maindroniidae [Zygentoma]) is a product of fusion of transverse connecting bars of the anterior and posterior arms, respectively (see **6 The hexapod orders [Dicondylia]**). In some groups an accessory bridge is formed by medially fused internally directed processes, the **laminatentoria.** This feature is sometimes erroneously referred to as "perforated corpotentorium". It is likely an autapomorphy of Dictyoptera but does also occur in other groups (e.g., some groups of beetles). The **anterior tentorial pits** or grooves are the invagination sites of the anterior arms. They arise at the proximolateral edge of the clypeus, close to the secondary mandibular joint, almost always on the transverse frontoclypeal strengthening ridge (epistomal ridge) or on the subgenal ridge. The **dorsal tentorial arms** originate from the anterior arms (most groups of Polyneoptera) or at the junction area of the posterior and anterior arms. They are not invaginations of the head capsule but usually connected to it by fibrillar structures (fused with the wall of the head capsule in some cases).

In apterygote hexapods the tentorium is held in its position by several muscles that originate on the head capsule. Only one muscle (M. tentoriofrontalis anterior) is preserved in lower neopteran insects. Additionally, the posterior arms almost always serve as attachment areas of extrinsic maxillary, labial and hypopharyngeal muscles, and the corpotentorium as an area of origin of ventral pharyngeal dilators. The dorsal and anterior arms are usually attachment sites of extrinsic antennal muscles, and the latter also of tentorial muscles of the mandibles.

Reductions and simplifications of the tentorium occur, especially but not only in groups with reduced or strongly modified mouthparts (e.g., Strepsiptera, Diptera).

1.2.4 Labrum and epipharynx

The **labrum** (Figs 1.2.1.1, 1.2.2.1, 1.2.4.1), is an unpaired anterior appendage of the head and forms the anterior closure of the **preoral cavity** between the paired mouthparts. It usually covers a considerable part of the mandibles. The preoral cavity or **cibarium** can be defined as the space between the inner labral wall (ventral[4] **epipharynx**)

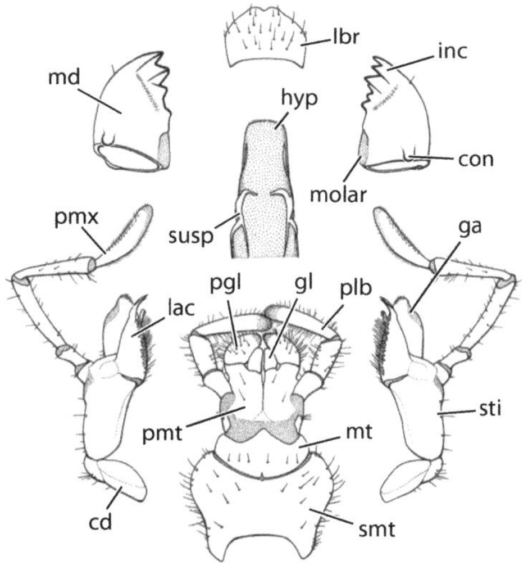

Fig. 1.2.4.1: Mouthparts of a generalized insect (*Periplaneta americana*, Blattodea), labrum (top), mandibles, hypopharynx (center), maxillae, labium (bottom). Abbr.: cd – cardo, con – condyle (primary mandibular joint), ga – galea, gl – glossa, hyp – hypopharynx, inc – mandibular incisivi, lbr – labrum, lac – lacinia, md – mandible, molar – molar region, mt – mentum, pgl – paraglossa, plb – palpus labialis, pmt – prementum, pmx – palpus maxillaris, smt – submentum, sti – stipes, susp – hypopharyngeal suspensorium. Courtesy of H. Pohl.

4 The terms referring to directions (e.g., anterior, ventral) apply to an orthognathous head here and in the following. The corresponding terms for a prognathous head are: anterior (orthognathous) = dorsal (prognathous); posterior = ventral; dorsal = posterodorsal; ventral = anterior.

(Figs 1.2.2.1A, 6.18.2B) and the inner surface of the clypeal region (dorsal epipharynx) on one hand, and the **hypopharynx** on the other (see below). A wider preoral space limited by the labrum and labium would also include the **salivarium** (see below). The proximal border of the epipharynx (and cibarium) is the **anatomical mouth**, the anterior opening of the **pharynx.**

The homology of the labrum is disputed. Its insertion below (or anterad in prognathous hexapods) the antennae suggests that it may belong to the first head segment, possibly homologous to the antennae of Onychophora and the "great appendages" of Cambrian fossils (e.g., †Anomalocarida). However, it was pointed out by Denis & Bitsch (1973) that a true preantennal segment (and the "acron") is not traceable in Hexapoda. Moreover, the innervation from the tritocerebrum and embryological evidence suggest that it may rather belong to the intercalary (= 3rd) segment.

At its base the labrum is usually moveably connected with the ventral clypeal margin by an internal membranous fold. However, it is sometimes immobilized and more or less completely fused with the clypeus (e.g., predacious beetle larvae: Adephaga, Hydrophiloidea, Cleroidea etc.). The product of fusion is called **clypeolabrum** or **frontoclypeolabrum** if the frontoclypeal strengthening ridge is absent. The anterior labral wall is almost always sclerotized and equipped with setae. The ventral (anterior in prognathous insects) and lateral margins are usually rounded. A more or less deep ventromedian emargination is present in different groups (e.g., Orthoptera, Blattodea partim). Sensilla and/or microtrichia are often present along the ventral edge. Basolateral sclerotized rods, the **tormae**, reinforce the epipharynx and their distal part serves as attachment area of the lateral labral retractors (see below).

The inner labral wall, i.e. the ventral epipharynx (Figs 1.2.2.1B, 6.18.2B), is usually largely membranous or semimembranous. It often bears a vestiture of specifically arranged microtrichia (non-articulated hairs), which are usually posteriorly directed, thus facilitating the food transport towards the anatomical mouth. In some groups of insects the lateral margin of the dorsal (clypeal) epipharyngeal part is fused with the lateral margin of the dorsal hypopharynx, thus forming a more or less elongated **prepharyngeal tube**, which is continuous with the pharynx.

M. labroepipharyngalis, which is present in most lineages, connects the anterior and posterior labral wall. Its contraction widens the anterior region of the preoral cavity. Other intrinsic labral muscles can occur in apterygote hexapods, ephemeropterans and in members of Polyneoptera. The labrum is retracted by two pairs of extrinsic muscles in the groundplan of hexapods (reduced in different groups). M. frontolabralis (Figs 6.14.3, 6.18.2B) usually originates on the median frontal region and inserts medially on the external labral wall. M. frontoepipharyngalis originates laterad the former and inserts on the tormae. Reduction of labral muscles occur in different groups. M. frontolabralis, for instance, is always absent in Coleoptera.

In some groups the labrum is strongly modified. In some basal lineages of Diptera (e.g., Culicidae) and in Siphonaptera it forms a part of the piercing-sucking apparatus (Figs 6.35.1, 6.35.2) with a food channel on its ventral (epipharyngeal) side. The labrum partly covers the labial rostrum in Hemiptera (Figs 6.21.2, 6.24.1, 6.24.4). It is almost always completely reduced in Strepsiptera but still present in the groundplan of the order.

1.2.5 Antennae

The **antennae** are the paired appendages of the 2nd segment and homologous to the 1st antennae (=antennulae) of crustaceans (Figs 6.1.1, 6.1.2, 6.4.1). The antennal nerve originates from the second part of the brain, the **deutocerebrum**. In Collembola and Diplura (and in Myriapoda and crustaceans) muscles are present in all antennomeres (often referred to as antennal segments) except for the apical one. This type of antenna (="Gliederantenne" in German) belongs to the groundplan of Hexapoda (Figs 6.1.1, 6.3.3) An autapomorphy of Insecta is the antenna of the flagellar type ("Geißelantenne") (Figs 6.4.1, 6.4.2), with muscles only in the basal antennomere, the **scapus** (Mm. scapopedicellares lateralis and medialis). A chordotonal organ is present in the **pedicellus** (2nd antennomere), which bears a more or less elongated **flagellum** without muscles, which is usually composed of many antennomeres (**flagellomeres**).

The antennae are inserted in the **antennal foramen** (see above), between or above the compound eyes in most groups (Fig. 1.2.1.1). They are almost always movable in all directions and more or less densely covered with setae and different sensilla, mostly chemo- and mechanoreceptors (Figs 4.4.1B, 4.9.1B). The scapus usually articulates with the **antennifer** and is moved by three or four extrinsic antennal muscles in most groups (Mm. tentorioscapales anterior, posterior, lateralis and medialis). The two intrinsic muscles (see above) are attached to the base of the pedicellus and move the remaining part of the antenna. The Johnston's organ in the pedicellus of Insecta is a complex chordotonal organ composed of numerous scolopidial sensilla. It functions as a mechanoreceptor and registers movements of the flagellum. It also perceives acoustic signals in some groups (e.g., Culicidae, Chironomidae).

The antennae can be modified in many different ways (e.g., Fig. 6.35.4). They are completely absent in Protura (Figs 6.2.1, 6.2.2), largely reduced in larvae of some groups of Holometabola (e.g., Strepsiptera) (Fig. 6.30.1), and shortened and more or less bristle-like in Odonata (Fig. 6.7.1A), Ephemeroptera (Fig. 6.6.2) and Auchenorrhyncha (Fig. 6.21.2). The ancestral type (groundplan of Insecta) is the filiform antenna with a multisegmented, slender flagellum. In moniliform antennae the flagellomeres are more or less globular (e.g., Isoptera, Embioptera partim), whereas they are sawtooth-shaped in the serrate type (e.g., Elateridae [Coleoptera]). The flagellomeres of pectinate antennae bear elongate extensions either on one or on both sides (e.g., different groups of Lepidoptera, Tipulidae [Diptera]). In capitate or clubbed antennae one or several apical segments are symmetrically or asymmetrically widened (e.g., different groups of Lepidoptera and Coleoptera). Geniculate antennae are hinged or bent (e.g., different groups of Hymenoptera [e.g., ants], Curculionidae [Coleoptera]). In "higher" dipterans (Cyclorrhapha) the first flagellomere (postpedicellus) is strongly enlarged whereas the remaining flagellum forms a bristle-like **arista** (aristate type) (Fig. 6.35.4).

1.2.6 Mandibles

The **mandibles** are the paired appendages of the 4th head segment (Figs 1.2.1.1, 1.2.4.1) and receive their innervation from the anterior part of the suboesophageal ganglion. They play a dominant role in the mechanical processing of food. Unlike the **maxillae** (Fig. 1.2.4.1) they are primarily a compact, undivided and strongly sclerotized element without appendages. In Dicondylia (Zygentoma + Pterygota) they are connected with the head capsule by a posterior primary mandibular joint (condyle on the mandible) and an anterior secondary joint (condyle at the clypeal base). The movements of the dicondylous mandible are restricted to a single level perpendicular to the axis between these two articulations. The mandible is almost always distinctly curved inwards, especially at its outer margin, and it is usually broadest at its base. One or several apical or subapical teeth (**incisivi**) (Fig. 1.2.4.1) are used to perforate and crush food substrate. Often, especially in hexapods feeding on plant matter or fungi, a mesally directed prominence is present at the mandibular base, the **mola**; usually its surface is modified for efficient grinding, with parallel ridges or densely arranged tubercles. In predacious forms (e.g., Odonata, Mantophasmatodea) the mesal surfaces of the mandibles form distinct cutting edges. An articulated mesal appendage, the **lacinia mobilis** or prostheca, occurs in several groups. It can be sclerotized or largely membranous and is often equipped with microtrichia.

In neopteran insects with well-developed mandibles a pair of antagonistic muscles with a cranial origin move the mandibles, the very large M. craniomandibularis internus (flexor=adductor) and the distinctly smaller M. craniomandibularis externus (extensor=abductor) (Figs 6.14.3, 6.18.2B). Both insert on strongly developed tendons attached to the mandibular base. In the groundplan of Hexapoda strongly developed muscles arising from the hypopharynx (M. hypopharyngomandibularis) or the anterior tentorial arms (Mm. tentoriomandibulares lateralis superior and inferior, Mm. tentoriomandibulares medialis superior and inferior) are present, and also a transverse muscle medially connected by a ligament (e.g., Zygentoma). They insert on the internal surface of the mandible. The hypopharyngeal and tentorial muscles persisting in many pterygote groups are almost always very thin. One of them is usually accompanied by a nerve and functions as a proprioreceptor.

Only the primary joint is present in Collembola, Protura and Archaeognatha and both joints are missing in Diplura. Mandibular sucking channels occur in larvae of some groups of beetles (e.g., Dytiscidae, Lampyridae). The mandibles are transformed into stylets in Hemiptera and biting flies (in addition to stylets formed by the laciniae, see below). The mandibles form a sucking apparatus together with the laciniae in neuropteran larvae.

1.2.7 Maxillae

The **maxilla** (Figs 1.2.4.1, 6.14.2, 6.27.3C), which is homologous to the maxillula (1st maxilla) of crustaceans, is usually composed of cardo, stipes, lacinia (=inner endite lobe), galea (=outer endite lobe), and palp. The relatively short **cardo** articulates with the head capsule, usually in a more or less deep maxillary groove (=**fossa maxillaris**). At its base it bears a lateral process for muscle attachment. The articulation point lies between the insertion of the extensor and the flexor, which originate on the lateral head capsule and the posterior tentorial arm, respectively. The **stipes** is connected with the cardo by a hinge. It is usually the largest part of the maxilla and bears the palp and endite lobes. The ventral and lateral sides are sclerotized and divided into a **basistipes** and **mediostipes** in some groups (e.g., Coleoptera). The **lacinia** is usually more or less firmly connected or fused with the stipes mesally. It is sclerotized and apically curved and pointed in most groups, and usually set with strong setae or spines along its mesal edge. In groups with sucking-piercing mouthparts (e.g., Anoplura, Hemiptera, biting flies) the lacinia is transformed into a piercing stylet (Figs 6.35.1, 6.35.2). The **galea** is usually less strongly sclerotized and equipped with chemoreceptors; it is composed of one or two segments. The maxillary **palp** is usually five-segmented and also equipped with chemoreceptive sensilla; its apex often bears a dense field of sensorial structures.

In most groups two extrinsic maxillary muscles originate from the head capsule (M. craniocardinalis and M. craniolacinialis) and three from the posterior tentorial arms (M. tentoriocardinalis, Mm. tentoriostipitales anterior and posterior). Additional muscles insert on the endite lobes, on the base of the palp, and on the base of the palpomeres.

1.2.8 Labium

The **labium** or lower lip (Figs 1.2.4.1, 6.14.2, 6.27.3C) forms the posterior (orthognathous head) or ventral (prognathous head) closure of the preoral space and of the salivarium. It is homologous to the maxilla (2nd maxilla) of crustaceans, but its proximal parts form a structural unit without a recognizable median line or suture. The proximal **postmentum** primarily forms the posterior closure of the head capsule and is adjacent with the foramen occipitale unless a gula or hypostomal bridge is present; it is often divided into a **mentum** and **submentum** (Fig. 1.2.4.1). The anterior element, the **prementum**, bears the appendages, i.e. the palps and the **paraglossae** and **glossae** (outer and inner endite lobes; serially homologous to the galea and lacinia); it may or may not be incised medially. The labial palp is usually three-segmented but otherwise very similar to the maxillary palp; it is either inserted on a more or less distinct **palpiger** or directly on the distolateral edge of the prementum. The musculature is similar to the muscle equipment of the maxilla. Three pairs of premental retractors are usually present in Pterygota, one of them originating on the postlabium (M. submentopraementalis) and two on the posterior tentorial arms (Mm. tentoriopraementales inferior and superior). In addition intrinsic premental muscles are associated with the glossae and paraglossae, the palps, and the salivarium. Intrinsic palp muscles are also usually present.

The labium is transformed into a rostrum in Hemiptera (Figs 6.24.1, 6.24.4A). It forms a sheath for the stylet-like mandibles and laciniae. The outer and inner endite lobes of the labium are distinctly reduced or missing in holometabolous insects with the notable exception of Hymenoptera. In bees, the labium forms a complex with the maxillae. Both elements of this maxillolabial complex are connected by a small rod-like sclerite, the **lorum**. The frontally fused, strongly elongated and pubescent glossae serve as uptake apparatus of liquid food.

1.2.9 Hypopharynx

The **hypopharynx** is an unpaired tongue-like structure between the paired mouthparts (Figs 1.2.4.1, 6.18.2B). It is likely a derivative of several segments and not serially homologous to the primarily paired appendages (Denis & Bitsch 1973).

The hypopharynx is usually largely semimembranous, but almost always reinforced by sclerites (Fig. 1.2.4.1: susp). It forms the posterior floor of the cibarium and a ramp for transporting food towards the anatomical mouth (Fig. 6.18.2B). Its anterior surface is often at least partly sclerotized and bears a vestiture of posteriorly directed **spinulae** which facilitate this process (see epipharynx). Between the hypopharynx and the anterior side of the distal part of the labium lies the salivarium, a pocket where the salivary ducts open (see **1.2.10 Salivarium**) (Fig. 6.18.2B). **Superlinguae** are lateral lobes of the hypopharynx (Fig. 6.6.5), flanking the median **lingua**. They are likely a groundplan feature of Hexapoda (present in Collembola, Diplura and Archaeognatha) but they are missing in Zygentoma and distinctly reduced or absent in Pterygota (with the exception of Ephemeroptera: Fig. 6.6.5). Reduced superlinguae are usually represented by lateral sclerotizations, the basal sclerites or basal plates, which sometimes extend to the salivarium. They serve as attachment areas of the tentorial retractors of the hypopharynx, M. tentoriohypopharyngalis. The **suspensoria** (=**fulturae**) are paired sclerites or groups of sclerites embedded in the proximolateral hypopharyngeal wall (Fig. 1.2.4.1: susp). They form long anterior projections, the oral arms which end laterad the anatomical mouth and serve as attachment area of M. oralis transversalis, M. frontooralis and M. tentoriooralis. The last two muscles were addressed as retractors of the mouth angle by Snodgrass (1935). Their antagonist is M. craniohypopharyngalis, a retractor originating from the head capsule (apterygote lineages) or posterior tentorium (Pterygota). The muscle is missing in different groups, especially in correlation with a reduced salivarium (e.g., Coleoptera). Several other muscles linked with the hypopharynx occur in different lineages.

The hypopharynx can be distinctly modified or reduced in different groups of hexapods. It is greatly reduced in size in Hymenoptera and fused with the distal part of the labium in Coleoptera, which consequently lack a salivarium. It forms one of the stylets in basal dipteran groups (e.g., Culicidae). In Hemipterans it is perforated by the canal of the salivary pump and continuous with the food canal of the laciniae.

1.2.10 Salivarium

The **salivarium** is the pocket between the posterior side of the hypopharynx and the anterior (internal) side of the distal part of the labium (Figs 1.2.2.2B, 6.18.2B). It is part of the preoral cavity in the wider sense and receives the openings of the **salivary ducts**. The salivary glands (Fig. 1.2.10.1) are usually well developed and located in the anterior part of the thorax. In some groups they are connected with a large reservoir (Fig. 1.9.2). The salivarium is usually equipped with three pairs of short muscles originating from the hypopharynx (M. hypopharyngosalivaris) and the prementum (M. praementosalivaris anterior and posterior), respectively. In certain groups a small ring muscle (M. annularis salivarii) is present where the salivary duct opens into the salivarium. Reductions of the salivarium (including muscles and glands) occur in different groups, notably in Coleoptera (completely absent) and in Megaloptera (vestigial).

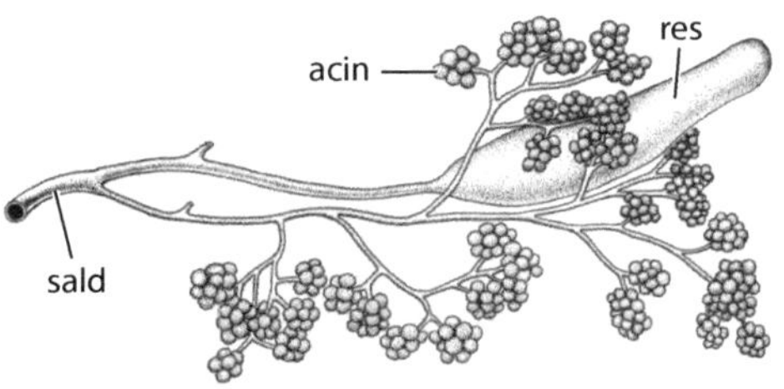

Fig. 1.2.10.1: Salivary gland. Abbr.: acin – acini (berry-like substructure) of salivary gland, res – reservoir, sald – salivary duct. Redrawn from Seifert (1995).

[Snodgrass (1935); Weber (1933, 1938); v. Kéler (1963); Matsuda (1965); Denis & Bitsch (1973); Weber & Weidner (1974); Chapman (1998); Wipfler et al. (2011)]

Table I: Generalized nomenclature for cephalic muscles (based on Wipfler et al. [2011] and von Kéler [1963], modified). The terms used for the orientation of structures refer to a prognathous head. For orthognathous heads: dorsal = anterior, ventral = posterior, anterior = ventral, posterior = dorsal.

Abbrev.	Name	Origin	Insertion
Antennal muscles			
0an1	M. tentorioscapalis anterior	anterior or dorsal tentorial arms	anteriorly on base of scapus
0an2	M. tentorioscapalis posterior	anterior or dorsal tentorial arms	posteriorly on base of scapus
0an3	M. tentorioscapalis lateralis	anterior or dorsal tentorial arms	lateral margin of scapus
0an4	M. tentorioscapalis medialis	anterior or dorsal tentorial arms	mesal margin of scapus
0an5	M. frontopedicellaris	frons	lateral edge of pedicellus
0an6	M. scapopedicellaris lateralis	posterior or posterolateral area of scapus	posteriorly or posterolaterally on base of pedicellus
0an7	M. scapopedicellaris medialis	mesal or posteromesal area of scapus	mesal, anterior or anteromesal base of pedicellus
0an8	M. intraflagellaris	base of pedicellus	close to tip of flagellum
Antennal heart muscles			
0ah1	M. interampullaris	antennal ampulla	antennal ampulla of opposite side
0ah2	M. ampulloaortica	antennal ampulla	cephalic aorta
0ah3	M. ampullopharyngalis	antennal ampulla	anterior pharynx
0ah4	M. ampullofrontalis	antennal ampulla	frons
0ah5	M. frontopharyngalis	frons (laterally of antennal ampulla)	dorsolaterally on pharynx
0ah6	M. frontofrontalis	frons (laterally of antennal ampulla)	opposite side of frons
Labral muscles			
0lb1	M. frontolabralis	central region of frons or rarely on antennal base	posterior labral margin
0lb2	M. frontoepipharyngalis	central region of frons or rarely on antennal base	posterolateral edge of labrum (labro-epipharyngeal border) or tormae
0lb3	M. epistoepipharyngalis	epistomal sulcus	posterolateral edge of labrum (labro-epipharyngeal border)
0lb4	M. labralis transversalis	ventrolateral edge of labrum	opposite side of ventrolateral edge of labrum
0lb5	M. labroepipharyngalis	centrally on labrum	epipharynx
0lb6	M. labrolabralis	centrally on labrum (between bundles of 0lb5)	medially on apical labral area

Mandibular muscles			
0md1	M. craniomandibularis internus	Ventral, posterior, lateral and/or dorsal parts of the head capsule and/or posterior and anterior tentorial arms	adductor (mesal) tendon of mandible
0md2	M. craniomandibularis externus anterior	gena (below compound eyes)	base of mandible (behind anterior mandibular articulation)
0md3	M. craniomandibularis externus posterior	gena and postgena	abductor (lateral) tendon of mandible
0md4	M. hypopharyngomandibularis	process of loral arm or separate sclerite close to it	dorsolateral area of mandible (inside of mandible)
0md5	M. tentoriomandibularis lateralis superior	anterior and dorsal tentorial arm	laterally on mandibular base (between processus paratentorialis and posterior mandibular articuation
0md6	M. tentoriomandibularis lateralis inferior	anterior tentorial arm	ventrally on mandibular rim (inside of mandible)
0md7	M. tentoriomandibularis medialis superior	anterior and dorsal tentorial arm	dorsomesal area of mandible (inside of mandible)
0md8	M. tentoriomandibularis medialis inferior	anterior tentorial arm	dorsal part of mandibular base (inside of mandible)
Maxillary muscles			
0mx1	M. craniocardinalis	gena and postgena	proximal part of cardo (cardinal process)
0mx2	M. craniolacinialis	posterior part of gena	proximal egde of lacinia
0mx3	M. tentoriocardinalis	tentorium	distal part of cardo (close to stipitocardinal suture)
0mx4	M. tentoriostipitalis anterior	tentorium	stipital ridge
0mx5	M. tentoriostipitalis posterior	tentorium	proximal rim of stipes (close to stipito-cardinal sulcus)
0mx6	M. stipitolacinialis	proximolaterally on stipes (dorsally of maxillary palp)	base of lacinia (in some cases common tendon with 0mx4)
0mx7	M. stipitogalealis	laterally on stipes (dorsally of maxillary palp)	base of galea
0mx8	M. stipitopalpalis externus	laterally on stipes (dorsally of maxillary palp) or stipital ridge	laterally on base of first maxillary palpomere
0mx9	M. stipitopalpalis medialis	stipital ridge	posteriorly on base of first maxillary palpomere
0mx10	M. stipitopalpalis internus	stipital ridge	mesally on base of first maxillary palpomere

Abbrev.	Name	Origin	Insertion
0mx11	M. stipitalis transversalis	lateral wall of proximal part of stipes (close to stipitocardinal suture)	stipital ridge
0mx12	M. palpopalpalis maxillae primus	laterally on base of first maxillary palpomere	laterally on base of second maxillary palpomere
0mx13	M. palpopalpalis maxillae secundus	mesally on base of second maxillary palpomere	mesally on base of third maxillary palpomere
0mx14	M. palpopalpalis maxillae tertius	base of third maxillary palpomere	mesally on base of fourth maxillary palpomere
0mx15	M. palpopalpalis maxillae quartus	mesally on base of fourth maxillary palpomere	mesally on base of fifth maxillary palpomere
Labial muscles			
0la1	M. postoccipitoglossalis medialis	postoccipital phragma	glossa
0la2	M. postoccipitoglossalis lateralis	postoccipital phragma (laterad 0la1)	dorsolaterally on base of glossa
0la3	M. postoccipitoparaglossalis	postoccipital phragma (laterad 0la2)	base of paraglossa
0la4	M. postoccipitopraementalis	postoccipital phragma (laterad 0la3)	posterolateral corner of prementum
0la5	M. tentoriopraementalis	posterior tentorial arm and/or cranium	proximal rim of prementum
0la6	M. tentorioparaglossalis	posterior tentorial arm and/or postoccipital ridge and/or cranium	paraglossae (close to base of labial palp)
0la7	M. tentorioglandularis	posterior tentorial arm	labial gland
0la8	M. submentopraementalis	central or proximolateral region of submentum or gula	medially on proximal premental margin or distal part (close to galea)
0la9	M. postmentomembranus	postmentum	internal (ventral) membrane of labium
0la10	M. submentomentalis	submentum	mentum
0la11	M. praementoparaglossalis	proximal rim of prementum (lateral half)	proximolaterally on paraglossa
0la12	M. praementoglossalis	central region of prementum	mesal wall of glossa
0la13	M. praementopalpalis internus	distal area of prementum (medially)	mesally on first palpomere
0la14	M. praementopalpalis externus	medially on proximal rim or proximolateral corner of prementum	laterally on first palpomere
0la15	M. praementomembranus	postmentum	internal (ventral) membrane of labium
0la16	M. palpopalpalis labii primus	proximomesally on first labial palpomere	mesally on base of second labial palpomere
0la17	M. palpopalpalis labii secundus	lateral wall of second labial palpomere	mesally on base of third labial palpomere

Hypopharyngeal muscles			
0hy1	M. frontooralis	frons	oral arm of suspensorial sclerite
0hy2	M. tentoriooralis	anterior tentorial arm or epistomal sulcus or frons close to it	oral arm of suspensorial sclerite
0hy3	M. craniohypopharyngalis	posterior tentorial arm or corpotentorium	suprasalivarial sclerite
0hy4	M. postoccipitalohypopharyngalis	postoccipital phragma	hypopharyngeal phragma
0hy5	M. tentoriosuspensorialis	anterior margin of corpotentorium	hypopharyngeal suspensorial sclerite
0hy6	M. postmentoloralis	postmentum	loral arm of suspensorial sclerite
0hy7	M. praementosalivarialis anterior	distolateral area of prementum (close to labial palp)	laterally on salivarium
0hy8	M. praementosalivarialis posterior	proximal or proximolateral part of prementum	laterally on salivarium
0hy9	M. oralis transversalis	oral arm of suspensorial sclerite	oral arm of suspensorial sclerite of opposite side
0hy10	M. loroloralis	loral arm of suspensorial sclerite	oral arm of suspensorial sclerite of opposite side
0hy11	M. lorosalivarialis	hypopharyngeal suspensorial sclerite	suprasalivarial sclerite
0hy12	M. hypopharyngosalivarialis	hypopharyngeal suspensorial sclerite or tentorium	salivary orifice
0hy13	M. annularis salivarii		ring muscle close to salivary orifice
Tentorial muscles			
0te1	M. tentoriofrontalis posterior	lateral edge of anterior part of corpotentorium	frons
0te2	M. tentoriofrontalis anterior	dorsally on anterior part of corpotentorium	frons
0te3	M. tentoriofrontalis dorsalis	dorsal tentorial arm	frons
0te4	M. posterotentorialis	dorsolaterally on metatentorium	external rim of tentorium
0te5	M. tentoriotentorialis longus	ventrolaterally on anterior part of corpotentorium	mesally on posterior tentorial part
0te6	M. tentoriotentorialis brevis	anterior part of tentorium	posterior part of tentorium
Cibarial muscles			
0ci1	M. clypeopalatalis	postclypeus	roof of cibarium
Buccal muscles			
0bu1	M. clypeobuccalis	postclypeus (posterad 0ci1)	roof of bucca (between anatomical mouth and frontal ganglion)
0bu2	M. frontobuccalis anterior	frons (behind epistomal sulcus)	dorsal wall of bucca (directly behind frontal ganglion)
0bu3	M. frontobuccalis posterior	posterior part of frons	dorsal wall of bucca (directly in front of brain)

Abbrev.	Name	Origin	Insertion
0bu4	M. tentoriobuccalis lateralis	dorsal tentorial arm	lateral wall of bucca
0bu5	M. tentoriobuccalis anterior	anterior tentorial arm or corpotentorium	ventral wall of bucca (directly behind anatomical mouth)
0bu6	M. tentoriobuccalis posterior	posterior tentorial arm or corpotentorium	ventral wall of bucca (directly in front of brain, ventrally of 0bu3)
Pharyngeal muscles			
0ph1	M. verticopharyngalis	vertex or occipitale	dorsally on postcerebral pharynx (directly posterad of brain)
0ph2	M. tentoriopharyngalis	posterior tentorial arm	ventrally on postcerebral pharynx (beneath 0ph1)
0ph3	M. postoccipitopharyngalis	postocciput (close to median line)	posterior pharynx
Stomodaeal muscles			
0st1	M. annularis stomadaei		ring muscle layer covering pharynx
0st2	M. longitudinalis stomadaei		longitudinal muscles along pharynx

1.3 Thorax

1.3.1 Segmentation and composition of segments

The presence of a **thorax** composed of three clearly defined segments is arguably the most important autapomorphy of Hexapoda. It is usually the second largest tagma of the body and structures related to locomotion are concentrated in this region. On each of the three segments a pair of legs is inserted ventrolaterally. Two pairs of wings with a dorsolateral articulation area are present in most lineages of Pterygota (groundplan autapomorphy). The three segments are more or less similar and relatively simple in the apterygote lineages. In pterygote insects the **mesothorax** and **metathorax** (together forming the **pterothorax**) are strongly modified and much more complex than the **prothorax**, especially on the dorsal side (tergal region) and dorsolaterally (wing base).

All three segments are composed of dorsal, lateral and ventral elements, the **tergum**, **pleuron** and **sternum** (Figs 1.3.3.1, 6.13.3, 6.18.3). The individual regions of sclerotized cuticle (**tergites**, **pleurites** and **sternites**) are connected by membranes and semimembranous areas. The membranous areas are exposed to varying degrees in almost all groups (Fig. 6.13.3), especially on the ventral side (not in Coleoptera [Figs 6.29.4, 6.29.5] and some groups of Heteroptera). The unsclerotized regions ensure the necessary flexibility within the segments, especially at the articulation areas of the locomotor organs. The three segments are connected by intersegmental membranes, which are often more or less concealed. On the dorsal and ventral side the membranous connection does not correspond with the true segmental border (see below). Laterally the thoracic **spiracles** are embedded in the intersegmental membranes (Fig. 1.3.3.1A), in most groups two pairs belonging to the meso- and metathorax, each pair located in front of the respective segment. Thoracic spiracles are usually larger than their abdominal equivalents, which are serially homologous. The two pterothoracic segments often form a more or less rigid functional unit, whereas the flexibility between the pro- and mesothorax is usually relatively high. Posteriorly the metathorax is broadly connected with the abdomen in most groups. Different types of fusion of metathoracic and abdominal elements occur in various groups (e.g., Hymenoptera).

The prothoracic tergum or **pronotum** is usually a simple, plate-like structure, whereas the dorsal and dorsolateral regions of the pterothorax of pterygote insects are modified in a complex way in correlation with the flight function (see **1.3.3 pterothoracic segments**) (Figs 1.3.3.1, 6.13.3). Dorsolateral tergal duplicatures (**paranota**, **laterotergites**) probably belong to the groundplan of Insecta (preserved in the original form in Archaeognatha and Zygentoma). They are possibly precursors of the meso- and metathoracic wings of Pterygota. The lateral sclerotized element of the thoracic segments, the **pleurite**, is located between the tergum and sternum. It is possibly formed by the subcoxa, an element supposedly separating from the coxal primordium in the embryonic development (Matsuda 1970), but this interpretation is uncertain. It is divided by a slanted dorsoventral **pleural ridge**, which is always present in the meso- and metathorax, but often reduced or absent in the prothorax (Figs 1.3.3.1A, 6.13.3). It stabilizes the lateral body wall and divides it in the anterior **episternum** and the posterior **epimeron**. It is also important in the functional context of the articulations of the coxae and wings. It connects the pleural wing joint with the pleurocoxal joint, the latter generally formed by a condyle at the posteroventral edge of the ridge and a corresponding coxal concavity. An additional coxal articulation is formed by the triangular anteroventral **trochantin** (Figs 1.3.3.1A, 6.13.3A), an additional pleural element below the episternum (small and crescent-shaped, reduced in some groups of Holometabola). The sternal region usually contains more extensive unsclerotized areas than the lateral and dorsal walls of the segments (Figs 1.3.3.1C, 6.13.3B). However, it is almost always reinforced by several sternites. A series of five elements is arguably present in the groundplan of Hexapoda, the short anterior **presternum**, the extensive **basisternum**, the **furcasternum** (or **sternellum**), the **spinasternum**, and the posterior **poststernite** (Matsuda 1970). In Pterygota, the presternum is almost generally absent (possibly represented by isolated sclerites in Plecoptera) and the spinasternum is reduced to varying degrees in many lineages. In Holometabola the true sternal elements are largely invaginated, mostly represented by an internal median longitudinal ridge (**discrimen**), and externally largely replaced by pleural elements (**preepisternum**, **katepisternum**).

More or less extensive ligamentous endoskeletal elements occur in the apterygote lineages. They are of subepidermal origin, usually branched, and serve as insertion points of muscles and tendons (Matsuda 1970). These structures are completely absent in most groups of Pterygota, at least in the adult stage. The pterygote endoskeletal elements are the **furcae** and the smaller **spinae** (Fig. 1.3.3.1D). The former arise from the furcasternum. Primarily their invagination sites are distinctly separated and connected by a transverse ridge, the **sternocosta**. In correlation with a narrowed or invaginated sternum (Holometabola) the furcal bases can be more or less closely adjacent or even arising from a common invagination site. The dorsolaterally extending furcal arms are often fused to the pleural ridge or connected with it by short muscles or fibrillae. They serve as muscle attachment area, especially for muscles attached to basal elements of the legs, the coxa and the trochanter, but also for the ventral intersegmental muscles. The spina is an unpaired invagination of the spinasternum and serves as attachment area of coxal and intersegmental muscles. It is partly or completely reduced in many groups, especially in the metathorax (distinctly developed in Grylloblattodea).

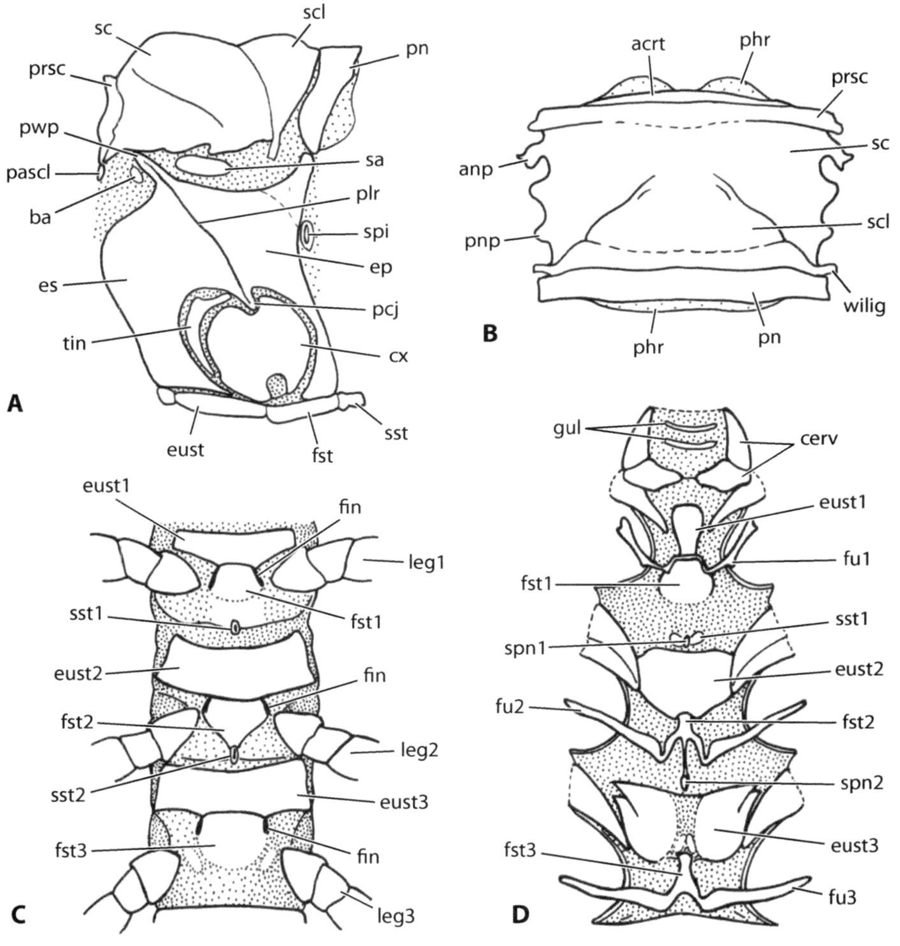

Fig. 1.3.3.1: Thoracic skeleton. A, pterothoracic segment, lateral view, schematized; B, pterothoracic tergal region, schematized; C, sternal region, ventral view *Isoperla* sp. (Plecoptera, Perlodidae); D, sternal region, internal view, *Periplaneta americana* (Blattodea, Blattidae). Abbr.: acrt – acrotergite, anp – anterior notal wing process, ba – basalare, cerv – cervical sclerite, cx – coxa, ep – epimeron, es – episternum, eust1–3 – pro-, meso-, metathoracic eusternum, fin – furcal invagination site, fst1–3 – pro-, meso-, metathoracic furcasternum (sternellum), fu1–3 – pro-/ meso-/ metafurca, gul – gularia, leg1–3 – fore-/ mid-/ hindleg, pascl – prealar sclerite, pcj – pleurocoxal joint, phr – phragma, plr – pleural ridge, pn – postnotum, pnp – posterior notal wing process, prsc – prescutum, pwp – pleural wing process, sa – subalare, sc – scutum, scl – scutellum, spi – spiracle, spn1/2 – pro-/ mesospina, sst1/2 – pro-, mesothoracic spinasternum, tin – trochantin, wilig – wing ligament. Redrawn from Seifert (1995).

1.3.2 Prothorax

The prothorax is anteriorly connected with the head by the cervical membrane, usually with embedded **cervical sclerites** (Figs 1.3.3.1D, 6.13.3A) forming articulations and serving as muscle attachment areas. One or two pairs of lateral cervical sclerites (**laterocervicalia**) are present in most groups. Unpaired ventral sclerites occurring occasionally are called **gularia**. Unpaired dorsal sclerites occur in several groups (e.g., Zoraptera, Coleoptera [Hydrophilidae]). Paired dorsal cervical sclerites are present in Dermaptera.

The pronotum is often divided by a median line or zone of weakness (ecdysial line in immatures) but otherwise forms a single, more or less plate-like structure, with or without a distinct lateral edge. It mainly serves as attachment area of large dorsoventral leg muscles and some of the extrinsic muscles of the head. The size of the pronotum varies greatly. It forms a conspicuous pronotal shield in Blattodea, Coleoptera and some other groups. In orthopterans it is saddle shaped and laterally covers extensive parts of the pleuron. Anteriorly the pronotum covers the posterior margin of the head in many groups. Posteriorly it often overlaps with the anterior region of the mesotergum. Distinctly developed lateral duplicatures are present in the primarily wingless Archaeognatha and Zygentoma (see above: paranota). Prothoracic winglets were present in some Paleozoic insects (e.g., †Palaeodictyoptera) but absent in all extant pterygote lineages. The prothoracic pleuron is generally less complex than the corresponding regions of the pterothoracic segments. All modifications or structures related to flight are lacking (e.g., **basalare** and **subalare**) (Fig. 6.13.3A). The pleural suture is often short, indistinct or lacking. In some groups more or less extensive propleural parts are invaginated below the pronotum, thus forming a **cryptopleuron** (e.g., Coleoptera, especially Polyphaga). Compared to the tergal and pleural elements, the prosternum differs less profoundly from its pterothoracic counterparts, as it is not directly connected to the flight organs. Modifications occur in different groups, for instance in relation with hypognathous heads or raptorial forelegs. The prosternum is often reduced in width in Holometabola but is broad in Megaloptera and Raphidioptera.

The prothoracic musculature differs distinctly from the pterothoracic muscle systems (Figs 1.3.3.2, 1.3.3.3, 6.13.4, see also **Table II**). A complex array of extrinsic head muscles (**cervical muscles, musculi cranii**) is always present, whereas muscles related to the flight apparatus are absent for obvious reasons. The leg muscles are similar to those of the pterothorax (see below). The cervical muscles (Fig. 6.13.4; Matsuda 1970: fig. 19, 22; Friedrich & Beutel 2008) form a complex system of dorsal and ventral longitudinal retractors and of oblique muscles, thus guaranteeing the movability of the head in all directions (Figs 1.3.3.2, 1.3.3.3C). They originate on the prothoracic **phragma**, different regions of the pronotum, the propleuron, the prosternum, and the profurca. They insert on the cervical sclerites, the tentorium and on different regions of the posterior head capsule, especially on the postoccipital ridge.

1.3.3 Pterothoracic segments

The meso- and metathorax are largely unmodified and similar to the prothorax in the apterygote lineages (e.g., Fig. 6.1.1). Far-reaching modifications have resulted from the acquisition of wings and functions related to flight, especially of the tergal and pleural elements (Figs 1.3.3.1–1.3.3.3, 6.13.3, 6.13.4).

The meso- and metaterga of pterygote insects are complex and specifically subdivided structures (Fig. 1.3.3.1B). The main division is into the large anterior **notum** and the transverse posterior **postnotum**. The notum is primarily subdivided into the short **prescutum**, the extensive **scutum**, and a **scutellum**, which is usually triangular. The part of the notum connected with the wing base is called the **alinotum**. In apterygote hexapods an **acrotergite** is separated from the prescutum by the **antecosta**, an intersegmental furrow. The acrotergite and antecosta are equivalent with the postnotum of the preceding segment of Pterygota. The antecosta (or internal part of the postnotum) forms a **phragma** for attachment of the dorsal intersegmental muscles. In most groups of Pterygota it is an extensive attachment site of large dorsal indirect flight muscles and plays an important role in the flight apparatus. The transverse prescutum is strongly narrowed in some groups. Laterally it forms the **prealare** (or **prealar sclerite**), an attachment area of short direct flight muscles. In some groups a process of the prescutum is bent downwards and linked with the episternum, thus forming a **prealar bridge**. The scutum is the largest element of the notum and the area of origin of large dorsoventral and oblique indirect flight muscles, and also of muscles attached to the basal elements of the leg. Some smaller muscles associated with the flight apparatus originate from its lateral regions. Laterally the scutum bears the anterior, the postmedian and the posterior **wing processes** (=**alar processes**), and occasionally also an anteromedian process. These small projections, usually more or less triangular in shape, are directly connected with the wing base and interact with its **axillary sclerites** (see **1.3.6 Wings**) (Fig. 1.3.3.1A, B). The scutal surface is subdivided into different regions partly corresponding with the sites of origin of the large dorsoventral muscles. The separating lines are frequently addressed as sutures (e.g., Matsuda 1970), but are in fact ridges locally increasing the rigidity of the sclerite. They also play a role in the context of the specific deformations resulting from alternating contractions of dorsal longitudinal and dorsoventral indirect flight muscles. A short anterolateral scutal line separates the area bearing the anterior alar process (occasionally also an additional anteromedian process) from the main scutal region. In a similar way, the oblique posterolateral scutal line delimits the area bearing the posterior alar process. The transverse transscutal line between the posterior margins of the anterior (or anteromedian) alar process divides the scutum in an anterior and a posterior half. Additional scutal lines can occur, such as the oblique notaulix on the anterior alinotum, or the more or less parallel-sided longitudinal "suturae parapsidiales" (Matsuda 1970). The anteromedially sinuate scutoscutellar line separates the scutum from the scutellum. Laterally it reaches the **axillary ligament** (Fig. 1.3.3.1A,

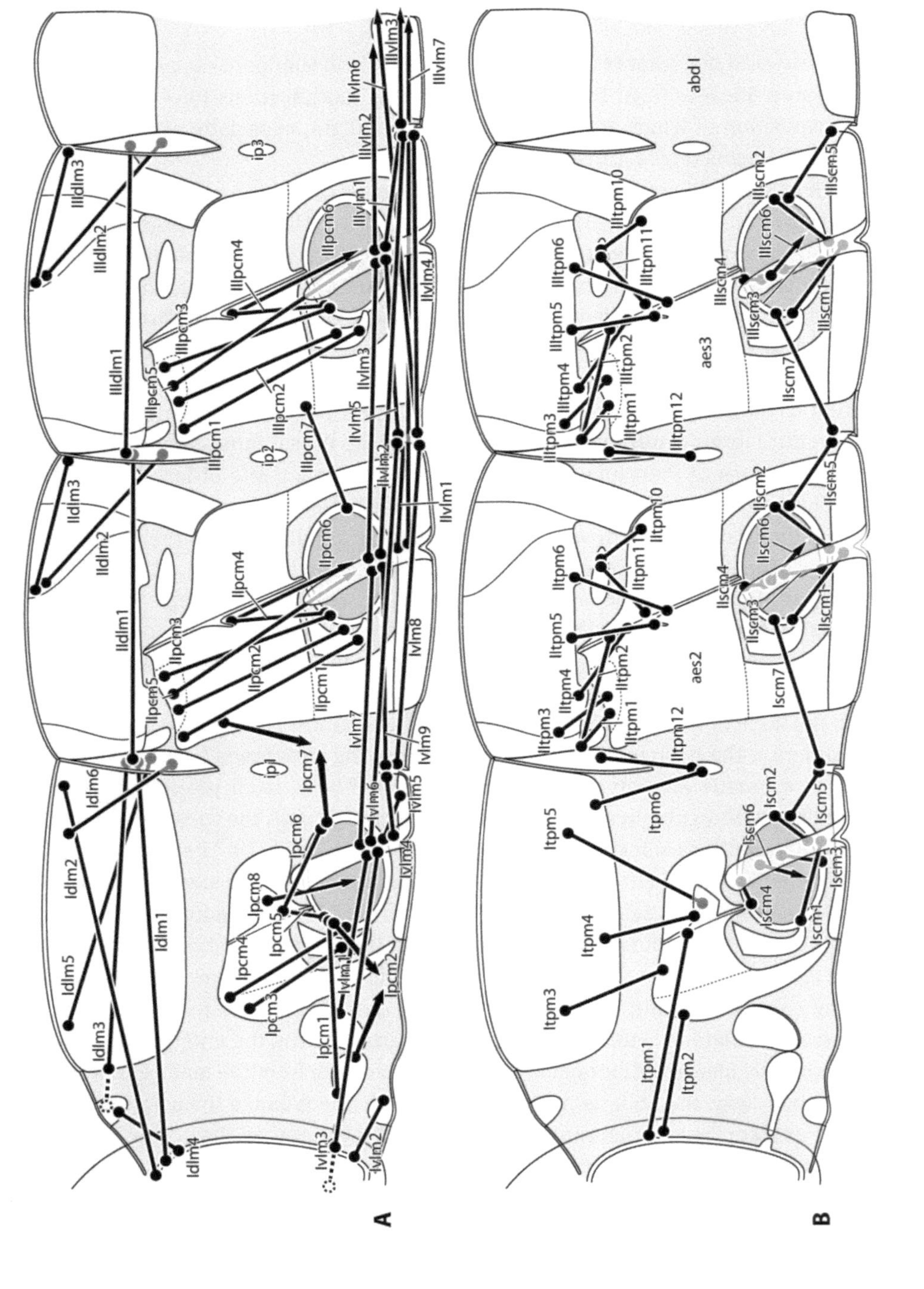
A
B
IdIm1
IdIm2
IdIm3
IdIm4
IdIm5
IdIm6
IIdIm1
IIdIm2
IIdIm3
IIIdIm1
IIIdIm2
IIIdIm3
Ipcm1
Ipcm2
Ipcm3
Ipcm4
Ipcm5
Ipcm6
Ipcm7
Ipcm8
IIpcm1
IIpcm2
IIpcm3
IIpcm4
IIpcm5
IIpcm6
IIpcm7
IIIpcm1
IIIpcm2
IIIpcm3
IIIpcm4
IIIpcm5
IIIpcm6
Ivlm1
Ivlm2
Ivlm3
Ivlm4
Ivlm5
Ivlm6
Ivlm7
Ivlm8
Ivlm9
IIvlm1
IIvlm2
IIvlm3
IIvlm4
IIvlm5
IIvlm6
IIvlm7
IIIvlm1
IIIvlm2
IIIvlm3
ipl
ip2
ip3
Itpm1
Itpm2
Itpm3
Itpm4
Itpm5
Itpm6
IItpm1
IItpm2
IItpm3
IItpm4
IItpm5
IItpm6
IItpm10
IItpm11
IItpm12
IIItpm1
IIItpm2
IIItpm3
IIItpm4
IIItpm5
IIItpm6
IIItpm10
IIItpm11
IIItpm12
Iscm1
Iscm2
Iscm3
Iscm4
Iscm5
Iscm6
Iscm7
IIscm1
IIscm2
IIscm3
IIscm4
IIscm5
IIscm6
IIscm7
IIIscm1
IIIscm2
IIIscm3
IIIscm4
IIIscm5
IIIscm6
aes2
aes3
abd I

C

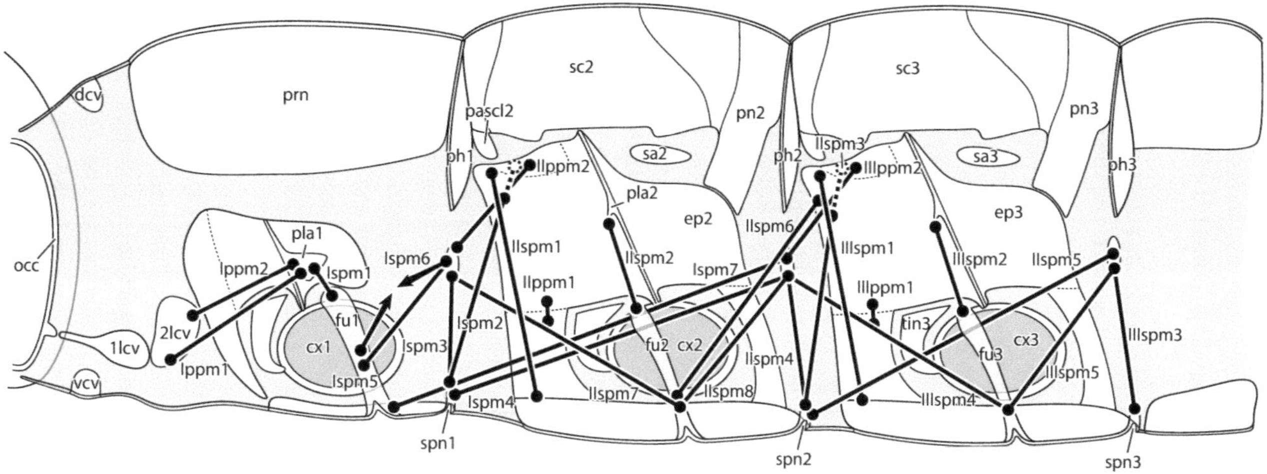

Fig. 1.3.3.2: Schematic drawings of thoracic muscle system. A, dorsal longitudinal (dlm), pleuro-coxal (pcm) and ventral longitudinal muscles; B, tergo-pleural (tpm) and sterno-coxal muscles (scm); C, pleuro-pleural (ppm) and sterno-pleural muscles (spm). Abbr.: 1/2lcv – 1st/2nd lateral cervical sclerite, abdI – abdominal segment I, aes2/3 – mes-/ metanepisternum, cx1/2/3 – pro-/ meso-/ metacoxa, dcv – dorsal cervical sclerite, ep2/3 – mes-/ metepimeron, fu1–3 – pro-, meso-, metafurca, ip1/2/3 – pro-/meso-/metathoracic interpleurite, occ – occipitale, pascl – prealar sclerite, ph1/2/3 – pro-/ meso-/ metaphragma, pla1/2 – pro-/ mesothoracic pleural arm, pn2/3 – meso-/ metapostnotum, prn – pronotum, sa2/3 – meso-/ metathoracic subalare, sc2/3 – meso-/ metascutum, spn1/2 – pro-/ meso-/ metaspina, tin3 – metatrochantin, vcv – ventral cervical sclerite. See Table II (end of 1.3) for detailed description of muscles. From Friedrich & Beutel (2008), modified.

B) which is continuous with the posterior margin of the wing base. The posteriormost part of the tergum is the postnotum (see above: antecosta and acrotergite). Similar to the prescutum it is bent downwards in some groups and connected with the epimeron, thus forming a **postalar bridge.**

The membranous region below the wing base is ventrally adjacent with the upper edge of the pleuron. The **epipleurites**, the anterior **basalare** and the posterior **subalare**, are embedded in this membrane (Figs 1.3.3.1A, 6.13.3A). They are likely small isolated elements of the anepisternum and epimeron, respectively (see below). Both are important attachment sites of direct flight muscles. The basalare is usually a small external sclerotisation connected with an extensive interior muscle disc by a tendon, whereas the subalare is a relatively extensive sclerotisation exposed at the surface of the membrane. In winged insects, the pleuron is always composed of an anterior episternum and a posterior epimeron. Both sclerites are separated by a very distinct oblique external line corresponding with the more or less extensive internal pleural ridge between the anterodorsal pleural wing joint and the posteroventral pleurocoxal articulation (see above and **1.3.4 Legs**). Its anterodorsal apex forms the **fulcrum** or **pleural wing process** in all winged insects. It articulates with the second axillary sclerite and plays a crucial role in the flight apparatus. An internal process of the pleural ridge is the pleural arm. It is almost always connected with the furcal arms by a muscle. The ventral end of the pleural ridge forms a ventromedially directed lower pleural arm in some groups (e.g., Orthoptera, Antliophora). The episternum is divided into several regions (Fig. 6.13.3A). The dorsal **anepisternum** is separated from the ventral **preepisternum** by the **anapleural cleft**. The paracoxal suture separates the katepisternum from the preepisternum. The relatively narrow **katepisternum** is adjacent with the **trochantin** and the coxa. The small, crescent-shaped trochantin articulates with the coxal base anterior to the pleurocoxal joint. The epimeron is usually undivided. In some groups a fairly indistinct horizontal suture incompletely divides the sclerite into a dorsal and a ventral part, the **anepimeron** and the **katepimeron** (Matsuda 1970). The meso- and metathoracic spiracles are usually embedded in the intersegmental membrane in front of the respective segment.

As pointed out above, the meso- and metasternal structures are primarily more or less similar to the corresponding elements of the prothorax (apterygotes and basal pterygote lineages) (e.g., Matsuda 1970: fig. 35). In most pterygote insects the entire sternal region is more extensive and often also the individual sternites (Figs 1.3.3.1C, D, 6.13.3B). The furcae (**mesendosternite, metendosternite**) are often extensive and stabilized by their lateral connection with the pleural arms (see above). The spina is often reduced in the mesothorax and almost always missing in the metathorax. Far-reaching modifications occur in acercarian lineages and a far-reaching internalization of the true sternal elements (**endosterny**) is an autapomorphy of Holometabola. The ventral body wall is then largely formed by the pleural preepisternum.

The pterothoracic muscle system is more complex than that of the prothorax (Figs 1.3.3.2, 1.3.3.3; see also **Table II**), except for the absence of extrinsic head muscles (cer-

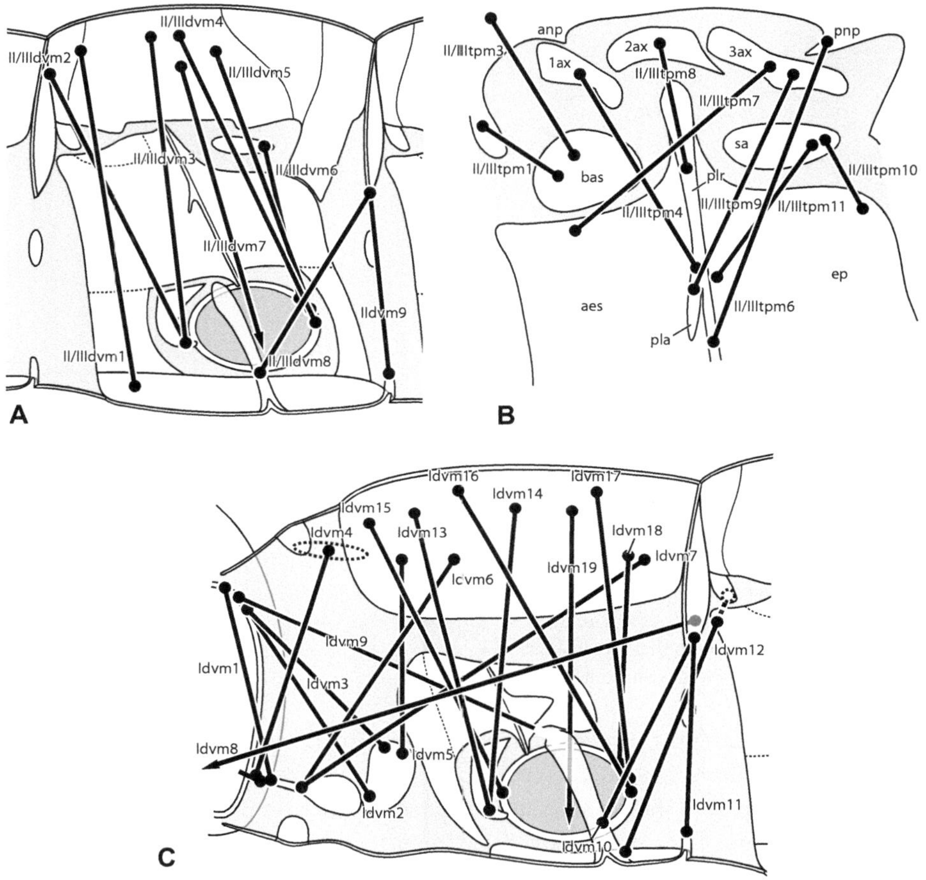

Fig. 1.3.3.3: Schematic drawings of thoracic muscle system. A, dorso-ventral muscles (dvm) of a pterothoracic segment; B, tergo-pleural wing base muscles (tpm); C, dorso-ventral muscles (dvm) of prothorax. Abbr.: 1/2/3ax – 1st / 2nd /3rd axillary sclerite, aes – anepisternum anp – anterior notal process, bas – basalare, ep – epimeron, pla – pleural ridge, plr – pleural ridge, pnp – posterior notal process, sa – subalare. See Table II (end of 1.3) for detailed description of muscles. From Friedrich & Beutel (2008), modified.

vical muscles). Ventral longitudinal muscles are present and the full sets of extrinsic and intrinsic leg muscles (Fig. 1.3.3.2). Additional muscles are related to the flight apparatus. Direct flight muscles insert on sclerites at the wing base (Fig. 1.3.3.3B). Their function is the fine adjustment during flight. Other muscles already present in the hexapod groundplan (dorsal longitudinal muscles, noto-sternal muscles, noto-coxal muscles) play an important role as indirect flight muscles (Figs 1.3.3.2, 1.3.3.3A, 1.3.3.4) in almost all pterygote groups (not in Odonata). The dorsal longitudinal muscles and dorsoventral muscles alternatively deform the pterothoracic tergites.

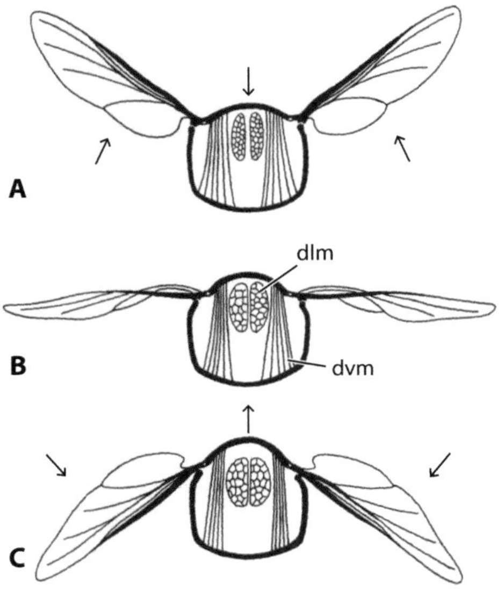

Fig. 1.3.3.4: Indirect flight muscle system. A, upstroke; B, intermediate position; C, downstroke. Abbr.: dlm – dorsal longitudinal muscles, dvm – dorsoventral muscles. Redrawn from Seifert (1995).

The resulting vibrations of the segments cause a rapid sequence of upstrokes and downstrokes of the wings (Fig. 1.3.3.4). Some muscles have alternative double functions, either as part of the flight apparatus or as leg muscles involved in the locomotion on the ground (e.g., M. notocoxalis, M. nototrochanteralis, M. coxobasalaris).

1.3.4 Legs

Three pairs of legs are present in the groundplan of Hexapoda. Reductions occur in very few specialized groups, for instance in parasitic females of Strepsiptera (Fig. 6.30.2) and semi-sessile females of Coccoidea. Different degrees of reduction occur in larvae of some holometabolan lineages (e.g., Mecoptera, Curculionidae [Coleoptera]), and the larval thoracic legs are completely absent in Diptera and Siphonaptera (Figs 6.34.4, 6.35.9, 6.35.10). Various kinds of abdominal appendages occur, especially in the apterygote lineages, but they are always distinctly modified and never function like thoracic walking legs. Secondarily evolved **abdominal prolegs** of holometabolan larvae may be involved in locomotion to a certain degree (e.g., symphytan larvae, lepidopteran caterpillars) (see **1.4 Abdomen**).

The uniramous thoracic legs of Hexapoda (Fig. 1.3.4.1) are likely derived from a biramous **arthropodium**, which is a groundplan autapomorphy of Euarthropoda

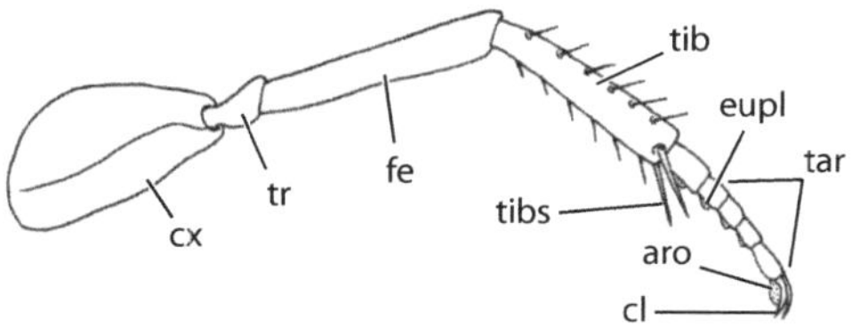

Fig. 1.3.4.1: Leg of a polyneopteran insect, schematized.
Abbr.: aro – arolium, cl – claws, cx – coxa, eupl – euplantulae, fe – femur, tar – tarsus, tib – tibia, tibs – tibial spur, tr – trochanter.

(e.g., †Trilobita, most crustacean lineages). Like in other terrestrial euarthropod lineages (Arachnida, Myriapoda) the **exopodite**, which is usually used as a swimming organ and accordingly set with fringes of hairs, is reduced or absent. The coxal **styli** of the middle and hindlegs of Archaeognatha (and on abdominal leglets) are possibly vestiges. Mesal endites of the leg base of the ancestral euarthropod legs, forming the main mechanical elements of the ventral food rim (present in the extinct †Trilobita, in Xiphosura, and in different crustacean groups) are also completely missing in Hexapoda, like in the other terrestrial groups of Arthropoda.

The proximal element of the 6-segmented hexapod leg (Fig. 1.3.4.1) is the **coxa.** It is ventrolaterally connected to the respective thoracic segment. Only one articulation is usually present in primarily wingless hexapods, but two or three in most insect lineages. The main lateral articulation is the pleurocoxal joint. Its condyle formed by the ventral end of the pleural ridge fits into a socket at the external coxal base. A second pleural articulation occurring in many groups is the trochantinocoxal joint. A pointed process of the trochantin articulates with a concavity at the anterior part of the coxal base. The mesal articulation is the sternocoxal joint. It is modified in holometabolan lineages (e.g., Amphiesmenoptera, Antliophora), with a distinctly elongated sternal process fitting into a medial coxal concavity. The coxa is relatively short but voluminous, providing attachment areas for different basal leg muscles and also containing intrinsic coxotrochanteral depressor and levator muscles. The latter originate exclusively from the coxa. Extrinsic and intrinsic depressor muscles are attached to a strong trochanteral apodeme. In many groups the coxa is divided into an anterior **basicoxa** and a posterior **meron** (or **merocoxa**). Both regions are separated by the merocoxal ridge, and an additional basicoxal ridge often separates a short proximal basicoxite from the main part of the coxa. The mesal surfaces of pterothoracic coxae are partly membranous in some groups of Holometabola (e.g., Trichoptera, Diptera). In these groups basi- and merocoxa are interconnected by a sclerotized bar, the **sternocoxale**, which also bears the socket of the sternocoxal joint. The **trochanter** is a small sclerite and usually of triangular shape. The joint connecting it with the coxa is always dicondylous, with the pivots placed anteriorly and posteriorly. A well-developed apodeme is mesally continuous with its base and reaches into the coxa. The oblique or transverse connection between the trochanter and **femur** is not a true articulation. It allows only a minimal movability between these two elements if at all. The femur is usually the largest element of the hexapod leg. It forms a monocondylous or dicondylous joint with the **tibia**. The absence of a patella between these two leg elements is a possible autapomorphy of Hexapoda. The tibia is often as long as the femur but always more slender and often cylindrical. It is usually equipped with more or less regular rows of strong setae and a pair of strong apical spurs is present in many groups. The tibiotarsal joint is usually dicondylous but monocondylous articulations occur in adults of some lineages and in holometabolan larvae. The **tarsus** is undivided in Collembola (Fig. 6.1.1), Protura (6.2.3) and Diplura, and also in †Monura (Fig. 6.A.2), an extinct lineage of Insecta (groundplan of Hexapoda). It is likely that either three or five tarsomeres belong to the groundplan of Insecta (excl. †Monura). The proximal tarsomere is often the longest and is called the **basitarsus**. The subdivision of the tarsus results in an increased flexibility of the distal part of the legs and is a precondition for the acquisition of tarsal attachment structures (see **1.3.5 Attachment structures**). Specific articulations between the tarsomeres are lacking. The **pretarsus** articulates with the apical tarsomere, the **distitarsus**. It is a short but well-defined tubular element (annulus) of the leg in Collembola, Protura and Diplura (groundplan of Hexapoda). It is distinctly reduced in Insecta (possible autapomorphy), where it is only represented by the **unguitractor** (**unguitractor plate**). A single pretarsal claw is present in Collembola and Protura, probably equivalent to the dactylopodite of crustaceans. It is distinctly modified (e.g., serrate) in some representatives of the former groups. The presence of double claws (**ungues**) is a potential synapomorphy of Diplura and Insecta. In Insecta they articulate with the unguitractor. Small sclerites, the **auxilia**, are usually present at the base of the claws. The unguitractor often bears a sclerotized **empodium**, which is usually spine-shaped. Various pretarsal attachment devices occur in pterygote insects. The most common of them are the unpaired **arolium** between the claws, and the paired **pulvilli** below them (see **1.3.5 Attachment structures**).

The legs are equipped with an extensive set of different types of sensilla (see also **1.1.4 Cuticular sensilla, 1.1.5 Scolopidia**). Mechanoreceptive sensilla trichodea occur on all parts. Contact chemoreceptors are concentrated on the distal parts, especially on the tarsomeres and on pretarsal attachment structures. One or two **subgenual organs** are present in the proximal part of the tibia in most groups (absent in Coleoptera and Diptera). These chordotonal organs are usually composed of 10–40 scolopidia. They detect substrate vibrations. Proprioreceptors in the leg include fields of sensilla trichodea, sensilla campaniformia, and chordotonal organs.

Extrinsic leg muscles (Figs 1.3.3.2A, B, 1.3.3.3A, C, see also Table II) insert on the coxal base and the trochanter. They originate on the notum, on different pleural parts including the basalare and subalare, and on sternal elements, i.e. the furca and spina. The intrinsic leg musculature is very conservative. Two antagonistic coxal muscles, sometimes subdivided into several subcomponents, insert on the trochanter. A short muscle originating in the trochanter inserts on the femoral base in some groups (femoral reductor). Two antagonistic femurotibial muscles are present, the tibial flexor and extensor. In some cases a subcomponent can originate in the trochanter. Depressors and levators of the tarsus originate in the tibia. They insert on the base of the basitarsus. Intrinsic tarsal muscles are lacking. Muscles originating in the femur and tibia insert on a long tendon which moves the claws.

1.3.5 Attachment structures

Attachment devices have evolved on different body parts of insects, but they are almost always located on the distal parts of the thoracic legs (Figs 1.3.5.1, 1.3.5.2), especially on the tarsus and pretarsus, and in rare cases on the distal region of the tibia. Adhesive structures not associated with legs occur only in some highly specialized forms, such as for instance as circular sucking discs on the ventral side of the body segments of larvae of Blephariceridae (Diptera), which live in rapidly flowing streams, or as prey catching devices on the mouthparts of predacious larvae and adults of Scydmaeninae (Coleoptera) (e.g., Jałoszyński & Beutel 2012). Other examples would be the pygidial lobes of pistilliferan larvae (Mecoptera) (Fig. 6.33.5) or the modified mouthparts of some pollen-feeding beetles. Their specific properties and functional background are clearly different from those of typical tarsal or pretarsal attachment pads, which will be treated here in more detail.

Specialized hairy attachment structures have evolved in some representatives of Archaeognatha (*Meinertellus*: distal tarsal scopulae), but they are absent in other apterygote hexapods, which usually live in leaf litter or soil substrate. Specialized claws and associated structures occurring in Collembola should not be addressed as attachment structures.

The evolution of attachment devices in insects was apparently mainly triggered by the evolution of flight. It is interesting that despite of a considerable number of different morphological types only two different mechanisms have evolved, hairy surfaces and smooth flexible pads (Fig. 1.3.5.3). In both cases, the high flexibility of the material, i.e. the pad surface or the distal parts of the adhesive microtrichia (tenent hairs, acanthae), guarantees a maximum contact with surfaces, regardless of the microsculpture. Molecular adhesion plays a decisive role even though adhesive secretions are involved in different groups (Beutel & Gorb 2001, 2006).

The terms attachment- or adhesive devices are somewhat misleading as the main function is in fact efficient locomotion, especially on plant surfaces. This means that not only the ability to attach but also to detach easily from the substrate is an important functional requirement (Beutel & Gorb 2001).

In the context of flight, insects have evolved attachment devices in order to be able to move efficiently on different plant surfaces. As a counterstrategy, plants have modified their surfaces in many different ways to prevent phytophagous insects from moving and feeding on them. Plant surfaces may be extremely smooth, covered with

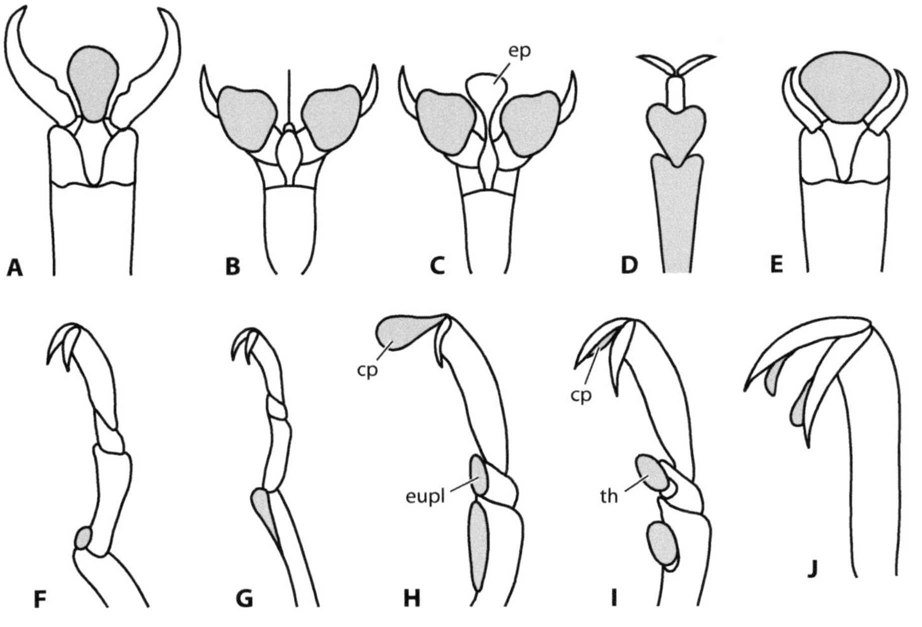

Fig. 1.3.5.1: Attachment structures (grey). A, pretarsal arolium (almost always smooth); B, pretarsal pulvilli (smooth or hairy); C, widened empodium with adhesive microtrichia; D, hairy adhesive soles of tarsomeres; E, pretarsal protrusible vesicle (arolium); F, eversible adhesive structure between tibial apex and tarsus; G, hairy fossula spongiosa (Reduviidae partim); H, tarsal euplantulae (almost always smooth); I, tarsal plantar lobes (basal groups of Hymenoptera); J, adhesive claw setae. Abbr.: ep – empodium, cp – claw pad, eupl – euplantulae, th – tarsal horns. Redrawn from Beutel & Gorb (2001) (see also Beutel & Gorb 2006).

waxy secretions, densely covered with fine hairs or thorns, or modified in other ways. Modifications of insect attachment structures and plant surfaces are complementary sides of an evolutionary arms race (e.g., Federle et al. 1997; Beutel & Gorb 2001).

As pointed out above, attachment structures are not restricted to one particular area of the leg. They may be located on claws, derivatives of the pretarsus, the tarsal apex, tarsomeres, or even on the distal tibia. The different structural types are described briefly in the following.

Arolium (Figs 1.3.5.1A, E, 1.3.5.2D, 1.3.5.4): the median hollow lobe of the pretarsus. It can be completely membranous or partly sclerotized. The arolium is morphologically and ultrastructurally completely different from a lobular median empodium (Fig. 1.3.5.1C). Its surface is almost always smooth (not in Mantophasmatodea and basal phasmatodeans; Beutel & Gorb 2008) and its cuticle consists of rod-like internal structures oriented perpendicularly or at some angle to the surface. The arolium is the most widespread attachment structure in insects and possibly a groundplan autapomorphy of Neoptera (Beutel & Gorb 2006).

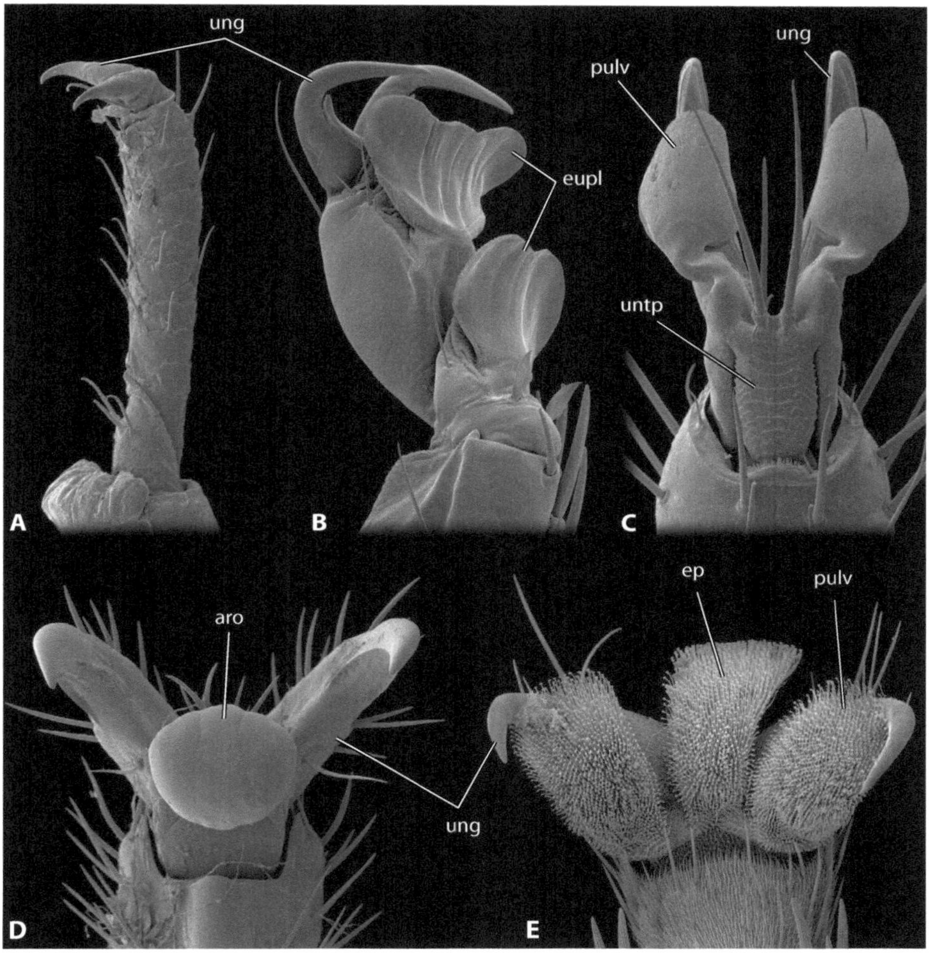

Fig. 1.3.5.2: Attachment structures, SEM micrographs. A, *Aphis sambuci* (Sternorrhyncha, Aphididae); B, *Trinoton anserinum* (Amblycera, Menoponidae); C, *Graphosoma lineatum* (Heteroptera, Pentatomidae); D, *Locusta migratoria* (Orthoptera, Acrididae); E, *Tabanus sp.* (Diptera, Tabanidae). Abbr.: aro – arolium, ep – empodium, eupl – euplantulae, pulv – pulvilli, ung– ungues (claws), untp – unguitractor plate. Courtesy K. Friedemann.

Empodium (Figs 1.3.5.1C, 1.3.5.2E): a median process between the claws and pulvilli (if present). It arises from the distal end of the unguitractor plate, is spine-shaped or lobe-like and is sometimes similar in form to the pulvilli ("empodial pulvillus"). In contrast to the (typical) arolium the empodium is composed of common layered cuticle and its surface is covered with acanthae. The empodium is strongly extended in many groups of Diptera (Beutel & Gorb 2001).

Auxiliae or auxiliary sclerites (=Basipulvilli s. Dashman 1953) (Fig. 1.3.5.4B): lateral sclerites beneath the bases of the claws.

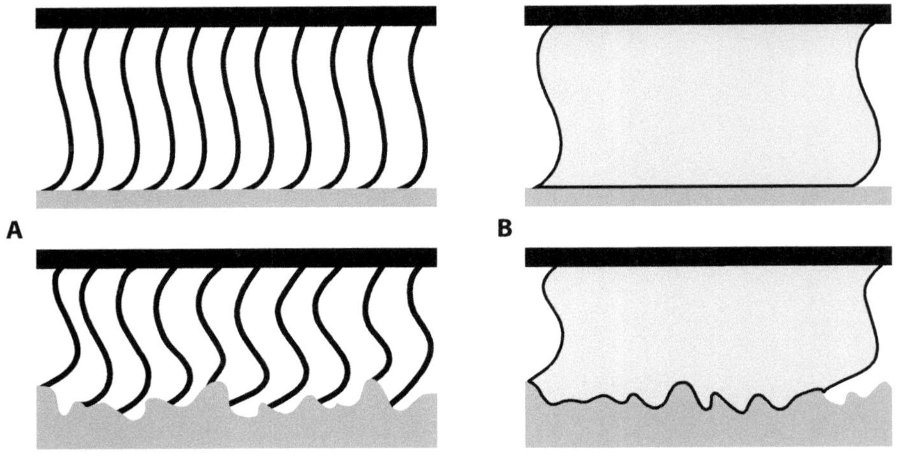

Fig. 1.3.5.3: Function of different types of tarsal or pretarsal attachment structures. A, hairy type; B, smooth type. Redrawn from Beutel & Gorb (2001).

Pulvilli (Figs 1.3.5.1B, C, 1.3.5.2C, E): smooth or hairy paired lateral membranous lobes below the claws (Bauchhenss 1979). They are located on the auxiliae, which are involved in the control of the movements of the pulvilli. Pulvilli of different shape have evolved in Acercaria and Holometabola. They are always absent in Polyneoptera.

Euplantulae (Figs 1.3.5.1H, 1.3.5.2B): flexible pad-like structures on the ventral side of one or more tarsomeres. Usually smooth but with a dense vestiture of spike-like acanthae in Mantophasmatodea (Beutel & Gorb 2006) and basal phasmatodeans (Timematodea). Euplantulae are a potential groundplan autapomorphy of Polyneoptera, and have likely evolved independently (and structurally different) in few subordinate taxa of Acercaria.

Claw pads (Fig. 1.3.5.1H, I: cp): one or both claws can be partly or completely membranized and then function as attachment devices. Typical claw pads occur in Ephemeroptera.

Fossula spongiosa (Fig. 1.3.5.1G): the distal region of the tibiae of Cimicomorpha (partim) bears a field of tenent hairs (acanthae) with spatulate or tapering apices interspersed with sensory setae. A unique feature of Reduviidae is the presence of a thick and flexible cuticle, underlain by a haemolymph cavity separated almost entirely from the interior of the remaining tibia by a cuticular invagination (Weirauch 2007).

Plantar lobes (Fig. 1.3.5.1I: th): plantar lobes occur in basal hymenopteran lineages ("Symphyta"). They are probably flat outgrowths of the tarsal cuticle (Beutel & Gorb 2006: fig. 5) and not derived from tarsal thorns as suggested in (Beutel & Gorb 2006). They are elongated in Xyelidae (sistergroup of the remaining Hymenoptera) and the surface is entirely smooth. Their presence is likely a groundplan autapomorphy of Hymenoptera with secondary reduction within the group.

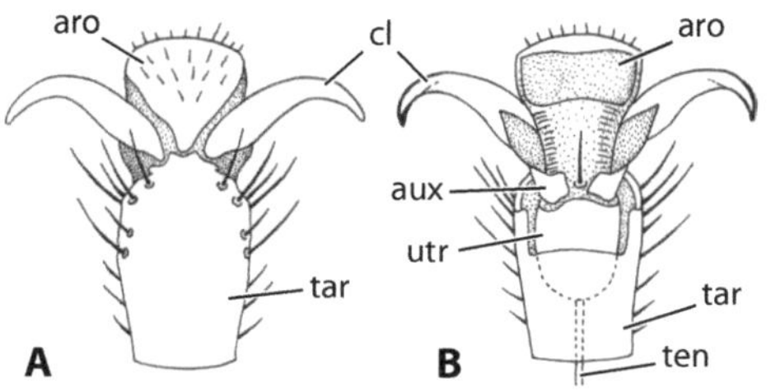

Fig. 1.3.5.4: Pretarsus, *Periplaneta americana*. A, dorsal view; B, ventral view. Abbr.: aro – arolium, aux – auxilium, cl – claw, tar – tarsus, ten – tendon of unguitractor, utr – unguitractor. Redrawn from Seifert (1995), after Snodgrass (1935).

1.3.6 Wings

The wings are doubtlessly an autapomorphy of Pterygota but their evolutionary origin is disputed (see for instance discussion in Grimaldi & Engel 2005). The traditional hypothesis (e.g., Snodgrass 1935; see also Hennig 1969) suggests that they originated from meso- and metathoracic paranota. The alternative exite or gill theory (e.g., Kukalová-Peck 1991) hypothesizes an origin from basal leg appendages (exites), which are considered as serial homologues of abdominal gills as they occur in aquatic larvae of some extant groups (e.g., Ephemeroptera, Megaloptera). In any case, the highly complex structure of the wings clearly indicates a single origin in the evolution of Hexapoda. Different modifications occur (see below) but the ancestral pattern is almost always still recognizable.

Fully developed wings (Fig. 1.3.6.1) occur only in mature insects, with the notable exception of the subimago of Ephemeroptera (possibly a groundplan feature of Pterygota). They are connected to the lateral edge of the scutum (alinotum) and the upper margin of the pleuron, involving three axillary sclerites of the wing base in most groups of Pterygota (Figs 1.3.6.1, 1.3.6.2). An atypical condition is found in Odonata, which like Ephemeroptera are unable to fold back their wings (Palaeoptera). They possess only two large plates (**humeral** and **axillary plate**) hinged to the scutum and supported by two arms of the pleural wing process. Three **axillary sclerites** (**pteralia** of German authors) are present in Ephemeroptera and most groups of Neoptera, but four in Orthoptera [partim], Hymenoptera and Diptera. They are embedded in the double membrane at the wing base and form a complicated 3-dimensional functional complex. Axillary 1 is a part of the dorsal membrane and articulates mesally with the anterior notal wing process, anterolaterally with the subcostal vein, and laterally with axillary 2. Axillary 2 is a sclerotisation of both, the dorsal and the ventral membrane (at least in some groups). Mesally it articulates with axillary 1, ventrally with the pleural wing process, and posteriorly with axillary 3. Anterolaterally it is connected with the third major longitudinal vein, the **radius** (R), and laterally it is adjacent

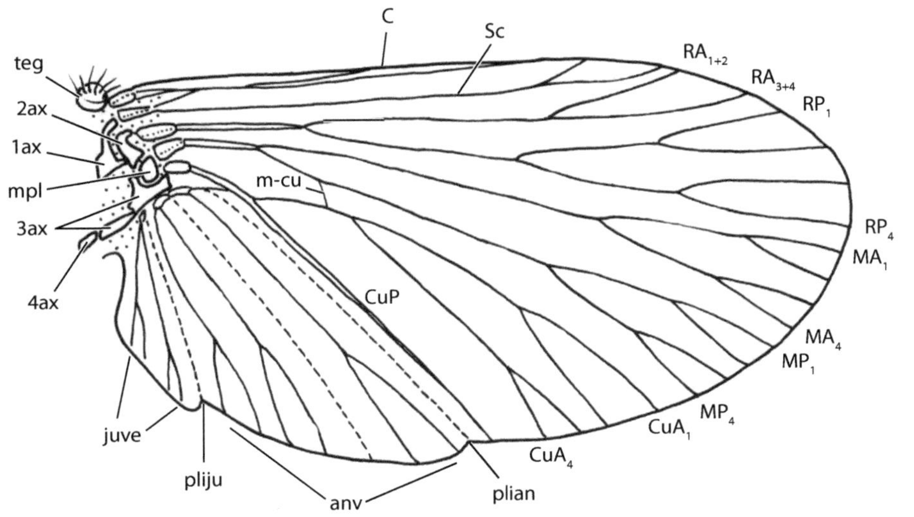

Fig. 1.3.6.1: Neopteran wing, generalized. Abbr.: 1–4ax – 1st – 4th axillary sclerite, anv – anal veins, C – costa, $CuA_{1/4}$ – branches of cubitus anterior, CuP – cubitus posterior, juve – jugal veins, $MA_{1/4}$ – branches of media anterior, m-cu – medio-cubital vein, $MP_{1/4}$ – branches of media posterior, mpl – median plate, plian – plica analis, pliju – plica jugalis, $RA_{1+2,\,3+4}$ – branches of radius anterior, $RP_{1/4}$ – branches of radius anterior, Sc – subcosta, teg – tegula. Redrawn from Gullan & Cranston (1994).

with the **median plates** (see below). Axillary 3 is usually Y-shaped. Its posterolateral margin is adjacent with the bases of the **anal** (A) and **jugal veins** (J; if present) and its posterior edge articulates with the posterior notal wing process. In Neoptera a muscle attached to axillary 3 triggers the rotation of this sclerite. As a consequence the wings can be folded back over the abdomen (autapomorphy of Neoptera).

A structure at the posterior wing base is the **axillary cord**. It connects the posterolateral edge of the alinotum with the posterior wing base. The **tegula** is a characteristic small lobe at the anterior wing margin. It is unsclerotized and covered with short sensilla trichodea. It is closely connected with the **humeral plate**, a small sclerotisation at the anterior wing base, which is laterally connected with the anteriormost longitudinal vein, the **costa** (C).

A simple criterion to distinguish between the wing base and the wing plate is that both membranes are completely fused in the latter, without epidermal cells between them. A proximal structure of the wing interacting with the axillaries 2 and 3 is the **median plate.** In the typical case it is diagonally subdivided into a proximal and a distal subunit (proximal and distal plate) and is in contact with the bases of three major longitudinal veins laterally, the **media** (M), the **cubitus** (Cu) and the **postcubitus** (PCu).

The very thin sheath formed by the connected two wing membranes is reinforced by a system of longitudinal and transverse veins. Only the longitudinal veins

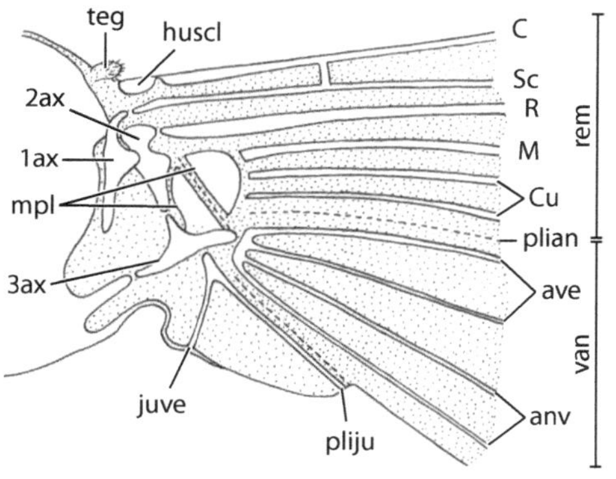

Fig. 1.3.6.2: Neopteran wing base. Abbr.: 1–3ax – 1st – 3rd axillary sclerite (4th usually missing), ave – anal veins, C – costa, Cu – cubitus, huscl – humeral sclerite, juve – jugal veins, M – media, mpl – median plates, plian – plica analis, pliju – plica jugalis, R – radius, rem – remigium (costal field), Sc – subcosta, teg – tegula, van – vannus (anal field). Redrawn from Dathe (2005).

contain haemolymph, tracheae and nerves. Their arrangement on alternating levels (costa convex, subcosta concave etc.) results in a structure resembling corrugated iron, which also increases the mechanical stability of the thin wing foil. The transverse veins are solid structures. They form a dense reticular network in some basal neopteran lineages, the **archedictyon** (probably a groundplan feature of Pterygota). This condition is preserved in the extant Ephemeroptera, Odonata (Palaeoptera) and Orthoptera, and also in the extinct †Palaeodictyoptera, which have also preserved prothoracic winglets with such a veinous pattern. In most extant groups the number of transverse veins is strongly reduced. They are completely absent in extant Strepsiptera and some other lineages.

The wing is subdivided into three main regions. The anterior **remigium** or **costal field** is the most extensive part and plays a dominant role in flight. The posteriorly adjacent **anal field** or **vannus** usually displays a fan-shaped pattern of veins. The **jugal field** at the posterior wing margin is small and completely absent in many groups. The remigium is usually subdivided by the median flexion line, which originates close to the media (M) and runs behind the radial sector (RP) (see below). The **plica vannalis** runs close to the postcubitus and separates the costal and anal fields. Additionally, a **plica jugalis** separating the anal and jugal fields can be present.

The first and strongest vein of the remigium is the unbranched **costa** (C). The following vein, the **subcosta** (Sc), often branches once. The **radius** (R) usually divides several times. Its anterior primary branch is referred to as **radius anterior** (RA) and the posterior one as **radial sector** or **radius posterior** (RP). The secondary branches are usually designated by subscripts (e.g., RA_1, RA_2). Connections by transverse veins resulting in closed radial cells occur frequently. The next vein, the **media** (M), also divides several times. The primary branches are the media anterior (MA) and the media posterior (MP). Similarly, the following **cubitus** (C) branches into the cubitus anterior (CuA) and cubitus posterior (CuP), both again dividing into further secondary and tertiary branches. The posteriormost longitudinal vein of the remigium is the **postcubitus** (PCu). The anal field is usually subdivided by several **anal veins** (1A, 2A, etc.). The small jugal field occasionally bears one or two short veins in some groups, but is often devoid of veins or reduced (see above).

The surface of the wing is usually more or less densely covered with very short microtrichia. The wing membrane can be partly or entirely transparent (e.g., Hymenoptera) but color patterns occur frequently, most conspicuously in Lepidoptera. A specialized structure of the anterior margin is the **pterostigma**, a pigmented spot occurring in Odonata and some other groups (e.g., Psocoptera, Megaloptera).

Different modifications and reductions of wings occur in pterygote insects. They are completely absent in all known species of Grylloblattodea, Mantophasmatodea and Siphonaptera and also in subordinate lineages of other groups (e.g., Phasmatodea). Alate and wingless morphs occur in Zoraptera and some subordinate lineages (e.g., Isoptera, Formicidae). Females of Embioptera are always wingless, whereas males have simplified wings which are shed after a short mating flight. The forewings are moderately sclerotized **tegmina** in different groups of Polyneoptera (e.g., Orthoptera, Blattodea). The anterior part of the forewings is sclerotized in Heteroptera (**hemielytra**) and the entire forewings in Coleoptera (**elytra**). The anal field of the hindwings of different polyneopteran lineages is enlarged, with a fan-like folding pattern (Plecoptera [partim], Orthoptera, Dictyoptera [excl. Isoptera], Dermaptera).

Aside from the ability to fold back the wings (autapomorphy of Neoptera) different transverse or longitudinal folding patterns occur. This is always linked with modifications of veins such as articulations (Scarabaeoidea) or bending zones (e.g., other groups of Coleoptera). A complicated mechanism of folding the greatly enlarged anal field has evolved in Dermaptera, which store the modified hindwings under the strongly shortened and sclerotized forewings. The hindwings of Diptera are modified as **halteres**, which are highly specialized gyroscopic sense organs and stabilize the flight by their rotating movements. A very similar condition has evolved independently in Strepsiptera, where the forewings are transformed into halteres. A typical modification of wings of very small insect is the reduction of the wing membrane to narrow straps with only one or two longitudinal veins, and fringes of long bristles along the margin. The air has an increased viscosity for these minute insects (low Reynolds number). Their modified wings enable them to “swim” through this medium. This condition has independently evolved in Thysanoptera and in some groups of Coleoptera (Sphaeriusidae [Myxo-

Table II: Generalized thoracic muscle nomenclature (from Friedrich & Beutel 2008, modified)

Prothorax

Abbrev.	Name	Origin	Insertion
Dorsal longitudinal muscles			
Idlm1	M. prophragma-occipitalis	middle region of prophragma	dorsal or dorsolateral area of occipitale
Idlm2	M. pronoto-occipitalis	posteromedian area of pronotum	dorsal or dorsolateral area of occipitale (close to Idlm1)
Idlm3	M. prophragma-cervicalis	prophragma (laterad of Idlm1, mediad of Idlm5)	dorsal cervical sclerite, cervical membrane or anterior margin of pronotum
Idlm4	M. cervico-occipitalis dorsalis	dorsal cervical sclerite or anterior margin of pronotum	dorsal or dorsolateral area of occipitale (laterad of Idlm2)
Idlm5	M. pronoto-phragmalis anterior	anterior half of central pronotal region	prophragma (laterad Idlm1, Idlm3)
Idlm6	M. pronoto-phragmalis posterior	posterior area of pronotum	ventrolateral part of prophragma or antero-dorsal area of mesanepisternum or mesobasalare
Dorsoventral muscles			
Idvm1	M. cervico-occipitalis anterior	anterior end of first lateral cervical sclerite	dorsal or dorsolateral part of occipitale
Idvm2	M. cervico-occipitalis medialis	posterior part of first or anterior area of second lateral cervical sclerite	dorsolateral area of occipitale (laterad of Idvm1)
Idvm3	M. cervico-occipitalis posterior	posterolateral part of second lateral cervical sclerite	dorsolateral area of occipitale (laterad of Idvm1)
Idvm4	M. pronoto-cervicalis lateralis	dorsal cervical sclerite, cervical membrane and/or anterior region of pronotum	posterior region of head or anterior part of first lateral cervical sclerite
Idvm5	M. pronoto-cervicalis anterior	anterior part of pronotum	posterior margin of second lateral cervical sclerite
Idvm6	M. pronoto-cervicalis medialis	central region of pronotum	central area of first lateral cervical sclerite
Idvm7	M. pronoto-cervicalis posterior	posterior half of pronotum	anterior region of first lateral cervical sclerite
Idvm8	M. prophragma-tentorialis	prophragma	tentorium or occipitale
Idvm9	M. profurca-occipitalis	tip of profurcal arm	dorsolateral area of occipitale (between Idlm1 and Idlm2)
Idvm10	M. profurca-phragmalis	dorsal face of proximal profurcal arm	ventral area of prophragma or anterior part of mesopleuron

Abbrev.	Name	Origin	Insertion
Idvm11	M. prophragma-spinalis	prophragma	prospina
Idvm12	M. profurca-mesonotalis	base of profurca or region behind it	anterior part of mesonotum or prealar sclerite
Idvm13	M. pronoto-trochantinalis anterior	anterior region of pronotum	trochantin (close to Idvm14)
Idvm14	M. pronoto-trochantinalis posterior	central region of pronotum	trochantin (close to Idvm13)
Idvm15	M. pronoto-trochantinocoxalis	anterior part of pronotum (anterior to Idvm13)	anterior coxal rim, rarely on trochantin
Idvm16	M. pronoto-coxalis anterior	paramedially on anterior area of pronotum	posterior procoxal rim
Idvm17	M. pronoto-coxalis posterior	posterior region of pronotum	posterolateral procoxal rim (laterad of Idvm16)
Idvm18	M. pronoto-coxalis lateralis	posterolateral part of pronotum	posterolateral procoxal rim (laterad of Idvm17)
Idvm19	M. pronoto-trochanteralis	pronotum (anterolaterad of Idvm17)	trochanter
Tergo-pleural muscles			
Itpm1	M. pleurocrista-occipitalis	dorsolateral or lateral area of occipitale	pleural ridge anterad of propleural arm
Itpm2	M. propleuro-occipitalis	dorsolateral or lateral area of occipitale or laterally on cervical membrane	anterior edge of propleuron
Itpm3	M. pronoto-pleuralis anterior	anterolateral part of pronotum	anterodorsal margin of propleuron
Itpm4	M. pronoto-apodemalis anterior	lateral part of pronotum (posterolaterad of Itpm3)	pleural apodeme, rarely tip of profurcal arm
Itpm5	M. pronoto-apodemalis posterior	posterolateral region of pronotum	posterior face of propleural arm and/or pleuron behind it
Itpm6	M. pronoto-intersegmentalis	posterolateral part of pronotum	intersegmental membrane or intersegmental sclerite between pro- and mesothorax
Pleuro-pleural muscles			
Ippm1	M. propleuro-cervicalis medialis	propleural arm or pleural ridge above	medial rim of (second) cervical sclerite
Ippm2	M. propleuro-cervicalis lateralis	base of propleural arm	posterolateral rim of cervical sclerite
Sterno-pleural muscles			
Ispm1	M. profurca-apodemalis	tip of profurcal arm	propleural arm
Ispm2	M. prospina-mesopleuralis	prospina	mesepisternum or mesobasalare or ventrolateral area of prophragma

Ispm3	M. prospina-intersegmentalis	prospina	intersegmental membrane or intersegmental sclerite between pro- and mesothorax
Ispm4	M. prospina-intersegmentalis posterior	prospina	intersegmental membrane or intersegmental sclerite between meso- and metathorax
Ispm5	M. profurca-intersegmentalis posterior	profurca	intersegmental membrane or intersegmental sclerite between pro- and mesothorax
Ispm6	M. profurca-mesopleuralis transversalis	profurca	mesanepisternum of opposite side
Ispm7	M. prosterno-intersegmentalis posterior	sternum posterad of profurcal stem	intersegmental membrane or intersegmental sclerite between meso- and metathorax
Pleuro-coxal muscles			
Ipcm1	M. procoxa-cervicalis	anterior rim of procoxa	first lateral cervical sclerite
Ipcm2	M. procoxa-cervicalis transversalis	anterior rim of procoxa	posterior face of first lateral cervical sclerite of opposite side
Ipcm3	M. propleuro-trochantinalis	anterodorsal area of propleuron	trochantin
Ipcm4	M. propleuro-coxalis superior	anterodorsal area of propleuron	anterior procoxal rim
Ipcm5	M. propleuro-coxalis inferior	ventral face of propleural apodeme and/or crista below it	anterior procoxal rim
Ipcm6	M. propleuro-coxalis posterior	propleural apodeme and/or crista below it	posterior procoxal rim
Ipcm7	M. mesanepisterno-procoxalis transversalis	anterior margin of mesanepisternum	posterior procoxal rim of opposite side
Ipcm8	M. propleuro-trochanteralis	dorsal margin of propleuron and /or propleural apodeme	trochanter
Ventral longitudinal muscles			
Ivlm1	M. profurca-cervicalis	anterior face of profurca	posterior rim of second lateral cervical sclerite
Ivlm2	M. submento-cervicalis	posterior submental edge	posteromedian rim of ventral cervical sclerite
Ivlm3	M. profurca-tentorialis	profurca	tentorium or occipitale
Ivlm4	M. profurca-spinalis	profurcal arm	prospina or profurcal arm of opposite side
Ivlm5	M. profurca-sternalis	profurca	prosternum between profurcal stem and prospina

Abbrev.	Name	Origin	Insertion
Ivlm6	M. profurca-mesospinalis	profurca	mesospina
Ivlm7	M. profurca-mesofurcalis	profurca	mesofurca
Ivlm8	M. prospina-mesospinalis	prospina	mesospina
Ivlm9	M. prospina-mesofurcalis	prospina	mesofurca
Sterno-coxal muscles			
Iscm1	M. profurca-coxalis anterior	lateral face of proximal profurcal stem and/or adjacent prosternum	anterior procoxal rim, close to trochantino-coxal articulation
Iscm2	M. profurca-coxalis posterior	posterolateral face of profurca	posterior procoxal rim
Iscm3	M. profurca-coxalis medialis	ventral face of proximal profurcal arm	mesal procoxal rim
Iscm4	M. profurca-coxalis lateralis	ventral area of distal profurcal arm	lateral coxal rim, close to pleuro-coxal articulation
Iscm5	M. prospina-coxalis	prospina	posterior procoxal rim
Iscm6	M. profurca-trochanteralis	profurca	trochanter
Iscm7	M. prospina-mesocoxalis	prospina	anterior mesocoxal rim

Mesothorax

Abbrev.	Name	Origin	Insertion
Dorsal longitudinal muscles			
IIdlm1	M. prophragma-mesophragmalis	median region of prophragma	median region of mesophragma
IIdlm2	M. mesonoto-phragmalis	posterolateral region of mesonotum	ventrolateral area of mesophragma (lateral of IIdlm1)
IIdlm3	M. mesoscutello-scutellaris	posterior mesoscutal rim	mesoscutellum
Dorsoventral muscles			
IIdvm1	M. mesonoto-sternalis	anteromedian region of mesonotum (laterad of IIdlm1)	anterolateral part of mesosternum
IIdvm2	M. mesonoto-trochantinalis anterior	prophragma and/or anterior region of mesonotum (laterad of IIdvm1)	trochantin, close to trochantino-coxal articulation

IIdvm3	M. mesonoto-trochantinalis posterior	central region of mesonotum	trochantin (partly close to trochantino-coxal articulation)
IIdvm4	M. mesonoto-coxalis anterior	central region of mesonotum (posterad of IIdvm3)	posterior mesocoxal rim
IIdvm5	M. mesonoto-coxalis posterior	central region of mesonotum (posterolateral of IIdvm4)	posterior mesocoxal rim (laterad of IIdvm4)
IIdvm6	M. mesocoxa-subalaris	posterolateral mesocoxal rim (lateral of IIdvm5)	subalare (or lateral region of mesonotum in wingless forms)
IIdvm7	M. mesonoto-trochanteralis	central region of mesonotum (lateral of IIdvm4)	trochanter
IIdvm8	M. mesofurca-phragmalis	dorsal face of proximal mesofurcal arm	ventrolateral area of mesophragma
IIdvm9	M. mesospina-phragmalis	mesospina	ventrolateral area of mesophragma (close to IIdvm8)
Tergo-pleural muscles			
IItpm1	M. prophragma-mesanepisternalis	prophragma, prealar sclerite or anterolateral region of mesonotum	anterodorsal edge of mesanepisternum or basalare
IItpm2	M. mesopleura-praealaris	pleural ridge below pleural wing process or mesopleural arm	subtegula, lateral part of prophragma, prealar sclerite or detached muscle disc
IItpm3	M. mesonoto-basalaris	anterolateral or lateral region of mesonotum	basalare
IItpm4	M. mesonoto-pleuralis anterior	mesopleural arm or pleural ridge above it	anterior notal process or first axillary sclerite
IItpm5	M. mesonoto-pleuralis medialis	mesopleural arm	lateral margin of mesonotum between anterior and posterior notal process
IItpm6	M. mesonoto-pleuralis posterior	pleural ridge below mesopleural arm	posterior notal process or fourth axillary sclerite
IItpm7	M. mesanepisterno-axillaris	dorsal part of mesanepisternum or basalare	third axillary sclerite
IItpm8	M. mesepimero-axillaris secundus	upper part of mesopleural ridge	second axillary sclerite
IItpm9	M. mesepimero-axillaris tertius	mesopleural arm or pleural ridge above it	third axillary sclerite
IItpm10	M. mesepimero-subalaris	posterodorsal part of mesepimeron	posterior region of subalare
IItpm11	M. mesopleura-subalaris	epimeral face of mesopleural ridge or arm	posterior region of subalare
IItpm12	M. prophragma-intersegmentalis	lateral part of prophragma or prealar sclerite	intersegmental membrane or intersegmental sclerite between pro- and mesothorax

Abbrev.	Name	Origin	Insertion
Pleuro-pleural muscles			
IIppm1	M. mesotransanapleuralis	mesopleuron dorsad of anapleural cleft	mesopleuron ventrad of anapleural cleft
IIppm2	M. mesobasalare-intersegmentalis	anteroventral area of mesobasalare	intersegmental membrane or intersegmental sclerite between pro- and mesothorax or anterior end of anapleural cleft
Sterno-pleural muscles			
IIspm1	M. mesopleura-sternalis	anterior part of mesanepisternum or mesobasalare	lateral region of mesosternum (laterad of IIdvm1)
IIspm2	M. mesofurca-pleuralis	tip of mesofurcal arm	mesopleural arm
IIspm3	M. mesospina-metanepisternalis	mesospina	anterior region of metanepisternum or metabasalare
IIspm4	M. mesospina-intersegmentalis	mesospina	intersegmental membrane or intersegmental sclerite between meso- and metathorax
IIspm5	M. mesospina-intersegmentalis posterior	mesospina	intersegmental membrane or intersegmental sclerite between metathorax and abdominal pleura
IIspm6	M. mesofurca-metanepisternalis	mesofurca	anterior region of metanepisternum
IIspm7	M. mesofurca-intersegmentalis anterior	mesofurca	intersegmental membrane or intersegmental sclerite between pro- and mesothorax
IIspm8	M. mesofurca-intersegmentalis posterior	mesofurca	intersegmental membrane or intersegmental sclerite between meso- and metathorax
Pleuro-coxal muscles			
IIpcm1	M. mesanepisterno-trochantinalis	anterodorsal edge of mesanepisternum	trochantin
IIpcm2	M. mesobasalare-trochantinalis	mesobasalare or anterodorsal part of mesanepisternum (posterior of IIpcm2)	trochantin or anterior mesocoxal rim
IIpcm3	M. mesanepisterno-coxalis anterior	dorsal part of mesanepisternum (anterior of pleural wing process)	anterolateral coxal rim (close to IIpcm4)

IIpcm4	M. mesanepisterno-coxalis posterior	ventral half of mesanepisternum (close to mesopleural arm) and/or paracoxal suture	anterolateral coxal rim (close to IIpcm3)
IIpcm5	M. mesanepisterno-trochanteralis	mesobasalare or anterodorsal part of mesanepisternum	trochanter
IIpcm6	M. mesopleura-trochanteralis	middle of mesanepisternum or mesopleural arm	trochanter (close to IIpcm4)
Ventral longitudinal muscles			
IIvlm1	M. mesofurca-spinalis	mesofurcal arm	mesospina or mesofurcal arm of opposite side
IIvlm2	M. mesofurca-metaspinalis	mesofurca	metaspina
IIvlm3	M. mesofurca-metafurcalis	mesofurca	metafurca
IIvlm4	M. mesospina-metaspinalis	mesospina	metaspina
IIvlm5	M. mesospina-metafurcalis	mesospina	metafurca
IIvlm6	M. mesospina-abdominosternalis	mesospina	lateral margin of second abdominal sternite
IIvlm7	M. mesofurca-abdominosternalis	mesofurca	median part of second abdominal sternite
Sterno-coxal muscles			
IIscm1	M. mesofurca-coxalis anterior	laterally on proximal mesofurcal stem and/or adjacent sternal region	anterior mesocoxal rim, close to trochantino-coxal articulation
IIscm2	M. mesofurca-coxalis posterior	posterior face of proximal mesofurca	posterior or posterolateral mesocoxal rim
IIscm3	M. mesofurca-coxalis medialis	ventral face of mesofurcal arm	mesal mesocoxal rim
IIscm4	M. mesofurca-coxalis lateralis	ventral face of mesofurcal arm (laterad of IIscm3)	lateral coxal rim, close to pleuro-coxal articulation
IIscm5	M. mesospina-coxalis	mesospina	posterior mesocoxal rim
IIscm6	M. mesofurca-trochanteralis	ventral face of mesofurca	trochanter
IIscm7	M. mesospina-metacoxalis	mesospina	anterior metacoxal rim

Metathorax

Abbrev.	Name	Origin	Insertion
Dorsal longitudinal muscles			
IIIdlm1	M. mesophragma-metaphragmalis	central region of mesophragma	central region of metaphragma
IIIdlm2	M. metanoto-phragmalis	posterior region of metanotum	ventrolateral part of metaphragma
IIIdlm3	M. metascutello-scutellaris	posterior metascutal rim	metascutellum
Dorsoventral muscles			
IIIdvm1	M. metanoto-sternalis	anterior region of metanotum	lateral area of metasternum
IIIdvm2	M. metanoto-trochantinalis anterior	mesophragma and/or anterior part of metanotum	trochantin, close to trochantino-coxal articulation
IIIdvm3	M. metanoto-trochantinalis posterior	central region of metanotum	trochantin (close to IIIdvm2)
IIIdvm4	M. metanoto-coxalis anterior	anterior half of metanotum (posterad of IIIdvm3)	posterior metacoxal rim
IIIdvm5	M. metanoto-coxalis posterior	posterolateral region of metanotum (posterolaterad of IIIdvm4)	posterior metacoxal rim (laterad of IIIdvm4)
IIIdvm6	M. metacoxa-subalaris	posterior metacoxal rim (laterad of IIIdvm5)	subalare (or lateral part of metanotum in wingless forms)
IIIdvm7	M. metanoto-trochanteralis	central region of metanotum (laterad of IIIdvm4)	trochanter
IIIdvm8	M. metafurca-phragmalis	dorsal face of proximal part of metafurcal arm	ventrolateral area of metaphragma
Tergo-pleural muscles			
IIItpm1	M. mesophragma-metanepisternalis	mesophragma, anterolateral part of metanotum or prealar sclerite	anterodorsal edge of metanepisternum or basalare
IIItpm2	M. metapleura-praealaris	metapleural ridge below pleural wing process or pleural arm	prealar sclerite, subtegula or lateral metaphragma
IIItpm3	M. metanoto-basalaris	anterolateral or lateral area of metanotum	metabasalare
IIItpm4	M. mesonoto-pleuralis anterior	metapleural arm or pleural ridge above	anterior notal process or first axillary sclerite
IIItpm5	M. metanoto-pleuralis medialis	metapleural arm	lateral part of metanotum between anterior and posterior notal process

IIItpm6	M. metanoto-pleuralis posterior	pleural ridge below metapleural arm	posterior notal wing process or fourth axillary sclerite
IIItpm7	M. metanepisterno-axillaris	dorsal part of metanepisternum or basalare	third axillary sclerite
IIItpm8	M. metepimero-axillaris secundus	dorsal part of metapleural ridge	second axillary sclerite
IIItpm9	M. metepimero-axillaris tertius	metapleural arm or pleural ridge above it	third axillary sclerite
IIItpm10	M. metepimero-subalaris	posterodorsal part of metepimeron	posterior half of subalare
IIItpm11	M. metapleura-subalaris	epimeral face of metapleural ridge (close to metapleural arm)	posterior half of subalare
IIItpm12	M. mesophragma-intersegmentalis	lateral part of mesophragma or prealar sclerite	intersegmental membrane or intersegmental sclerite between meso- and metathorax
Pleuro-pleural muscles			
IIIppm1	M. metatransanapleuralis	metapleuron dorsad of anapleural cleft	metapleuron ventrad of anapleural cleft
IIIppm2	M. metabasalare-intersegmentalis	anterior part of metabasalare	intersegmental membrane or intersegmental sclerite between meso- and metathorax or anterior end of anapleural cleft
Sterno-pleural muscles			
IIIspm1	M. metapleura-sternalis	anterior region of metanepisternum or basalare	lateral part of metasternum (laterad of IIIdvm1)
IIIspm2	M. metafurca-pleuralis	tip of metafurcal arm	metapleural arm
IIIspm3	M. metaspina-intersegmentalis posterior	metaspina	intersegmental membrane or intersegmental sclerite between metathorax and abdominal pleura
IIIspm4	M. metafurca-intersegmentalis anterior	metafurca	intersegmental membrane or intersegmental sclerite between meso- and metathorax
IIIspm5	M. metafurca-intersegmentalis posterior	metafurca	intersegmental membrane or intersegmental sclerite between metathorax and abdominal pleura
Pleuro-coxal muscles			
IIIpcm1	M. metanepisterno-trochantinalis	anterior margin of metanepisternum	trochantin
IIIpcm2	M. metabasalare-trochantinalis	basalare or anterodorsal part of metanepisternum	trochantin or anterior metacoxal rim
IIIpcm3	M. metanepisterno-coxalis anterior	dorsal part of metanepisternum (anterad of pleural wing process)	anterolateral metacoxal rim (close to IIIpcm4)

Abbrev.	Name	Origin	Insertion
IIIpcm4	M. metanepisterno-coxalis posterior	ventral half of metanepisternum (close to pleural arm) and/or paracoxal suture	anterolateral metacoxal rim (close to IIIpcm3)
IIIpcm5	M. metanepisterno-trochanteralis	basalare or anterodorsal metanepisternum (close to IIIpcm1)	trochanter
IIIpcm6	M. metapleura-trochanteralis	posteroventral part of metanepisternum and/or metapleural arm (close to IIIpcm4)	trochanter
IIIpcm7	M. metanepisterno-mesocoxalis	anteroventral margin of metanepisternum	posterior mesocoxal rim
Ventral longitudinal muscles			
IIIvlm1	M. metafurca-spinalis	metafurcal arm	metaspina or metafurcal arm of opposite side
IIIvlm2	M. metafurca-abdominosternalis	posterior face of metafurca	second abdominal sternite (usually on anterolateral margin)
IIIvlm3	M. metaspina-abdominosternalis	metaspina	second abdominal sternite (usually on anteromedial part)
Sterno-coxal muscles			
IIIscm1	M. metafurca-coxalis anterior	lateral face of proximal metafurcal stem and/or metasternum around	anterior metacoxal rim (close to trochantino-coxal articulation)
IIIscm2	M. metafurca-coxalis posterior	posterior face of metafurcal stem	posterior or posterolateral metacoxal rim
IIIscm3	M. metafurca-coxalis medialis	ventral face of metafurcal arm	mesal metacoxal rim
IIIscm4	M. metafurca-coxalis lateralis	ventral face of metafurca (lateral of IIIscm3)	lateral metacoxal rim (close to pleuro-coxal articulation)
IIIscm5	M. metaspina-coxalis	metaspina	posterior metacoxal rim
IIIscm6	M. metafurca-trochanteralis	ventral face of metafurcal arm	trochanter

phaga], Ptiliidae [Staphylinoidea]) and Hymenoptera (Mymaridae, Trichogrammatidae).

[Snodgrass (1935); Dashman (1953); Hennig (1969); Matsuda (1970); Bauchhenss (1979); Kukalová-Peck (1991); Brodsky (1994); Gullan & Cranston (1994); Seifert (1995); Federle et al. (1997); Chapman (1998); Beutel & Gorb (2001, 2006, 2008); Grimaldi & Engel (2005); Weirauch (2007); Friedrich & Beutel (2008); Jałoszyński & Beutel (2012); Gorb (2012)]

1.4 Abdomen

1.4.1 General organization

The abdomen of Hexapoda (Fig. 1.4.1.1) is primarily 11-segmented (groundplan), but there is a distinct tendency to reduce or retract the posterior segments. It is almost always broadly connected with the metathorax. The segments are usually linked by short intersegmental membranes, which are exposed (many groups) or internalized (e.g., Coleoptera). The **postabdomen** (see below **1.4.2 External structures of the male and female postabdomen**), which includes the genital segments (VIII and IX in females, IX in males) is tapering towards its apex in almost all groups. The abdomen contains the largest part of the digestive tract, the main excretory organs (**Malpighian tubules**) (Fig. 1.9.2), the main part of the **fat body** (corpus adiposum) (Fig. 1.12.1), the genital organs with accessory glands, the heart with the associated wing muscles, the dorsal and ventral **diaphragm** (if present), large parts of the tracheal system (usually with eight **spiracles**) and eight of 11 paired ganglia of the postcephalic central nervous system.

The skeletal elements of the abdomen (Fig. 1.4.1.1) are much less complex than those of the thoracic segments, which are characterized by their locomotor functions. Endoskeletal elements like furcae or spinae (sternal apophyses) are lacking. The pregenital segments I–VII (**preabdomen**) are usually composed of a simple plate-like **tergum**, a corresponding **sternum**, and a pleural membrane, which almost always bears the spiracles. In few groups the abdominal spiracles are shifted to the tergites. **Laterotergites** separated from the main part of the tergal sclerotization occur in few groups. Frequently the posterior parts of the tergites and sternites overlap with the corresponding elements of the following segments. Sternite I is reduced to varying degrees in different groups. It is only present as a membranous fold attached to the hind margin of the enlarged metacoxae in Coleoptera. Subdivisions of the anterior abdominal sclerites are characteristic of Neuropterida (potential autapomorphies). The sternite of the postgenital segment X is usually more or less strongly reduced. Segment XI is primarily composed of a dorsal unpaired **epiproct** and ventrolateral paired **paraprocts**, but it is often largely or completely reduced or fused to varying degrees with segment X. Complete or partial fusions of elements of anterior or inter-

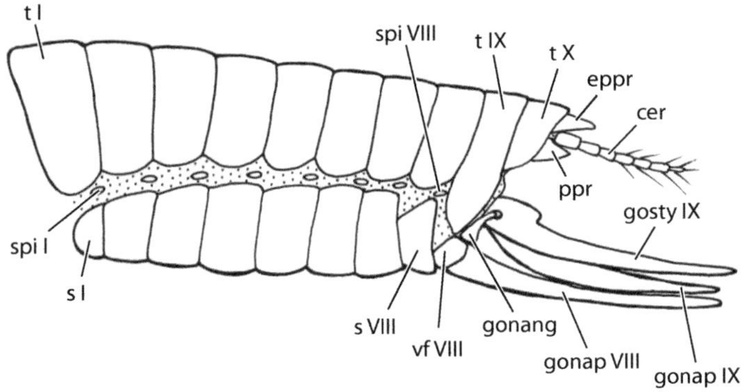

Fig. 1.4.1.1: Abdomen of a pterygote insect, lateral view, schematized. Abbr.: cer – cerci, eppr –epiproct, gonang – gonangulum, gonap VIII/IX – gonapophyses VIII/IX, gosty IX – gonostylus IX (3rd valve), ppr – paraprocts, spi I/VIII – spiracles I/VIII, s I/VIII – sternites I/VIII, t I/IX/X– tergites I/IX/X, vf VIII – valvifer VIII.

mediate segments do also occur. In Hymenoptera segment I forms a structural and functional unit with the pterothorax (autapomorphy) (Fig. 6.25.3). A terminal "telson" is generally absent as a defined sclerotized structure. It is a membranous ring around the anus ("periproct") if anything at all.

The appendages of the abdomen are strongly modified, reduced or completely absent. Appendages are always completely missing on segment X (Fig. 1.4.1.1). Plate-like abdominal **coxites** I–IX probably belong to the groundplan of Hexapoda (Fig. 6.4.3). They are referred to as **coxosternites** when they are fused with the sternites (Zygentoma excl. *Tricholepidion*). In the typical case, these structures are equipped with unsegmented **styli** and one or two pairs of eversible **coxal vesicles**. The styli are possibly homologous to the crustacean exopodites. The main function of the coxal vesicles is probably water uptake, but they are probably also hygroreceptors and involved in osmoregulation. The coxites are paired in Diplura and Archaeognatha but medially fused in Zygentoma (and with the sternum). They are absent in all pterygote lineages. The styli and coxal vesicles are missing on most abdominal segments (I–VI) in Zygentoma (excl. *Tricholepidion*) and are never present in Pterygota. Appendages of segment XI are the **cerci** (Figs 1.4.1.1, 6.3.1, 6.4.1), which are primarily multisegmented, equipped with mechanoreceptive sensilla, and inserted between the epiproct and the paraprocts. It is unclear whether they belong to the groundplan of Hexapoda (implying secondary loss in Collembola and Protura) or a possible clade comprising Diplura and Insecta (=Cercophora). In Japygoidea (Diplura) and Dermaptera they are modified as 1-segmented and strongly sclerotized structures functioning like a pair of forceps. They are reduced to varying degrees in different groups of Pterygota and completely absent in Acercaria and Coleoptera. The **terminal filament**, a groundplan autapomorphy of Insecta, is an unpaired, multisegmented appendage of the abdominal apex. It is present in Archaeognatha and Zygentoma, and usually also in Ephemeroptera, but absent in Odonata and in Neoptera with the exception of some plecopteran larvae. Highly specialized abdominal appendages occur in Collembola. The strongly shortened 6-segmented abdomen bears the unpaired **ventral tube** on segment I, the small **retinaculum**, an arresting device, on segment III, and the large, jumping device on segment IV, the distally paired **furculum (furcula, furca)** (Figs 6.1.1, 6.1.2).

Different abdominal appendages or **prolegs** occur in holometabolan larvae, notably in basal groups of Hymenoptera ("Symphyta") and in Lepidoptera. They are short and indistinctly segmented and apically smooth in Hymenoptera, but equipped with distal groups of small hooks or crochets in many groups of Lepidoptera. The prolegs are usually present on abdominal segments II–VIII and X in Tenthredinidae and related hymenopteran groups, and on segments III–VI and X in many lepidopteran caterpillars. Terminal prolegs (**pygopods**) of segment X are present in larvae of different groups of Holometabola (e.g., Carabidae, Staphylinidae [Coleoptera]). They are equipped with hooks in the aquatic larvae of Corydalidae (Fig. 6.27.1B) (Megaloptera), Trichoptera (Fig. 6.31.7), Nannochoristidae (Mecoptera), Gyrinidae, and Hydraenidae (Coleoptera), probably as an anchoring device in swift running water. Lateral abdominal gills have evolved independently in Ephemeroptera (Fig. 6.6.2) and Megaloptera (Fig. 6.27.1), and in some groups of Coleoptera (e.g., Gyrinidae, *Coptotomus* [Dytiscidae], *Berosus* [Hydrophilidae]).

The abdominal spiracles (Fig. 1.4.1.1) are usually smaller than those of the thorax and almost always embedded in the pleural membranes of segments I–VIII. Spiracle I is often larger than II–VII (see **1.7 Tracheal system**).

Like the skeletal elements the preabdominal muscle system is less complex than that of the thoracic segments. The dorsal and ventral longitudinal muscles and the tergosternal muscles are much less strongly developed. The former are subdivided into median, paramedian and lateral subunits. The unambiguous identification of pleural longitudinal muscles and diagonal muscles is not possible without sternal apophyses. The tergal and sternal longitudinal muscles often do not reach the segmental borders. They are often attached to the lateral segmental walls. They can shorten the segment or result in torsions. The flattening of the abdomen is achieved by contractions of tergosternal, tergopleural and sternopleural muscles, which are all referred to as transverse muscles, and also by intersegmental muscles connecting the lateral segmental borders with the tergite of the following segment. All these muscles are antagonists of the longitudinal muscles on the dorsal and ventral side. The combined activities can result in pumping movements of the abdomen (e.g., ventilation of the tracheal system). The spiracles I–VIII are also equipped with closing muscles in almost all groups of insects. They are missing in Ephemeroptera and the apterygote hexapods. The muscles associated with the male and female genital apparatuses are usually complex and highly variable. As the presently available data are insufficient a table of the abdominal muscles is not presented in this volume.

1.4.2 External structures of the male and female postabdomen

The postabdomen comprises the terminal segments X and XI, and the genital segments VIII and IX which are usually characterized by external genital appendages and generally by the openings of the genital ducts (Fig. 1.4.1.1). The genital openings usually lie on or behind segment VIII in females, and on or behind segment IX in males (arguably on segment X in the groundplan of Hexapoda). The postabdominal segments are almost always very distinctly modified in both sexes.

A male copulatory organ is absent in the groundplan of Hexapoda. A simple membranous **penis** is present in Archaeognatha (Fig. 6.4.3B) and Zygentoma. It is formed by a paired anlage on the ventral side of the male genital segment IX. Paired penes are present in Ephemeroptera (Fig. 6.6.4) and a secondary copulatory organ on the ventral side of the abdominal base in Odonata. In almost all neopteran groups an unpaired **ductus ejaculatorius** enters a complex copulatory organ, which is often composed of a sclerotized **phallobase** (**basal piece**), paired **parameres**, an unpaired median **aedeagus**, and an eversible membranous **endophallus**, which often bears an armature of small sclerotized spines. Additional structures can be present on the distal part of the aedeagus, such as for instance paired **paraphyses** (**titillators**) or an unpaired **virga**. That these structures function as stimulating organs was suggested but lacks confirmation. The parameres, which articulate with the phallobase, usually function as clasping organs. They can be divided into basal basimeres and distal telomeres.

As a whole the male postabdominal segments and their substructures are highly variable. A lock-and-key mechanism with species-specific male and female genitalia fitting precisely into each other occurs in some groups such as for instance Geometridae (Lepidoptera) or Carabidae (Coleoptera). However, in many cases sexual selection (cryptic female choice) seems to be the main driving force behind the formation of complex and highly variable male genitalia (Fig. 6.13.5). The high variability and complexity of the male genitalia in many groups of hexapods play a crucial role in taxonomic work on the species level. Species in many groups can only be defined using male genital characters. Female genitalia are usually more conservative and are rarely used by taxonomists (e.g., **spermatheca**).

The female genital opening lies on or posterior to sternite VIII in most groups, rarely posterior to sternite VII. The ovipositor formed by appendages of the female genital segments VIII and IX is a groundplan autapomorphy of Insecta (=Ectognatha incl. †Monura) (Figs 1.4.1.1, 6.A.2). It is composed of the elongated (Archaeognatha) or relatively short proximal **coxopodites** VIII and IX (=**valvifers** or **gonocoxites**) and by elongate, stiff **gonapophyses** VIII and IX (=1st and 2nd **valves** or **valvulae**) (Fig. 1.4.1.1).

Gonostyli are inserted on both pairs of coxopodites in Archaeognatha (groundplan of Insecta?). A small and often triangular sclerite connecting the coxopodites VIII and IX is the **gonangulum** (Fig. 1.4.1.1). It is present as a well-defined element in the groundplan of Dicondylia and retained in many groups of Pterygota. In the typical orthopteroid ovipositor (groundplan of Neoptera?) a 3rd valve is present, a more or less elongate and broadened appendage inserted on the posterior part of coxopodite IX. The upper edges of the gonapophyses form a tongue and groove connection with the ventral edges of the gonapophyses IX and the 3rd valves. The latter form a sheath enclosing the gonapophyses. The valves of both sides can be moved alternatively by retractor muscles and thus move the egg released from the genital opening towards the apex of the ovipositor.

The ovipositor (Figs 1.4.1.1, 6.12.3A) is modified or reduced to varying degrees in different insect lineages. It is distinctly shortened in Caelifera (Orthoptera) (Fig. 6.12.3B), where it functions as a burrowing device, and also in Dictyoptera in correlation with the presence of a genital atrium where the egg cocoons (**oothecae**) are formed. Within Holometabola it is well-preserved with its typical elements in the basal lineages of Hymenoptera (groundplan) (Fig. 6.25.5), but it is strongly modified in Neuropterida and Coleoptera (completely absent in Strepsiptera) and largely or completely reduced in Mecopterida. In most amphiesmenopteran and antliophoran groups the telescoping postabdomen is used to deposit eggs. An ovipositor modified as a stinging device is an autapomorphy of Aculeata (e.g., wasps, ants and bees). Secondary ovipositors occur in Boreidae (Mecoptera) and some groups of Diptera.

1.4.3 Male and female internal genital organs

The main elements of the male internal genital organs are the **testes** (Fig. 1.4.3.1). They are always paired but in few lineages secondarily connected (e.g., Orthoptera [partim]). The testes are usually composed of a series of sperm producing **follicles** (tubes). More than 100 of these subunits can be present per testis in some groups (e.g., Acridiidae), but only one tubular follicle is present in non-polyphagan beetles, and only one sac-shaped structure in Diptera and some entognathous hexapods. The arrangement of the follicles is comb-like in males of Zygentoma (*Lepisma*) and Archaeognatha (*Machilis*), but usually botrydoid (grape-shaped) or brush-like in the pterygote lineages. The wall of the follicles is a thin epithelium, in some groups with two layers of cells on a basal lamina. The primordial germ cells are usually located in the apical region. The entire structure is enclosed in a sheath of connective tissue in many cases. The shape of the testes can vary considerably among hexapod lineages. They can be fan-shaped (e.g., Byrrhidae [Coleoptera]), divided into several small lobes (e.g., Dascillidae [Coleoptera]), or tubular (e.g., Campodeoidea [Diplura], Phasmatodea). Strongly elongated and coiled tube-shaped testes are present in males of archostematan and adephagan beetles.

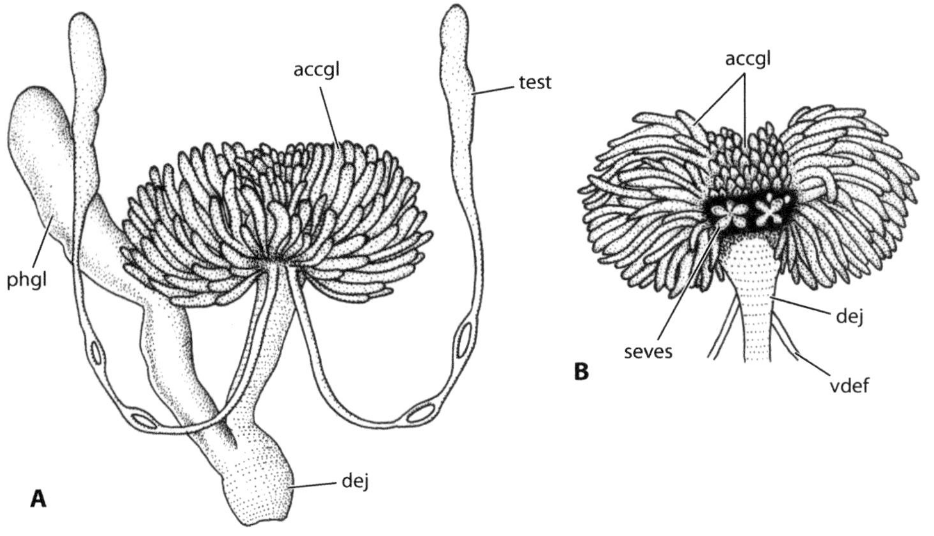

Fig. 1.4.3.1: Male internal reproductive organs of *Periplaneta americana* (Blattodea, Blattidae). A, dorsal view; B, accessory gland, ventral view. Abbr.: accgld – accessory gland, dej – ductus ejaculatorius, phgl – phallic glands, seves – seminal vesicles, test – testes, vdef – vas deferens. Redrawn from Seifert (1995).

The sperm cells are released into the paired mesodermal **vasa deferentia**, which are often enlarged proximally, thus forming **vesiculae seminales** functioning as sperm storage organs. In almost all groups they unite to form an unpaired ectodermal **ductus ejaculatorius** (Fig. 1.4.3.1). In most cases it leads to an intromittent organ (absent in entognathous hexapods, simple in Archaeognatha and Zygentoma), with the gonopore usually placed in the membrane connecting sternites IX and X. The epithelium of the ductus ejaculatorius is one-layered and as a structure of ectodermal origin covered by a thin intima. In some groups it is equipped with a strongly developed muscularis and forms a simple sperm pump for spermatozoa contained in liquid (e.g., Zoraptera [partim], Strepsiptera). In Ephemeroptera the ductus ejaculatorius is lacking. The vasa deferentia lead directly to the paired gonopores and copulatory organs. In Dermaptera the ductus is paired, but vestigial on one side in some species (Chapman 1998).

In some groups the epithelium of the vasa deferentia has a glandular function, but they usually form tubular diverticula differentiated as mesodermal **accessory glands** (mesadenia) (Fig. 1.4.3.1: accgl). In some nematoceran groups (Diptera) parts of the ectodermal ductus ejaculatorius are enlarged and modified as accessory glands (ectadenia). Ectodermal accessory glands also occur in Coleoptera and Heteroptera, and both types can be present in some species of these orders. The secretions of the accessory glands play an important role in formation of the spermatozoa (sperm packages). Their shape and also the number can vary strongly. They are absent in

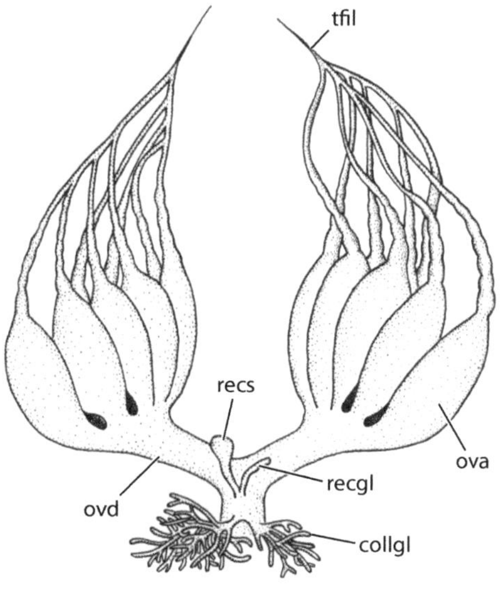

Fig. 1.4.3.2: Ovaries and associated structure. Abbr.: collgl – collateral glands, ova – ovarioles, ovd – oviduct, recs – rectal sac, recgl – rectal gland, tfil – terminal filament. Redrawn from Seifert (1995).

apterygote lineages, Odonata, and Ephemeroptera (hexapod and pterygote groundplan?) and also in muscoid dipterans. A single pair is present in Lepidoptera and other groups, two pairs in *Tenebrio* (Coleoptera), and 15 pairs or more can occur in Orthoptera (Chapman 1998).

The main elements of the internal female genital apparatus are the paired ovaries, each of them composed of several or many **ovarioles** or **egg tubes** in almost all groups (Figs 1.4.3.2, 1.4.3.3). The number and arrangement varies. Only one pair of sack-shaped elements with unclear homology is present in Collembola, Campodeoidea (Diplura) and Protura (groundplan of Hexapoda?). The ovaries of viviparous dipterans (e.g., Hippoboscidae) contain only two ovarioles. Four per ovary are present in smaller orthopterans but more than 100 in larger species. The ovarioles are usually arranged as a bundle (e.g., Orthoptera, Blattodea, Heteroptera, Hymenoptera, Coleoptera, Lepidoptera), but a serial arrangement occurs in some lineages, such as for instance Archaeognatha, Zygentoma and Ephemeroptera (groundplan of Insecta and Pterygota?). Botrydoid ovaries with many small ovarioles are characteristic for Coccoidea (Sternorrhyncha) and also occur in nematoceran groups of Diptera (e.g., Culicidae). Atypical horseshoe-shaped ovaries (distally connected) with numerous finger-shaped ovarioles are present in Plecoptera (autapomorphy). A common sheath of the ovaries is usually absent, but occurs in Diptera.

The typical ovarioles (Fig. 1.4.3.3) are tube-like, apically tapering structures forming a cluster on both sides. They are enclosed by an external ovariole sheath or **tunica externa**, and an internal **tunica propria**, an elastic basal lamina. In the **germarium** the primordial germ cells are formed and the first meiotic divisions take place. It is almost always located in the distal part but laterally in the atypical ovaries of Collembola. When the **oocytes** (immature egg cells) leave the germarium they are clothed by **follicle cells**. In the proximal **vitellarium** yolk is produced and the **oocytes** increase in size. Apically the ovarioles bear thin terminal filaments, with a cellular core covered by the tunica propria. They are often connected distally, thus forming a suspensory ligament. In some groups the ligaments of both sides merge, or they are attached to the body wall or the dorsal diaphragm, suspending the ovaries in the mixocoel.

Three different types of ovarioles can be distinguished (Fig. 1.4.3.3). The **panoistic** ovariole (e.g., Archaeognatha, Zygentoma, Ephemeroptera, Plecoptera [groundplan Insecta]) lacks nurse cells (Fig. 1.4.3.3A). All primordial germ cells develop as **oocytes**. The nutrition is provided by the **follicle epithelium** of the ovarioles. In polytrophic or telotrophic **meroistic** ovarioles some priordial germ cells form nurse cells (**trophocytes**) and others oocytes. In the **polytrophic** type (Fig. 1.4.3.3B) each oocyte is enclosed by follicle cells. In **telotrophic** type (Fig. 1.4.3.3C) trophic tissue is present and oocytes, and prefollicular tissue in the apical region. The trophocytes are confined to the germarium and enclose a central trophic core in this region. The advancing oocytes are connected to the trophic core by thin nutritive cords.

At their base the ovarioles are usually connected with a common **calyx** by narrowed stalk-like pedicels. From there the mature oocytes enter the paired **oviducts**. Their epithelium of cubic or columnar cells on a basal lamina is usually enclosed by a thin muscularis. In almost all groups (not in Ephemeroptera) they join an unpaired median **oviduct**. Due to its ectodermal origin its inner surface is coated by a thin intima. Externally it is usually covered by thin layers of circular and longitudinal muscles. The **gonopore** (usually unpaired, paired in Ephemeroptera) at the posterior end of the oviduct releases the eggs into a more or less extensive genital chamber in most groups. This can form a specifically shaped diverticulum or pouch, the **bursa copulatrix**, which receives the male intromittent organ during copulation. In other cases only a more or less narrow tube-like **vagina** is present, with the **vulva** as its terminal opening. In viviparous dipterans, the genital chamber forms an "uterus". From the female genital tract eggs enter the basal part of an ovipositor in most groups (groundplan of Insecta). A **spermatheca** or **receptaculum seminis** is present in many groups and usually unpaired (two in *Blaps* [Coleoptera], three in the groundplan of Diptera [Hennig 1973]). Its insertion is a landmark for separating the median oviduct from the vagina. Its internal surface is coated with a thin intima. Spermatozoa can be stored in the spermatheca until they are used for fertilization. In most cases it comprises a more extensive pouch-like part and a narrow **spermathecal duct**, connecting it to the bursa copulatrix or vagina. Tubular ectodermal **sperma-**

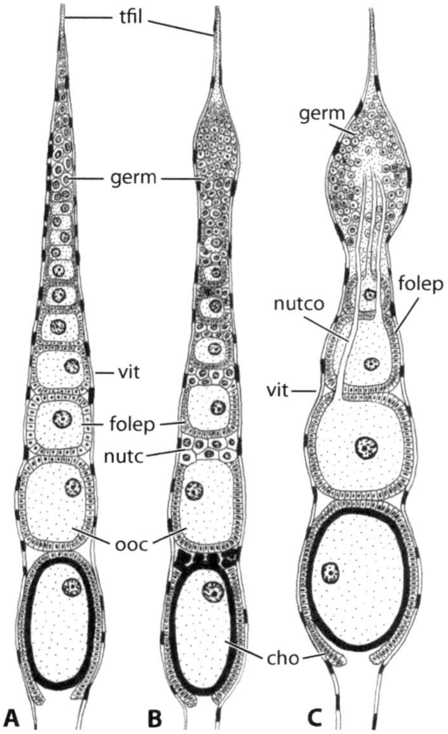

Fig. 1.4.3.3: Ovariole types. A, panoistic; B, polytrophic (meroistic); C, telotrophic (meroistioc). Abbr.: cho – chorion, folep – follicle epithelium, germ – germarium, nutc – nutritive cell, nutco – nutritive cord, ooc – oocyte, tfil – terminal filament, vit – vitellarium. Redrawn from Seifert (1995).

thecal glands are connected to the duct occur in different groups. Paired ectodermal **accessory glands** are almost always associated with the female genital tract (absent in Grylloblattodea) and more than one pair is present in some groups. They usually produce substances coating and protecting the eggs or enabling the females to glue them on specific substrates. The function of the secretions varies in different lineages.

[Weber (1933); Snodgrass (1935); v. Kéler (1963); Hennig (1973); Matsuda (1976); Martoja (1977); Gullan & Cranston (1994); Seifert (1995); Chapman (1998)]

1.5 Nervous system

In its basic design the complex nervous system of Hexapoda is similar to that occurring in other groups of arthropods (Fig. 1.5.1.1). It integrates sensory input and internal physiological information and controls the complex locomotor apparatus. It is subdivided into three major elements, the **central nervous system (CNS)**, the **visceral (sympathetic) nervous system**, and the peripheral nervous system (Fig. 1.5.1.1). The main subdivision is the CNS comprising the complex **brain (cerebrum**, supraoesophageal ganglion), the **suboesophageal complex (suboesophageal ganglion)**, and the postcephalic ventral nerve cord or ganglionic chain. The visceral nervous system comprises three subcomponents, the **stomatogastric** (stomodaeal) **nervous system**, the **ventral nerve (ventral visceral nervous system)**, and the **caudal visceral nervous system**. The peripheral nervous system establishes the connections to the appendages, the body wall, and the organs. A ventral nerve cord with segmental ganglia is arguably an autapomorphy of Arthropoda (with possible reversal in Onychophora). A similar condition occurs in Annelida (with great variation within the group), but not in Cycloneuralia (Nematoda, Priapulida, etc.), the presumptive sistergroup of arthropods.

The **ganglia** of hexapods and euarthropods in general are primarily paired segmental nerve centers. They are a concentration of perikarya (cell bodies) of motoneurons and interneurons (see below) and projections connecting different types of nerve cells. The perikarya are always placed at the periphery. The internal layer only consisting of axons is called the **neuropil**. The ganglia are connected by transverse **commissures** and longitudinal **connectives**. Different types of concentration of the ventral nerve cord occur in many groups. The primarily paired ganglia can be more or less completely fused along the median line and also in the longitudinal direction (see below), involving reductions of the commissures and connectives, respectively.

The neurons (nerve cells) are the basic components of the nervous system. They are composed of the perikaryon and two types of projections, a short dendrite which receives stimuli, and a long axon (neurite) which transmits information to another neuron or to effector organs such as for instance muscles. The cell bodies and other parts of the neurons are enclosed by folds of one or several glial cells. Synapses occur only where these folds are absent. The glial cells can enclose an individual axon or several smaller ones. They provide the neurons with nutrients and play an important function in the development of the CNS (Chapman 1998). The entire system of glial cells is referred to as **neuroglia**. The sheath enclosing the whole nervous system is called the **neurilemma** (perilemma). Its inner cellular layer (**perineurium**) contains glycogen in higher concentration. The external layer or neural lamella is formed by the basement membrane and contains collagen.

Most hexapod neurons are monopolar, i.e. the dendrite and the axon share a single common projection arising from the perikaryon. Peripheral sense cells are bipolar (Fig. 1.1.4.1), with a short and usually unbranched distal dendrite and a long proximal axon connected to the CNS. Multipolar neurons occur in ganglia and can be associated with stretch receptors and thermoreceptors (Chapman 1998). The three main functional types of neurons are the afferent sensory neurons, the efferent motor neurons, and the interneurons (association neurons) which transmit information

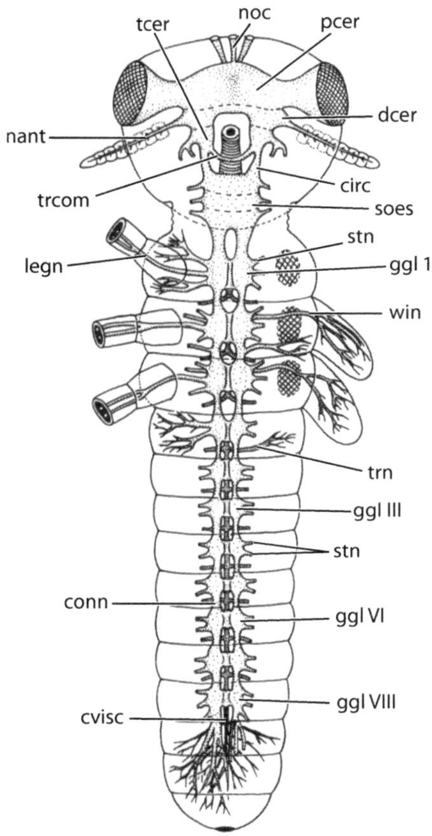

Fig. 1.5.1.1: Nervous system, schematized, ventral view. Abbr.: circ – circumoesophageal commissure, cvisc – caudal visceral nervous system, conn – connective, dcer – deutocerebrum, ggl 1/III/VI/VIII – thoracic and abdominal ganglia, legn – leg nerve, nant – nervus antennalis, noc – ocellar nerve, pcer – protocerebrum, soes – suboesophageal ganglion, stn – stem nerve, tcer – tritocerebrum, trcom – tritocerebral commissure, trn – tracheal nerve, win – wing nerve. Redrawn from Seifert (1995).

between other nerve cells. Specialized neuroendocrine cells (**neurosecretory cells**) (Fig. 1.5.1.2) produce secretions, especially neurohormones.

1.5.1 Brain

The **brain** (**cerebrum**) of hexapods is the central element of the CNS (e.g., Strausfeld 1976) (Figs 1.5.1.1–1.5.1.3). It is composed of three divisions associated with the first three head segments, the **protocerebrum**, the **deutocerebrum** and the **tritocere-**

brum. These main elements are usually only indistinctly separated from each other (Figs 1.5.1.2, 1.5.1.3). The external layer of the brain, the **cell body rind** or **cortex**, is formed by numerous perikarya of neurons. The inner **neuropil**, which appears largely homogenous in histological sections, is formed by axons, dendrites and glial cells, and is free of perikarya. Neuropil areas with an increased density of synapses are referred to as **glomeruli**.

The protocerebrum is the largest and most complex part of the brain (Figs 1.5.1.2, 1.5.1.3). It is associated with the compound eyes and ocelli. Its symmetrical halves are referred to as hemispheres and its frontal and ventral convexities as protocerebral lobes. The nerves of the ocelli arise from the middle region between the hemispheres, the **pars intercerebralis** (Fig. 1.5.1.2). The lateral **optic lobes** form the connection between the central elements of the protocerebrum and the compound eyes that are usually strongly developed. They are connected by the optic commissure and composed of the **lamina ganglionaris**, which is adjacent with the retinulae, the **medulla** (externa), and the **lobula** (=**medulla interna**) (Fig. 1.5.1.2). The axons leading to the retinula cells form chiasmata, i.e. they are intercrossing between the three optic neuropils. The lateralmost part of the protocerebrum adjacent with the lobula is the **lateral horn** (equivalent to the medulla terminalis with hemiellipsoid bodies of crustaceans; Fanenbruck et al. 2004). It is also connected with the deutocerebral antennal lobes (see below).

The **central body** complex (Strausfeld 2009) in the central protocerebral region (Fig. 1.5.1.2) is composed of the **protocerebral bridge** and the central body which lies below it. The latter is subdivided into an upper and a lower part and a pair of globular noduli. The functions of the central body complex are not entirely clarified but they likely comprise the perception of polarized light and the control of movements. The most conspicuous protocerebral structures are the paired **corpora pedunculata** or mushroom bodies (Strausfeld et al. 1998). They are composed of a large, more or less hemispherical **calyx** and a narrower, stalk-like basal portion referred to as **pediculus** (Fig. 1.5.1.2). The total number of intrinsic neurons (=Kenyon cells) is about 230,000 in cockroaches and ca. 2,500 in *Drosophila*. The distal part of the calyx, which can be subdivided in large insects (alpha-, beta-, gamma lobes), is a globulus with densely packed cell bodies (Fig. 1.5.1.2). The mushroom bodies are associative centers and very highly developed in insects with complex patterns of behavior, especially in social species such as for instance the honey bee. Important functions are related to learning and memory, especially in the context of olfaction, but also motor control and sensory filtering. The protocerebral or first commissure lies posterior to the mushroom bodies. It connects the ventral bodies, two small glomeruli. Further elements of the protocerebrum are two paired groups of **neurosecretory cells** (Fig. 1.5.1.2). One of them lies in the pars intercerebralis. Its axons form the nervi corporis cardiaci I which enter the **corpora cardiaca** (**neurohaemal organ**; see **1.11 Endocrine organs and the hormone system**) and deliver neurohormones, like the nervi corporis cardiaci II, which originate from the second group of neurosecretory cells located laterad the

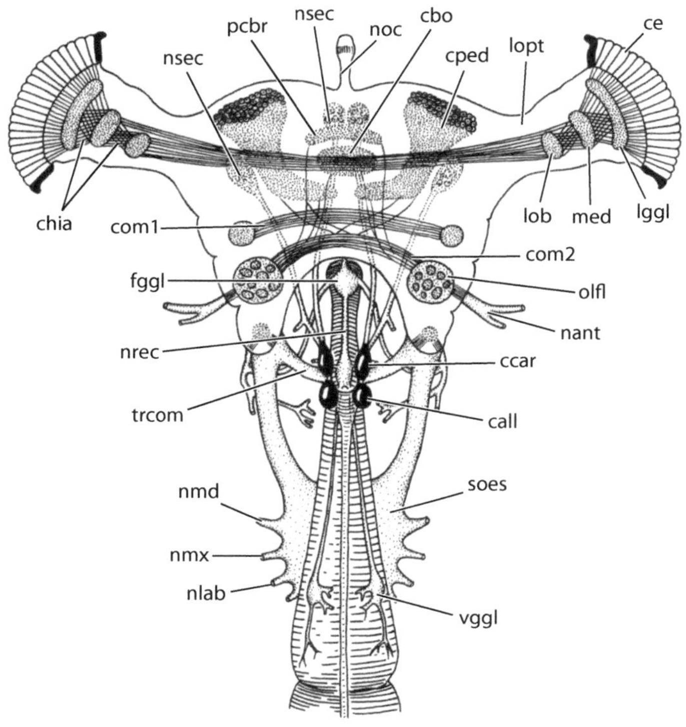

Fig. 1.5.1.2: Brain and suboesophageal ganglion with associated organs, dorsal view. Abbr.: call – corpora allata, cbo – central body, ccar – corpora cardiaca, ce – compound eye, chia – chiasmata, com 1/2 – 1st/2nd commissure, cped – corpora pedunculata, fggl – frontal ganglion, lggl – lamina ganglionaris, lob – lobula, lopt – lobus opticus, med – medulla, nant – nervus antennalis, nlab – nervus labialis, nmd – nervus mandibularis, nmx – nervus maxillaris, noc – ocellar nerve, nrec – nervus recurrens, nsec – neurosecretory cells, olfl – olfactory lobe, pcbr – protocerebral bridge, soes – suboesophageal ganglion, trcom – tritocerebral commissure, vggl – ventricular ganglion. Redrawn from Seifert (1995).

corpora pedunculata (Fig. 1.5.1.2). In some groups an unpaired nervus connectivus originates on the frontal side of the protocerebrum and connects it directly with the **frontal ganglion**, the central element of the **stomatogastric nervous system** (see below) (Figs 1.5.1.2, 1.5.1.3).

The deutocerebrum is associated with the antennae and the second head segment. It is distinctly smaller than the protocerebrum and usually not distinctly separated from it externally (Figs 1.5.1.2, 1.5.1.3). It controls antennal movements and receives input from chemo-, mechano- and thermoreceptors. The **antennal lobes** with glomeruli are important elements of the deutocerebrum. They are connected with the

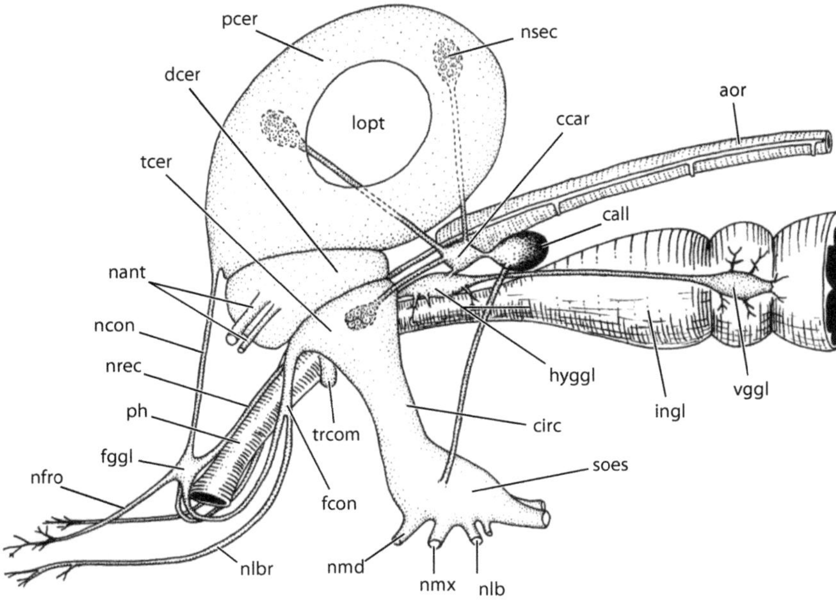

Fig. 1.5.1.3: Brain and suboesophageal ganglion with associated organs, lateral view. Abbr.: aor – aorta, call – corpora allata, ccar – corpora cardiaca, circ – circumoesophageal connective, dcer – deutocerebrum, fcon – frontal connective, fggl – frontal ganglion, hyggl – hypocerebral ganglion, ingl – ingluvies, lopt – lobus opticus, nant – nervus antennalis, ncon – nervus connectivus, nfro – nervus frontalis, nlbr – nervus labralis, nlb – nervus labialis, nmd – nervus mandibularis, nmx – nervus maxillaris, nrec – nervus recurrens, nsec – neurosecretory cells, pcer – protocerebrum, ph – pharynx, soes – suboesophageal ganglion, tcer – tritocerebrum, trcom – tritocerebral commissure, vggl – ventricular ganglion. Redrawn from Seifert (1995).

protocerebral lateral horn by the antenno-protocerebral tracts, and also with the dorsal lobe or antennal mechanosensory and motor center (homologue to the lateral antennal neuropil of crustaceans). They receive input from the sensory neurons of the antenna. The second commissure connects the olfactory lobes of both sides. The antennal nerves originating from the deutocerebrum is composed of motoneurons and a thicker sensory branch. The deutocerebral cell body rind contains additional neurosecretory cells (Seifert 1995).

The tritocerebrum on both sides of the foregut is the smallest part of the brain (Figs 1.5.1.2, 1.5.1.3). It belongs to the intercalary segment 3 of the head, which bears the 2nd antenna in crustaceans. In contrast to the proto- and deutocerebrum its commissure lies below the pharynx, and is thus distinctly separated from this part of the brain. The paired labral nerve originates together with the paired **frontal connectives** on the anterior side of the tritocerebrum. The latter connect this part of the brain with the frontal ganglion (Fig. 1.5.1.3). The glomeruli of the tritocerebrum are less strongly developed than in the anterior parts of the brain, but it also contains cells producing neurohormones. In some groups their axons form nervi corporis cardiaci III (Seifert 1995). The posteriormost part of the tritocerebrum is the origin of the **circumoesophageal connectives**, which connects it with the anteriormost part of the suboesophageal complex.

The relative size and the shape of the brain depend strongly on the total body size of the hexapod (e.g., Beutel et al. 2005). It appears very small and dumbbell-shaped in large hexapods (e.g., cockroaches, locusts), but large and deformed and fitting in available narrow spaces in very small hexapods (e.g., primary larvae of Ptiliidae [Coleoptera]). As an effect of miniaturization the brain can be more or less completely shifted to the thorax, especially in holometabolan larvae. In the extremely small primary larvae of Strepsiptera (average total size ca. 200 μm) it forms a compact unit around the midgut with all other elements of the central nervous system and is located in the middle region of the postcephalic body (Beutel et al. 2005).

Far-reaching structural affinities between the brains of Hexapoda, Malacostraca and Remipedia have been pointed out in a recent study, the presence of a medulla terminalis in the protocerebrum, and of a lateral antenna 1 neuropil and olfactory lobes with glomeruli in the deutocerebrum (see above; Fanenbruck et al. 2004).

1.5.2 Suboesophageal complex

The **suboesophageal complex** (=suboesophageal "ganglion") (Figs 1.5.1.1–1.5.1.3) is connected with the tritocerebrum by the circumoesophageal connectives, which are primarily more or less elongate structures encompassing the posterior pharynx, but short and broad in very small hexapods and generally in Hemiptera and Antliophora, where the brain and suboesophageal complex form a compact unit around the anterior foregut. Connective nerves often originating from the circumoesophageal connectives mainly innervate the hypopharynx.

The suboesophageal complex is a moderately elongate structure composed of three primary pairs of ganglia belonging to the posterior three head segments (mandibular-, maxillary- and labial segments). The three segmental subcomponents (ganglia) form a compact structural unit but can appear as distinctly separated subdivisions in the embryonic development (Figs 1.5.1.1–1.5.1.3). The complex usually lies in the posteroventral head region but is partly or completely shifted to the prothorax in different groups, especially as a result of miniaturization (e.g., Beutel et al. 2005). It supplies the mouthparts with sensory nerves and motor neurons (Figs 1.5.1.2, 1.5.1.3). The first two parts innervate the mandibles and maxillae, respectively, including the associated muscles. The third part innervates the labium and associated structures, muscles and the distal parts of the salivary ducts, and also large parts of the cervical region. This part is connected with the prothoracic pair of ganglia by connectives which are often medially fused.

1.5.3 Postcephalic ganglionic chain

A pair of ganglia is present in each segment of the thorax and in the abdominal segments I–VII (Fig. 1.5.1.1). The complex in abdominal segment VIII is a product of fusion of ganglia of the terminal segments, but separate segmental units are often recognizable in the embryonic development. Fusions occur in many groups (see above). The ganglia of the abdominal segment I are often more or less completely fused with the metathoracic ganglia. The eight abdominal ganglia are concentrated in two complexes in Zoraptera and the presence of only one compact complex is an autapomorphy of Acercaria. An extreme degree of concentration is found in the cyclorrhaphan Diptera (e.g., *Calliphora*), where the ventral ganglionic chain does not even reach the posterior thorax (Seifert 1995).

The thoracic ganglia are generally larger than those of the abdomen (Fig. 1.5.1.1). The prothoracic pair can be closely connected or fused with the suboesophageal complex. A pair of efferent leg nerves arises from each of the thoracic ganglia and one or two pairs of stem nerves. The latter comprise motor and sensory neurons. In pterygote insects a pair of sensory wing nerves is present in both pterothoracic segments. Additional nerves arising from the connectives are called **connective neurons**. In the abdominal ganglia the leg nerves and wing nerves are missing for obvious reasons. Two pairs of **stem nerves** are usually present and connective nerves also occur.

1.5.4 Visceral nervous system (VNS)

The anterior and most complex subdivision of the **VNS** is the **stomodaeal (stomatogastric) nervous system** located in the head and the anterior thorax (Fig. 1.5.1.3). Its central element is the unpaired **frontal ganglion**, which is also a landmark for the anatomical mouth. It is usually triangular or rhomb-shaped in frontal view, and always connected with the tritocerebrum by the paired frontal connectives, and in some groups with the anterior side of the protocerebrum by the unpaired nervus connectivus (see above). The **nervus procurrens** from the epipharyngeal region enters the frontal ganglion anteromedially. It is mainly a sensory nerve receiving input from the preoral region, but also contains efferent fibers innervating intrinsic and extrinsic labral muscles. The **nervus recurrens**, which originates from the frontal ganglion posteromedially extends along the anterior foregut dorsomedially below the brain and enters the small **hypocerebral ganglion**, which is closely connected with the **corpora cardiaca** (see **1.11**) (Fig. 1.5.1.3). In some groups the nervus recurrens continuous as an unpaired nerve and branches shortly before the posterior end of the foregut, usually after forming a small unpaired **ventricular ganglion**. In other groups it divides into two branches posterior to the hypocerebral ganglion or two separate nerves originate from it. These paired branches extend along the side walls of the oesophagus and form paired ventricular ganglia towards their posterior end.

The sensory neurons and motor neurons connected to the stomatogastric nervous system innervate different regions and structures of the preoral cavity, the salivary ducts, and the foregut and its musculature. Proprioreceptors in the pharyngeal wall, the frontal ganglion, and the nervus recurrens are also involved in the ecdysis.

The second part of the visceral nervous system is the **unpaired nerve** (**ventral visceral nervous system**, **ventral nerve**, **Leydig's nerve**). It contains sensory and motor fibers as well as axons of neurosecretory cells. In the thorax it originates posteromedially from the commissure or the median region between the paired ganglia and divides into two branches anterior to the following ganglion. The paired branches turn laterad and function as motor neurons of the thoracic spiracles. In the abdomen the ventral nerve extends from one ganglion to the next in the median line. Like in the thorax, motor neurons of the spiracles originate from it.

A nerve originating ventromedially from the ganglionic complex of segment VIII (Fig. 1.5.1.1) ramifies and forms the **caudal visceral nervous system** in the posterior abdomen. It innervates the posterior part of the digestive tract and its musculature and also the genital organs.

[Strausfeld (1976, 2009); Gullan & Cranston (1994); Seifert (1995); Strausfeld et al. (1998); Chapman (1998); Dettner & Peters (2003); Fahnenbruck et al. (2004); Beutel et al. (2005)]

1.6 Photoreceptor organs

1.6.1 Extraocular photoreception

Extraocular (non-visual, non-image-forming) **photoreception** was described for several hexapod orders (e.g., Arikawa et al. 1980). This can be either direct photoreception by the central nervous system, often related to the circadian clock (Bennett 1979), or function via light-sensitive areas of the body surface involving epidermal cells with pigments and a connection to the nervous system (**dermal photoreception**). Photoreceptors were for identified found in the brain and the abdominal ganglionic complex VIII of *Periplaneta americana* (Arikawa et al. 1980). Simple dermal photoreception is apparently often related to the perception of the length of the photoperiod, which is known to affect certain developmental processes. Known examples are beetle larvae (e.g., *Dytiscus marginalis*, *Tenebrio molitor*), butterfly caterpillars, roaches, eye-less beetles, and adults of certain species of Lepidoptera. Interestingly, photoreceptive areas occur on the genitalia of both sexes of Papilionidae (Lepidoptera). Apparently, the specialized light-sensitive cells (**phaosomes**) monitor the genitalia during copulation (Arikawa et al. 1980; Miyako et al. 1993).

1.6.2 Compound eyes

Compound eyes are complex and highly efficient photoreceptor organs (Fig. 1.6.2.1). They occur in extant and extinct lineages of Euarthropoda (e.g., †Trilobita, Xiphosura, †Eurypterida, Pancrustacea [incl. Hexapoda], Myriapoda [usually strongly modified or reduced]) and are arguably an autapomorphy of this extremely successful lineage. Apparently, compound eyes evolved in the Early Cambrian (ca. 550 Ma), when ancestral euarthropods (and other groups of organisms) ceased to live within soft marine substrates such as sand or mud. The presence of highly developed eyes is likely linked with the evolution of complex appendages, which allow efficient locomotion on the substrate surface. Compound eyes of hexapods usually possess a large visual field and they are suitable to detect fast movements, and in some groups polarized light.

The paired compound eyes are almost always placed at the lateral sides of the head. They are usually distinctly convex and can cover extensive areas, in some cases almost the entire cephalic surface. The shape varies strongly. In most groups they are round, oval or kidney-shaped. Typical compound eyes are composed of numerous functional units of equal architecture, the **ommatidia** (Fig. 1.6.2.2), each of which is equipped with an individual external **corneal lens**. These units are radially arranged and usually adjacent with each other, but in some groups separated externally by chitinous bridges (e.g., Strepsiptera). The number of ommatidia varies strongly. Approximately 30,000 are present in some species of Odonata, ca. 300 in females of fireflies (Lampyridae [Coleoptera]), 12 in Lepismatidae (Zygentoma), and eight is the maximum number in Collembola. A complete reduction of the compound eyes occurs in many groups. They are constantly absent in extant Protura and Diplura ("Non-oculata"), in wingless morphs of Zoraptera (and other groups), and in many cave-dwelling (e.g., *Anophthalmus* [Carabidae]) or parasitic species.

The diameter of the ommatidial facets may differ in different regions of the compound eyes. They are larger in the anterior and upper region in Tabanidae (Diptera). Partially or completely divided compound eyes occur in different groups, such as for instance Ephemeroptera (males of many species) and Gyrinidae ("whirligig beetles" [Coleoptera]). The upper "turban eye" of ephemeropteran males appears pedunculate and contains an elongate haemolymphatic space in its "stalk". Its facets are distinctly larger than those of the smaller ventral part of the compound eye. It is adapted to low light intensity.

The hexapod ommatidium (Fig. 1.6.2.2) is composed of a dioptric apparatus and a proximal sensory part. The **corneal lens** is the external part of the light-gathering apparatus. It usually has a hexagonal shape and is also referred to as facet. In many cases the facets are biconvex, and they are always formed by a portion of transparent

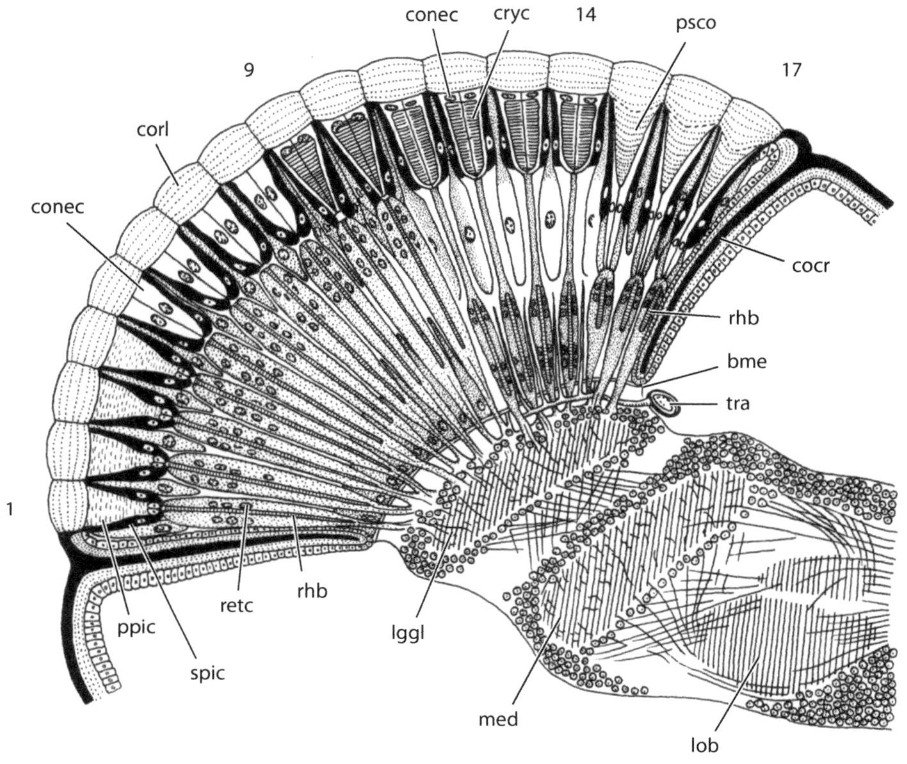

Fig. 1.6.2.1: Compound eye, displayed with different types of ommatidia, 1–4, 15–17 pseudocone, with extracellular crystalline cone directly adjacent with cornea, crystalline cone cells shifted towards retina; 5–8 acone, crystalline cone cells (Semper cells) form no cone; 9–14 eucone, crystalline cone formed within cone cells. Abbr.: bme – basement membrane, cocr – circumocular ridge, conec – crystalline cone cells, corl – corneal lens, cryc – crystalline cone, lggl – lamina ganglionaris, lob – lobula (medulla interna), med – medulla (externa), ppic – primary pigment cell, psco – pseudocone, retc – retinula cell, spic – secondary pigment cell, rhb – rhabdom, tra – trachea. Redrawn from Seifert (1995).

cuticle. Each corneal lens is formed by a pair of epidermal cells, the **corneagenous cells.** Below it four **cone cells (Semper cells)** form a second lens in many groups, the tetrapartite **crystalline cone.** This condition and the presence of four proximal processes of the cone cells (connected to the basal matrix) was considered as an autapomorphy of Pancrustacea (=Tetraconata) (e.g., Richter 2002). However, a similar configuration occurs in Scutigeromorpha (Myriapoda) (Müller et al. 2003). **Acone** compound eyes lack a crystalline cone (e.g., in Tipulidae [Diptera], Hemiptera). In the **eucone** type it is formed by intracellular secretions of the cone cells, with the nuclei always located between the corneal lens and the cone. The **pseudocone** type is characterized by the extracellular formation of the cone, which is adjacent with the

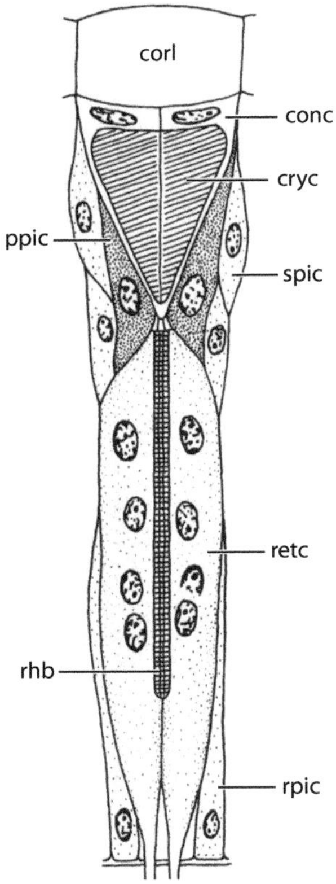

Fig. 1.6.2.2: Ommatidium. Abbr.: conc – corneagenous cell, corl – corneal lens, cryc – crystalline cone, ppic – primary pigment cell, retc – retinula cell, rhb – rhabdom, rpic – retinula pigment cell, spic – secondary pigment cell. Redrawn from Seifert (1995).

corneal lens and sometimes fused with it, thus forming a pseudocone. In this type the cone cells are shifted proximally towards the retinula.

The sensory elements are elongate photoreceptor cells arranged along the longitudinal axis of the ommatidium. Together they form the club-shaped **retinula**. Eight retinula cells are present in almost all groups of Hexapoda (Fig. 1.6.2.2), but six or nine occur in some lineages, and the number can be distinctly increased in Coleoptera (Scarabaeoidea) (Gokan & Meyer-Rochow 2000). The retinula cells contain screening pigment granules and the part oriented towards the longitudinal axis of the ommatidium is densely packed with a set of microvilli showing a strict parallel arrangement, the **rhabdomere**. In the typical case, the rhabdomeric microvilli are set at an angle to those of the adjacent cell, but aligned with those of the retinula cell on the opposite side (Chapman 1998). The unit formed by the rhabdomeres of each ommatidium is called the **rhabdom**. In some groups, the rhabdomeres of a given ommatidium are separated from each other. This configuration is called an open rhabdom. It occurs in Dermaptera and Diptera, and also in some groups of Heteroptera and Coleoptera. However, in most groups the rhabdomeres are connected along their longitudinal axis within their ommatidium, thus forming a fused rhabdom. The retinula cells of ommatidia of this type have the same visual field, whereas those of hexapods with open rhabdoms have separate fields of view, shared with retinula cells of other ommatidia.

The ommatidia are more or less completely isolated from each other by **secondary pigment cells**, which contain numerous screening pigment granules (Fig. 1.6.2.1). The **primary pigment cells** enclose the crystalline cone with narrow proximal processes. The secondary pigment cells (12–18 per ommatidium in most groups) usually cover the proximal part of the primary pigment cells and at least the distal region of the retinula cells. The proximal parts of the photoreceptor cells are usually separated by retinula pigment cells. Their number varies between taxa.

There are several functional sub-types of the hexapod compound eyes (e.g., Land & Nilsson 2002) (Fig. 1.6.2.1). In the **apposition eye**, the ommatidia are optically isolated from each other by a pigment sheath comprising both the secondary pigment cells and screening pigment granules present in the retinula cells. In the light-adapted state, when apposition is performed, it is the longitudinal migration of retinular screening pigment granules along the entire rhabdom that enables a complete optical isolation of the ommatidia. Then, each ommatidium functions as an independent unit. Only light passing through the dioptric apparatus parallel to the longitudinal axis of the ommatidium (or at a small angle) reaches the rhabdom. The apposition eye is considered as an adaptation to perceive light at high intensities and ensures high spatial resolution. This is apparently linked with a diurnal lifestyle. The counterpart is the **superposition eye**, which usually possesses a shorter rhabdom. Between the crystalline cone and the distal tip of the rhabdom, there is a zone devoid of any screening pigment granules, termed the clear zone. Consequently, the light channeled through one cone may be spread not only to the retinula of the same ommatidium but also to the retinulae of neighboring ommatidia (Seifert 1995; Chapman 1998).

It is important to note that optical types are never fixed, but may change from apposition into superposition at night (and vice versa before daytime). This is due to photoreceptor dynamics (photoperiodic elongation or shortening of rhabdomeric microvilli), the displacement of entire pigment cell bodies and, most importantly, the withdrawal or migration of screening pigment granules from/into the clear zone (Meyer-Rochow 1999).

The axons of the retinula cells pass through the basal matrix of the compound eye and are connected with the optic neuropils (Fig. 1.6.2.1). Most of them end in the lamina ganglionaris (1st to 6th retinula cell) but some reach the medulla (7th and 8th

retinula cells). In hexapods with a fused rhabdom the axon bundle originating from each ommatidium usually remains connected and associated with visual interneurons originating in the lamina ganglionaris and medulla to form a **cartridge.** In contrast to that, cartridges of hexapods with open rhabdoms receive axons from retinula cells with the same field of vision but belonging to different ommatidia. This recombination of retinular axons in the optic neuropil represents is referred to as **neuronal superposition** (Nilson 1989; Chapman 1998).

Up to five different types of photopigments may occur in hexapod compound eyes (Odonata, few Lepidoptera). A visual pigment with a maximum absorption in the green range of the spectrum (maximum 490–540 nm) is always present (Chapman 1998).

Thin tracheae reach into the compound eyes and enter the space between the ommatidia. In some lepidopteran lineages, they form a densely packed light-reflecting inner layer, the **tapetum.** This structural modification is a characteristic feature of the superposition eye (see above). The taenidia in the tapetum are enlarged and flattened (Chapman 1998).

The entire compound eye is enclosed by a more or less extensive internal circumocular ridge with a central opening for the protocerebral optic lobes (Fig. 1.6.2.1). This endoskeletal structure increases the mechanical stability of the head capsule (see **1.2.2. Head capsule**). An additional low external ridge occurs in different lineages.

Compound eyes are usually absent in larvae of Holometabola. However, simplified types with a single corneal lens and without a crystalline cone occur in some basal lineages of Hymenoptera. Simplified compound eyes also occur in most groups of Mecoptera, but they are absent in Boreidae (see below) and strongly reduced in Nannochoristidae.

1.6.3 Ocelli

Dorsal **ocelli** (Fig. 1.6.2.3) are present in adults of most groups of Hexapoda. They also occur in immature stages but are absent in nymphs of Acercaria and in larvae of Holometabola (potential synapomorphy). In most cases, three ocelli are present and arranged in a triangle, with the unpaired ocellus in front of or below the paired ones. The loss of the unpaired median ocellus has occurred several times independently (rarely of the paired ones). Complete reduction occurs in different lineages, especially in wingless forms (e.g., Grylloblattodea, wingless morphs of Zoraptera, extant Dermaptera, most groups of Coleoptera). The main function of the ocelli is the perception of changes of the light intensity.

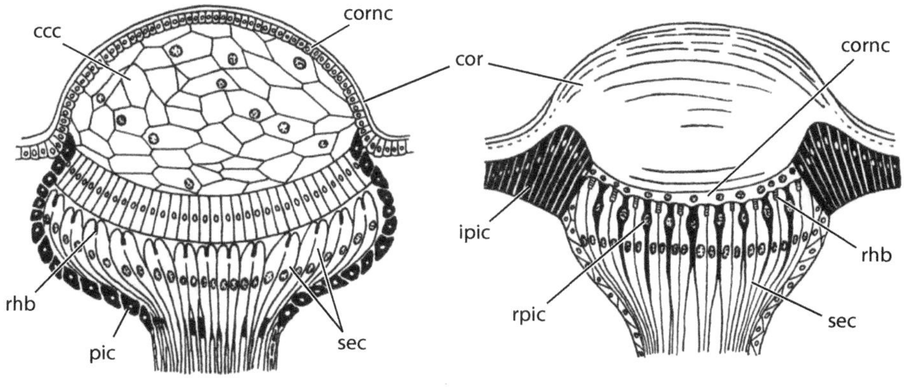

Fig. 1.6.2.3: Two types of ocelli. Abbr.: cornc – corneagenous cell, ccc – cellular crystalline cone, cor – cornea, ipic – iris pigment cells, pic – pigment cell, rhb – rhabdom, rpic – retinula pigment cell, sec – sensory cell. Redrawn from Seifert (1995).

In most groups, the ocelli are covered by a thickened (not in some orthopterans), transparent and undivided corneal lens. Below it retinula cells (ca. 800 in *Locusta*) are densely packed with a much less regular arrangement than in the compound eyes. A rhabdomere is present at least on one side of the retinula cells (unidirectional type). Rhabdoms are formed involving between two and seven cells. Accessory pigment cells may be present or absent (e.g., Blattodea). A reflecting tracheal **tapetum** occurs in some groups (Chapman 1998). The nerves of the ocelli originate on the dorsal (or frontal) region of the protocerebrum.

1.6.4 Stemmata

Stemmata are specific lateral eyes (the term "lateral ocelli" is inappropriate) of larvae of most groups of Holometabola. They are replaced by the adult compound eyes during the metamorphosis. The number ranges between seven and one and they are absent in different groups (e.g., Hymenoptera, Mecoptera excl. Boreidae, Siphonaptera, Cyclorrhapha [Diptera]). Structurally, they resemble ommatidia and are suitable to perceive movements, directions and distances. It was shown that specialized stemmata of some predacious beetle larvae (Cicindelinae, Dytiscinae) are image forming lens eyes (Buschbeck & Friedrich 2008).

Each stemma is covered by a single biconvex corneal lens and a crystalline cone is also present in most groups (eucone type). It is formed by three or four Semper cells in most groups but the number is increased in Neuropterida. The crystalline cone is usually missing in Coleoptera (acone type). Below it, a regularly arranged cluster of retinula cells is present. Their number is distinctly increased in Neuropterida (up to

40 in Megaloptera) (Paulus 1986). Only one rhabdom is present in the stemmata of most groups (e.g., Boreidae [Mecoptera], Neuroptera [partim], Trichoptera, Lepidoptera, Coleoptera [major part]). Up to 5,000 cells can be contained in the extended and cup-shaped retina of the two largest stemmata of tiger beetle larvae (Cicindelinae). In the two largest stemmata of larvae of *Thermonectes* (Dytiscidae) several morphologically distinct retinae are arranged on different layers (Buschbeck & Friedrich 2008).

The optic tracts connecting the stemmata with the protocerebrum are much less developed than the optic neuropils of the adults. They are greatly elongated in the extremely miniaturized first instar larvae of Strepsiptera due to the position of the brain (and other parts of the central nervous system) in the middle region of the postcephalic body.

[Bennett (1979); Paulus (1979, 1986); Arikawa et al. (1980); Nilsson (1989); Miyako et al. (1993); Seifert (1995); Chapman (1998); Meyer-Rochow (1999); Gokan & Meyer-Rochow (2000); Land & Nilsson (2002); Richter (2002); Müller et al. (2003); Buschbeck & Friedrich (2008)]

1.7 Tracheal system

A tracheal system has evolved several times independently in different groups of arthropods as an adaptation to the terrestrial environment. It is likely that it appeared for the first time in Onychophora in the Silurian, and then in several euarthropod groups including several lineages of Arachnida (Chelicerata). It was a considered as an autapomorphy of a presumptive monophylum Tracheata (=Atelocerata, Antennata) comprising Hexapoda and Myriapoda. However, the tracheal system in several myriapod subgroups (Notostigmophora, Craterostigmomorpha, Diplopoda) differs strongly from the condition in most groups of Chilopoda (Lithobiomorpha, Scolopendromorpha, Geophilomorpha) and Hexapoda (Fig. 1.7.1). Moreover, the tracheal system of the entognathous Collembola and Protura is profoundly different from the typical condition found in Insecta.

Tracheae are formed as invaginations of the external body wall, as hollow apodemes. Consequently their lumen is coated with a very thin cuticle, the intima. The external openings are the **spiracles** (Fig. 1.7.2). They are usually located in membranous pleural areas of the pterothorax and in the pleural membranes of the abdominal segments I–VIII. However, this typical pattern of pterygote insects is very likely not part of the hexapod groundplan. Only one pair of spiracles is present in the cervical membrane in Collembola, and even this is missing in some subgroups. The usual postcephalic spiracles are completely lacking in this order, and abdominal spiracles are also absent in Protura. In both basal hexapod lineages the tracheal system is more or less restricted to the anterior body region, and sometimes even completely absent. Consequently it cannot be completely excluded that a tracheal system was missing in the hexapod groundplan, and the primary pattern in Insecta is also quite unclear. The

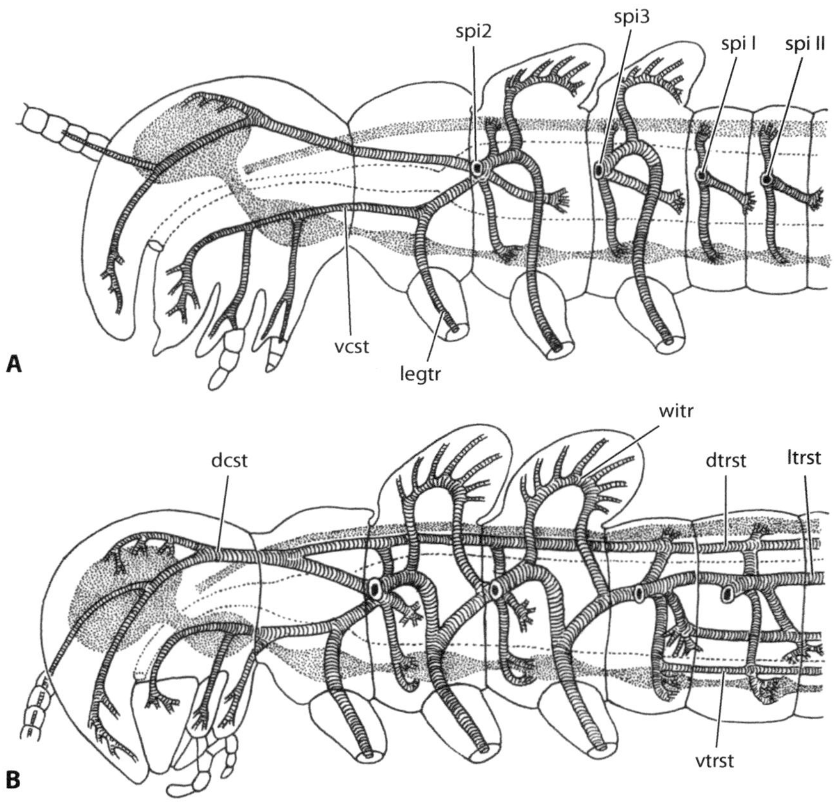

Fig. 1.7.1: Tracheal system, schematized, lateral view. A, apterygote hexapod; B, pterygote insect. Abbr.: dcst – dorsal cephalic stem, dtrst – dorsal tracheal stem, legtr – leg trachea, ltrst – lateral tracheal stem, spi2/3,I/II – thoracic and abdominal spiracles, vcst – ventral cephalic stem, vtrst – ventral tracheal stem, witr – wing trachea. Redrawn from Seifert (1995).

configuration of spiracles appears more complex in the apterygote Diplura, Archaeognatha and Zygentoma than in Pterygota.

The external openings of the tracheal system, the **spiracles**, are usually embedded in the pleural membranes as pointed out above, but occasionally also in small sclerotized plates called **peritremes** (Fig. 1.7.2: petr). They are round in some groups but usually more or less elliptical or even fissure-shaped. An extended chamber is usually present below the body surface. This atrium is often cone-shaped and almost always equipped with an opening-and-closing mechanism, a valve which is either present at its external or internal opening (Fig. 1.7.2). Closing muscles are usually part of this apparatus, which is in most cases formed by a rigid and a flexible cuticular element. The spiracular muscles are controlled by the transverse nerves of the unpaired ventral visceral nervous system. The abdominal spiracles of apterygote

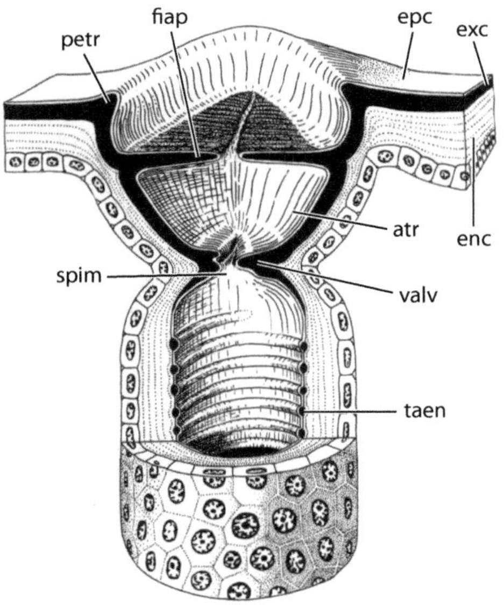

Fig. 1.7.2: Spiracle. Abbr.: atr – atrium, enc – endocuticle, epc – epicuticle, exc – exocuticle, fiap – filter apparatus, petr – peritreme, spim – spiracular mouth, taen – taenidium, valv – valve. Redrawn from Seifert (1995).

hexapods and Ephemeroptera lack closing muscles, but they are generally present in Odonata and the neopteran lineages. The atrium is usually also equipped with a filter mechanism with microtrichia, which prevents particles from entering the tracheal system. The main tracheal branches are segmentally arranged in Diplura and Insecta and connected by longitudinal branches and anastomoses in the typical case (Zygentoma and Pterygota) (Fig. 1.7.1). Large air-sacs are developed in insects with very good flying abilities (e.g., bees, cyclorrhaphan flies).

The thin ectodermal intima of the tracheae is reinforced by a helical cuticular thickening, the **taenidium**. The short horizontal proximal branch ramifies within the body, dividing into a dorsal, a ventral and a visceral trachea. The dorsal one supplies the heart, the dorsal muscles and the dorsal integument, the ventral one the ventral ganglionic chain, ventral muscles and the ventral body wall, and the visceral branch the gut, the Malpighian tubules, the internal genital organs and the fat body. The oxygen is directly transported to muscles and different organs. The main mechanism is diffusion, in many hexapods supported by breathing movements resulting from by contractions of the abdomen. The terminal parts of the tracheae (Fig. 1.7.3) are called **tracheoles**. These very thin tubes differ distinctly from the proximal sections, not only by their very narrrow lumen (less than 1 μm), but also by their lack of an intima. Moreover, they are not filled with air but with liquid. They end blindly in contact with

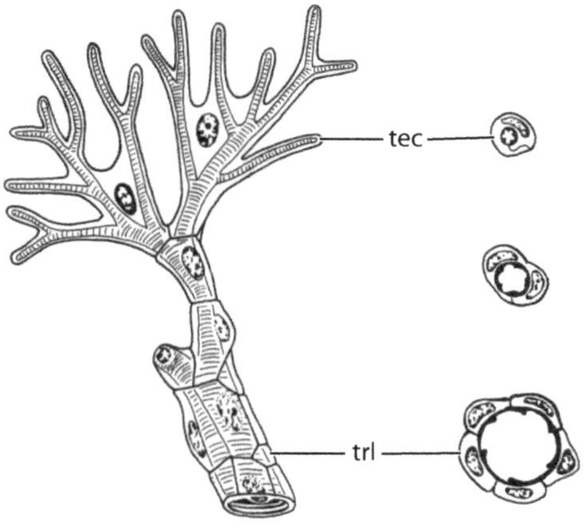

Fig. 1.7.3: Distal part of trachea: tracheole and terminal cells. Abbr.: tec – terminal cell, trl – tracheole. Redrawn from Seifert (1995).

the tissues they supply with oxygen, sometimes even indenting into the cells (Gullan & Cranston 1994). In contrast to the wider proximal parts of the tracheal system the tracheoles stay intact throughout an hexapod's live, whereas the proximal parts are shed during ecdysis with other parts of the cuticle.

The presence of two thoracic and eight abdominal functional spiracles is called a holopneustic condition. This is a groundplan feature of Pterygota and maintained in the vast majority of lineages. A hyperpneustic condition occurs in the dipluran family Japygidae, where additional spiracles are present in the meso- and metathorax. Other variations occur in different groups, especially in immature stages with a modified lifestyle, such as for instance in aquatic larvae. In the hemipneustic type some or all spiracles are formed in the development but closed at a later stage. Apneustic larvae lack spiracles completely.

Different types of spiracles occurring in groups of hexapods can be of systematic value. The simplest type is the annular spiracle. Other types are annular-biforous, biforous, or cribriform spiracles (see e.g., Stehr 1987, 1991). Modified breathing organs occur in aquatic larvae such as for instance lateral abdominal tracheal gills (e.g., Ephemeroptera, Gyrinidae [Coleoptera]), tubular gill tufts (Plecoptera), spiracular gills (Myxophaga [Coleoptera]), anal gills (e.g., larvae of Anisoptera [Odonata] or Scirtidae [Coleoptera])

A plastron is a layer of air held by a dense vestiture of microtrichia, either on most parts of the body (e.g., Dryopidae [Coleoptera]) or only on the ventral side (Hydrophilidae [Coleoptera]). It functions as an additional air supply and in some cases also as a physical gill.

[Stehr (1987, 1991); Gullan & Cranston (1994); Seifert (1995)]

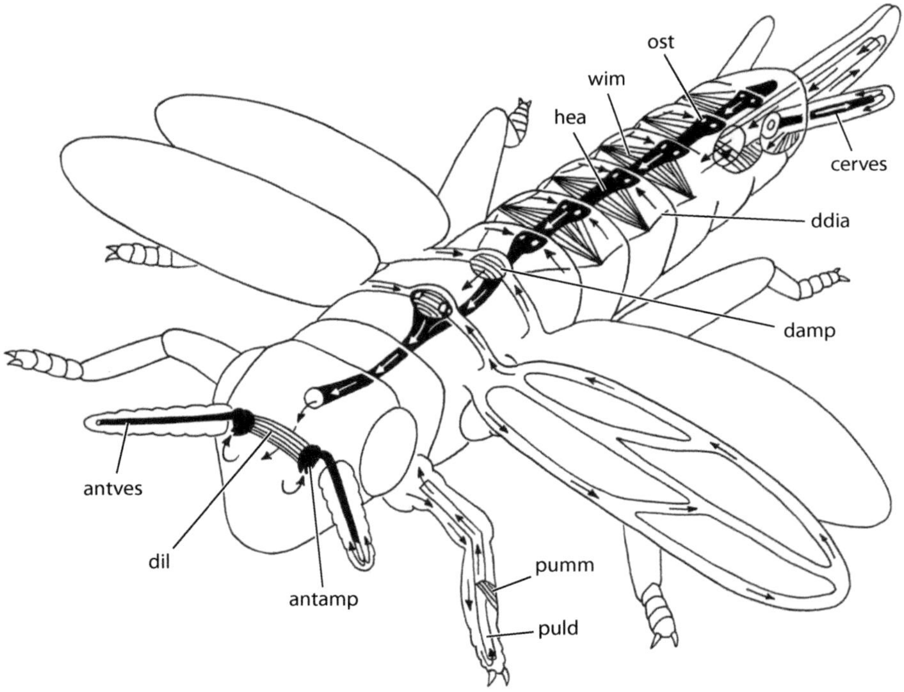

Fig. 1.8.1.1: Circulatory system, schematized. Abbr.: antamp – antennal ampulla, antves – antennal vessel, cerves – cercal vessels, damp – dorsal ampulla, ddia – dorsal diaphragm, dil – dilator muscle, hea – heart, ost – ostium, puld – pulsatile diaphragm, pumm – pumping muscle, wim – wing muscles. Redrawn from Pass (1998).

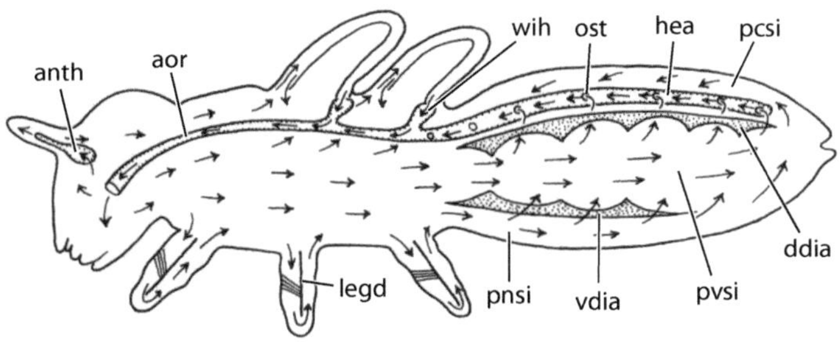

Fig. 1.8.1.2: Circulatory system, schematized. Abbr.: anth – antennal heart (ampulla), aor – aorta cephalica, ddia – dorsal diaphragm, hea – heart, legd – leg diaphragm, ost – ostium, pcsi – pericardial sinus, pnsi – perineural sinus, pvsi – perivisceral sinus, vdia – ventral diaphragm, wih – wing heart. Redrawn from Seifert (1995).

1.8 Circulatory system

Hexapoda have an open circulatory system with a main **dorsal vessel** with lateral openings (**ostia**) (Figs 1.8.1.1, 1.8.1.2), as it is also present in all other arthropod lineages with very few exceptions (e.g., absent in the strongly miniaturized Tardigrada). Its complete absence in all subgroups of Cycloneuralia (e.g., Nematoda, Priapulida) suggests that the presence is an autapomorphy and important evolutionary novelty of Arthropoda in the widest sense. This implies that a similar but more complete circulatory system in Annelida (and Mollusca) has evolved independently.

The body lumen of arthropods is often referred to as **mixocoel**, implying the fusion of the primary and secondary body cavity (coelom). However, the strong and consistent support for a clade Ecdysozoa (=Cycloneuralia + Arthropoda) in molecular studies suggests that a true coelom as it occurs in Annelida and other lophotrochozoan lineages (e.g., Sipunculida, Mollusca) was never present in Arthropoda. Therefore, the neutral term **haemocoel**, i.e. the space containing the haemolymph (blood + fluid of the primary body cavity) is preferable. Whether cavities formed in the embryonic development of hexapods (see **2.2.1 Cleavage**) are homologous to true coelomic cavities is questionable.

Compared to some crustacean lineages, especially in Malacostraca ("higher crustaceans"), the circulatory system of most hexapods is distinctly simplified, largely or completely lacking an arterial system. Paired segmental vessels occur in some groups with large species (e.g., Mantodea, Blattodea, Orthoptera [partim]). However, they are short, and apparently de novo formations (Hertel & Pass 2002) not belonging to the hexapod groundplan. An important functional aspect of the hexapod circulatory system is that the haemolymph does not transport oxygen in the body (see **1.7 Tracheal system**).

1.8.1 Haemocoel, diaphragmata and alary muscles

The haemocoel in the postcephalic body is usually divided by two extensive connective tissue membranes, the dorsal **diaphragm** below the heart and the ventral diaphragm above the ventral ganglionic chain (Fig. 1.8.1.2). The resulting three compartments are the dorsal **pericardial sinus**, the **perivisceral sinus** around the alimentary tract, and the ventral **perineural sinus**. The dorsal diaphragm is a fenestrated membrane and usually incomplete laterally, thus leaving open a connection to the perivisceral sinus. In some groups it propels the haemolymph flow actively. In the honeybee it contributes to the ventilation of tracheal sacs, together with the **alary muscles**. These muscles either stretch from one side of the body to the other just below the heart, or they are directly connected to it. Usually they fan out from the attachment points at the terga towards a broad median connecting zone (Chapman 1998). The alary muscles are controlled by visceral nerves. In Orthoptera two muscle pairs can be present in the thorax and up to ten in the abdomen, but the number is usually lower (e.g., 4–7 pairs in Heteroptera) (Chapman 1998). The ventral diaphragm is a horizontal septum above the ventral ganglionic chain and usually restricted to the abdomen (not in some orthopterans and hymenopterans). It supports the backflow of the haemolymph in the ventral abdominal region, in some cases by peristaltic movements. The structure can vary in different body parts and life stages. In the thorax of some orthopterans it is membranous and largely or completely lacking muscles, whereas it is a solid muscular sheet in the abdomen. In roaches only a small rudiment of the ventral diaphragm is retained as a thin muscle band, the hyperneural muscle. The ventral diaphragm is a fenestrated membrane in adults of *Corydalus* (Megaloptera) but a muscular sheet in the larvae (Chapman 1998). In groups without segmental ganglia in the abdomen (concentration of the ventral ganglionic chain in the posterior thorax) the ventral diaphragm is lacking (Wasserthal 2003).

In most hexapods the haemocoel is restricted to narrow spaces between muscles, endoskeletal elements, tracheae, fat body lobes, and other structures and organs. The haemolymph is pressed between the narrow sinuses, which are thus functionally partly equivalent to blood vessels. The lumen of the legs is divided by a septum (Fig. 1.8.1.2). The haemolymph pressed into the legs by a **pulsatile organ** flows down on one side and up on the other (Chapman 1998).

1.8.2 Dorsal vessel

The main element of the circulatory system is the tube-like **dorsal vessel** (Figs 1.8.2.1, 1.8.2.2), which functions as the main haemolymph pump in the hexapod body. It is placed dorsomedially closely adjacent to the terga and connected to the dorsal body wall or suspended by elastic filaments (Chapman 1998). The contractile wall of the vessel usually consists of one or two layers of muscle cells with a circular or spiral arrangement, often with two cells forming a closed ring. The basal laminae of the cells form an outer (**adventitia**) and an inner layer (**endocard**) covering the external and internal wall of the tube, respectively.

In some hemimetabolous lineages and most groups of Holometabola the dorsal vessel is distinctly subdivided into a weakly contractile **anterior aorta** (or **aorta cephalica**) and a posterior pumping region referred to as **heart** (Hertel & Pass 2002). The tube-like, unperforated aorta reaches into the head anteriorly and usually opens below or slightly anterad the brain. Its muscularis is weakly developed. It is usually straight, but aortic loops in the thorax occur in some groups of Lepidoptera and Hymenoptera. They may have a function in the context of thermoregulation. A small element in close vicinity with the dorsal aorta is the **circumoesophageal vessel ring** around the anterior foregut. It occurs in the apterygote lineages (Pass 2000) and is probably a groundplan feature of Hexapoda. The secondary loss is a potential autapomorphy of Pterygota.

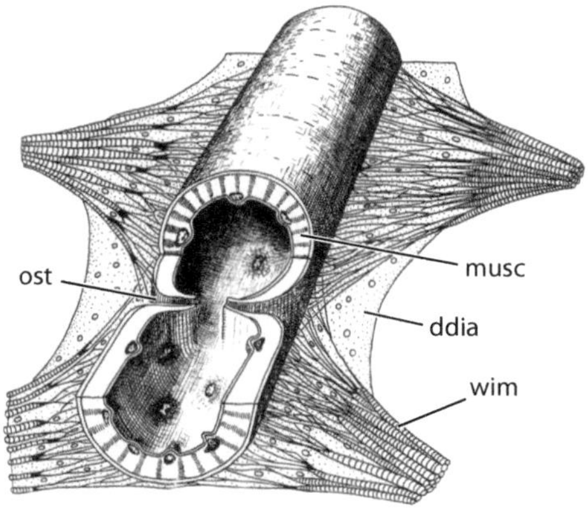

Fig. 1.8.2.1: Heart, dorsolateral view, partly opened horizontally. Abbr.: ddia – dorsal diaphragm, musc – muscularis, ost – ostium, wim – wing muscles. Redrawn from Seifert (1995).

The heart in the postcephalic body region rarely reaches the prothorax anteriorly (e.g., in Blattodea) (Figs 1.8.1.1, 1.8.1.2). In different lineages it is completely restricted to the abdomen (e.g., Anoplura). It is suspended by filaments formed by suspensory cells (**pericardial cells**) and by the flat and more or less horizontal alary muscles (Fig. 1.8.2.1). It is placed above the dorsal diaphragm. Its muscularis is well-developed and its peristaltic contractions (systole) result in the flow of the haemolymph. The dilation of the vessel (diastole) resulting in the inflow through the lateral ostia (see below) is usually achieved by the elastic suspending filaments. An additional simultaneous contraction of the alary muscles occurs in few groups of insects (Hertel & Pass 2002). The heart beat rhythm in hexapods is caused by a myogenic automatism of the muscle fibers of the dorsal vessel. This is likely a groundplan feature of Hexapoda (Hertel & Pass 2002). In most groups this mechanism can be modified by a neuronal control effected by a pair of lateral heart nerves and segmental nerves from the ventral nerve cord.

The heart is posteriorly closed in most pterygote insects (not in Ephemeroptera). Paired incurrent lateral ostia are always present. The anterior and posterior lips of the vertical slit-like openings are reflexed into the heart, thus forming a valve allowing only the flow of haemolymph into the vessel at diastole, but preventing its outward passage (Fig. 1.8.2.2). In some groups (e.g., Orthoptera) the heart appears chambered, especially in the thorax, due to extended ampullae at the sites of the ostia (Chapman 1998). The maximum number of lateral incurrent ostia is 13 (some Blattodea). Two or three are often located in the thorax and the remaining ones in the abdomen. The number is distinctly reduced in different holometabolan lineages (e.g., 7–8 in many groups of Lepidoptera, five in aculeate hymenopterans). Only two or three pairs are present in Anoplura and many groups of Heteroptera (Chapman 1998). Additional

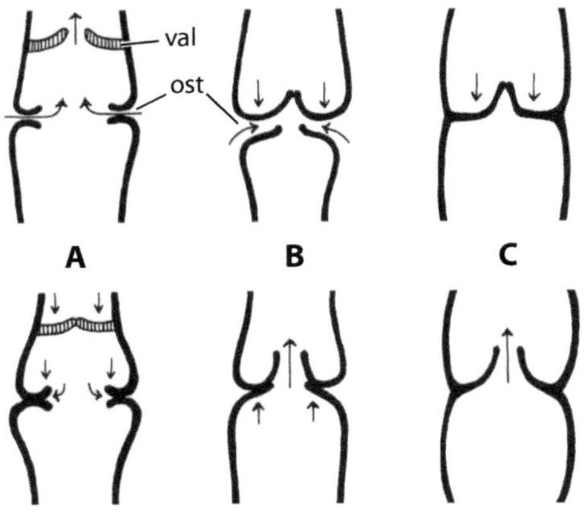

Fig. 1.8.2.2: Heart with valves and flow of haemolymph (A–C), schematized. Abbr.: ost – ostium, val – valve. Redrawn from Seifert (1995).

excurrent ostia occur in few groups, paired in Zygentoma, Orthoptera and Phasmatodea, and unpaired in Plecoptera and Embioptera. In Orthoptera two thoracic and five abdominal ventrolateral excurrent ostia are present. Externally the openings are surrounded by a papilla of spongiform multinucleate cells, which expand during systole when haemolymph is pressed out, and again contract during diastole (Chapman 1998). They penetrate the dorsal diaphragm in orthopterans but not in Phasmatodea.

In Odonata and in most groups of Neoptera the flow of the haemolymph is unidirectional towards the head, with specialized valves preventing the backflow. This is apparently not a groundplan feature of the entire Hexapoda. The heart is posteriorly open in apterygotes and Ephemeroptera, and the flow of the haemolymph in the dorsal vessel is bidirectional. In Diplura the haemolymph flow anterior to abdominal segment VIII is towards the head, and posterior to it towards the abdominal apex. This is regulated by an intercardiac valve posterior to the last pair of ostia (Hertel & Pass 2002). A unique characteristic of Diplura are arteries directly supplying the antennae and cerci. This feature arguably belongs to the groundplan of Hexapoda as similar conditions occur in other arthropod lineages (Hertel & Pass 2002). In Archaeognatha, Zygentoma and Ephemeroptera a vessel of the unpaired terminal filament (groundplan autapomorphy of Insecta) is present in the posterior abdomen. The posterior closure of the dorsal vessel and the resulting unidirectional flow in Odonata and Neoptera (groundplan) is possibly correlated with the loss of the terminal filament (Hertel & Pass 2002).

Resulting from periodic heartbeat reversal the direction of the flow alternates in some groups of Holometabola (Hertel & Pass 2002). Corresponding anatomical prerequisites, are either a secondarily opened posterior end of the dorsal vessel or two-way ostia (Hertel & Pass 2002).

1.8.3 Accessory pulsatile organs

Accessory pulsatile organs are evolutionary novelties of pterygote insects (Pass 2000; Hertel & Pass 2002) (Fig. 1.8.1.1). These pumping devices occur at the bases of the antennae (antennal hearts, antenna-hearts), the legs (leg hearts, leg-hearts) and the wings (wing circulatory organs, wing-hearts). Different muscles can be recruited for operating them. An unpaired transverse cephalic muscle can be integrated in the antennal hearts (e.g., Dictyoptera), or paired muscles posteriorly inserted on the pharynx (e.g., Mecoptera). The leg hearts of Heteroptera are operated by a portion of the pretarsal flexor, in some cases detached from the tendon of the main part of the muscle. Innervations have been described for antenna-hearts and leg-hearts. The wing circulation is ensured by modified portions of the dorsal vessel in basal pterygote lineages, whereas independent organs – muscular plates with associated elastic suspending strands – are present in more advanced groups (Hertel & Pass 2002). Pumping organs related to wing circulation have evolved several times independently in Pterygota (Pass 2000).

1.8.4 Haemolymph

The **haemolymph** of hexapods is composed of plasma and different types of **haemocytes** (Fig. 1.8.4.1). As it is usually not involved in the transport of oxygen its volume is relatively small, varying between 10% and 40% of the total body volume (Wasserthal 2003). The main functions are the transport of nutrients, metabolites, excretory products, hydrogen carbonate, neuromodulators and hormones. It is also involved in wound repair, defense responses to intruding organisms, the ventilation of the tracheal system, and thermoregulation. It also carries defensive substances against predators in different groups of hexapods (e.g., Coccinellidae). Another functional complex is a variety of hydraulic mechanisms occurring in the hexapod body, for instance in the context of copulation and molting.

The haemolymph plasma is usually more or less transparent or yellowish. Respiratory pigments are almost always lacking. Haemoglobin occurs in larvae of some species of Chironomidae (Diptera). The composition of the plasma varies among taxa and depending on the developmental stage and the nutritional status. It contains high levels of free amino acids, different proteins (especially storage proteins but also lipid transport proteins and enzymes), protease inhibitors (at least in some groups), carbohydrates (especially glucose and the disaccharide trehalose), end products of the nitrogen metabolism, lipids (especially as diacylglycerols), and organic acids. The most abundant inorganic anion in many groups of hexapods is chloride (low concentration in Holometabola). Sodium is usually the most abundant cation (Chapman 1998).

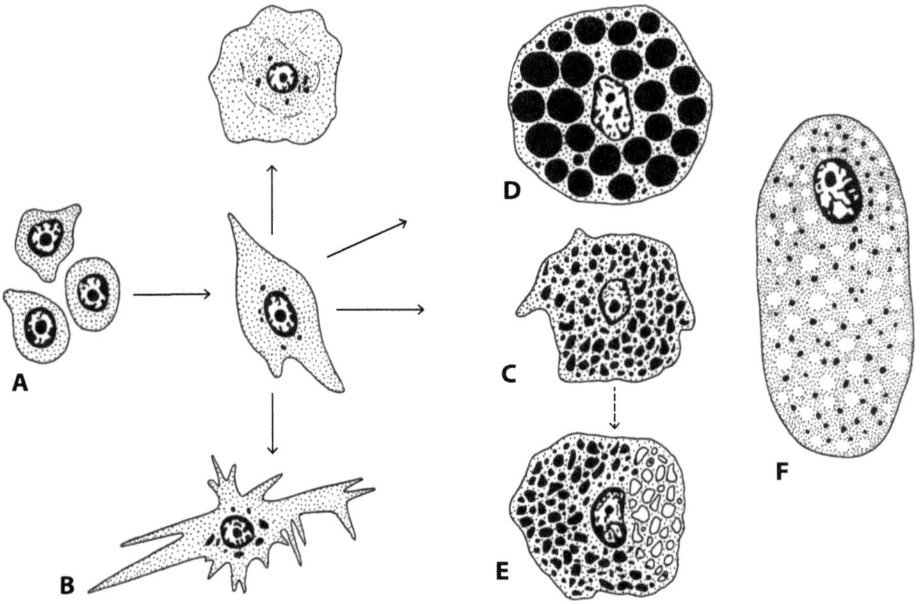

Fig. 1.8.4.1: Haemolymph cells. A, prohaemocytes; B, plasmatocyte; C, granulocyte; D, sphaerulocyte; E, cystoctyte; F, oenocytoid. Redrawn from Seifert (1995).

A considerable variety of different haemocytes occurs in hexapods. Their concentration in the haemolymph varies between taxa and developmental stages. The stem cells of most of them are mesodermal **prohaemocytes**, small and globular cells with a large nucleus in relation to the cytoplasm. Exceptions are the **oenocytoids** which are usually derived from hypodermal cells. Haemocytes can be formed by mitoses of free cells or in **haemocytopoietic** (haemopoietic) **organs**, which are more or less closely associated with the heart (e.g., Orthoptera).

Plasmatocytes are abundant in the haemolymph (usually more than 30% of the haemocytes) and highly variable in shape. They are involved in phagocytosis and the encapsulation of intruding organisms. **Granulocytes**, which are likely derived from plasmatocytes, are also abundant and characterized by large amounts of endoplasmatic reticulum and Golgi complexes. They discharge their contents on intruding organisms. **Cystocytes** are similar to the granulocytes and also common in different groups. They contain more granules but less endoplasmatic reticulum and Golgi vesicles. **Adipohaemocytes** are characterized by a small and eccentrically placed nucleus and lipid droplets. Their function is unclear, like in the case of the **oenocytoids**, which mainly occur in Lepidoptera (Chapman 1998).

[Seifert (1995); Chapman (1998); Pass (2000); Hertel & Pass (2002); Wasserthal (2003)]

1.9 Digestive tract

The digestive tract of hexapods is divided into three main sections with a different ontogenetic origin (Figs 1.9.1, 1.9.2). The **foregut** and the **hindgut** are formed by the ectoderm and therefore coated by an intima, whereas the **midgut** is of entodermal origin. The epithelium is single-layered. Its external basal lamina is thickened in some groups and then referred to as **membrana propria** or **tunica**. It is enclosed by a muscularis formed by circularly arranged fibers (external layer) and longitudinal fibers (internal layer).

1.9.1 Preoral cavity

The **preoral cavity** is the space enclosed between the labrum and labium and the paired mandibles and maxillae (Fig. 1.9.1). It is not part of the digestive tract in the strict sense as it lies below or anterior to the anatomical mouth opening (depending on the orientation of the head), i.e. the opening of the pharynx. The area between the internal wall of the labrum (epipharynx) and the hypopharynx is the **cibarium**. In this region the food is manipulated, processed mechanically by the mandibles and to a lesser degree by the laciniae, and diluted with secretions of the salivary glands. The hypopharynx usually forms a ramp leading towards the **anatomical mouth** opening. The laciniae often play an important role in the transport of particles together with the epipharyngeal and hypopharyngeal surfaces which are often covered with posteriorly directed microtrichia. The process is usually supported by upward (backward) movements of the labrum (Mm. frontolabralis and frontoepipharyngalis) and the hypopharynx (M. frontohypopharyngalis). The cibarium is widened by contractions of the intrinsic labral muscle (M. labroepipharyngalis) and especially M. clypeobuccalis, which is often strongly developed (e.g., Acercaria) and can be divided into a more or less elongated series of bundles (e.g., Mecoptera excl. Nannochoristidae). In some groups of hexapods, especially with predacious feeding habits, the proximal epi- and hypopharynx fuse along their lateral margins, thus forming a closed tube, which is then referred to as **prepharynx**. In some cases it can be strongly elongated (e.g., larvae of Carabidae and Gyrinidae [Coleoptera]). The external opening of the prepharyngeal tube is the **functional mouth opening**.

1.9.2 Foregut

The foregut is ontogenetically derived from the ectodermal stomodaeum. Its anteriormost part is the **pharynx** (Figs 1.9.1, 6.18.2B: ph). Its opening is the **anatomical mouth** which is always clearly defined by the position of the frontal ganglion, the insertion of the anterior frontopharyngeal muscle (M. frontopharyngalis anterior),

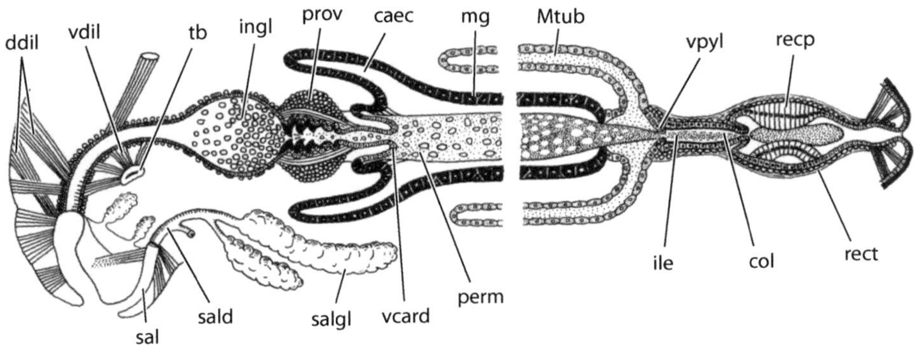

Fig. 1.9.1: Digestive tract with associated organs, schematized. Abbr.: caec – caeca, col – colon, ddil – dorsal pharyngeal dilators, ile – ileum, ingl – ingluvies, mg – midgut, Mtub – Malpighian tubules, perm – peritrophic membrane, prov – proventriculus, recp – rectal papillae, rect – rectum, sal – salivarium, sald – salivary duct, salgl – salivary gland, tb – tentorial bridge, vcard – valvula cardiaca, vdil – ventral pharyngeal dilators, vpyl – valvula pylorica. Redrawn from Seifert (1995).

and usually by a strongly developed ring muscle. The lumen of the pharynx can be rather wide or strongly narrowed. The inner surface is coated by a thin cuticular intima. The ring muscles and longitudinal muscles are usually well-developed. Dorsolateral, ventrolateral and ventral longitudinal folds are almost always present. They serve as attachment sites of dorsal, lateral and ventral **dilators** which widen the pharynx, and the longitudinal fibers of the muscularis are enclosed between them. The dorsal dilators originate from the wall of the head capsule, anterior (precerebral) and posterior (postcerebral) to the brain. The ventral dilators usually arise from the posterior tentorium (posterior arms and lateral parts of the tentorial bridge) but can also arise from the posteroventral region of the head capsule, especially in forms with reduced tentorial elements. The pharynx is posteriorly continuous with the **oesophagus** in the postoccipital region. The border can be marked by oesophageal dilators, thin bundles originating from the postoccipital ridge and posterior tentorium, respectively. However, these inconspicuous muscles are often missing. Generally, the muscularis of the oesophagus is less strongly developed than that its pharyngeal equivalent and the longitudinal folds are indistinct (anteriormost oesophagus) or entirely missing. Longitudinal muscles are present or absent. The anterior oesophagus in the thoracic region is extended as a voluminous crop or **ingluvies** (Fig. 1.9.2: ingl) in different groups (e.g., roaches). It is strongly dilatable and mainly serves for the storage of large amounts of food. The fibers of its thin muscularis form anastomoses and as a whole a reticulate meshwork. The posteriormost region of the foregut is often differentiated as a **proventriculus** (Fig. 1.9.2: prov) with rows of cuticular teeth and acanthae and strongly developed layers of ring muscles and longitudinal muscles. This device is used for a final mechanical processing of the food. The passage into the midgut is regulated by a sphincter muscle at the posterior margin of the proventriculus.

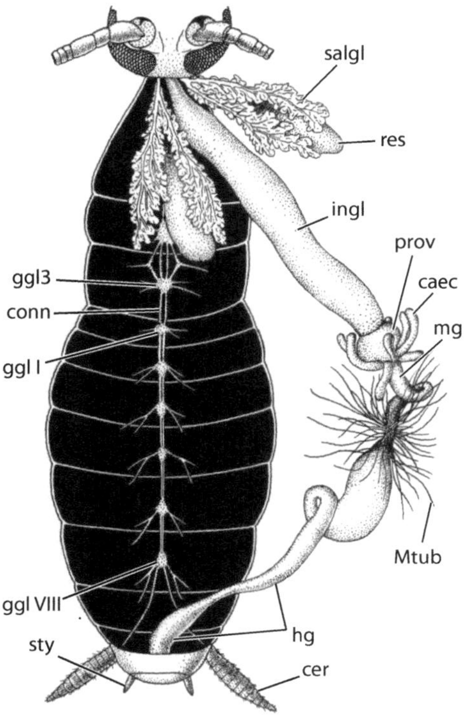

Fig. 1.9.2: *Periplaneta* sp. (Blattodea), digestive tract and other internal organs. Abbr.: caec – caeca, cer –cercus, conn- connective, ggl3/I/VIII – metathoracic/1st/8th abdominal ganglion, hg – hindgut, ingl – ingluvies, mg – midgut, Mtub – Malpighian tubules, prov – proventriculus, res – reservoir of salivary duct, salgl – salivary gland, sty – stylus. Redrawn from Seifert (1995).

1.9.3 Midgut

The border between the ectodermal foregut and the entodermal midgut is clearly marked by the absence of an intima in the latter, by the **valvula cardiaca**, a ring-shaped fold functioning as a valve, and in some groups by the insertion of a group of more or less extensive midgut **caeca** (Figs 1.9.1, 1.9.2), which enlarge the surface of the area involved in secretion and resorption. The caeca can be short papillae or elongate tubes and the number varies between one pair and up to 30. The epithelium is similar to that of the remaining midgut and the main functions are likely surface enlargement and secretion of hydrolases (Fig. 1.9.3.1). The midgut is usually distinctly shorter than the foregut. The cells are cuboidal or cylindrical (columnar) and apically lined by **microvilli**, which greatly enlarge the surface oriented towards the gut lumen (**rhabdorium, brush border**). Externally a basement membrane is present and a very thin muscularis. Midgut cells are regularly discarded and replaced. The undifferentiated replacement cells lie at the base of the epithelium without contact to the lumen. Sometimes they form large groups, regenerative crypts, which can penetrate the basement membrane and muscularis and reach into the mixocoel.

The secretion is usually apocrinous or merocrinous, but in some cases entire cells disintegrate (holocrinous secretion). The enzymatically digested food is resorbed by way of endocytosis between the microvilli of the midgut cells, usually between cycles of secretion, but in some cases also by specialized non-secretory cells. Another function of the midgut is to metabolize resorbed substances before they are released to the haemolymph (e.g., 20-hydroxy-ecdysone).

The vulnerable midgut epithelium is usually protected by a **peritrophic membrane** (Figs 1.9.1, 1.9.3.1), which is in fact not a membrane but a dense netting of thin cuticular fibrillae. It is continuously produced at the posterior foregut border and dismantled at the posterior end of the midgut by the **valvulae pylorica** (Fig. 1.9.1). It can be absent in groups exclusively feeding on fluids (e.g., aphids). The main functions of the midgut are secretion of digestive enzymes and resorption of food between the microvilli. The active form of ecdysone (20-hydroxy-ecdysone) is also occasionally produced by midgut cells. The hind border is marked by the valvula pylorica and the insertion of the Malpighian tubules.

1.9.4 Hindgut

The hindgut (Figs 1.9.1, 1.9.2) is ontogenetically derived from the proctodaeum and consequently coated with a thin intima like the foregut. Its muscularis of longitudinal

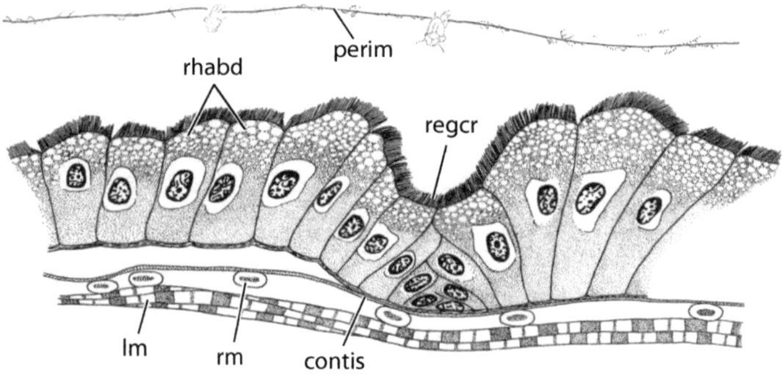

Fig. 1.9.3.1: Midgut wall. Abbr.: contis – connective tissue, perim – peritrophic membrane, lm – longitudinal muscle, regcr – regenerative crypt, rhabd – rhabdorium (microvilli seam), rm – ring muscle. Redrawn from Seifert (1995).

fibers and an outer layer of ring muscles is usually well-developed. The anteriormost part of the hindgut is the more or less globular **pylorus**, the area of origin of the **Malpighian tubules** (see **1.10. Excretory organs**). The ring-shaped **valvula pylorica** is equipped with a strongly developed muscularis at its posterior margin. Analogous to the valvula cardiaca it controls the passage of the gut contents and it also degrades the peritrophic membrane. The posteriorly adjacent **ileum** and **colon** are often not clearly distinguishable and not characterized by specific histological properties. The posterior margin of the colon is marked by a ring-shaped **valvula rectalis** in some groups. The **rectum** is usually distinctly expanded, with its epithelium forming three or six voluminous one- or bilayered **rectal papillae** (Fig. 1.9.1). The epithelium of the papillae is always columnar. The lateral membranes of the high cells are intensively folded and these areas contain large mitochondria and are specialized on the uptake of potassium cations in a labyrinth of cellular interspaces. A strong gradient is formed resulting in a very efficient absorption of water, which then re-enters the mixocoel via the labyrinth. A **sphincter** is present at the posterior end of the rectum. The anus is flanked by anal valves.

[Seifert (1995); Lehane (1998)]

1.10 Excretory organs

Metabolic waste products such as urates can be stored to a certain degree in specific cells of the fat body (Fig. 1.12.2). Labial glands (Collembola, Diplura) or midgut cells can be involved in the excretion process (Collembola, Protura) and the rectum plays an important role in osmoregulation (see **1.9.4. Hindgut**). However, the **Malpighian tubules** are clearly the main excretory and osmoregulatory organs (Figs 1.9.1, 1.9.2, 1.10.1). They play an essential role in removing nitrogenous waste from the haemolymph.

It is disputable whether the Malpighian tubules belong to the groundplan of Hexapoda as they are completely missing in Collembola (and aphids) and only present as short papillae in Protura and Diplura (partim, missing in *Japyx*) (Dunger 2005). The number varies strongly between the orders. Ten or 12 are present in Archaeognatha and usually 4–8 in Zygentoma. High numbers occur in some polyneopteran groups (ca. 100 in Mantodea and roaches) (Fig. 1.9.2) and ca. 50 are present in some groups of Hymenoptera. A low number (4–8) is characteristic for Acercaria and Holometabola (excl. Hymenoptera).

The Malpighian tubules of Insecta are more or less thin, tube-like structures originating from the pylorus at the midgut-hindgut border (Fig. 1.9.2). They are usually unbranched and floating freely in the haemolymph. The apical parts are branched, inflated or connected pairwise in few groups. The ontogenetic origin is disputed. The traditional view is that they are of ectodermal origin. However, they lack an intima

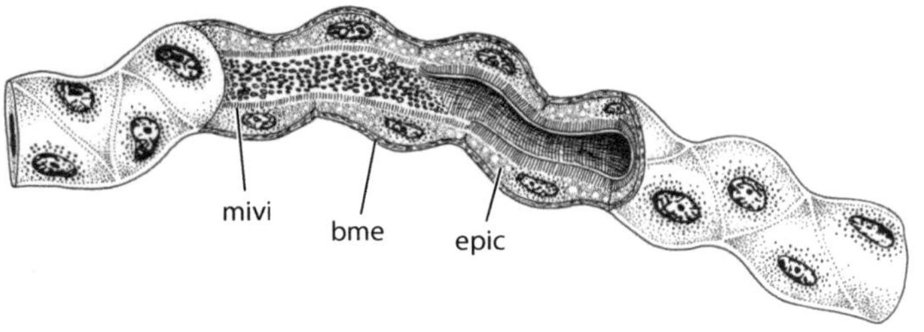

Fig. 1.10.1: Part of Malpighian tubule, partly opened. Abbr.: bme – basement membrane, epic – epithel cell, mivi – microvilli. Redrawn from Seifert (1995).

and recent investigations of the ontogenesis of *Drosophila* (Denholm et al. 2003) have shown that mesodermal cells can be involved. The single-layered epithelium is similar to that of the midgut (Figs 1.9.3.1, 1.10.1). Often two cell types occur, primary cells and stellate cells. A basement membrane covers the external (basal) surface of both types and the internal (apical) surface is covered by microvilli (rhabdorium) as it is the case in the midgut. The distribution and density can vary in different regions and is often lower and more scattered in the case of the stellate cells. An additional surface enlargement is caused by multiple folds of the basal cell membrane, the basal labyrinth.

The passage of excretion products from the haemolymph to the lumen of the Malpighian tubules is partly a passive diffusion (urea, amino acids) but active pumping mechanisms using adenosine triphosphate (ATP) are also involved (sodium, potassium). ATP is produced by numerous mitochondria in the folds of the basal labyrinth. Clear pre-urine is formed in the distal area of the tubules. Concrement vacuoles are produced by the endoplasmatic reticulum of the cells and are concentrated in the proximal region of the tubules. The cells of the proximal regions re-absorb water from the lumen which likely results in the formation of crystals of uric acid and other substances. A **basal ampulla** lacking typical microvilli is present in different groups. Its elongated cells are extend into the lumen of the pylorus and are also involved in the water re-absorption.

The excretions of most hexapods mainly consist of uric acid, but they can also contain ammonia (mainly aquatic insects and carnivorous flies), urates, urea, pteridines, amino acids, hypoxanthine, allantoin and allantoic acid. The rectal papillae play an important role in the water re-absorption before the waste is deposited via the anus (see **1.9.4 Hindgut**).

[Gullan & Cranston (1994); Seifert (1995); Denholm et al. (2003); Dunger (2005)]

1.11 Endocrine organs and the hormone system

The endocrine system of hexapods (Fig. 1.11.1) is well-developed even though less complex than in vertebrates. Hormones are substances produced by endocrine glands, by **neurosecretory cells**, or by **neurohaemal organs** (Figs 1.5.1.2., 1.5.1.3, 1.11.2). They are transported to other body regions by blood or haemolymph where they can have different physiological effects, in hexapods mainly related to development and molting, but also to other important body functions.

Groups of neurosecretory cells (Nc) of the brain (see **1.5.1 Brain**) are important elements of the endocrine systems (Figs 1.5.1.2., 1.5.1.3). The Nc are characterized by numerous grana (ca. 100–400 nm), an unusually high content of rough endoplasmatic reticulum, and extended terminal vesicles of the axons, where grana are stored. They can release their secretions directly at the perikaryon or via the axon. Lateral protocerebral Nc produce the **prothoracicotropic hormone** (**PTTH**) and the group in the pars intercerbralis of the protocerebrum the neurohormone D (synonyms: Pea-CAH I, myotropin I). The eclosion hormone, which plays an important role in the molting process, is produced by Nc in the tritocerebrum.

Neurohaemal organs are elements of the endocrine system associated with secretory neurons and the haemolymph system. The **corpora cardiaca** (**Cc**) and the **corpora allata** (**Ca**) are closely connected with the brain (Figs 1.5.1.3, 1.11.2), whereas the segmental **perisympathetic organs** (**PSO**) are associated with the postcephalic ganglionic chain. The Cc are small organs below the posterior brain region and usually close to the anterior opening of the cephalic aorta. They are primarily paired but fused in few groups. In some cases they are fused with the hypocerebral ganglia and they are ontogenetically derived from these neuronal structures. The Cc receive different neurohormones (e.g., **PTTH**, neurohormone D) from neurosecretory cells in the brain and store them or deliver them to the haemolymph. They also produce several neurohormones (e.g., adipokinetic hormone) which are involved in the regulation of the water balance and the lipid- and carbohydrate concentration of the haemolymph. The Ca are usually well-defined, small, round or ovoid organs posterior to the brain and the Cc, above the foregut and immediately laterad the cephalic aorta. Ontogenetically they are formed as ectodermal invaginations. They are usually paired but like the Cc medially fused in different groups. Their intrinsic cells produce the **juvenile hormone** which suppresses the molt to the adult stage and regulates developmental processes. Its secretion is suppressed by **allatostatin** and stimulated by **allatotropin**. Both substances are produced by neurosecretory cells in the brain. The **perisympathetic organs** are located in the postcephalic body and closely linked to the segmental ganglia of the thorax and abdomen. They contain neurosecretory axons ensheathed with glia cells and irregularly scattered gland cells. They produce numerous multifunctional neuropeptides.

The cerebral groups of neurosecretory cells are connected with the Corpora cardiaca (Cc) and Corpora allata (Ca) (see below) by the nervi corporis cardiaci and nervi

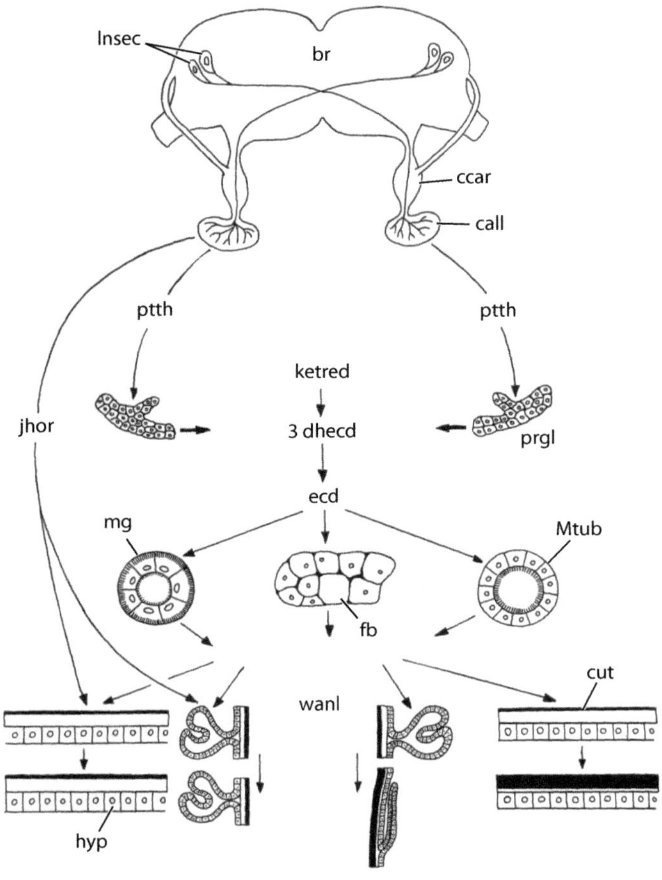

Fig. 1.11.1: Hormononal control of molting and metamorphosis in *Manduca sexta* (Lepidoptera, Sphingidae). Abbr.: br – brain, 3dhecd – 3 dehydroxy ecdysone, ecd – ecdysone, call – corpora allata, ccar – corpora cardiaca, cut – cuticle, fb – fat body, hyp – hypodermis, jhor – juvenile hormone, ketred – ketoreductase, lnsec – lateral neurosceretoric cells, mg – midgut, Mtub – Malpighian tubules, prgl – prothoracic gland, ptth – prothroacotropic hormone, wanl – wing buds (anlagen). Redrawn from Seifert (1995).

corporis allati (Fig. 1.11.2). The nervus corporis cardiacis II extends from the lateral protocerebrum through the Cc and reaches the Ca posteriorly. A short nerve connects the Cc with the hypocerebral ganglion in most groups.

An important element of the endocrine system of immature hexapods is the paired **prothoracic gland**, which is usually located in the anterior thoracic region, but occasionally also in the posterior region of the head (ventral gland). Ontogeneti-

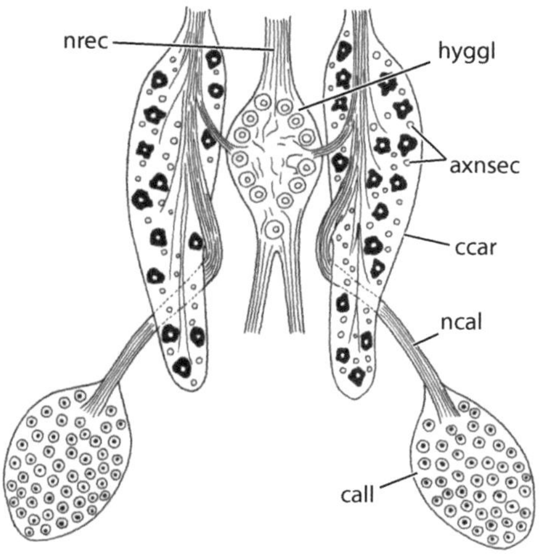

Fig. 1.11.2: Neurohaemal organs. Abbr.: axnsec – axon of neurosecretory cells, call – corpora allata, ccar – corpora cardiaca, hyggl – hypocerebral ganglion, ncal – nervus corporis allati, nrec – nervus recurrens. Redrawn from Seifert (1995).

cally it develops in the lateral area of the labial segment. It is usually elongated and branched in some groups. It is formed by rather diffuse tissue and often extends along tracheae or is embedded between fat body lobes. It receives its innervation either from the suboesophageal complex or from the pro- or mesothoracic ganglion. It produces the molting hormone **ecdysone**, which is later transformed into the active form 20-hydroxy-ecdysone by cells of the fat body, midgut or Malpighian tubules (Fig. 1.11.1). The prothoracic glands are stimulated by the **prothoracicotropic hormone (PTTH)**, which is produced by neurosecretory cells in the brain (see above) (Fig. 1.5.1.2). They degenerate after the adult stage is reached. At least in some adult hexapods, production of small amounts of ecdysone takes place in the ovaries and testes, respectively.

Hormone producing cells interspersed between digestive cells of the midgut are referred to as the **diffuse endocrine system**. They produce peptide hormones also occurring in vertebrates, as for instance gastrin, insulin, glucagon, vasopressin or β-endorphin (Dettner & Peters 2003). However, the functions of these substances in hexapods are still unknown.

[Seifert (1995); Dettner & Peters (2003); W. Hertel (pers. comm.)]

1.12 Fat body

The **fat body** or **corpus adiposum** occurs as a diffuse organ in members of Myriapoda and in crustaceans, but its presence as a voluminous organ mainly concentrated in the abdomen (Fig. 1.12.1) is likely an autapomorphy of Hexapoda. With its capacity to mobilize energy-rich substances efficiently it was probably an important prerequisite for the evolution of flight, likely the most important innovation in the evolution of Hexapoda.

The fat body is a diffuse, ill-defined organ. It is highly variable in its shape, extension and histological properties among taxa, and also depending on the developmental stage and physiological condition of an hexapod. It is composed of small whitish lobes of sheets or ribbons of cells (Fig. 1.12.2) which are immersed in the haemolymph (Gulan & Cranston 1994; Seifert 1995). Other organs, especially in the abdomen, are embedded in the complex 3-dimensional system formed by the fat body lobes (e.g., salivary glands, tracheae, ovarioles). Sometimes, especially in caterpillars and other holometabolan larvae, it shows a division into an inner visceral layer around the digestive tract and an outer parietal layer more closely associated with the integument. It is a multifunctional organ and plays a highly important role in the metabolism of hexapods. It is primarily designed to synthesize and store energy-rich substances such as glycogen, fat and protein. An important function, as pointed out above, is to mobilize these products very efficiently, for instance in the context of flight or specific development stages (e.g., metamorphosis).

The dominant cell type in the fat body is the **trophocyte**. These cells are responsible for the main functions of storage, mobilization of energy-rich substances and synthesis. Their size and contents (e.g., small droplets of lipids, albuminoids) are indicators of the developmental stage, active tissue formation or starvation. They are involved in the metabolism of carbohydrates, lipids and nitrogenous compounds. They synthesize the active form of ecdysone (20-hydroxy-ecdysone) and also sugar and proteins (e.g., vitellogenin, storage proteins) of the haemolymph. The fat body can also play a role in excretion, especially in Collembola, which lack Malpighian tubules. Metabolic waste is stored as spherules of urates (including uric acid) in specific urate cells, the **urocytes.** Fat body cells can also contain endosymbiotic microorganisms, mostly bacteria, but sometimes also yeasts or fungi. In certain aphids (Aphidoidea) endosymbionts can occur in all fat body cells, whereas they occur only in specific **mycetocytes** in roaches and certain scale insects (Coccoidea). The specialized mycetocytes are free of reserve substances and excretion products and can be scattered over the entire fat body or concentrated in small groups (mycetomes) in the central region of fat body lobes. The symbionts provide the hosts with vitamins and other substances the hexapods cannot produce themselves. Some symbionts can make use of uric acid for their own protein synthesis (Seifert 1995).

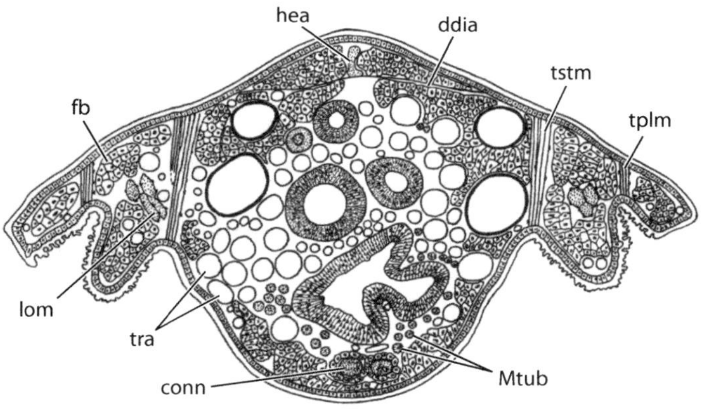

Fig. 1.12.1: Abdomen with fat body lobes and other internal organs. Abbr.: conn – connective of ventral nerve cord, ddia – dorsal diaphragm, fb – fat body, hea – heart, lom – longitudinal muscle, Mtub – Malpighian tubule, tplm – tergopleural muscle, tra – trachea, tstm – tergosternal muscle. Redrawn from Seifert (1995).

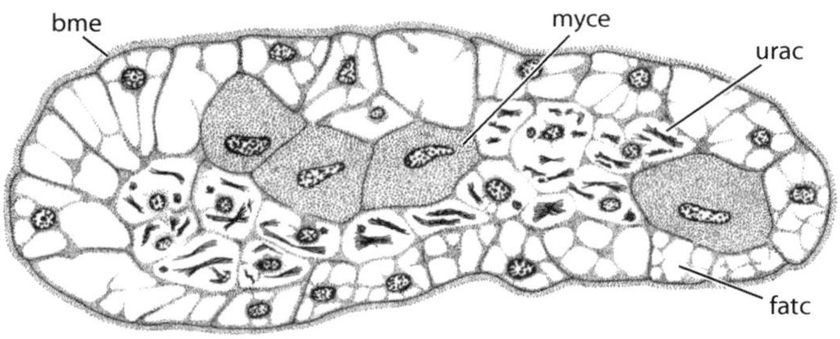

Fig. 1.12.2: Fat body lobe. Abbr.: bme – basement membrane, fatc – fat cell, myce – mycetome, urac – urate cell. Redrawn from Seifert (1995).

An important function of the fat body in immatures is the production of energy-rich yolk proteins as a preparation for the molt to the adult stage. In some holometabolan groups (Lepidoptera, Diptera) the larval fat body dissolves completely during the metamorphosis and is replaced by an adult fat body formed by cells kept at an embryological stage in the larva. In the highly specialized beetle family Lampyridae and some closely related elateroid groups the fat body or specialized regions of it form luminous organs. Light emission is caused by oxidation of luciferin involving the enzyme luciferase and adenosine triphosphate (ATP) as an energy source. The species specific pattern of light flashes is triggered by nervous impulses.

[Gullan & Cranston (1994); Seifert (1995)]

2 Reproduction, development and immature stages

2.1 Fertilization and egg structure

External sperm transfer is a plesiomorphic groundplan feature of Hexapoda. This applies to all primarily wingless groups. A **spermatophore** (sperm package) is either deposited on a stalk formed of secretions (e.g., Collembola) or on silk threads (Archaeognatha, Zygentoma [partim]). Direct sperm transfer using an intromittent organ (penis, aedeagus) is an important evolutionary novelty of Pterygota, but it is unclear whether this is a groundplan feature of this lineage. The mechanism is completely different in Ephemeroptera and Neoptera on one hand, and Odonata on the other (secondary copulatory organ at the base of the abdomen; Fig. 6.7.4B). Females of many groups possess a bursa copulatrix (expansion of the vagina) which receives the aedeagus. Sperm is often stored in a receptaculum seminis or spermatheca, which can have a species specific shape and/or number. There the sperm cells can be kept in a functional condition for a longer period in some groups and then used for fertilization repeatedly (e.g., Hymenoptera). Males of most groups of Pterygota still produce spermatophores (groundplan of Hexapoda) (Fig. 6.13.5). However, more or less complex sperm pumps have evolved independently in several lineages (Strepsiptera, Mecoptera [excl. Boreidae], Diptera [partim], Siphonaptera). In these groups the males transmit fluid containing spermatozoa. A secondary external sperm transfer with a spermatophore attached to the female postabdomen occurs in one species of Zoraptera.

Parthenogenesis, i.e. the development of eggs without fertilization, occurs in some groups of hexapods as the normal mode of reproduction. It is usually found in phytophagous lineages (e.g., Sternorrhyncha, Chrysomelidae [Coleoptera]) where it enables populations to increase rapidly during favorable conditions. In groups with facultative **arrhenotoky** eggs may or may not be fertilized and haploid eggs produce males (Hymenoptera, Aleyrodidae [partim], Thysanoptera, Micromalthidae [Coleoptera]). **Thelytoky** with extremely rare males and prevailing parthenogenesis is known from some species of Orthoptera and some other groups. Parthenogenetic and bisexual races co-occur in some cases (e.g., some species of Psychidae [Lepidoptera] and Coccoidea [Sternorrhyncha]). Facultative parthenogenesis as a result of lacking opportunity to mate is probably widespread in hexapods (Chapman 1998).

The eggs are formed in the follicles of the ovarioles. The primary oocyte usually undergoes the 1st and 2nd **meiotic division** shortly before, during or shortly after egg deposition. Eggs ready for deposition comprise the egg cell proper, the thin **vitelline envelope**, and the external **chorion**. They are usually large relative to body size and **macrolethical**, i.e. containing a large amount of yolk. Notable exceptions are the small and microlethical eggs of Collembola, Protura and a few other groups, notably many parasitoid Hymenoptera. The cytoplasm of the egg cell is divided into three components, a layer around the nucleus, a more or less thin peripherical **periplasm** beneath the **oolemma**, and the **cytoplasmic reticulum**, a fine meshwork which connects both and also suspends the nucleus in the yolk mass. The ratio of cytoplasm and yolk mass varies considerably. Most groups of hemimetabolous insects have large eggs with lots of yolk, a thin periplasm and a poorly developed cytoplasmic reticulum, whereas the periplasm and cytoplasmic reticulum are usually well-developed in eggs of holometabolous insects (Anderson 1972a, b; Heming 2003).

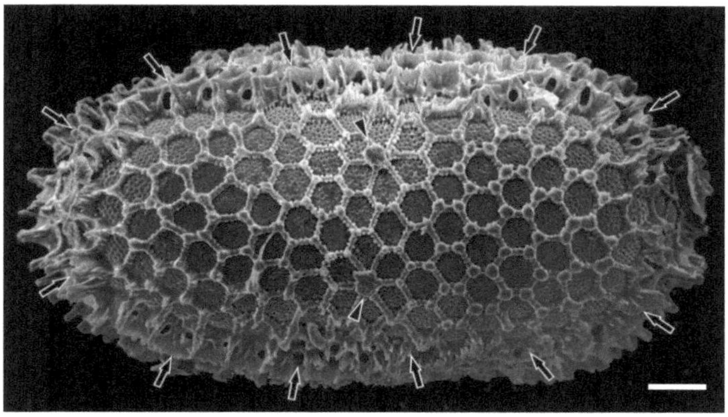

Fig. 2.1.1: *Zorotypus caudelli* (Zoraptera, Zorotypidae), egg, ventral view, SEM micrograph. Arrows and arrowheads show fringes and pairs of small polygons with micropyles, respectively. Scale bar: 50 μm. From Mashimo et al. (2011), modified.

The shape of hexapod eggs is usually oval, ellipsoid or almost cylindrical and more or less elongated (Fig. 2.1.1). They are often slightly curved, with a convex and a concave side. Rounded eggs occur in Heteroptera, Phasmatodea, and some basal lepidopteran lineages, and conical eggs in *Pieris* (Pieridae [Lepidoptera]). Eggs with long stalks are characteristic for some groups of Neuroptera (e.g., Chrysopidae). An anterior cap referred to as **operculum** occurs in different groups (e.g., Phasmatodea, Acercaria [see below]). It is pushed off by the hatching first instar nymph. In some species of Phasmatodea the operculum bears a knob-shaped or conical **capitulum**, which facilitates the transport of the eggs by ants. The chorion is produced by the follicle cells in the ovarioles. It is usually divided into an external **exochorion**, a trabecular **endochorion**, and an internal wax-layer. The exochorion forms a meshwork mainly involved in gas exchange. The wax layer is usually placed on the surface of the vitelline membrane and reduces desiccation (Heming 2003). In some groups the chorion it is very thin and more or less smooth (e.g., Psocodea, parasitic Hymenoptera, Strepsiptera, many groups of Coleoptera) but usually it is well-developed and distinctly sculptured (e.g., Orthoptera, Phasmatodea). A honey comb pattern (Fig. 2.1.1) occurs in many groups (e.g., Zoraptera, Gyrinidae, Archostemata [both Coleoptera]). The surface structure of the chorion reflects the surface of the follicle cells which deposited it to a certain degree. One or more openings for passage of spermatozoa are often present. The position of these **micropyles** varies strongly. In several lineages they are arranged around the anterior pole of the egg. In Zoraptera a pair is present at the equator of the eggs, and a single micropyle is located near the posterior pole in phlaeothripid Thysanoptera. In most groups of Acercaria an operculum is present at the anterior pole with one or up to a dozen micropyles. There is little relation between the position of the micropyle and the site of fertilization. To unite, both pronuclei migrate over a considerable distance towards the cleavage center within the yolk mass, the female one from the maturation island, and its male counterpart from the micropyle. Additional perforations of the chorion are **aeropyles** for gas exchange and chorionic **hydropyles** for the uptake of water. Numerous of these small openings are usually present. The layer of air in the endochorion has direct access to the egg through pores or canals (Heming 2003).

2.2 Embryonic development

2.2.1 Cleavage and germ band formation

Cleavage is meroblastic and superficial in most groups of Hexapoda and in Arthropoda in general (Fig. 2.2.1.1). However, within each major arthropod lineage the holoblastic type occurs, usually in basal groups (Machida et al. 1990). This suggests that superficial cleavage may have been acquired independently several times. The superficial cleavage pattern is apparently linked with the large size and specific configuration of the eggs, with a thin superficial **periplasm** covering a large central yolk mass. Cleavage is (secondarily?) holoblastic in some groups with small eggs containing limited amounts of yolk, such as Collembola (until the 8-cell stage), Protura, Aphididae (viviparous species), or Hymenoptera (parasitic species) (see above). However, holoblastic cleavage also occurs in Archaeognatha, which produce normally sized eggs (ca. 1 mm).

In the typical superficial **meroblastic** (=partial) cleavage the diploid zygote nucleus divides several time mitotically but without cytokinesis. This takes place in the **cleavage center**, a species specific region in the egg cell. The nuclei lie within the central yolk mass after the first divisions (Fig. 2.2.1.1A). With a small body of **ovoplasma** enclosing them they form **energids**. After several division cycles most of the nuclei migrate towards the periphery and then enter the peripherical cytoplasm (**periplasma**), where they form the **plasmodial preblastoderm** (**syncytial blastoderm**) (Fig. 2.2.1.1C). The cleavage divisions can be synchronous throughout the process or become desynchronized at a certain stage (parasynchrony or asynchrony). At the periphery, mitoses continue at a progressively slower rate. After the 256 cell

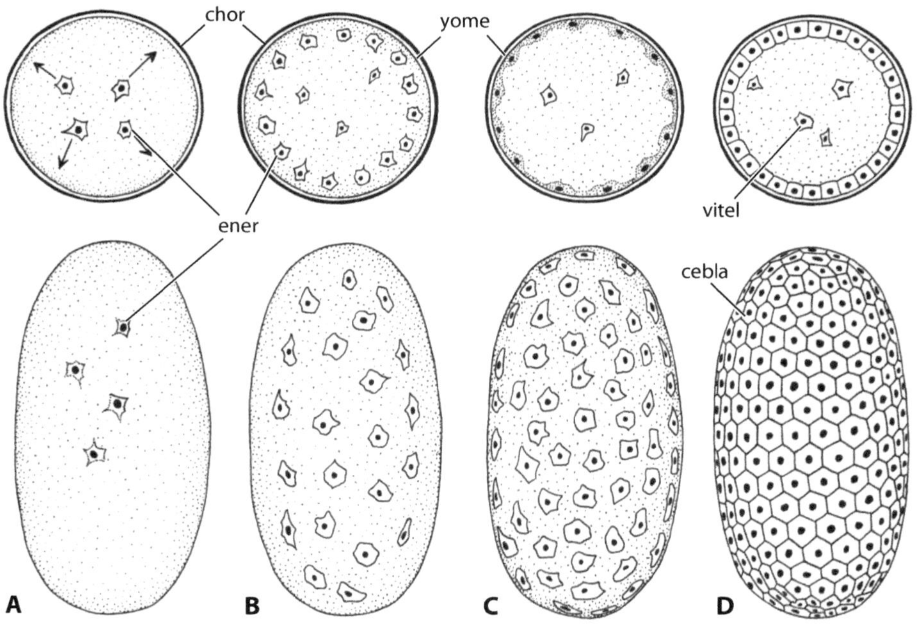

Fig. 2.2.1.1: Early stages of superficial cleavage, cross sections (upper row) and ventral view. A, B, energid stages; C, syncytial preblastoderm, D, undifferentiated blastoderm. Abbr.: cebla – cellular blastoderm, chor – chorion, ener – energids, vitel – vitellophages, yome – yolk membrane. Redrawn from Seifert (1995).

stage (in *Drosophila*) is reached, external membrane grows inwards and separates the energids (Fig. 2.2.1.1D). The lateral membranes form before the basal membranes, which separate the cells from the yolk mass. The external layer of cells is referred to as **cellular blastoderm**. Some nuclei remain in the central yolk mass and help to mobilize the nutrients (**vitellophages** or **yolk cells**) (Fig. 2.2.1.1D: vitel). In many groups other nuclei migrate to the pole plasm at the posterior region of the egg, where they absorb granules containing RNA and differentiate as primordial germ cells. The single-layered blastoderm is thin and loosely arranged in the early stage of its formation. Regularly shaped cubic cells are formed after further mitoses, and in the following process the blastoderm differentiates into the embryonic and extraembryonic areas (Fig. 2.2.1.2). The former proliferates resulting in relatively small cells crowding to form a thickened plate, the **germ band** or **germ anlage** (**"Keimstreifen"**). The larger cells of the extraembryonic region form the **serosa** and adjacent to the embryo the **amnion**.

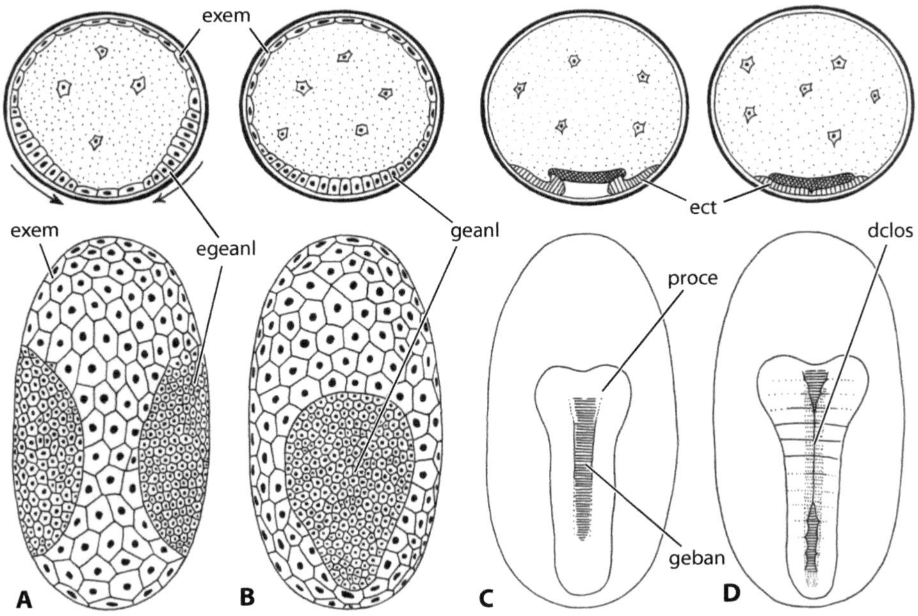

Fig. 2.2.1.2: Blastoderm stages, formation of the embryo, cross sections (upper row) and ventral view. A, B, later blastoderm stages, formation of the germ anlage; C, germ band, formation of the embryo, D, early stage of segment formation in the embryo. Abbr.: dclos – dorsal closure, ect – ectoderm, egeanl – early germ anlage, exem – extraembryonal area, geanl – germ anlage, geban – germ band, proce – procephalon. Redrawn from Seifert (1995).

2.2.2 Short germ and long germ embryos

The size ratio of the germ band to the entire egg is group specific. The proportions of the regions developing into the head, thorax and abdomen also differ considerably. In short germ eggs (hemimetabolous insects, basal holometabolan lineages) the germ band is heart- or keyhole-shaped and small, in some cases occupying less than 10% of the entire egg surface (Figs 2.2.2.1A, 6.13.6A). The embryo is divided into the **protocephalon** (prospective head) and **protocorm** (postcephalic body). The former is distinctly larger than the latter in the protopod stage, which is characterized by the absence of segments and appendages. A short growth zone is present in the anterior region of the small protocorm. The semi-long germ egg is an intermediate condition. In the long germ egg (Collembola, Diplura, "intermediate and higher" holometabolan orders) the proportions of the three presumptive tagmata are about the same as in the hatching first instars (Fig. 2.2.2.1B). The germ band can occupy up to 90% of the egg surface and a protopod stage and a growth zone are missing. Appendage formation often begins in the prothoracic segments and proceeds anteriorly and posteriorly. Long germ eggs are probably an adaptation for rapid embryogenesis (Heming 2003).

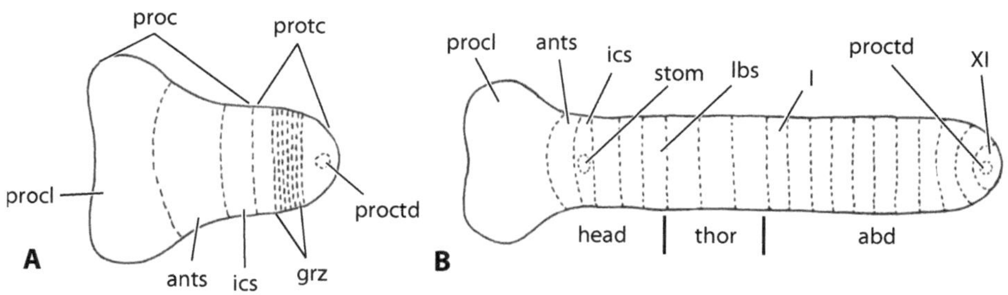

Fig. 2.2.2.1: Different forms of embryos. A, short germ embryo; B, long germ embryo. Abbr.: abd – abdomen (I,XI), ants – antennal segment, grz – growth zone, ics – intercalary segment, lbs – labial segment, proc – proecephalon, procl – protocephalic lobes, protc – protocorm, proctd – proctodaeum, stom – stomodaeum, thor – thorax. Redrawn from Heming (2003), modified.

2.2.3 Germ layer formation and blastokinesis

During the proliferation of cells of the germ anlage and the elongation of the embryo the formation of the three germ layers takes place. The ectoderm, mesoderm and entoderm are formed in a process which may be considered as a very atypical **gastrulation.** Longitudinally arranged midventral cells in the germ band move upward into a space between the embryo and the yolk, thus forming an inner layer of mesoderm. The ectoderm is formed by the remaining cells on the surface. The inner layers can be formed in three different ways: by formation of a midventral furrow that subsequently closes over (most orders), by lateral overgrowth of a midventral plate (Hymenoptera, e.g., *Apis*), or by ingression of individual ventral midline cells (Lepidoptera) (Heming 2003).

In groups with short or semi-long germ eggs the embryo is shifted to the central region containing the yolk mass by a rotation (**anatrepsis**) (Fig. 2.2.3.1). This results in the formation of the amniotic cavity. Later in development a reverse movement called **katatrepsis** results in return to the original external position of the embryo (Fig. 2.2.3.2); both movements collectively being referred to as **blastokinesis**. The process differs in groups with a long germ egg. The embryo remains on the ventral side of the egg throughout embryogenesis. The amnion is formed by amnioserosal folds which arise at the margins of the embryo and extend inward and fuse midventrally, thus enclosing an amniotic cavity and the embryo (Heming 2003) (Fig. 2.2.3.3). Formation of an **amnionic cavity** is a groundplan apomorphy of Insecta. An opening is retained in Archaeognatha, but the cavity is closed in Zygentoma and Pterygota (synapomorphy).

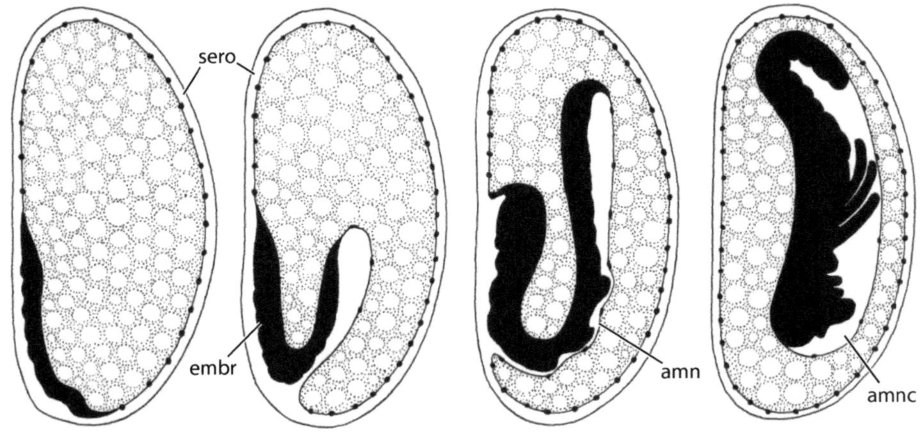

Fig. 2.2.3.1: Blastokinesis, anatrepsis, embryo shifted to the interior of the developing egg, formation of the amniotic cavity, groups with short or semi-long germ eggs. Abbr.: amn – amnion, amnc – amniotic cavity, embr – embryo, sero – serosa. Redrawn from Seifert (1995).

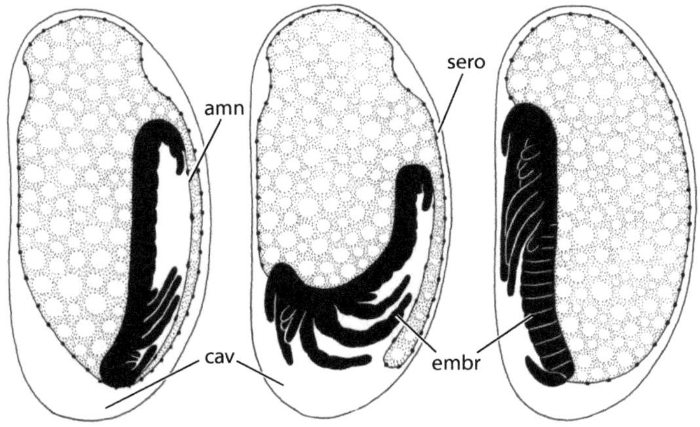

Fig. 2.2.3.2: Blastokinesis, katatrepsis, embryo shifted to the external surface of the developing egg, reversal of anatrapsis. Abbr.: amn – amnion, cav – coelomic cavity, embr – embryo, sero – serosa. Redrawn from Seifert (1995).

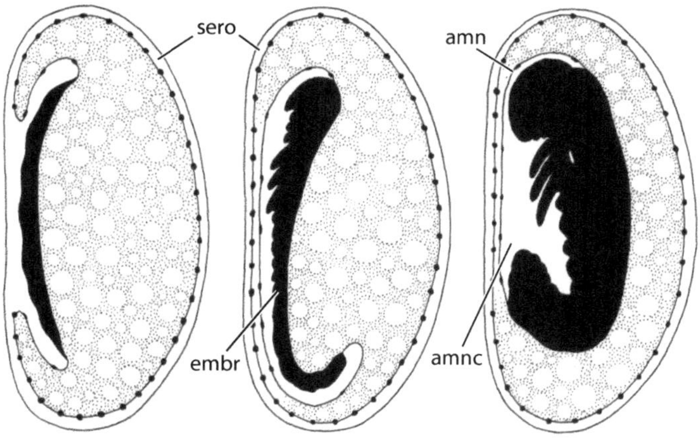

Fig. 2.2.3.3: Blastokinesis, internalization of the embryo and formation of the amniotic cavity, groups with long germ eggs. Abbr.: amn – amnion, amnc – amniotic cavity, embr – embryo, sero – serosa. Redrawn from Seifert (1995).

2.2.4 Segmentation

After the inner layer is formed the embryo is divided into segments by transverse furrows (Figs 2.1.1.2, 6.13.6). In members of most orders each of the segments develops a pair of appendages, the antennae and three gnathal appendages (prospective mouthparts) in the cephalic region, three pairs of thoracic legs, ten pairs of abdominal appendages, and cerci on abdominal segment XI. Those of segment I develop into **pleuropodia** (Fig. 2.2.4.1). Those of segments II–IX develop into simplified abdomi-

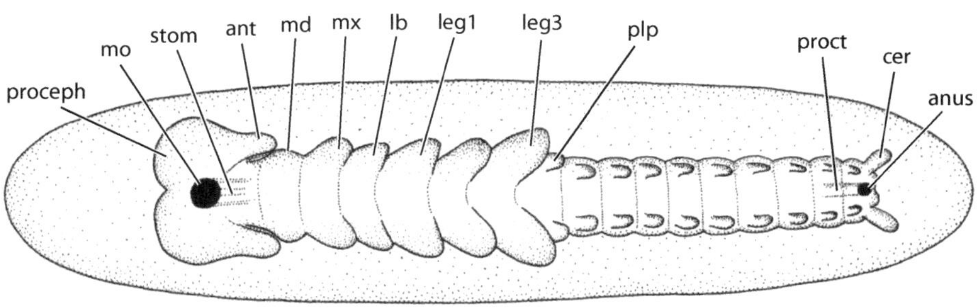

Fig. 2.2.4.1: Short germ embryo at the polypod stage, after germ band elongation, gastrulation and segmentation. Abbr.: ant – antenna, cer – cerci, lb – labium, leg1/3 – fore- / hindleg, md – mandible, mx – maxilla, mo – mouth opening, plp – abdominal pleuropodium, proceph – procephalon, proct – proctodaeum, stom – stomodaeum. Redrawn from Seifert (1995).

nal legs (coxites bearing styli and coxal vesicles) in apterygote lineages and into prolegs in some groups of Holometabola (symphytan larvae, lepidopteran caterpillars), but disappear before hatching in most hexapod orders.

Traditionally the protocephalon (Fig. 2.2.2.1A) was considered as a unit composed of an acron and three pregnathal segments, the first of them bearing the labrum. A recent developmental study with a focus on the segment polarization gene *engrailed* (Haas et al. 2001) has shown that it is in fact composed of an anterior ocular segment, an antennal segment, and an intercalary segment 3 with the labrum. The protocephalic lobes (Fig. 2.2.2.1) give rise to the protocerebrum including the optic lobes and lateral eyes. The ectodermal stomodaeum (prospective foregut) invaginates medially on the intercalary 3rd segment.

2.2.5 Organogenesis

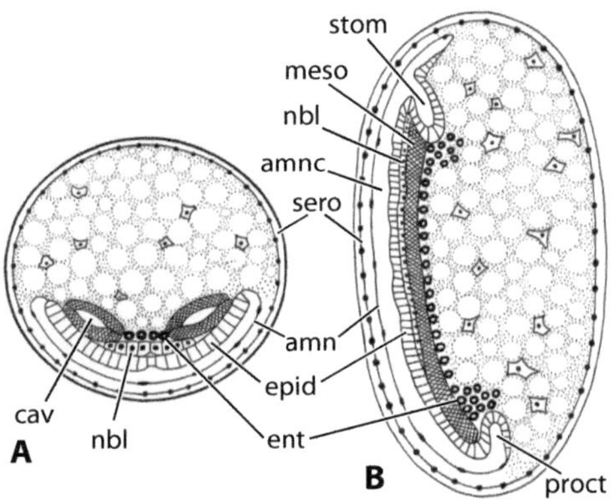

Fig. 2.2.5.1: Differentiation of the germ layers. A, cross section; B, sagittal section. Abbr.: amn – amnion, amnc – amniotic cavity, cav – coelomic cavity, epid – epidermis, ent – entoderm, meso – mesoblasts, nbl – neuroblasts, proct – proctodaeum, sero – serosa, stom – stomodaeum. Redrawn from Seifert (1995).

In late postembryonic development the external ectoderm (Fig. 2.2.5.1) forms the body wall, the foregut (stomodaeum) and hindgut (proctodaeum), and invaginations such as tracheae and apodemes or apophyses. The entoderm forms the midgut, and the mesoderm the musculature, the circulatory system, and elements of the reproductive organs (Fig. 2.2.5.2). The primarily paired midgut primordia at the apices of the stomodaeum and proctodaeum spread around the yolk to form the midgut epithelium. The Malpighian tubules are mainly formed by the ectoderm, but cells of mesodermal origin (mesoblasts) can be involved, at least in *Drosophila*. **Coelomic cavities** as they

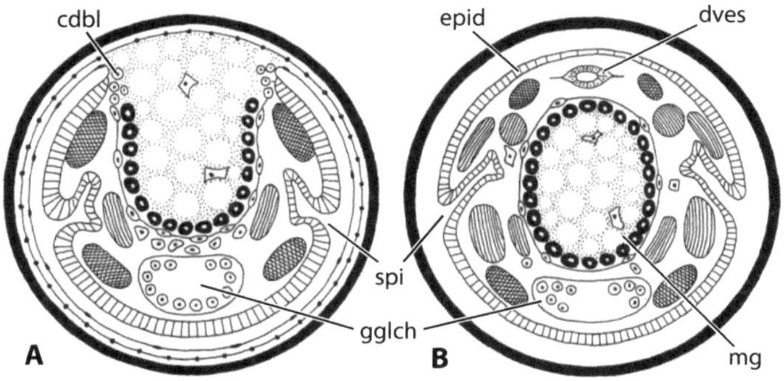

Fig. 2.2.5.2: Organogenesis, cross sections. A, before dorsal closure; B, after dorsal closure. Abbr.: cdbl – cardioblasts, gglch – ganglionic chain, dves – dorsal vessel, epid – epidermis, mg – midgut, spi – spiracle. Redrawn from Seifert (1995).

occur in most groups of Bilateria are also formed in many Hexapoda and in other groups of Arthropoda, but after the middle developmental stages, they collapse and their lumina fuse with the **primary body cavity**, thus forming a **mixocoel** (preserved and associated with metanephridia in adult Onychophora).

Specific regions in the developing hexapod embryo can be assigned to presumptive organs at an early stage. The determination processes are triggered by intensified protein syntheses. In hemimetabolous insects this takes place during the segregation of the germ anlagen, in Holometabola already during the formation of the blastoderm (potential autapomorphy). Blastoderm formation also takes place at an earlier stage in this lineage (1.5 h in *Drosophila* [Diptera] versus 30 h in *Gryllus* [Orthoptera]). The timing, like the developmental rate in general, strongly depends on temperature.

2.3 Postembryonic development

Three different types of **postembryonic development** are traditionally distinguished, the ametabolous, the hemimetabolous (Fig. 2.3.1) and the holometabolous modes (Fig. 2.3.2). The former two cannot be clearly defined. In both cases the adult stage is reached after an **incomplete metamorphosis**. In ametabolous development (apterygote lineages) the immature stages differ only in size, absence of sexual maturity and in minor morphological features (e.g., number of antennomeres) from the adults, and molting continues after reaching sexual maturity. In hemimetabolous lineages compound eyes and external wing buds (Anlagen) occur already in **nymphs**. The morphological configuration gradually converges towards the adult condition with fully developed wings and genital organs (Fig. 2.3.1). The immatures are addressed as **larvae** if specific features are present, such as for instance the labial mask in Odonata or the abdominal tracheal gills in Ephemeroptera.

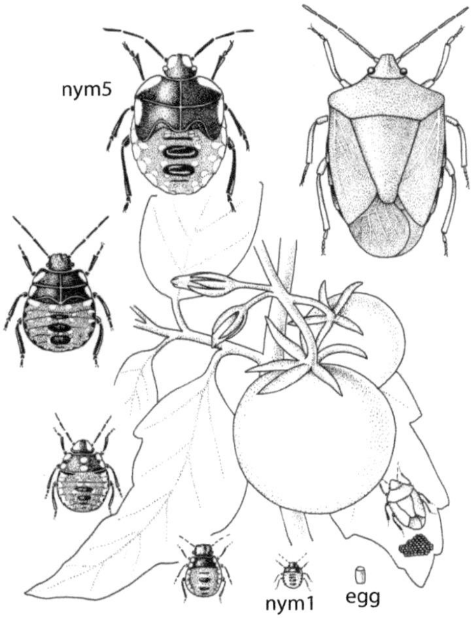

Fig. 2.3.1: Hemimetabolous development, life cycle of Nezara (Heteroptera, Pentatomidae). Abbr.: nym1/5 – 1st and 5th instar nymph. Redrawn from Gullan & Cranston (1994).

Postembryonic development is fundamentally different in Holometabola, which are mainly characterized by **complete metamorphosis** (autapomorphy). Larvae and adults differ more or less completely in their morphology (Fig. 2.3.2) and usually also in their microhabitats and feeding habits. Compound eyes and external wing buds are absent in the larvae (Endopterygota; autapomorphy [reversal in Strepsiptera]). During a preimaginal pupal stage the entire morphology is completely remodeled. The pupa is largely immobile (except for those of certain Raphidioptera, Megaloptera, Trichoptera, and some groups of Neuroptera and Diptera [e.g., mosquitos]) and non-feeding, but should not be referred to as a resting stage. Pupation, which normally takes place in a cryptic or protected microhabitat (e.g., in soil or under bark), is characterized by increased physiological activity. Larval structures disintegrate, and the compound eyes, wing buds and genital organs appear for the first time in postembryonic development. In the typical case, the wing buds (and other adult structures) are present as **imaginal discs** below the larval cuticle.

An advantage of holometabolous development is the potential to make use of different food sources and microhabitats as larvae and adults, resulting in reduced intraspecific competition. Moreover, the absence of external wing buds in larvae (the endopterygote condition) improves their ability to penetrate narrow crevices such as

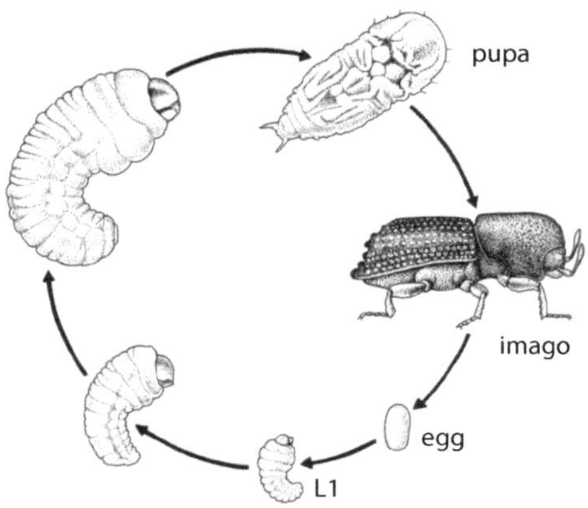

Fig. 2.3.2: Holometabolous development, life cycle of *Ips* (Coleoptera, Curculionidae, Scolytinae). Redrawn from Gullan & Cranston (1994).

for instance under bark (subcortical habitats) or to burrow in various substrates (e.g., plant tissue). Evolutionary costs of holometabolous development are the reduced sclerotization in the pupal stage and the strongly restricted movability, which makes it very vulnerable to predators.

2.4 Larval and pupal forms

Different types of larvae occur in holometabolous insects. The **oligopod** larvae are characterized by a well-developed head capsule and thoracic legs, and the absence of leglets on abdominal segments I–IX. Three main subtypes occur. The **campodeoid** larvae are slender and agile, with distinct sclerites of the postcephalic segments (e.g., Carabidae, Staphylinidae [Coleoptera], Neuroptera, Raphidioptera) (Figs 6.26.1, 6.29.8, 6.29.9) The terminal abdominal segment X usually forms an unpaired conical or cylindrical pygopod, which facilitates locomotion. **Asseliform** (isopod-like) larvae (e.g., Silphidae [Coleoptera]) are dorsoventrally flattened and their tergites form distinct lateral duplicatures (paranotal lobes). **Scarabaeiform** larvae ("Engerlinge") are characterized by an unsclerotized, sausage-shaped postcephalic body, with an enlarged terminal segment IX. **Oligopod** (**erucoid**) larvae (lepidopteran caterpillars, larvae of basal hymenopteran lineages [sawflies etc.]) possess a well-developed head (usually orthognathous) and serially arranged abdominal prolegs. **Maggots** are weakly sclerotized and legless (Diptera), with (**eucephalic**) or without (**acephalic**) a well-developed head capsule.

In the pupal stage movable mandibles are present in the **pupa dectica** of Neuropterida, basal lepidopteran lineages, Trichoptera and Mecoptera. The mandibles are immobilized in the **pupa adectica** of the remaining lineages. In the adecticous **pupa libera** (e.g., Coleoptera [partim], Hymenoptera [partim], Siphonaptera) the body appendages (wing sheaths, legs) are free, whereas they are glued to the body in the **pupa obtecta** ("Mumienpuppe"; e.g., Lepidoptera [partim]) (Fig. 6.32.5). The adecticous **pupa pharata** ("Tönnchenpuppe") is enclosed in a barrel-shaped, sclerotized puparium (Fig. 6.35.12) formed by the last larval exuvia (e.g., Strepsiptera, some Staphylinidae [Coleoptera], Cyclorrhapha [Diptera]).

[Anderson (1972a, b]; Machida et al. (1990); Gullan & Cvanston (1994); Seifert (1995); Chapman (1998); Haas et al. (2001); Gilbert (2003); Heming (2003); Mashimo et al. (2011)]

3 Glossary

3.1 Morphology

3.1.1 General terms

Muscularis: muscle layer around different organs.

Tagma: functional group of segments

3.1.2 Integument

Acanthae: hairs without sockets, one per hypodermal cell.

Apodeme: solid cuticular invagination serving as muscle insertion area.

Apolysis: the separation of the old cuticle from the epidermis

Apophysis: hollow cuticular invagination.

Arthropodin: flexible water-soluble matrix protein of the procuticle and endocuticle.

Cuticulin: matrix protein in the intermediate layer of the epicuticle, similar to sclerotin.

Dense lamina: homogenous inner layer of the epicuticle; highly robust mechanically.

Ecdysial space: expanding space below the exuvia

Ecdysis: molting, shedding and replacement of the cuticle.

Endocuticle: inner layer of cuticle, with flexible protein matrix and chitin.

Epicuticle: multilayered external cuticular layer without chitin.

Exocuticle: cuticular layer below epicuticle, with hardened protein matrix and chitin.

Exuvia: ecdysial membrane and part of the cuticle separating from the epidermis in the ecdysis.

Hypodermis (=epidermis): layer of epidermal cells below cuticle.

Intima: thin cuticular layer lining internal structures of ectodermal origin.

Membrane: very flexible, unsclerotized area of the integument.

Mesocuticle: cuticle where the tanning process terminates at an intermediate stage.

Microtrichium (trichome): hair without articulation, usually solid; several microtrichia are formed by one hypodermal cell.

Phragma: large, ridge-shaped internal apodeme.

Pore canals: thin perforations of the cuticle.

Procuticle: soft and undifferentiated precursor of the endo- and exocuticle.

Prosclerotin: intermediate form of matrix protein, before the hardening process is completed.

Resilin: rubber-like matrix protein.

Ridge: cuticular strengthening ridge.

Sclerite: distinctly sclerotized region of the integument, with thick layer of exocuticle.

Sclerotin: hardened matrix protein of the exocuticle.

Seta (=macrotrichium): articulated hair with socket, formed by trichogen cell.

Spinulae: minute spines on the cuticular surface.

Suture: ecdysial suture, zone of weakness.

Thecogen cell: epidermal cell involved in the formation of a sensorial seta.

Tormogen cell: specialized epidermal cell, forming membranous area between seta and surrounding cuticle.

Trichogen cell: specialized epidermal cell, forming seta.

3.1.3 Sensilla and sensory organs associated with the integument

Chordotonal organs: scolopidium (or groups of scolopidia) connecting articulated sclerotized elements.

Scolopale: element of the scolopidium consisting of fibrous material containing actin arranged in ring or a series of rods.

Scolopale cap cell (=attachment cell): cell of scolopidia, placed on top of scolopale cell and the dendrite

Scolopale cell: thecogen cell of scolopidia

Scolopidium: specialized subcuticular mechanoreceptor; consisting of scolopale cell, scolopale cap cell, and one or several sensory neurons.

Scolops (scolopale cap): cuticular structure associated with scolopidium.

Sensillum basiconicum: sensillum with rounded, conical seta.

Sensillum chaeticum: shorter than sensillum trichodeum, thick-walled.

Sensillum coeloconicum: sensillum with modified seta in a cavity, with narrowed external opening of surrounding cuticle.

Sensillum placodeum: flat sensillum with strongly modified plate-like seta.

Sensillum trichodeum: hair-like sensilla, seta associated with sensory cell.

Trichobothrium: very long and fine sensillum, vibroreceptor.

3.1.4 Head capsule

Alaforamen: part of foramen occipitale above tentorial bridge; passage of oesophagus and tracheal stems.

Antennifer: small cephalic process articulating with the antenna.

Anterior tentorial arm: paired tentorial element anterior to tentorial bridge; originates at anterior tentorial pit.

Anterior tentorial pit (or groove): externally visible invagination area and origin of anterior tentorial arm; adjacent with dorsolateral clypeal margin.

Anteclypeus: transparent distal part of clypeus.

Antecosta: dorsal intersegmental furrow separating thoracic segmens, attachment site of dorsal longitudinal indirect flight muscles (see phragma).

Antennal foramen: insertion site of the antenna.

Cephalon: head

Circumocular ridge: internal ridge enclosing compound eye.

Clypeolabrum: area comprising the fused clypeus and labrum.

Clypeus: anteroventral part of head capsule; primarily separated from frons by transverse frontoclypeal strengthening ridge between anterior tentorial pits; usually bearing the labrum at its distal margin.

Corpotentorium: see tentorial bridge.

Fossa maxillaris: articulatory groove of the maxilla.

Frontoclypeal (strengthening) ridge (=frontoclypeal “suture”, epistomal “suture”): internal ridge separating clypeus from frons, between anterior tentorial pits.

Cervical membrane: membrane connecting the head and prothorax.

Cervical muscles (=musculi cranii): extrinsic muscles moving the head.

Cervical sclerite: articulatory sclerite embedded in the cervical membrane.

Coronal suture: unpaired median ecdysial sutures dividing vertex; anteriorly continuous with frontal sutures.

Cryptopleuron: more or less completely invaginated propleura.

Dorsal tentorial arm: paired dorsal element of tentorium, originating at anterior arm or connecting area of anterior and posterior arm; often connected with head capsule by fibrillae.

Epicranial suture: Y-shaped ecdysial sutures dividing vertex and frons, composed of coronal suture and paired frontal sutures.

Foramen occipitale (=foramen magnum): posterior opening of head capsule; divided by tentorial bridge into alaforamen and neuroforamen.

Frons: frontal part of head capsule, delimited by ecdysial lines (frontal sutures); often triangular and bearing unpaired (median) ocellus.

Frontal suture: paired ecdysial sutures enclosing frons; anterior parts of epicranial suture.

Frontoclypeolabrum: area comprising the fused frons, clypeus and labrum.

Gena: lateral region of head capsule; not separated from vertex by a ridge or suture.

Gnathencephalon: posterior part of the head with the mandibles, maxillae and labium.

Gula: sclerite partially closing foramen occipitale, formed by sclerotized region of ventral cervical membrane.

Gular ridges: lateral internal ridges enclosing the gula.

Gularia: unpaired ventral cervical sclerites.

Epipharynx: semimembranous or membranous posterior wall of labrum and roof of cibarium.

Hypognathous head: head with posteriorly directed mouthparts.

Hypostoma: posterior part of subgena.

Hypostomal bridge: partial closure of foramen occipitale, formed by ventromedially adjacent extensions of lateral head capsule.

Hypostomal ridge: posterior part of subgenal ridge.

Laterocervicalia: paired lateral cervical sclerites.

Occipital ridge (="occipital suture"): ridge on posterior head region; present in some groups, oblique and short or more or less parallel to postocciptal rigde (Orthoptera). No true segmental border.

Orthognathous head: head with ventrally directed mouthparts.

Neuroforamen: part of foramen occipitale below tentorial bridge; passage of cervical connectives between suboesophageal complex and prothoracic ganglion.

Phragma (=antecosta): dorsal thoracic transverse ridge, attachment site of dorsal longitudinal indirect flight muscles.

Postclypeus: sclerotized and pigmented proximal part of clypeus.

Posterior tentorial arm: paired posterior element of tentorium; originates at posterior tentorial pit.

Posterior tentorial pit (or groove): externally visible invagination of posterior part of head capsule and origin of posterior tentorial arm; adjacent with foramen occipitale or gula.

Postgena: posterior genal region anterior to postoccipital ridge.

Postgenal bridge: partial closure of foramen occipital formed by postgenal extensions.

Postocciput: posteriormost head region.

Postoccipital ridge: strengthening ridge enclosing the foramen occipitale dorsally and laterally; also serving as muscle attachment area.

Prognathous head: head with anteriorly directed mouthparts.

Rostrum: snout-like prolongation of the head (e.g., Mecoptera).

Subgena: narrow lateral region of head capsule, above external side of mandibular base, usually separated from gena by subgenal ridge.

Subgenal ridge: strengthening ridge separating gena from subgena.

Tentorial bridge (=corpotentorium): transverse element connecting posterior tentorial arms, often plate-like.

Tentorium: cephalic endoskeleton, usually composed of paired anterior, posterior and dorsal tentorial arms and a transverse tentorial bridge (corpotentorium).

Vertex: posterodorsal part of head capsule, usually medially divided by ecdysial line (coronal suture).

3.1.5 Head appendages

Antenna: head appendage of segment 2, innervated by deutocerebrum; in Insecta composed of scapus, pedicellus and flagellum; primarily multisegmented and filiform.

Antennomere: antennal segment.

Arista: bristle-like part of a modified antenna.

Basistipes: basolateral part of stipes.

Cardo: basal element of maxilla, articulating with head capsule; usually moved by one cranial muscle and one tentorial muscle.

Flagellomere: segment of the flagellum.

Flagellum: third element of antenna of Insecta, usually multisegmented; lacking muscles.

Galea: outer endite lobe of maxilla, usually semimembranous and equipped with sensilla (mainly chemoreceptors).

Glossa: paired inner endite lobes of the prementum.

Haustellum: snout-like extrusible complex formed by labium and hypopharynx (Trichoptera)

Hypopharynx: median unpaired lobe between mouthparts and epipharynx; anterior (dorsal) surface continuous with anterior pharynx.

Incisivi: apical and subapical teeth of mandible.

Labellae: modified lobe-like labial palps of Diptera.

Labrum: unpaired upper lip, connected with distal edge of clypeus.

Lacinia: inner endite lobe of maxilla, usually firmly connected or fused with stipes; usually sclerotized, with mesally curved apex and articulated spines arranged along mesal edge.

Lacinia mobilis: movable mesal appendage of the mandible.

Lingua: median part of hypopharynx.

Lorum: small rod-like sclerite connecting the maxilla and labium in bees.

Maxilla: head appendage of segment 5, innervated by maxillary nerve originating from intermediate part of suboesophageal complex; primarily composed of cardo, stipes, endite lobes (lacinia, galea) and maxillary palp.

Mediostipes: mesal part of stipes.

Mentum: intermediate part of labium, anterior part of postmentum; bears the prementum at its distal edge.

Mola: proximomesal grinding area of mandible.

Palp: tactile appendage of maxilla and labium.

Palpifer: small separate element of stipes, bears the maxillary palp.

Palpiger: small separate element of prementum, bears the labial palp.

Paraglossa: paired outer endite lobes of the prementum (labium).

Pedicellus: second antennal segment in Insecta; contains Johnston's organ and lacks intrinsic muscles.

Postmentum: posterior main element of labium, posteriorly adjacent with foramen occipitale in hexapods without gula or hypostomal bridge; often separated into two sclerites, the anterior mentum and the posterior submentum.

Postpedicellus: enlarged 1st flagellomere (brachyceran Diptera).

Prementum: distal main element of labium, bearing the endite lobes (glossa and paraglossa) and the labial palp.

Prostheca (=lacinia mobilis): articulated mesal lobe of the mandible.

Retinaculum: fixed prominent tooth of mesal mandibular edge.

Scapus: basal element of antenna, with two muscles attached to the pedicellus.

Stipes: second element of maxilla, connected with cardo by a hinge; bearing endite lobes (galea and lacinia) and maxillary palp.

Submentum: posterior part of a divided postmentum.

Superlinguae: lateral appendages of hypopharynx.

Suspensorium (=fulturae, hypopharyngeal arms, oral arms): sclerotized strengthening elements of proximal hypopharynx.

Tormae: lateral apodemes of labral base.

3.1.6 Thorax

Acrotergite: sclerite separated from the prescutum by the antecosta (apterygote groups).

Alinotum: part of the notum connected with the wing base.

Anapleural suture (=paracoxal suture): suture horizontally delimiting dorsal and ventral parts of episternum and (partially) epimeron; sometimes anteriorly widened and membranous (=anapleural cleft).

Anepimeron: dorsal part of epimeron (present as separate element in some groups).

Anepisternum: dorsal part of episternum, ventrally delimited by anapleural suture.

Antecoxa: lower part of episternum, anterad the trochantin.

Axillary ligament: connects the scutoscutellar line with the posterior margin of the wing base.

Axillary sclerites (=axillaria, pteralia): small sclerites of wing base (usually 3); articulating with notal wing processes.

Basalare: sclerite associated with wing base, more or less distinctly separated from anepisternum; usually with internal muscle disc; attachment site of direct flight muscles.

Basisternum: main part of sternum, between presternum and furcasternum.

Discrimen: longitudinal ridge medially dividing the sternum

Endosterny: pterothoracic internalization of sternal elements (Holometabola).

Epimeron: part of pleuron posterior to pleural ridge.

Epipleurites: basalare and subalare.

Episternum: part of pleuron anterior to pleural ridge.

Fulcrum (=pleural wing process): anterodorsal apex of the pleural ridge.

Furca (=sternal apophysis, endosternite): endoskeletal element (entapophysis) of furcasternum; paired furcal arms often connected with pleural ridge by short muscle.

Furcasternum (=sternellum): sternal element posterior to basisternum, area of origin of furca.

Humeral plate: sclerite of the wing base (pteralium); articulates with costal vein.

Katepimeron: ventral part of epimeron (present as separate element in some groups).

Katepisternum: posterior part of ventral episternal region; separated from preepisternum by precoxal suture.

Median plate: sclerite of the wing base (pteralium); undivided and firmly connected with main longitudinal veins in the pterygote groundplan; subdivided and detached from main longitudinal veins in Neoptera.

Mesothorax: 2nd thoracic segment, anterior segment of pterothorax; bearing forewings (groundplan of Pterygota).

Metathorax: 3rd thoracic segment, posterior segment of pterothorax; bearing hindwings (groundplan of Pterygota).

Notal wing processes (=alar processes): anterior, postmedian and posterior processes of the lateral scutal margin; articulate with the axillary sclerites.

Notum: larger anterior part of tergum, composed of presutum, scutum and scutellum.

Paranota (=laterotergites): dorsolateral tergal duplicatures.

Pleural ridge: internal ridge, corresponding to externally visible pleural "suture"; separates episternum and epimeron; anterodorsally forming the pleural wing process and posteroventrally the pleuro-coxal joint.

Pleurite: sclerotized element of pleuron.

Pleuron: lateral wall of thoracic segments, almost always divided by pleural ridge into anterior episternum and posterior epimeron.

Postalar bridge: lateral postnotal margin bent downwards and connected with the epimeron.

Postnotum: smaller posterior part of tergum, forming internal intersegmental phragma; attachment area for dorsal indirect flight muscles.

Poststernite: posteriormost sternal element

Prealare (=prealar sclerite): small attachment area of short direct flight muscles.

Prealar bridge: prescutal process, bent downwards and linked with episternum.

Precoxal suture: suture separating katepisternum and preepisternum.

Precoxal ridge: mediolaterally directed ridge in front of coxa, corresponding with externally visible precoxal suture.

Preepisternum: anterior part of ventral episternal region; delimited from katepisternum by precoxal suture; largest part of ventrite in Holometabola.

Prescutum: anteriormost part of notum.

Presternum: anteriormost part of sternum.

Pronotum: prothoracic tergum.

Prothorax: 1st thoracic segment, always completely wingless in extant hexapods.

Pteralia: small sclerites of wing base (axillary sclerites + plates).

Pterothorax: 2nd and 3nd thoracic segments; bearing wings (groundplan of Pterygota), meso- and metathorax.

Scutum: large anterior part of notum.

Scutellum: smaller posterior part of notum.

Spina: smaller posterior endoskeletal element of thoracic segments, unpaired, arising on spinasternum.

Spinasternum: small posterior sternal element, determining the posterior segmental border; area of origin of spina.

Sternum: ventral wall of thoracic (or abdominal) segment, usually divided into presternum, basisternum, furcasternum and spinasternum in unspecialized hexapods; largely or completely invaginated in Holometabola (cryptosterny).

Sternite: sclerotized element of sternum.

Sternocosta: transverse ridge connecting paired furcal arms.

Subalare: sclerite associated with wing base, usually embedded in membrane between tergum and pleuron; attachment site of direct flight muscles.

Tergum: dorsal wall of a thoracic (or abdominal) segment.

Tergite: sclerotized element of tergum.

Trochantin: small triangular or sickle-shaped ventral pleural sclerite; anteriorly articulating with coxa (trochantino-coxal joint).

3.1.7 Legs

Arolium: unpaired pretarsal attachment device; largely or completely membranous, almost always smooth.

Arthropodium: leg with true instrinsic articulations (autapomorphy of Euarthropoda).

Auxilia: small sclerites at the base of the claws.

Basicoxa: anterior part of coxa.

Basitarsus: proximal tarsomere.

Coxa: basal element of leg; articulated with pleural, sternal and usually also trochantinal elements of thoracic segment.

Distitarsus: apical tarsomere.

Empodium: spine-like pretarsal process.

Euplantulae: attachment pads on the ventral side of the tarsalia, almost always smooth.

Femur: 3rd element of leg, often largest part.

Meron (=merocoxa): posterior part of coxa.

Pretarsus: apical element of leg bearing ungues; distinct and cylindrical in entognathous hexapods, strongly reduced in Insecta.

Pulvilli: paired pretarsal attachment pads, below claws; hairy or smooth, in some groups filament-like.

Sternocoxale: sclerotized bar interconnecting basi- and merocoxa in some groups.

Stylus: coxal appendages (Archaeognatha) (see also abdomen).

Subgenual organ: complex chordotonal organ near the femuro-tibial joint.

Tarsomere: subelement of tarsus in Insecta; number often taxon-specific, sometimes differing on legs of different segments (diagnostic value: tarsal formula).

Tarsus: 5th part of leg; undivided in entognathous orders, composed of 3–5 tarsomeres in most groups of Insecta.

Tibia: 4th element of leg; usually cylindrical and equipped with paired terminal spurs, often with rows of spines.

Trochanter: 2nd element of leg, usually small and triangular.

Ungues: pretarsal claws; unpaired in Collembola and Protura, usually paired in Diplura and Insecta.

Unguitractor (=unguitractor plate): pretarsal sclerite bearing the claws.

3.1.8 Wings

Anal field (=vannus): posterior part of wings, usually enlarged in Polyneoptera.

Anal veins: longitudinal veins of the anal field proximally articulated with 3rd axillary sclerite; usually unbranched and arranged like a fan.

Axillary cord: connects the posterolateral edge of the alinotum with the posterior wing base.

Costa: main longitudinal vein along anterior wing margin, articulated with humeral plate.

Costal field (=remigium): anterior part of wing; usually largest region.

Cubitus: 5th main longitudinal vein.

Elytra: sclerotized forewings with completely reduced or strongly modified venation.

Halteres: wings transformed into drumstick-like gyroscopic sense organs.

Hemielytra: forewings with leathery anterior corium and posterior membrane.

Humeral and axillary plate: plate-like sclerites of the wing base.

Jugal field: small wing region posteriorly adjacent with anal field.

Jugal veins: veins of the jugal field.

Media: 4th main longitudinal vein.

Median plate: sclerotization of the wing base.

Plica jugalis: separates the anal and jugal fields.

Plica vannalis: separates the costal and anal fields.

Postcubitus: posteriormost longitudinal vein of the remigium.

Pterostigma: haemolymph sinus located in the anterodistal wing region; usually with conspicuous coloration.

Radius: 3rd main longitudinal vein.

Subcosta: 2nd main longitudinal vein, following costal vein.

Tegmina: leathery forewings.

Tegula: small lobe at the anterior wing margin.

3.1.9 Abdomen

Cerci: tactile appendages of segment XI, primarily multisegmented and tactile bearing setae.

Coxal vesicles: small eversible vesicles on coxites.

Coxite: proximal part of vestigial abdominal appendages (only apterygote groups).

Coxosternite: coxite fused with sternum (Zygentoma excl. *Tricholepidion*).

Epiproct: unpaired dorsal element (tergum) of segment XI.

Furculum: appendage of segment IV, jumping device (Collembola).

Laterotergite: separate lateral portion of abdominal tergum (only few groups).

Paraproct: paired lateral element of segment XI.

Pleural membrane: lateral unsclerotized part of abdominal segments, connecting tergum and sternum; usually bearing spiracles I–VIII.

Postabdomen: posterior segments (VIII–XI) bearing genital appendages.

Preabdomen (=pregenital segments): anterior abdominal segments (I–VII).

Proleg: short leg-like abdominal appendage of some holometabolan larvae.

Pygopod: terminal proleg of segment X (e.g., larvae of Trichoptera).

Retinaculum: small unpaired appendage of segment III, arresting device of furcula (Collembola).

Sternum: plate-like ventral sclerite of abdominal segment.

Styli: appendages of coxites or coxosternite.

Tergum: plate-like dorsal sclerite of abdominal segment.

Terminal filament: unpaired multisegmented appendage of the abdominal apex (epiproct); present only in Archaeognatha, Zygentoma, and Ephemeroptera.

Urogomphi: appendages of tergite IX of beetle larvae.

Ventral tube: unpaired ventral appendage of segment I (Collembola).

3.1.10 Male genital organs

Accessory glands: usually formed as diverticula of vasa deferentia; secretions form spermatophores and seminal fluid, and activate the spermatozoa.

Aedeagus (penis is often used as a synonym): male copulatory organ.

Ductus ejaculatorius: unpaired ectodermal duct connecting vasa deferentia (i.e. vesiculae seminales) and aedeagus.

Endophallus: membranous extrusible internal part of aedeagus.

Follicle: thin tube-like subelement of the testis, proximal to vas efferens.

Parameres: paired lateral parts of aedeagus.

Paraphyses (=titillators): paired appendages of aedeagus.

Phallobase (=basal piece): part of segment IX bearing copulatory organ.

Penis: main element of the male copulatory apparatus.

Testes: male gonads, usually composed of a series of testis tubes or follicles.

Vas deferens: paired ducts connecting vasa efferentia with the ductus ejaculatorius.

Vas efferens: narrowed section of testes connecting testicular follicles with vas deferens.

Vesicula seminalis: widened section of vas deferens, sperm storage area.

Virga: unpaired appendage of aedeagus.

3.1.11 Female genital organs

Accessory glands (=colleterial glands, cement cells): glands associated with female genital tract, secretions usually protect eggs or cement them to surfaces.

Bursa copulatrix: evagination of the vagina, receives the copulatory organ.

Coxopodite (=valvifer, gonocoxa, gonocoxite): short basal piece of ovipositor elements of female genital segments VIII and IX.

Follicle cells: cells enclosing the oocytes.

Follicle epithelium: nutritive epithelium of ovarioles.

Genital chamber: cavity formed by an inflection of the body wall, receiving the common oviduct.

Germarium: distal area of ovarioles where oocytes are formed.

Gonapophysis (=valves or valvulae 1–3): distal piece of ovipositor elements of female genital segments VIII and IX, inserted on coxopodites, usually elongated.

Gonopore: external opening of the common oviduct.

Gonostyli: appendages of gonocoxae (usually only on coxopodite IX)

Meroistic ovarioles: produce oocytes and specialized nutrient cells (trophocytes).

Oocyte: female germ cell, immature egg cell.

Ootheca: egg cocoon.

Ovarioles: tube-like, distally tapering subelements of the paired ovaries.

Oviduct: female genital duct; originating as paired structures from ovarioles, uniting as common (median) oviduct.

Ovipositor: paired postabdominal apparatus (segments VIII and IX) primarily used for depositing eggs.

Panoistic ovarioles: produce only oocytes, lacks specialized nutrient cells (trophocytes).

Pedicel: stalk of ovarioles.

Polyotrophic ovarioles: each follicle surrounded by several trophocytes.

Spermatheca (=receptaculum seminis): unpaired or paired evagination of female genital tract; storage of sperm; connected by spermathecal duct (=ductus receptaculi) with vagina or bursa copulatrix.

Spermathecal gland: provides nourishment for spermatozoa stored in spermatheca.

Telotrophic ovarioles: trophocytes confined to germarium, connected to oocytes by cytoplasmatic strands.

Terminal filament: thread-like apical part of ovarioles.

Trophocytes: nutrient cells of ovarioles.

Tunica externa: external ovariole sheath.

Tunica propria: elastic internal basal lamina of ovariole.

Vagina: unpaired terminal part of female genital duct, tube-like vulva.

Valvula 1: appendage of gonocoxa VIII.

Valvula 2: mesal appendage of gonocoxa IX.

Valvula 1: lateral appendage of gonocoxa IX.

Vitellarium: proximal part of ovarioles; growth of oocytes, production and deposition of yolk.

Vulva: external opening of the female genital tract, often forming a vagina.

3.1.12 Nervous system and associated structures

Antennal lobe: deutocerebral neuropil receiving input from olfactory sensory neurons of the antenna.

Brain (=cerebrum, supraoesophageal ganglion): main element of the CNS, above pharynx; comprising proto-, deuto- and tritocerebrum.

Cell body rind: outer part of central nervous system, mainly formed by perikarya.

Central body: central element of protocerebrum, ellipsoid glomerulus.

Central nervous system (=CNS): brain, suboesophageal complex and postcephalic ganglionic chain.

Chiasmata: intercrossings of nerves connecting the optic neuropils.

Circumoesophageal connectives: connect tritocerebrum and suboesophageal ganglion, laterally enclosing pharynx.

Commissure: transverse connection between ganglia.

Connective: longitudinal connection between ganglia.

Calyx: widened distal part of corpora pedunculata.

Caudal visceral nervous system: median visceral nerve originating from the ganglionic mass of segment VIII.

Cell body rind (=cortex): outer layer of brain, consisting of perikarya.

Central body complex: central ellipsoid glomerulus of protocerebrum.

Circumoesophageal connective: connects the tritorebrum and suboesophageal complex laterad the pharynx.

Connective neurons; nerves arising from connectives.

Corpora pedunculata: mushroom bodies of protocerebrum; associative centers.

Deutocerebrum: middle part of brain, associated with antennae.

Frontal connective: nerve connecting frontal ganglion and tritocerebrum.

Frontal ganglion: central element of stomatogastric nervous system, above anatomical mouth.

Ganglion: compact group of neurons, with external cell body rind and internal neuropil.

Globulus: region with densely packed perikarya.

Glomerulus: densely packed neuropil with increased number of synapses.

Hypocerebral ganglion: small ganglion formed by nervus recurrens; closely connected with corpora cardiaca.

Lamina ganglionaris: lateral optic neuropil of optic lobes; adjacent with retina.

Lateral horn: lateralmost part of the protocerebrum adjacent with lobula.

Lobula (medulla interna): mesal optic neuropil, connected with medulla (externa) by nerves forming chiasmata; mesally connected with medulla terminalis of protocerebrum.

Medulla (externa): intermediate optic neuropil of optic lobe, connected with lamina ganglionaris and lobula by nerves forming chiasmata.

Nervus connectivus: unpaired nerve connecting protocerebrum and frontal ganglion.

Nervus procurrens: unpaired anteriorly directed nerve originating from frontal ganglion.

Nervus recurrens: posteriorly directed nerve originating from frontal ganglion; anterior part unpaired.

Neurilemma (=perilemma): glial sheath, perineurium + neural lamella (basement membrane).

Neuroglia: entire system of glial cells (sheath cells).

Neurohaemal organ: element of the nervous system releasing neurohormones to the haemolymph, neurosecretory cells and glia cells.

Neuropil: inner part of the central nervous system, without perikarya.

Optic lobes: lateral lobes of the protocerebrum, containing the three optic neuropils.

Pars intercerebralis: median region between protocerebral lobes.

Perineurium: cellular sheath of elements of the nervous systems.

Protocerebral bridge: glomerulus between pars intercerebralis and central body.

Protocerebral lobes: dorsal and frontal convexities of protocerebrum.

Protocerebrum: anteriormost part of brain, associated with compound eyes and ocelli.

Stem nerve: nerves originating from postcephalic ganglia.

Stomatogastric nervous system (=stomodaeal nervous system): visceral nervous system of foregut.

Suboesophageal ganglion (=suboesophageal complex): tripartite cephalic part of central nervous system, below pharynx; associated with mandibles, maxillae and labium.

Tritocerebrum: posterior part of brain, with commissure below pharynx; origin of circumoesophageal connectives.

Unpaired nerve (unpaired ventral nerve, Leydig's nerve): series of unpaired visceral nerves originating on commissures.

Ventricular ganglion: ganglion formed by paired posterior part of stomatogastric nervous system.

Visceral (sympathetic) nervous system (=VNS): subdivision of the nervous system, comprising the stomatogastric system, the unpaired nerve, and the caudal visceral system.

3.1.13 Photoreceptor organs

Acone: crystalline cone missing.

Apposition eye: ommatidia optically isolated by pigment sheath; suitable for high light intensity.

Cartridge: unit containing axon bundles from single (fused rhabdom) or several ommatidia (open rhabdom) and visual interneurons originating in the lamina ganglionaris and medulla.

Compound eye: paired main light sense organs; almost always composed of numerous ommatidia.

Cone cells (Semper cells): four cells forming the crystalline cone.

Corneal lens: specialized transparent cuticle at the surface of the ommatidia.

Corneagenous cells: cells producing the corneal lens.

Crystalline cone: distal element of the ommatidium below the corneal lens, formed by four cells; part of the dioptric apparatus.

Dermal photoreception: photoreception at the body surface via epidermis cells with photopigments.

Eucone: crystalline cone formed by intracellular secretions of the cone cells.

Extraocular photoreception: non-image-forming photoreception not involving specific visual organs.

Neuronal superposition: recombination of retinular axons of different ommatidia in the optic neuropil.

Ocellus: relatively simple median eye with single cornea lens and always lacking a crystalline cone: three are probably present in the groundplan of Insecta, but reductions occur in many groups.

Ommatidium: single unit of compound eye; usually composed of cornea, crystalline cone, primary and secondary pigment cells, and eight retinula cells.

Optic lobe: extension of protocerebrum, containing lamina ganglionaris, medulla and lobula.

Phaosomes: light-sensitve cells on the genitalia of certain butterflies.

Primary pigment cells: enclose crystalline cone with narrow proximal processes.

Pseudocone: extracellular formation of the crystalline cone.

Retinula: perceptive element of ommatidium, formed by retinula cells.

Rhabdom: unit formed by the rhabdomeres of each ommatidium, open or fused.

Rhabdomere: set of microvilli in a retinula cell, with a strict parallel arrangement.

Secondary pigment cells: cover proximal part of primary pigment cells and at least distal region of retinula cells.

Stemmata: simple lateral eyes of holometabolan larvae; often arranged in groups of six or five on both sides.

Superposition eye: ommatidia incompletely isolated by pigment sheath; suitable for low light intensity.

Tapetum: tracheae forming a densely packed light-reflecting inner layer in compound eyes or ocelli.

3.1.14 Tracheal system

Air sac: large extension of tracheae.

Anastomosis: longitudinal or transverse connection between primarily segmental tracheae.

Atrium: extended chamber of a spiracle; below the body surface.

Peritreme: small sclerite bearing a spiracle.

Spiracle: external opening of the tracheal system.

Taenidium: spiral-shaped strengthening ridge of the tracheae.

Trachea: tube-shaped elements of the tracheal system.

Tracheole: thin terminal portion of the tracheae, filled with liquid.

3.1.15 Circulatory system

Accessory pulsatile organs (=accessory hearts): accessory pumping devices at the base of the antennae (antennal hearts), in the legs (leg hearts),at the wing base (wing circulatory organs), and at the base of abdominal appendages (e.g., cercal hearts).

Adipohaemocytes: haemocytes characterized by lipid droplets.

Adventitia: basal lamina of heart cells forming an external layer.

Alary muscles: flat visceral muscles keeping the heart in position.

Antennal heart (=antennal ampulla): pulsatile organ associated with the antenna.

Anterior aorta (=aorta cephalica): anterior part of dorsal vessel, usually reaching into the head anteriorly.

Circumoesophageal vessel ring: ring-shaped element of the circulatory system around the anterior foregut.

Cystocytes: haemocytes similar to granulocytes.

Diaphragm: septum dividing haemocoel; present below heart (dorsal), above ganglionic chain (ventral), and in legs.

Dorsal vessel: heart and cephalic aorta.

Dorsal ampulla (wing heart): accessory heart at the wing base.

Endocard: basal lamina of heart cells forming an internal layer.

Granulocyte: haemocyte with increased number of granula.

Haemocoel (=mixocoel): body cavity containing the haemolymph.

Haemocyte: haemolymphatic cell of mesoblastic origin.

Haemocytopoietic (=haemopoietic) organs: organs producing haemocytes, closely associated with the heart.

Haemolymph: body fluid in arthropods.

Heart: dorsomedian main pumping element of the circulatory system.

Muscularis: ring muscle layer around heart (or other organs).

Nephrocyte: sessile cell in the haemocoel, high metabolic activity, carrying out pinocytosis.

Oenocytoid: large cell of hypodermal origin, with unclear function.

Ostia: openings of the heart; usually paired and lateral, but also present as unpaired ventromedian openings in few groups.

Pericardial cell: nephrocyte in the pericardial sinus or the lumen of the heart.

Pericardial sinus: part of haemocoel around heart, above dorsal diaphragm.

Perivisceral sinus: part of haemocoel around gut, between dorsal and ventral diaphragm.

Perineural sinus: ventral part of haemocoel around ganglionic chain, below ventral diaphragm.

Plasma: body fluid without haemocytes.

Plasmatocyte: haemocyte with few or without granula, highly deformable; functions are phagocytosis and encapsulation.

Prohaemocyte: small and globular mesodermal stem cell of most haemocytes.

Pulsatile organ: see: accessory heart

Protohaemocyte: undifferentiated haemocyte.

Sphaerulocyte: large haemocyte with globular vacuoles.

3.1.16 Digestive tract

Anatomical mouth: opening of pharynx, always below frontal ganglion; with attachment of the anterior precerebral frontopharyngeal dilator, often enclosed by strongly developed ring muscle.

Basal ampulla: thickened basal part of Malpighian tubules; without typical microvilli.

Caecum: appendage of anterior midgut, short papillae or long tubes.

Cibarium: anterior part of preoral space between epipharynx and hypopharynx.

Colon: posterior region of hindgut, anterior to rectum.

Dilator: muscle widening the pharynx (or oesophagus); origin on head capsule or tentorium.

Foregut: ectodermal anterior section of digestive tract, lined by an intima.

Functional mouth opening: external opening of prepharynx.

Ingluvies: crop-like extended part of the posterior foregut.

Ileum: anterior region of hindgut, posterior to pylorus.

Hindgut: ectodermal posterior section of digestive tract, lined by an intima.

Malpighian tubules: tube-like excretory organs inserted at the midgut-hindgut border.

Membrana propria (=tunica): thickened external basal lamina of the gut epithelium.

Microvilli: microscopic cellular membrane protrusions increasing the surface area.

Midgut: entodermal middle section of digestive tract.

Oesophagus (=esophagus): part of foregut following the pharynx; with one pair of thin dilators or completely lacking dilator muscles; with thin ring muscle layer.

Peritrophic membrane: dense netting of thin cuticular fibrillae; protecting midgut epithelium.

Pharynx: anteriormost section of foregut, between anatomical mouth opening and oesophagus; located in the head; with well-developed ring muscles, longitudinal muscles and dilators.

Preoral cavity: space between mouthparts anterior to anatomical mouth.

Prepharynx (=prepharyngeal tube): preoral tube formed by a fusion of the lateral margins of the proximal epi- and hypopharynx; usually shaped like a flattened U in cross section.

Proventriculus: specialized posteriormost part of foregut, with rows of cuticular teeth and acanthae and strongly developed musculature.

Pylorus: anteriormost section of hindgut, area of origin of Malpighian tubules.

Rectal papillae: thickened cylindrical epithelium of rectum; specialized on water reabsorption.

Rectum: terminal part of ectodermal hindgut, usually with rectal papillae.

Regenerative crypts: large groups of replacement cells in the midgut.

Rhabdorium (=brush border): microvilli-bearing inner surface of midgut cells.

Salivarium: posterior part of preoral space between hypopharynx and labium, with opening of salivary ducts.

Salivary ducts: connect the salivary gland with the salivarium.

Sphincter: closing muscle at the end of the rectum.

Valvula cardiaca: ring-shaped fold at anterior end of foregut, functioning as a valve.

Valvula pylorica: ring-shaped fold at posterior end of pylorus, functioning as a valve.

Valvula rectalis: ring-shaped fold between the colon and the rectum.

3.1.17 Excretory organs

Basal ampulla: thickened basal part of Malpighian tubules; without typical microvilli.

Malpighian tubules: tube-like excretory organs inserted at the midgut-hindgut border.

3.1.18 Endocrine organs and the hormone system

Allatostatin: suppresses the secretion of juvenile hormone.

Allatotropin: stimulates the secretion of juvenile hormone.

Corpora allata (=Ca): small paired neurohaemal organs posterior to brain; produce juvenile hormone; connected with corpora cardiaca.

Corpora cardiaca (=Cc): small paired neurohaemal organs anterior to corpora allata; connected with neurosecretory cells in the brain.

Diffuse endocrinous system: hormone producing cells interspersed between digestive cells of the midgut.

Ecdysone: molting hormone; produced in the prothoracic gland.

Juvenile hormone: produced in the corpora allata; suppresses the molt to the adult stage.

Perisympathetic organs (=PSO): segmental neurohaemal organs associated with the postcephalic ganglionic chain.

Neurosecretory cells: specialized nerve cells producing neurohormones.

Prothoracic glands: glands in the prothorax producing ecdysone.

Prothoracotropic hormone (=PTTH): produced by neurosecretory cells in the brain, stimulates the prothoracic gland.

3.1.19 Fat Body

Corpus adiposum (=fat body): large diffuse multifunctional organ; usually concentrated in the abdomen.

Mycetocyte: cell containing symbionts. No storage function.

Mycetomes: aggregates of mycetocytes.

Trophocyte: storage cell, dominant cell type in the fat body.

Urocyte: specialized urate cell.

3.2 Reproduction, development and immature stages

3.2.1 Fertilization and egg structure

Aeropyles: openings in the chorion for the gas exchange.

Arrhenotoky: unfertilized eggs develop into males.

Capitulum: knob-shaped or conical structure on the operculum.

Chorion: external cover of the egg, often distinctly sculptured.

Endochorion: internal layer of the chorion.

Exochorion: external layer of the chorion.

Macrolecital: with large amount of yolk.

Meroblastic cleavage: incomplete cleavage.

Microlecital: with small amount of yolk.

Micropyles: openings in the chorion for spermatozoa.

Operculum: cap-like surface structure of eggs; pushed off during hatching.

Periplasm: thin peripherical cytoplasmic layer of the egg.

Spermatorphore: sperm package.

Thelytoky: females are produced from unfertilized eggs.

Vitelline envelope: thin layer on the egg surface, below the chorion.

3.2.2 Cleavage and embryonic developement

Amnion: extraembryonic area of the developing blastoderm; close to the embryo.

Amnionic cavity: hollow space within the germ formed during the blastokinesis.

Anatrepsis: first stage of the blastokinesis; the embryo is shifted to the central part; formation of the amniotic cavity.

Blastoderm: closed superficial epithelium in the superficial cleavage pattern.

Blastokinesis: rotation of the embryo.

Cellular blastoderm: blastoderm after the formation of cell borders.

Cleavage: first divisions (of cells or nuclei) in the embryonic development.

Cleavage center: central portion of the egg cell.

Coelomic cavities: secondary body cavities filled with liquid and enclosed by mesodermal coelothel. Reduced in the later embryonic development in hexapods.

Energids: nuclei formed after the first divisions in the central part of the egg.

Germ band (=germ anlage, "Keimstreifen"): differentiated region of the blastoderm; prospective embryo.

Holoblastic: complete cleavage.

Katatrepsis: reversal of the anatrespsis; the embryo is moved back to the surface.

Mixocoel: product of fusion of the primary body cavities and the dissolved coelomic cavities.

Ovoplasm: yolk-rich central cytoplasm, enclosing energids.

Periplasm: peripherical cytoplasm of the egg.

Plasmodial preblastoderm (=syncytial blastoderm): early stage of the blastoderm formation, without cell borders.

Primary body cavity: body lumen between the integument and digestive tract; contains the haemolymph.

Protocephalon: large prospective head of the short germ embryo.

Protocorm: small prospective postcephalic body of the short germ embryo.

Serosa: extraembryonic part of the developing blastoderm.

Superficial cleavage: special type of meroblastic cleavage, division of nuclei close to the surface.

Vitellophages (=yolk cells): nuclei remaining in the central yolk; mobilize nutrients.

3.2.3 Postembryonic development

Ametabolous development: instars increase only in size.

Complete metamorphosis: profound structural transformation leading to the adult stage after the pupation.

Hemimetabolous development: without complete metamorphosis.

Holometabolous development: with complete metamorphosis and pupa.

Imaginal discs: groups of undifferentiated cells in holometabolan larvae; primordia of adult structures (e.g., wings).

Incomplete metamorphosis: gradual transformation leading to the adult morphology.

Larva: immature stage differing strongly from adults; with specific larval structures.

Naiads: aquatic immature stages of Ephemeroptera, Odonata and Plecoptera, with hemimetabolous development but specific larval structures (e.g., gills).

Nymph: immature stage without specific larval organs or structures.

Postembryonic development: developmental processes after the hatching of the first instar.

Pupa: non-feeding preimaginal stage of Holometabola; usually largely immobilized.

3.2.4 Larval and pupal forms

Acephalic: head capsule reduced.

Asseliform (isopod-like) larvae: dorsoventrally flattened, tergites with lateral duplicatures.

Campodeoid larvae: slender, with distinct sclerites on postcephalic segments.

Erucoid larvae: with serially arranged abdominal prolegs.

Eucephalic: head capsule distinctly developed.

Maggots: weakly sclerotized, legless larvae.

Oligopod larvae: head capsule and thoracic legs well-developed, prolegs on abdominal segments I–IX absent.

Pupa adectica: mandible immobilized.

Pupa dectica: mandible movable.

Pupa libera: body appendages free.

Pupa obtecta (=“Mumienpuppe”): body appendages glued to body.

Pupa pharata (=“Tönnchenpuppe”): enclosed in barrel-shaped, sclerotized puparium.

Puparium: protective case formed by last larval exuvia.

Scarabaeiform larvae (=“Engerling”): unsclerotized, sausage-shaped postcephalic body, enlarged terminal segment IX.

4 Traditional and modern techniques in insect morphology

4.1 Fixation

Traditionally insects are killed (usually using ethyl acetate), dried, pinned and labeled. The long established procedure of pinning and storing the mounted specimens in entomological boxes or insect cases has the advantage of saving space and easy accessibility. Dried material is useful for taxonomic investigations and the study of exo- and endoskeletal features including genitalia. However, their use for detailed anatomical study (and extraction of DNA) is very limited.

A simple fixative frequently used by entomologists is 70–80% ethanol. It preserves internal soft parts to a certain degree and the specimens remain relatively pliable. This material is suitable for the study of external features and for long term storage. Usually it can be also used for anatomical study including histological sections. However, the quality of the tissue preservation is distinctly lower compared to material treated with other fixatives (e.g., Bouin's fluid, see below) and in older specimens muscles tend to become detached from their attachment sites. In many cases specimens loose or change their coloration.

For the extraction of DNA (for single gene sequencing) specimens are usually fixed in 98–100% ethanol. This material can be used for histological section series, but the specimens become very brittle due to strong dehydration.

Formaldehyde – acetic acid – ethanol (FAE): FAE, a solution composed of six parts concentrated formaldehyde, ten parts absolute ethanol and one part glacial acetic acid, is frequently used by insect morphologists. A mixture containing a higher portion of ethanol (17 parts) is called Landowsky's fluid. Specimens should be fixed for 1 to 24 hours depending on size. Both solutions are less toxic than fixatives containing picric acid and therefore less complicated in their application. They are good fixatives for anatomical work based on dissection, histological section series and μ-CT.

Kahle's fluid: This fixative is a slightly modified version of FAE containing distilled water. A standard mixture is composed of 15 parts ethanol, 5 parts concentrated formaldehyde, 1 part glacial acetic acid and 30 parts distilled water. It is traditionally used by North American insect morphologists.

Bouin's fluid: Bouin is a saturated aqueous solution of picric acid (15 parts), concentrated formaldehyde (5 parts) and glacial acetic acid (1 part). The fixative should be freshly mixed before use und stored in a refrigerator. Depending on size, specimens may be fixed from two hours to several days. After fixation they should be washed several times in 70% ethanol until no yellow coloration of the fluid can be observed.

Duboscq Brazil fluid: This alcoholic version of Bouin is preferred by some insect morphologists. Nine parts of saturated alcoholic solution of picric acid (10 g in 100 ml absolute ethanol) is mixed with four parts concentrated formaldehyde and one part glacial acetic acid. It is applied and stored like Bouin's fluid.

Carnoy's fluid: Due to the high alcohol content the mixture of six parts absolute ethanol, three parts chloroform and one part glacial acetic acid penetrates specimens very fast, but also results in strong dehydration. This can result in shrinking and hardening of tissues if applied to long. Fixation times of 1–5 hours are appropriate. After fixation the specimens have to be extensively washed in absolute ethanol.

4.2 Dissection

Simple dissection of fixed specimens is often an appropriate method to obtain morphological results very efficiently, especially when larger insects are under investigation (Fig. 4.2.1). Different simple tools such as razor blades or hand-sharpened pins can be used. In specimens preserved in fluid the connection of muscles and sclerites is preserved and can be used for detailed investigations of attachment sites and the interpretation of movements by pulling muscles with fine forceps. Dissections of wet specimens are carried out in glycerin or alcohol. For the visualization of the tracheal system preparation in water has proved suitable. The contrast between the air-filled, silvery-white tracheae and the musculature can be increased by applying methylene blue solution. Specimens longitudinally sectioned in the sagittal plane mounted on plasticine or other adhesive substances provide a good impression of the spatial arrangement of structures. In critical-point-dried or freeze dried specimens muscles can be easily detached from the cuticle in a stepwise process, without damaging the exo- and endoskeletal attachment areas. This is a convenient method for obtaining an overview of the muscle arrangement (Fig. 4.2.1).

It has to be kept in mind that simple dissection techniques have their limitations. Small or very small structures can be easily overlooked, especially if small insects are examined. Useful results cannot be obtained when the size of specimens is less than 2 mm. The histological properties of tissues cannot be assessed or only very tentatively. Moreover, the specimens are usually more or less destroyed after the procedure.

A specific approach of dissecting and drawing is recommended by K. Klass (Museum of Zoology, Dresden). The cuticle is considered as a continuous plane that is extensively folded and includes areas that are more or less strongly invaginated or evaginated. In drawings, the geometry of cuticular folding is strictly followed. Black lines are only used for "edges", i.e. lines along which the cuticle bends away from the observer's view. It is a geometric requirement that there are always two edges of opposite orientation starting in the same point, though in a particular drawing the starting point of one of them (at least) is always covered. Initially drawings include only struc-

Fig. 4.2.1: Sagittal section of a critical point dried specimen of *Sphecius spheciosus* (Hymenoptera, Crabronidae), gut removed. Courtesy I. Mikó.

tures displayed in the exposed cuticular surface. Deeper layers are successively made visible by removing parts of the cuticle via (virtual) cutting lines. These also follow the geometry of the cuticular foldings and are represented in a style different from the "edges". Sclerites are represented as darker patches. Shading is applied wherever the cuticle disappears beneath an edge (K. Klass, pers. comm.; see also Klass & Matushkina 2012; Schneider & Klass 2013).

[Klass & Matushkina (2012); Schneider & Klass (2013)]

4.3 Maceration

Maceration is frequently used in insect morphology to dissolve internal soft parts for the detailed examination of skeletal elements (Fig. 4.3.1). The most frequently used agent is potassium hydroxide (KOH, approximately 5% aqueous solution) which macerates soft tissue very efficiently, especially at higher temperatures. Good results are obtained when specimens are kept at 60° C for few hours. Prolonged treatment results in reduced coloration and sclerites are rendered more or less transparent. The latter

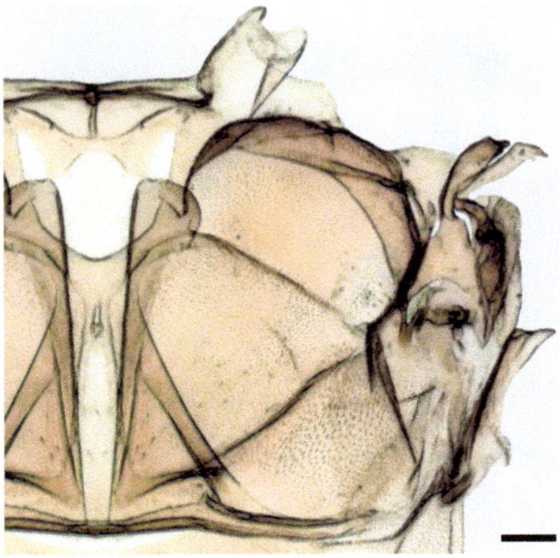

Fig. 4.3.1: Lightmicroscopical image of the macerated metanotum and wing base sclerites of *Elodes pseudominuta* (Coleoptera, Scirtidae). Scale bar: 100 µm.

effect can be compensated by applying chlorazol black or pyrogallol staining, which renders sclerotized structures blackish. The staining is reversible and can be removed by washing with ethanol. The cleared specimens can be stored in glycerin or 70% ethanol or critical-point-dried and also used for scanning electron microscopy (SEM).

Alternatives are sodium hydroxide (NaOH), lactic acid ($C_3H_6O_3$), lactophenol (lactic acid + distilled water + glycerin + phenol; 1:1:2:1) or Diaphanol (chlorine dioxide – acid acid). A less agressive method is to let the insects rot slowly in distilled water at room temperature. After some days or few weeks (depending on size) the decaying soft tissue can be removed using fine forceps. An advantage of this simple technique is that the cuticle retains its original coloration.

4.4 Scanning electron microscopy (SEM)

Scanning electron microscopy (SEM) (Fig. 4.4.1 and figures in chaper 6) is an approach frequently used by entomologists since the 1970ies. It is comparatively easy to handle and an excellent tool to examine and visualize cuticular surface structures (Figs 4.4.1B, 6.1.3). Arguably SEM has even contributed to a decline of detailed anatomical study (e.g., of the skeletomuscular system) as visually attractive results can be obtained within short time and without intensive training. Internal soft parts can also be visualized to a certain degree. In some cases SEM is useful to obtain a general overview of internal structures. However, it is clearly insufficient for a detailed ana-

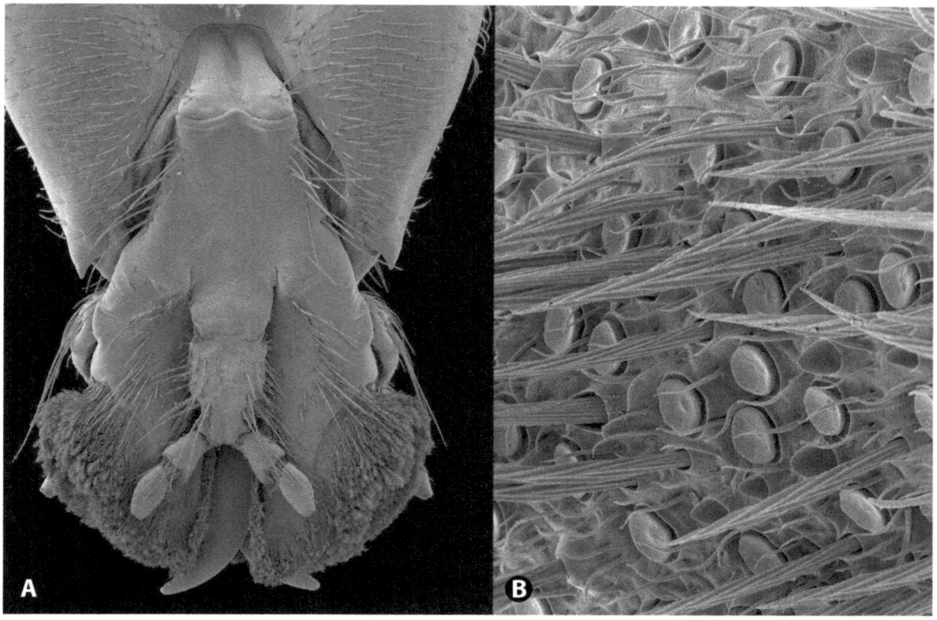

Fig. 4.4.1: SEM micrographs. A, *Merope tuber* (Mecoptera, Meropeidae), mouthparts, posterior view; B, *Philopotamus ludificatus* (Trichoptera, Philopotamidae), sensilla trichodea and fungiform pseudoplacoid sensilla of antenna.

tomical reconstruction. Furthermore the degree of sclerotisation, transparency and pigmentation of the cuticle cannot be observed using SEM, except for parts with distinct fine structure (e.g., membranes vs. sclerites).

Cleaning of the specimens is necessary if the cuticular surface is covered with secretions, soil particles, food substrates or other materials. Especially in the mouthpart region (e.g., Fig. 6.5.3) structural details can be obstructed by adhering substrates. An efficient method for cleaning relatively robust specimens is to expose them to ultrasonic sound. This approach should not be used for very fragile insects (e.g., small apterygotes, small larvae, poorly preserved specimens). To remove surface contamination of strongly encrusted specimens (e.g., dung feeders, fossorial insects) the application of KOH (ca. 1–5%) or cleaning agents (e.g., 3% Triton®) for few minutes prior to ultrasonic treatment is very useful.

After cleaning the insects are dehydrated in an ethanol series (final medium: 100% ethanol, acetone or isoamyl acetate). Air drying can be sufficient for strongly sclerotized specimens (e.g., beetles), but critical point drying is preferable in most cases to avoid shrinking artifacts. Equally good results are obtained by gradually transferring the specimens to distilled water and then freeze drying them. However, this procedure is comparatively time-consuming. Alternatively, chemical drying with

HMDS (hexamethyldisilazane) yields good results in short time (Brown 1993). Specimens are dehydrated with an ethanol series and kept in pure HMDS for two periods of 30 min. Finally the specimens are placed in an open Petri dish under a laboratory hood to allow the fluid to vaporize.

After the drying procedure the specimen has to be mounted on a stub via carbon adhesive pads or on a specimen holder with glue or nail polish. The holder can be a simple thin wire fixed to a stub. A much more sophisticated tool is a rotatable specimen holder (Pohl 2010). The insect is fixed on a thin pin on the rotatable arm of the device. This allows for an exact orientation of the specimen and its complete 360° rotation. Thus all standard views (usually dorsal, lateral and ventral view) can be obtained with a single specimen. Moreover, the hollow basal part of the brass specimen holder absorbs the electrons that pass the specimen. This results in a homogenous black background and distinctly reduces charging, which is often a problem, especially when insects are covered with a dense vestiture of setae (e.g., Diptera, Hymenoptera). Before the specimen is examined it has to be sputter coated with a very thin layer of a conductive element (e.g., gold, platinum, carbonate).

A modern approach which is sometimes used in entomology is the ESEM mode (environmental pressure). Moist or even live insects can be examined in low vacuum without conductive coating. The resolution is lower than in high vacuum but charging of the unprocessed specimens is reduced to a minimum. This technique also allows the examination of museum specimens and even type material as the surface is not affected by the procedure.

[Pohl (2010); Brown (1993)]

4.5 Transmission electron microscopy (TEM)

Transmission electron microscopy (Fig. 4.5.1) is an excellent tool for investigations on the ultrastructural level. It has played a pivotal role in the understanding and interpretations of the structure and functions of cells and tissues (Stirling & Woods 2002). Its role in insect phylogeny is relatively limited. However, it is essential for the study of specific character systems such as for instance the sperm ultrastructure (e.g., Dallai et al. 2012b).

The fundamental advantage of TEM is the enormously increased resolution compared to light microscopy (Woods & Stirling 2002). Magnifications of several hundred thousand times are possible. A disadvantage is the complicated and time consuming preparation of the material in part using highly toxic substances. In the following only a brief description of the fixation and preparation is given. More specific literature should be used by researchers intending to use TEM, and satisfying results cannot be obtained without intensive training.

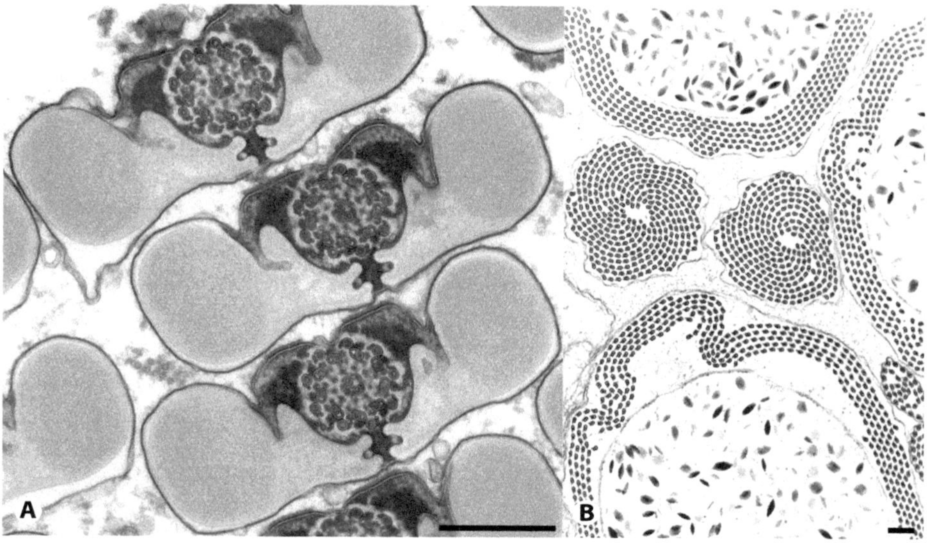

Fig. 4.5.1: TEM micrographs of spermatozoa. A, *Chrysoperla carnea* (Neuroptera, Chrysopidae); B, *Massalongia bachmaieri* (Diptera, Cecidomyiidae). Scale bar: 250 nm. Courtesy R. Dallai.

As the penetration of the material is essential for an effective fixation, the organs or structures to be investigated (e.g., testes, gland tissue) should be isolated. Smaller insects can be fixed as a whole, but opening the body cavity (e.g. by removing legs etc.) is recommended to improve the penetration. Glutaraldehyde (2.5%) is usually applied as primary fixative to cross-link proteins. For larger specimens paraformaldehyde is also an option. Both is prepared in a phosphate or natriumcacodylat buffer to maintain pH and with various additives (e.g., sucrose) to control osmolarity. After 1–4 h of primary fixation the specimens are washed overnight in buffer. This is followed by the secondary fixation in osmium tetroxide (2%) for 1–4 h to stain lipids (Hayat 1981) and washing in buffer for few minutes.

Epoxy resins (e.g., Spurr, Araldit, Eppon) are used as embedding materials. As they are completely immiscible with water, careful dehydration is essential. It is necessary to use a graded ethanol series but the time of dehydration should be also kept as brief as possible to minimize the risk of extracting cellular constituents (Woods & Stirling 2002). Gradual infiltration of the resin is required beginning with a 50 : 50 mixture of resin and solvent. Finally each specimen is transferred into fresh pure resin in a separate silicon mold. For polymerization a heating cabinet is necessary.

The polymerized resin block has to be tightly trimmed as a small pyramid around the black specimen. For sectioning the prepared specimen is fixed in an ultra-microtome. Glass knifes or diamond knifes can be used for ultrathin sectioning. The former are inexpensive but go blunt after short usage, whereas the latter are expensive but very durable and therefore preferable for continuous series of ultrathin sections. The sections of ca. 70 nm thickness are transferred to small round cupric holders (grids). Mesh grids give the sections the highest stability, but the balks of the mesh may hide structures of interest. A slot grid possesses a large single opening allowing to view the whole section. To carry the sections the slot has to be coated with an extremely thin and very fragile foil (e.g., formvar, pioloform).

To enhance the contrast of the sections in the electron beam they are treated with two heavy metal agents (see Reynolds 1963). Uranyl acetate is applied to stain proteins and nucleic acids and lead citrate is used to contrast cytoplasm, membranes and glycogen-rich structures. After drying the sections can be examined using the TEM.

The 3D-aplicabibility of TEM is limited. Slot grids have to be used and the preparation of continuous section series is very difficult. However, for detailed investigations below the cellular level there is no alternative. For lower magnifications SFBSEM is a possible alternative (see 4.6.).

[Reynolds (1963); Hayat (1981); Stirling & Woods (2002); Woods & Stirling (2002); Dallai et al. (2012b)]

4.6 Histology

The preparation of semithin sections is still widely used as standard technique for the investigation of internal features including the histological properties of tissues. In the first half of the 20th century, specimens were usually embedded in paraffin or celloidin (see e.g., Heddergott 1939). These embedding media are soft compared to the exo- and endoskeletal structures of most insects and do not rigidly interconnect with the cuticle. The use of teneral (freshly-molted) specimens or the application of specific chemicals reduces the effects, but the resulting sections are necessarily comparatively thick (5–50 μm) and show a high rate of artifacts as for instance deformations, loss of structures or interfolding of parts of the sections. An advantage is that after the final removal of the paraffin very contrast-rich staining protocols can be applied (e.g., methylene blue + basic fuchsine, Masson's trichrome stain). Towards the end of the 20th century traditional embedding media were more and more replaced by hard epoxy resins as already previously used in transmission electron microscopy (see above). Using these resins distinctly thinner sections (0.5–1.5 μm) could be produced. Furthermore, deformations and the loss of sections were greatly reduced.

Araldite has turned out as very useful for sectioning insects (and other arthropods) and is today the most often used embedding medium in entomology (Fig. 4.6.1A). However, other resins used for TEM are also suitable. The preparation of specimens is almost identical to the procedure described for TEM (see above). They should be well-fixed using FAE or buffered glutaraldehyde, but samples directly stored in 70%

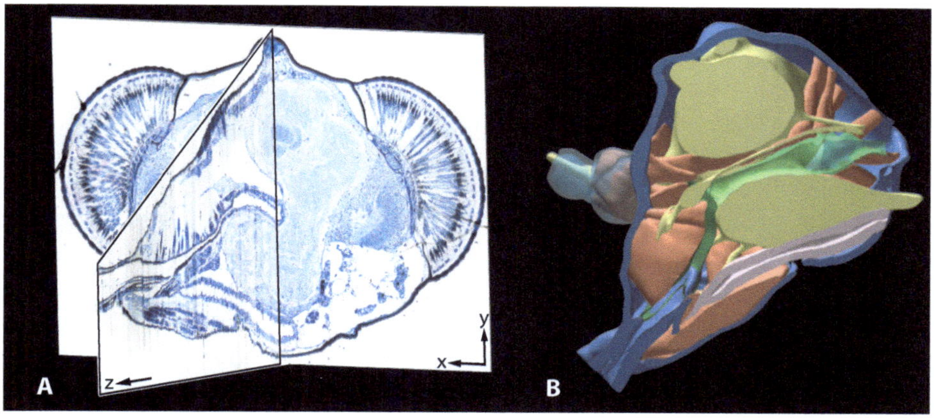

Fig. 4.6.1: *Nannochorista* sp. (Mecoptera, Nannochoristidae). A, image stack of histological sections. x, y represents original cross sections, z the computer-based virtual sagittal section; B, 3-dimensional reconstruction based on serial sections. Surface model, sagittal section.

ethanol are also suitable. For the latter, a postfixation with FAE prior to embedding is advisable. In order to increase the permeability for fixatives and the resin, removing appendages (e.g., legs, antennae) is recommendable. The specimens are gradually dehydrated in an ethanol series (up to pure ethanol) and transferred over several steps of ascending acetone-Araldite mixtures into pure resin. The infiltration can be facilitated by applying vacuum. Finally the samples are separately placed in silicon molts filled with resin. After a minimum of two days hardening at 60°C the resin blocks can be removed from the molts and trimmed as described for TEM (see above).

Ultra-microtomes used for TEM-sectioning are usually also suitable for the production of semithin sections (up to 1 µm tickness), but specific microtomes are used in most cases. Sections are prepared using glass- or diamond-knifes. The latter are much more durable and produce less artifacts. After cutting, every section is transferred to a drop of water placed on a microscope slide using a thin needle or eyelash glued on a glass-pipette. The slides, which are often slightly compressed after sectioning, are placed on a heating plate (60°C) for stretching. After the sections are completely dried, they can be stained using basic solutions. In contrast to paraffin embedding, the resin is usually not removed after sectioning, which impedes several traditional staining protocols. However, a combination of toluidine blue, borax (each 4 parts of 1% watery solution) and pyronin G (1 part of 1% watery solution) applied for about one minute at 40–60°C has turned out as an excellent staining method. After drying, the sections are sealed using cover glasses mounted with agents as Pertex or Eukit. Protected this way sections can be stored for a long time and investigated also using oil immersion microscopy at high magnification (100X).

In the last century the analysis of hundreds of sections via microscopy was very time consuming and the reconstruction of structures required strong imaginational skills. Today, motorized microscopes (e.g., slide scanner) semiautomatically digitalize sections in high-resolution and in very short time. Computer-based 3-dimensional alignment- and visualization-functions (see below) greatly facilitate a fast analysis and documentation of anatomical data (Fig. 4.6.1A).

Despite a considerable progress in histology the sections are never completely free of artifacts. The loss of terminal structures (e.g., tips of legs or palps) cannot be avoided in most cases. Deformations of the sections almost always occur, even though they are minimized compared to paraffin sections. The production of high-quality serial sections requires special training and the entire process takes at least two weeks. Therefore this technique is not practicable for a large taxon sampling from the economic perspective. Gross-anatomical data can be much more efficiently acquired using micro-computer tomography (see below). Nevertheless the very high optical resolution and the reliable and fast discrimination of tissues provided by semithin sections are still essential for the investigation of anatomical details. Today the combination of fast, artifact-free CT-scans for broader sets of taxa combined with histological sections of selected specimens is the best way to obtain an extensive and high quality anatomical data set.

[Heddergott (1939)]

4.7 Serial Block-Face Scanning Electron Microscopy (SBFSEM)

This recently developed technique combines SEM surface analysis with ultramicrotomy similar to what is used for TEM. It allows an efficient production of ultrathin serial sections perfectly suitable for 3D-reconstruction (Zankel et al. 2009). Advantages of SBFSEM are the low degree of artifacts, the very high 3D-resolution, and the almost automatic data acquisition. Compared to TEM the maximum magnification is distinctly lower. A disadvantage linked to the technical setup is the limitation to very small specimens (every dimension below 1 mm; Hörnschemeyer et al. 2012).

Specimen preparation is very similar to what is described for TEM (see 4.5) but contrasting is applied prior to resin embedding (block contrasting). Primary fixation in glutaraldehyde (2.5%) is followed by an extensive secondary treatment with osmium tetroxide (1–2 h). As for TEM uranyl acetate and lead citrate are used for contrasting but applied distinctly longer to generate high content of heavy metals in the specimen. This is not only necessary to distinguish between different tissue types but also to reduce charging artifacts during the scanning procedure (Hörnschemeyer et al. 2012). The contrasted specimens are dehydrated in an ethanol series, transferred into two stages of pure acetone and finally embedded in resin. Durcupan has proved

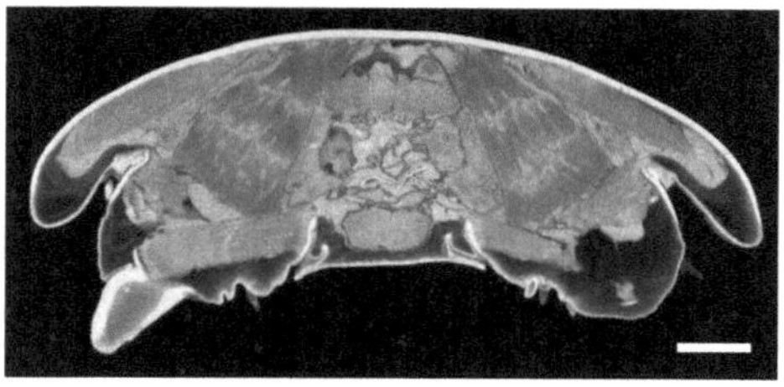

Fig. 4.7.1: Volume rendering of SBFSEM data. Primary larva of *Stylops melittae* (Strepsiptera, Stylopidae), virtual cross section through prothoracic region. Scale bar: 5 μm.

as most durable in exposure to electron beams. Other resins as Araldite or Spurr can also be used, but are less beam-stable. Softening during high resolution scans results in section artifacts.

The polymerized resin blocks are maximally trimmed to fit into the ultramicrotome mounted in the SEM chamber. In an automatic process the microtome cuts ultrathin sections (10–200 nm) which are discarded. After every section an image of the plane surface of the block is recorded by the SEM (Hörnschemeyer et al. 2012). The resulting large image stack is fully aligned and therefore perfectly suitable for 3D-reconstruction (Fig. 4.7.1).

[Zankel et al. (2009); Hörnschemeyer et al. (2012)]

4.8 Focused Ion Beam (FIB)

FIB technology is manly used in material sciences, but was also applied in insect morphology in recent years (e.g., Di Giulio et al. 2012). Dual beam Scanning Electron Microscopes are equipped with an additional, strongly focused ion beam (e.g., helium, gallium). This allows to remove precisely defined parts of the specimen to investigate structures below the surface or to open small compartments (e.g., surface receptors). The technique is also very useful for the preparation of TEM specimens. Due to the highly precise milling process structures can be removed layer by layer (Di Giulio et al. 2012). The freshly created surfaces are recorded using the electron beam.

The specimen preparation is identical to standard SEM procedures. Proper fixation (e.g., glutaraldehyde) and an ultrastructure-preserving drying process (e.g., critical point drying) are essential for good results.

[Di Giulio et al. (2012)]

4.9 Confocal laser scanning microscopy (CLSM)

Confocal laser scanning microscopy (Fig. 4.9.1) is a very efficient technique to study specific structural details of insects, but is rarely used in morphology based systematic investigations. Without applying specific staining methods (e.g., immunostaining) results can be obtained using the autofluorescence of the cuticle and unsclerotized body parts (e.g., muscles). No specific fixation is necessary. Specimens preserved in 70% ethanol are fully suitable. With the combined use of laser with different excitation wavelengths soft and hard parts can be easily differentiated (Fig. 4.9.1). Even the grade of sclerotisation of the cuticle and its resilin content can be visualized (e.g., Michels & Gorb 2012).

Very small or flat insects and small isolated body parts as for instance genitalia can be visualized (e.g., Klaus & Schawaroch 2006). Deeper layers of larger specimens cannot be detected without clearing the cuticle. Suitable agents are hydrogen peroxide (35% H_2O_2) (Stüben & Linsenmair 2008), methyl salicytate, lactic acid (Michels 2007) and Murray's clear (= BABB; 1 part benzyl alcohol + 1 part benzyl benzoate) (Zucker 2006). The specimens are kept in the solution for one hour to several days for bleaching and clearing. Soft parts are not drastically affected by this procedure and can be visualized after the process (Deans et al. 2012).

Depending on the thickness of the specimen, the number of scan channels, and the settings of the confocal microscope, the recording of high-quality scans can take several hours. It is important to keep the specimen fixed in its position during the scan. Even very slight movements strongly degrade the obtained image stack. Scans can be obtained with a minimum of preparation time by fixing the specimen in a drop of glycerine between the slide and the (high precision) cover glass, which should be supported by small spacers (e.g., wax or self-adhesive rings) to avoid squeezing. Ethanol, buffer and distilled water are also suitable media, but evaporation can induce movements of the specimen. For longer scans embedding media with a higher viscosity are recommended. Agarose (1%), glycerine jelly, and mowiol are useful, but also traditional media such as Canada balsam or Euparal (e.g., Schawaroch & Li 2007). They also facilitate the controlled adjustment of the specimen without deformation.

Autofluorescence of insect cuticle is induced by a laser wavelength of 488 nm (see e.g., Michels 2007). The emitted light is recorded in two separate spectra: green (ca. 500–570 nm) and red fluorescence (ca. 580–690 nm). The overlay of both channels allows for a detailed imaging of the grade of sclerotisation of the cuticle, with membranes appearing green and sclerites brown (see Fig. 4.9.1 and e.g., Deans et al. 2012). Additionally, resilin-rich structures can be visualized by applying UV-light (405 nm) and recording the emitted blue light (420–480 nm; e.g., Michels & Gorb 2012). Internal features can be visualized if the cuticle is weakly sclerotized and largely unpigmented, or if the specimen was bleached. The treatment with some bleaching agents can result in changes of the autofluorescence and affect the visualization of the grade of sclerotisation described above. The application of glutaraldehyde as fixa-

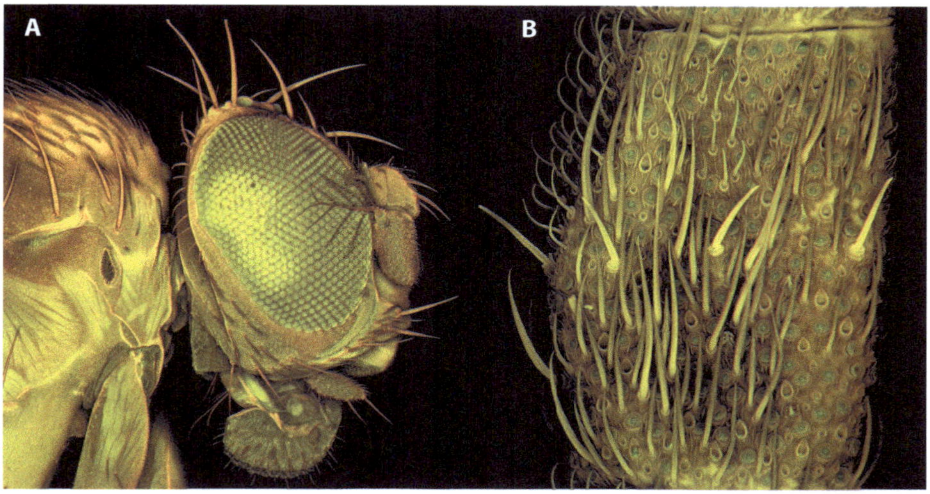

Fig. 4.9.1: Volume rendering of CLSM image stacks. Autofluorescence induced by 488 nm laser. Overlay of two recorded fluorescence channels: green (500–570 nm) and red (580–590 nm) A, *Drosophila melanogaster* (Diptera, Drosophilidae), head, lateral view; B, *Philopotamus ludificatus* (Trichoptera, Philopotamidae), antennal flagellomere.

tion agent increases the fluorescence at 488 nm. An efficient fluorescence staining of the integument is Congo red, which is excited at a wavelength of 561 nm (emission spectrum: 570–670 nm; Michels & Bützow 2010).

Data obtained with CLSM are useful for volume rendering in the first place. Maximum projections of the image stack usually result in less valuable visualizations. The reconstruction of surface-based 3D-models is possible for flat or small isolated objects such as genitalia or mouthparts (see e.g., Klaus & Schawaroch 2006; Michels 2007). The visualization of larger specimens (including softparts) is limited as the fading greatly increases in deeper layers (Fig. 4.9.1A).

[Zucker (2006); Klaus & Schawaroch (2006); Michels (2007); Schawaroch & Li (2007); Michels & Büntzow (2010); Deans et. al. (2012); Michels & Gorb (2012)]

4.10 Micro-computer tomography (μ-CT)

This highly efficient method to gain anatomical data is used in insect morphology since about ten years (Fig. 4.10.1). A pioneer study was published by Hörnschemeyer et al. (2002). The improved hard- and software of modern desktop scanners produce high resolution scans and make the technique more and more attractive for the anatomical investigation of small objects. The maximum resolution is presently ca. 0.5 μm. High end scanners produce data with a resolution better than 0.1 μm (nanotomography).

The greatest advantage of micro-computer tomography is the highly accelerated acquisition of almost artifact-free anatomical data. It is highly efficient compared to histology, the images stacks are perfectly aligned, and it is non-destructive. The specimens can be used for SEM or histological sectioning after the µ-CT scans are obtained. Unlike histology no structures are lost during the process. This allows a very detailed morphological documentation with a minimum of material. As a non-invasive technique µ-CT can also be applied to extremely rare species or even type material. High density resolution based on the specific absorption of different tissue types (e.g. skeleton, muscles, nervous system; Fig. 4.10.1), usually recorded with low beam energy, facilitates the discrimination of structures in the data set (e.g., Friedrich et al. 2008).

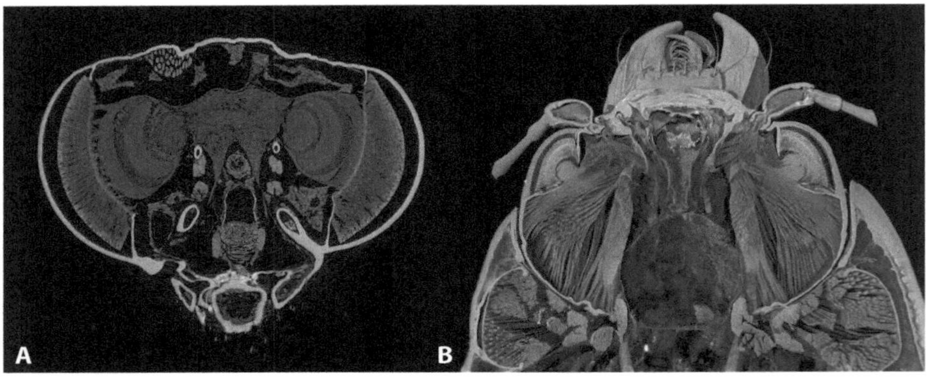

Fig. 4.10.1: Synchrotron-radiation-based µCT data. A, *Tenthredo vespa* (Hymenoptera, Tenthredinidae), head, virtual cross section; B, *Omophron* sp. (Coleoptera, Carabidae), volume rendering of whole data set, horizontal section. Note the high density resolution allowing easy discrimination of different tissue types.

The preparation of specimens is simple. They are usually dehydrated in an alcohol series and dried at the critical point to avoid shrinking artifacts. With this approach a high contrast between the tissues and the surrounding medium (air) is generated. The specimen can either be mounted on a specific holder with superglue or fixed in a small vessel (e.g. Eppendorf tube, pipette tip) glued to the holder. In order to obtain the maximum spatial resolution, the rotation axis of the sample should correspond to the longitudinal axis of the specimen. If drying the specimen is not an option (e.g., museum material stored in liquid) scanning in alcohol or distilled water is also possible. Due to the similar electron-density of these media and the insect tissues the obtained contrast will be very low. To compensate this reversible staining can be applied. The best results are obtained with alcoholic iodine solution (I2E; 1% iodine in pure ethanol) for one to five days followed by washing in alcohol (see e.g., Metscher 2009). The stained specimens have to be fixed mechanically to avoid movements caused by circulations of the liquid medium. Cotton and foamed plastic are suitable materials. The iodine can be removed by extensively washing in alcohol.

Specimens prepared for TEM investigations can be documented using μ-CT. The osmium-tetroxide stained and resin-embedded samples show a good contrast between tissues and embedding media. This is usually not the case in samples embedded for microtome sectioning, which are not specifically stained. However, machines optimized for phase contrast are able to visualize even such specimen. This technique is also an excellent approach for the investigation of fossils embedded in Amber (see e.g., Tafforeau et al. 2006; Pohl et al. 2010). Prior to the scan a dispensable part of Amber should be tested in the beam to estimate the darkening of the material during the procedure. A disadvantage of phase contrasted μ-CT scans is that different tissue types are not easy to discriminate because the coloration is almost identical.

The perfect alignment and high resolution of the resulting data set allows for fast and precise 3-dimensional analysis and documentation of anatomy (Figs 4.10.1B, 6.14.3).

[Hörnschemeyer et al. (2002); Tafforeau et al. (2006); Friedrich et al. (2008); Metscher (2009); Pohl et al. (2010)]

4.11 Computer-based 3-dimensional reconstruction

Computer-based 3-dimensional reconstructions can be based on a broad variety of high-quality data sets (Figs 4.6.1B, 4.7.1, 4.10.1B). Most common sources are histological serial sections (Figs 4.6.1B, 6.33.3B), μ-CT-image stacks (Figs 4.10.1B, 6.14.3, 6.33.3A) and CLSM-data (Fig. 4.9.1). 3D data sets are highly suitable for the analysis of the spatial arrangement of morphological structures by using virtual section planes computed by the software (Fig. 4.6.1A). Smaller data sets can be visualized on standard desktop computers, but powerful graphic workstations are needed for high resolution data and extensive imaging procedures.

Perfectly aligned data as for instance μ-CT- and SBFSEM image stacks can be immediately used for 3D reconstruction and visualization (Figs 4.7.1, 4.10.1B). Data based on other sources (e.g., images of histological or TEM sections) have to be aligned prior to the reconstruction (Fig. 4.6.1A). This can be done semiautomatically using specific functions implemented in commercial software packages (e.g., VSG Amira©, AutoAligner of Bitplane Imaris) or open source tools (e.g., ImageJ).

Volume rendering is a simple way to visualize the complete data stack and yields results similar to low magnification SEM images (Fig. 4.10.1B). An advantage is that the thickness of the body wall can be determined easily, which is not possible with SEM. Volume models can be used to create 3D-images (e.g., for red-cyan goggles) or videos (e.g., with ImageJ). In order to show internal features cutting planes can be applied to the volume rendering (Fig. 4.10.1B). The procedure is fast but limited in its potential to reveal the complexity of internal structural configurations. To obtain more specific 3D-reconstrutions (e.g., of the skeleto-muscular or nervous system) segmentation of the data set is necessary. Dependent on the data source and software, structures of interest can be manually outlined or semiautomatically labeled throughout the stack using commercial software packages (VSG Amira©, Bitplane Imaris, VGStudio MAX) or open source tools (e.g., Reconstruct™). Data prepared that way can be used to produce segmented volume renderings of selected structural complexes and greatly facilitate the coloration of volume data (Fig. 6.14.3).

The segmented image stacks are also the basis for the automatic creation of surface objects of discrete structures. Unlike volume renderings surface objects are hollow and represent only the outline of their structural basis, usually resulting in a more simplified 3D-model (Figs 6.13.5, 6.33.3). Main advantages of this visualization technique are the multiple opportunities of modification (e.g., simplification, minimizing artifacts), coloration, illumination and animation of anatomical structures using software (e.g., Autodesk® Maya®, Luxology® Modo, open source: Blender). Surface models can be used to calculate the volume of structures, to carry out finite element analyses (FEA) and to print magnified, solid models of the structures via rapid prototyping. Furthermore, surface-based 3D-models can be included in scientific presentations and publications using the common pdf-file format, allowing for an easy exchange of 3D contents.

4.12 Geometric morphometrics (Ming Bai)

Geometric morphometrics is a collection of approaches for the multivariate statistical analysis of Cartesian coordinate data, usually limited to landmark point locations. "Geometric" refers the geometry of Kendall's shape space: the estimation of mean shapes and the description of sample variation of shape using the geometry of Procrustes distance. "Morphometrics" is from the Greek "morph," meaning shape, and "metron," meaning measurement.

The analysis of shape is a fundamental element of biological research. The traditional comparative morphological approach (morphology as such is comparative as defined by J.W. v. Goethe) has a very long history in taxonomy and systematics. As statistics developed, the transition of biology from a descriptive discipline to a quantitative science began. Traditional morphometrics was the application of multivariate statistical analyses to sets of quantitative variables such as width, length, depths, volumes and areas (Bookstein 1998). This approach was limited in different ways, as for instance by the difficulty to assess the homology of linear distances. Another inherent problem is that the geometric relationships among the variables are not preserved. Consequently, alternative methods of quantifying and analyzing morphological shape were explored. In the 1980s, the nature of the data gathered and analysed changed fundamentally, with a focus on the coordinates of landmarks and the geometric information about their relative positions. The developing novel approach was referred to as geometric morphometrics (Adams et al. 2004).

The multivariate part of geometric morphometrics is usually carried out in a linear tangent space to the non-Euclidean shape space in the vicinity of the mean shape. Traditionally, morphometrics was mainly focused on size, which doubtlessly plays an important role. However, geometric morphometrics is better suited to assess information about shape. More generally, it is the class of morphometric methods that preserves complete information about the relative spatial arrangements of the data throughout an analysis. As such, these methods allow for the visualization of group and individual differences, sample variation, and other results in the space of the original specimens.

4.12.1 Terminology and principles

Shape. Shape is the geometrical information that remains when location, scale and rotational effects are filtered out from an object (Kendall 1977).

Kendall's shape space. Kendall's shape space provides a complete geometric setting for analyses of Procrustes distances among arbitrary sets of landmarks. Each point in this shape space represents the shape of a configuration of points in some Euclidean space, irrespective of size, position, and orientation. In shape space, scatters of points correspond to scatters of entire landmark configurations, not merely of single landmarks. Most multivariate methods of geometric morphometrics are linearizations of statistical analyses of distances and directions in this underlying space.

Homology. The concept of homology bridges the gap between the language of geometric morphometrics and the language of its biological or biomathematical applications. In theoretical biology, only the explicit entities of evolution or development, such as molecules, organs, or tissues, can be "homologous". Nevertheless, researchers using morphometrics often apply the concept to discrete geometric structures, such as points or curves, and, by a further extension, to the multivariate descriptors (e.g., partial warp scores) that arise as part of most multivariate analyses. In this context, the term "homologous" has no meaning other than that the same name is used for corresponding parts in different species or developmental stages. Consequently to declare something "homologous" means simply to address processes affecting such structures as if they would have a consistent biological or biomechanical meaning. Similarly, to declare an interpolation (such as a thin-plate spline) a "homology map" means that one intends to refer to its features as if they would be related with valid biological explanations pertaining to the regions between the landmarks, for which no data are available.

Landmarks. Landmarks are discrete points that are recognized as identical in all specimens. Landmarks with the same name, homologues in the purely semantic sense, are presumed to correspond in some sensible way over the forms of a data set. Landmarks are discrete, individually identifiable points. Each point on an individual can be matched to the corresponding point on another individual. By sampling landmarks of specimens shape variations can be quantified. There is no conceptual or mathematical distinction between 2D and 3D landmark analyses except for the number of coordinates at each landmark. However, there is an enormous technological difference between collecting 2D and 3D landmarks, and a great graphical difference between depicting the results.

Curves and semi-landmarks. The term curve is used to represent the shape information when it is difficult to select a landmark. Curves are resampled as semi-landmarks (sliding landmarks). Semi-landmarks are points (arbitrarily) spaced along a curve or over a surface. Each curve on one individual can be matched to a curve on another individual, but the points themselves might not correspond exactly. Semi-landmarks cannot be recognized as the "same" (homologous) points on all specimens, but they sample the "same" (homologous) curves.

Coordinates. A set of parameters locating a point in a geometrical space. Cartesian coordinates, for instance, locate a point on a plane or in physical space by a projection onto perpendicular lines through one single point, the origin. The elements of any vector may be considered as coordinates in a geometric sense.

Superimposition. The transformation of one or more figures to establish some geometric relationship with another figure. The transformations are usually affine transformations or similarities. They can be computed by matching two or three landmarks, by least-squares optimization of squared residuals at all landmarks, or in other ways. Sometimes this is informally referred to as a "fit" or "fitting," e.g., as a resistant fit.

Partial warps. Partial warps are an auxiliary structure for the interpretation of shape changes and shape variation in sets of landmarks. Geometrically, partial warps are an orthonormal basis for a space tangent to Kendall's shape space. Algebraically, the partial warps are eigenvectors of the bending energy matrix that describes the net local information in a deformation along each coordinate axis. Except for the maximum-scale partial warp, the one for uniform shape change, they have an approximate location and an approximate scale.

Thin-plate spline. In continuum mechanics, a thin-plate spline models the form taken by a metal plate that is constrained at some combination of points and lines and otherwise free to adopt the form that minimizes bending energy. In morphometrics, the same interpolation provides a unique solution to the construction of deformation grids for data in the form of two landmark configurations.

Bending energy. Bending energy is a metaphor from the mechanics of thin metal plates borrowed for use in morphometrics. It is assumed that a configuration of landmarks has been printed on an infinite, infinitely thin, flat metal plate, and that the differences in the coordinates of these same landmarks in another picture are taken as vertical displacements of this plate perpendicular to itself, one Cartesian coordinate at a time. The bending energy of one of these out-of-plane "shape changes" is the (idealized) energy that would be required to bend the metal plate so that the landmarks were lifted or lowered appropriately.

While in physics bending energy is a real quantity, measured in appropriate units, there is an alternate formula that remains meaningful in morphometrics: bending energy is proportional to the integral of the summed squared second derivatives of the "vertical" displacement – the extent to which it varies from a uniform tilt. The bending energy of a shape change is the sum of the bending energies that apply to any two perpendicular coordinates in which the metaphor is evaluated. The bending energy of an affine transformation is zero since it corresponds to a tilting of the plate without any bending. The value obtained for the bending energy corresponding to a given displacement is inversely proportional to the scale. Such quantities should not be interpreted as measures of dissimilarity (e.g., taxonomic or evolutionary distance) between two forms.

Cluster analysis. A method of analysis representing multivariate variation in data as a series of sets. In biology, the sets are often constructed in a hierarchical manner and shown in the form of a tree-like diagram called a dendrogram.

4.12.2 Geometric morphometrics analysis

Taxon sampling and data acquisition. The crucial first step is the selection of specimens suitable for the scientific question addressed. The second step is to collect photos (using a camera, drawing, SEM micrograph, etc.) based on the same direction rule, which ensures the comparability of **shape**. Although on principle there is no limitation for the photo size, it should be as small as possible as long as it is still clear enough. The third step is to built a data file (such as a .tps file) which links all photos involved. The fourth step is to gather landmark or semi-landmark data when the photos are loaded in software, such as TPS-Dig (Rohlf, 2006). The procedure of 3D data collecting is similar, but different in the 3D photo reconstruction and landmark gathering.

Landmark and curve selection. Criteria for selecting **landmarks** and **curves** are **homology**, adequacy of coverage, consistency of relative position, coplanarity and repeatability.

In the context of landmarks, homology means that the point is arguably the "same" in all specimens. The alternative is not "homoplasy" but rather "located at a different place". Homology (in a broad sense) means that landmarks are comparable. The second criterion, adequacy of coverage, is to avoid excess landmarks which are not related with the research interest. Consistency of relative position means that a switch in landmarks positions will result in an enormous change in shape (even including a twist of the object or a turn upside down). Coplanarity (for 2D data) implies that what appears like displacements within a plane could actually be displacements outside of the plane, if landmarks are on different planes, and foreshortening looks like shortening.

Superimposition. For most practical applications, the parameters describing the shapes for a sample of homologous landmark configurations are estimated by a Procrustes superimposition. This procedure is a least-squares oriented approach involving three steps. The position, scale and orientation of the specimens are removed, retaining only the shape for which the same position, size and orientation is assumed. After this the shapes are placed on top of each other. Two alternative methods can be used: Procrustes superimposition and Bookstein shape coordinates.

Plot and tree. The greatest strength of geometric morphometric methods is that graphical representations of results are possible as configurations of landmark points rather than as customary statistical scatterplots. The analysis of landmark data can be estimated by a superimposition procedure, followed by the projection of the aligned **coordinates** on a linear tangent **Kendall's shape space** for multivariate analyses, and the graphical visualization of results in terms of the configurations of landmarks (Kendall 1984; Rohlf 1999; Slice 2001). In Kendall's shape space, distances between pairs of points (specimens) approximate the Procrustes distances between the corresponding pairs of landmark configurations. **Partial warps**, which are the eigenvectors of the **bending energy** matrix ordered, from the **thin-plate spline** plus the uniform shape components (Rohlf & Bookstein 2003), are a convenient set of shape variables that can be interpreted as axes for this space. Scores on these axes can then be treated as multivariate data representing shape, and can be used in conventional multivariate analyses (Caldecutt & Adams 1998; Bookstein et al. 1999; Adams & Rohlf 2000; Gharaibeh et al. 2000; Rüber & Adams 2001; Klingenberg & Leamy 2001; Bai et al. 2010, 2011, 2012). Differences in shape among objects can be described not only as a plot, but also in tree form via **cluster analysis.** Different models, which include Single linkage (the distance between two units is the distance between the two closest members of those clusters), UPGMA (the distance between two clusters is the average of the distances between units in one cluster and units in the other cluster), Ward's (minimizes the variance of intra-cluster distances), etc., are used to calculate the similarity between clusters.

Combined statistics. Geometric morphometrics data can be applied to many research fields, such as for instance phylogeny, development, and ecology. This often requires a sophisticated quantitative representation of the phenotype that captures the functional, genetic or developmental attributes that are biologically important for the hypothesis to be tested, by linking the data or a communication of results. As existing coordinate-based geometric morphometric methods cannot be easily extended to 3D data. Geometric methods for their analysis is still a developing area. Further improvements are needed to address an even broader field of problems with greater sophistication than it is possible today.

[Kendall (1977, 1984); Bookstein (1998); Caldecutt & Adams (1998); Bookstein et al. (1999); Rohlf (1999, 2006); Adams and Rohlf (2000); Gharaibeh et al. (2000); Klingenberg & Leamy (2001); Rüber & Adams (2001); Slice (2001); Rohlf & Bookstein (2003); Adams et al. (2004); Bai et al. (2010, 2011, 2012)]

5 Phylogenetic reconstruction based on morphology

The purpose of the following chapter is to give a brief introduction into modern phylogenetic approaches. An in-depth coverage of the theoretical background and current phylogenetic methods would be clearly beyond the aim of this contribution which is focused on the morphology and phylogeny of a specific group. The main aim of the following account of Hennigian and cladistic principles is to facilitate the understanding of the phylogenetic chapters in the second part of this book and to give a brief introduction into the cladistic evaluation of morphological data sets.

Taxonomy has a long tradition in biology. It was consolidated by the groundbreaking work of the Swedish botanist Carolus Linnaeus in the 18th century. However, the classification of insects and other groups of organisms above the species level lacked a sound scientific basis for a long time (e.g., Beutel et al. 2009). In the 1950s and 1960s the German dipterist Willy Hennig (Hennig 1950, 1966) introduced **phylogenetic systematics** ('Phylogenetische Systematik'), which profoundly revolutionized the methods for reconstructing the relationships of organisms. In its modern numerical form – **cladistics** – it still plays an outstanding role in phylogenetics, even though other approaches are now widely used for analyzing data sets predominantly or exclusively consisting of molecular data (e.g., maximum likelihood [ML], Bayesian inference [BI]). As this volume is dedicated to morphology, we will only treat the Hennigian and cladistic principles and terminology in this chapter. A parsimony-based cladistic approach is almost exclusively used in morphology based phylogenetic analyses, even though Bayesian inference was also applied in some recent studies (e.g., Beutel et al. 2011).

5.1 Hennigian terminology and principles

A crucial advantage of the Hennigian approach is the use of precisely defined terms (Richter & Meier 1994). This greatly reduced or eliminated the subjectivity of earlier systematic approaches. The most important terms will be defined and briefly explained in the following.

– **Monophyly**. A group is monophyletic if it consists of an ancestor (**last common ancestor: lca**, =**stem species**) and all of its descendant species.

In the traditional pre-Hennigian systematics (**"Typologie"**; see Mayr 1975: p. 67), a group was considered as monophyletic if its members share a common ancestor. This definition is inadequate as any two species would form monophyletic groups. For example, *Homo sapiens* and the fruitfly *Drosophila melanogaster* share the stem species of Bilateria as a common ancestor. Yet, under a pre-Hennigian concept of monophyly the two species could have been considered a monophyletic group.

A non-monophyletic group in outdated entomology textbooks is "Apterygota" (=primarily wingless insects) comprising Collembola, Protura, Diplura, Archaeognatha, and Zygentoma. These groups share a common ancestor, but it is also the ancestor of Pterygota.

Monophyletic groups (=**clades**, **monophyla**) always originate from a single stem species. A scenario with supraspecific taxa rising from a "basal" or "ancestral group" (or „stock") are misleading and based on an imprecise concept of evolution (e.g., Beutel & Komarek 2007). This does also imply that no species or group can be the ancestor of another species or group existing at the same time (e.g., the present).

– **Paraphyly** and **polyphyly**. Group based definition: a group is paraphyletic if all descendants of a common ancestor are included except one (Hennig 1950). All other non-monophyletic groups are polyphyletic.

Character based definition: a group is paraphyletic if it is only based on plesiomorphies. Polyphyletic groups are based on homoplasious characters.

"Apterygota" is one example of a paraphyletic group. The Pterygota, which share a common ancestor (lca) with the apterygote lineages, are not included in this assembly. "Halteria" (Diptera and Strepsiptera) are polyphyletic. The common ancestor of Diptera and Strepsiptera did not have halteres and is excluded from the group. In addition, it is also the stem species of all other holometabolan lineages, except for Hymenoptera (Niehuis et al. 2012).

– **Stem species.** The stem species is the last common ancestor (**lca**) of a monophyletic group. This term is closely related with the term **groundplan** (see below).

– **Sister groups or sister taxa.** Two species or monophyletic supraspecific taxa with a common ancestor (=stem species) not shared by any other species. Sister taxa form a monophyletic group.

– **Groundplan.** The groundplan is the complete set of plesiomorphic and apomorphic features (=character states) of the stem species (lca). The groundplan concept differs fundamentally from the 'Bauplan' used by typologists. The groundplan is the reconstructed sum of all features of a stem species that really existed. In contrast, the 'Bauplan' is an idealistic concept, i.e. the attempt to identify features which are "typical" ('Typologie') or "essential" for a group (see Mayr 1975). It lacks objectivity and should not be used in systematics.

– **Apomorphy** and **plesiomorphy**. An essential feature of Hennigian systematics is the exclusive use of derived features (apomorphies) for establishing systematic units, i.e. monophyletic groups. In contrast to earlier concepts ('Typologie'), non-homologous similarity and plesiomorphic ("primitive") features are irrelevant for phylogenetic reconstruction. An autapomorphy is a derived feature characterizing only a single taxon. Synapomorphies indicate that two or several species or supraspecific taxa form a monophyletic group (clade). Symplesiomorphies are shared "primitive" features. They are phylogenetically irrelevant.

An apomorphy is a feature which has been modified in evolution from the groundplan condition of the immediate ancestor, i.e. the character state (see **5.2 Cladistics** for a definition of character and character state) found in the stem species (lca). The plesiomorphic condition is identical with what is found in the last common ancestor. The terms apomorphic and plesiomorphic are relative. A feature can be plesiomorphic on a certain taxonomic level (e.g., absence of wings in Hexapoda), and apomorphic on another (e.g., absence of wings in some species of Orthoptera). In order to avoid confusion, it may be noted that a feature, which is a synapomorphy of several species constituting a higher ranking taxon (e.g., a genus), is generally an autapomorphy of this more inclusive group.

The paraphyletic "Apterygota" share the absence of wings as a common feature. As this is plesiomorphic (if Hexapoda as a whole are under consideration) this is no argument for a monophyletic origin of the primarily wingless hexapods. The "Thysanura" are also paraphyletic. Archaeognatha and Zygentoma are very similar in their appearance and share a considerable number of features such as ectognathous mouthparts, the absence of wings, and the presence of coxal vesicles, styli, cerci, and the terminal filament. However, all these features are plesiomorphic, i.e. they belong to the groundplan of Insecta and are therefore irrelevant as arguments for a monophyletic origin.

– **Crown group** (Fig. 5.1.1): monophyletic group containing the last common ancestor of the extant representatives of a taxon and all its extinct and extant descendants.

– **Stem group** (=**stem lineage**) (Fig. 5.1.1): all extinct members of a taxon branching off before the last common ancestor of the crown group. None of them is more closely related with an extant member of the taxon. Stem groups or stem lineages are usually not monophyletic.

The extinct Carboniferous †Monura (Fig. 6.A.2) belong to the stem group of Insecta. The presence of a specific type of antenna and an ovipositor formed by specific elements of the abdominal segments VIII and IX are apomorphies shared with Insecta, whereas undivided tarsi are a plesiomorphy compared to the condition in the groundplan of the latter group, i.e. tarsi with at least 3 tarsomeres.

Phylogenetic reconstruction was a stepwise procedure in pre-cladistic phylogenetic systematists (Hennigian approach). Based on putative apomorphies sistergroup relationships were established for successively more inclusive groups, finally leading to a phylogenetic tree or **cladogram** ("argumentation plan of phylogenetic systematics"; Hennig 1966). The principle of **parsimony** was only applied informally, and Hennig and earlier systematists did not use a formal procedure to determine the

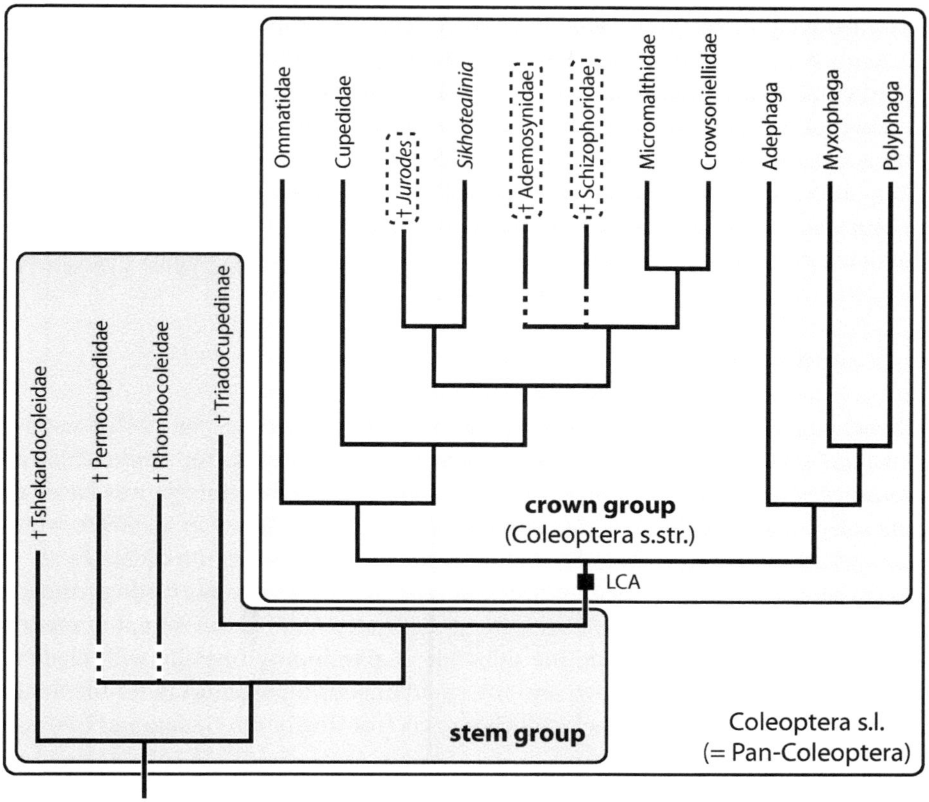

Fig. 5.1.1: Illustration of crown group/stem group concept (following Sudhaus & Rehfeld 1992) using the example of Coleoptera (tree based on Beutel et al. 2008). Crown group comprises all extant and extinct (brocken lined boxes) descendants of the last common ancestor (LCA) of the extant representatives. The stem group summarizes all non-crown group fossils.

polarity of characters, i.e. to identify plesiomorphic and apomorphic character states.

– **Polarity determination.** The outgroup comparison method was formally introduced by Watrous & Wheeler (1981) (see also Maddison et al. 1984; Nixon & Carpenter 1993). It became the by far most important method to determine the polarity of morphological characters ('Lesrichtung'), i.e. to determine whether a state is plesiomorphic or apomorphic. Note that in principle the **outgroup** comprises all taxa not belonging to the group under consideration (**ingroup**, see also below). Today character polarities are no longer determined a priori. Instead, selected outgroup taxa are included in numerical analyses and trees are rooted between the ingroup and outgroup (Nixon & Carpenter 1993).

In principle the ontogenetic method (see e.g., Minelli 1993) can also be used to root trees (Meier 1997). However, it is rarely applied as the practical application requires information on ontogenetic series. The paleontological or stratigraphical criterion is of very limited use for the polarity determination. The assumption that older forms exhibit more plesiomorphic character states is supported by common sense (Minelli 1993), but this criterion is unreliable as pointed out by Eldredge & Crawcraft (1980) (see also Minelli 1993; Wägele 2001). This is partly due to gaps in the known fossil record.

5.2 Cladistics

Phenetic methods or numerical taxonomy (e.g., Mitchener & Sokal 1957; Sneath & Sokal 1973) were an early approach to use quantitative methods for a more objective classification of organisms (see e.g., Mayr 1975). Whereas this concept was based on quantified overall similarity, the cladistic approach (e.g., Forey et al. 1992), which emerged in the 1970s, is rooted in the Hennigian principles, i.e. the exclusive use of apomorphies to establish monophyletic groups. The main features distinguishing it from the classical approach ('Phylogenetische Systematik') is the formal numerical character evaluation following the principle of **parsimony** (usually with cladistic computer programs, see below) and the simultaneous a posteriori (after the analysis) assessment of the polarity for all characters (**rooting** of the cladogram) (see e.g., Forey et al. 1992).

The cladistic evaluation of morphological characters can be divided into several successive steps which are briefly outlined in the following:

5.2.1 Taxon sampling

The crucial first step is the selection of ingroup and outgroup taxa. In contrast to the Hennigian approach precisely defined taxa (**terminals, terminal taxa**) are chosen for the analysis. Terminals can be species (exemplar approach) or supraspecific taxa which have to be monophyletic (groundplan approach) (Bininda-Emonds et al. 1998; Prendini 2001).

The ingroup is the taxon under consideration. Erroneously assumed monophyly can have serious negative consequences on obtained cladograms. The delimitation is usually based on previously published results, but may be unclear, especially on a lower taxonomic level. If this is the case, all taxa potentially belonging to the ingroup and multiple outgroups should be included in the sampling. A complete coverage of all species is the ideal case. However, for practical reasons this is usually not possible in Hexapoda, which contain many groups with a very high diversity. Selected terminals should not display a high number of autapomorphic features ("specialized taxa"). Long branches indicating a large number of evolutionary changes can cause problems in phylogenetic analyses of morphological and molecular data (**long branch attraction**; see e.g., Bergsten 2005).

The choice of outgroup taxa is a crucial step in cladistic investigations. An inappropriate selection can have serious negative effects on the outcome of the analyses. The outgroup terminals should be closely related to the ingroup to ensure that many features can be homologized between in- and outgroup taxa. The presumptive sister group is usually included but this is not mandatory, and sometimes the closest relative of the ingroup is disputed or unknown. Like with ingroups, taxa with many apomorphies should be avoided, and multiple outgroups should be included, which allows testing the monophyly of the ingroup. An "artificial outgroup" consisting of all putatively ancestral states for the ingroup (groundplan) is not recommended because it requires an intuitive a priori polarity assessments for all characters, which is in contrast to the cladistic concept.

5.2.2 Selection of characters

Given equal quality, a larger number of characters leads to a better resolution and higher support values (like a denser taxon sampling). A broad spectrum covering different body parts such as the adult head, thorax, wing venation, abdomen, genitalia and also external and internal larval features usually leads to good result in morphology-based phylogenetic analyses of hexapod groups (e.g., Beutel et al. 2011). The selection of suitable character systems plays a crucial role. Larval chaetotaxy (patterns of setae, pores and sensilla) and other cuticular surface patterns are fast evolving systems comparable with genes with a high substitution rate (e.g., COI). These characters may work well at the species and generic level in some groups (splitting events in the Cenozoic), but a high degree of homoplasy can impede their use at higher taxonomic levels. The phylogenetic signal will be largely or completely obliterated by "noise". In contrast, features of the musculature or the general configuration of the mouthparts tend to be more conservative, comparable to genes with a slow rate of evolution (e.g., protein-encoding nuclear genes). They are usually suitable for the reconstruction of the phylogeny on higher levels (e.g., interordinal relationships, splitting events in the earlier Mesozoic and Paleozoic). On a low taxonomic level these characters would be largely constant and not provide phylogenetic signal.

Normally only informative characters should be used, i.e. no characters with the same state for all terminals (constant characters) and no highly variable characters with no identical states shared by two or several of the included taxa (only autapomorphies).

An important aspect of morphology-based phylogenetic analyses is the documentation of data, which is often neglected when characters are mainly or exclusively extracted from the literature. New techniques which became available in the last decade (e.g., Confocal laser scanning microscopy [CLSM], Micro-computer tomography [µ-CT], computer-based 3D reconstruction; see **4. Traditional and modern techniques in insect morphology**) have greatly accelerated the acquisition and documentation of high quality morphological data. It should be attempted to store images of all entries (character states) in a data matrix (see below) and in a suitable data bank (e.g., MorphDBase, www.morphdbase.de/). These data should be made freely accessible to other scientists.

5.2.3 Character state coding and building a list of characters

After determining topological homology (Brower & Schawaroch 1993), the primary homology statements (de Pinna 1991) are coded as characters and character states. A character and its states codify the primary homology hypotheses (e.g., hindleg; forewing) and the character states are different variants of it (e.g., [0] cylindrical/ [1] broadened; [0] transparent/ [1] sclerotized). A careful definition of characters and character states is essential in morphology-based phylogenetic reconstruction.

Similarity (or congruence) of structures in different terminal taxa play a crucial role in phylogenetic investigations and there are three different explanations for similarity: synapomorphy, symplesiomorphy, and **homoplasy**. Only synapomorphy reflects relationship (or affinity) in a phylogenetic sense (monophyly). Homoplasy can be due to reversal (e.g., secondary loss of wings in a pterygote taxon, e.g., Grylloblattodea) or convergence (piercing sucking mouthparts in hemipterans and biting flies).

Characters can be discrete (qualitative) or continuous (quantitative). Only the former type will be treated in the following. Binary characters have only two states (e.g., forewings: [0] absent; [1] present) whereas multistate characters have three or more states (e.g., number of labial palpomeres: [0] 1; [1] 2; [2] 3; [3] 5). All states coded for a single character have to be directly comparable and refer to potentially homologous structures / conditions. Mixed codings (e.g., tegula: [0] absent; [1] enlarged, rounded; [2] reduced; [3] sickle-shaped, bearing long hairs) are inapproprioate.

Multistate characters can be broken down into binary characters (e.g., 2. Forewings: [0] present; [1] absent. 3. Sclerotisation of forewings: [0] unsclerotized; [1] sclerotized [absence coded as inapplicable (–)]). However, this approach is only appropriate if the characters are ordered (see below) (Meier 1994).

Characters in cladistic analyses are usually treated as equally weighted, i.e. each character is assigned the same weight. A priori weighting is problematic as there is no objective basis for determining the weight. However, with implied weighting there are numerical techniques for an unequal treatment of characters based on observed homoplasy (Goloboff 2005). A more objective way of weighting is to break down complex character systems into several binary characters (see above). This procedure reflects the complexity of a character.

Multistate characters are usually treated as unordered or non-additive, i.e. transformations between each of the states are possible. If an intermediate state '2' can be identified linking character states '1' and '3' characters can be treated as ordered or additive. Subsequently, the transition from state '1' to state '3' would require two steps (instead of one in unordered coding).

A phenomenon analyzed recently using basal pterygote insects ("the Palaeoptera problem" is **concerted convergence** (Blanke et al. 2013). It was shown that entire series of characters have evolved convergently under the same or similar selective pressures. This phenomenon can lead to serious errors in phylogenetic reconstruction. A formal approach to detect the occurrence of concerted convergence was developed by Holland et al. (2010).

The **consistency index** (**CI**) of a character indicates the degree of observed homoplasy. Non-homoplasious apomorphies appearing only once on a cladogram have a consistency index of 1, those appearing twice on separate branches 0.5, and those appearing four times 0.25 etc. (the CI decreases when homoplasy increases). Other measures for homoplasy are the retention index (RI) or the rescaled consistency index (RCI) (Farris 1989). Note that the comparison of CI values is problematic. It is strongly correlated with the number of the included taxa.

5.2.4 Data matrix

Characters and their states are arranged in a **data matrix**. Conventionally the taxa (terminals) are arranged in rows and the characters in columns. The character states are primary homology hypotheses which are tested in the analyses. Several software applications are available for entering and editing the data (see below).

Data matrices are a central element of the cladistic approach but are also very useful in other contexts, such as for instance in taxonomic studies without numerical character evaluation. They generally greatly increase the transparency by assigning features precisely to specific taxa.

5.2.5 Parsimony analyses

Even though likelihood methods (**ML, BI**) based on evolutionary models are widely applied in phylogenetic analyses based on molecular data, maximum parsimony (**MP**) is still the most commonly used approach for analyses of morphological character sets. It is based on the assumption of parsimony in evolution and basically goes back to Ockham's razor: the explanation requiring the lowest number of changes (the simplest theory) is regarded as the most likely one and should be preferred.

In the cladistic approach parsimony analyses of data matrices are carried out using computer software. Cladistic algorithms are used to calculate the branching pattern(s) (trees, cladograms) explaining the distribution of a given set of character states with a minimum number of character state transformations (or evolutionary steps) (see e.g., Forey et al. [1992] for different parsimony criteria such as Wagner-, Fitch-, Camin-Sokal-, polymorphism-, generalized parsimony). For smaller data sets (approximately 20 terminal taxa) exact methods (**exhaustive search**, **branch-and-bound search**) can be used. They guarantee the discovery of all trees with a minimal number of steps (optimal trees). For larger data sets **heuristic methods** are used. They include the possibility that not all or none of the optimal trees are found but only a local minimum.

Earlier cladistic computer software are **Hennig86** (Farris 1986), **PAUP** (Macinthosh, PowerMac, Windows and Unix/OpenVMS versions) (Swofford 1998), and **NONA** (Goloboff 1995). However, they are all superseded by **TNT** (Goloboff et al. 2003, 2008). NONA and TNT are available at the homepage of the **Willi Hennig Society**: http://www.cladistics.org/). TNT is suitable for analyzing very large data sets. Programs for entering data and analyzing trees are **MacClade** (Macintosh), **WinClada**, and **Mesquite** (Maddison & Maddison 2003, 2005).

In contrast to the traditional Hennigian procedure, the character state polarity is not determined a priori of (before the analysis). Apomorphic states are instead identified through **rooting** of the phylogenetic network after the analysis (a posteriori). In a first step an **unrooted network** is obtained. The polarity becomes implicit simultaneously for all characters through the position of the root, i.e. the selection of the outgroup.

5.2.6 Trees, their presentation and statistics

In morphology-based systematic investigations the phylogenetic branching patterns are usually presented as rooted **cladograms**, which do not represent the relative amount of evolutionary changes (apomorphies) or time. The branch length is proportional to the amount of character change in **phylograms** and to evolutionary time in **chronograms.** The general term for phylogenetic trees is **dendrogram.**

Characters and their derived states (apomorphies) should be mapped on trees in morphology-based phylogenetic studies. Characters can be optimized using software that is also used for entering and editing data (e.g., WinClada, see above). In addition to that, lists of apomorphies (with corresponding plesiomorphies) should be presented. Non-homoplasious changes (CI: 1) should be highlighted. Different categories of apomorphies can occur depending on whether there are different ways to optimize the same character onto a tree. **Unambiguous changes** are sometimes highlighted in bold compared to changes with ambiguous optimizations. The most commonly inspected optimizations are based on the accelerated (**ACCTRAN**) (preferring reversal) or delayed transformation mode (**DELTRAN**) (preferring convergence). It has to be noted that these optimizations do not affect the length (=number of steps) or topology of the resulting tree(s) but help to solve ambiguous changes and to map apomorphies on the cladogram.

Consensus trees summarize the results of cladistics analyses, which often result in several or even numerous cladograms of equal length (most parsimonious trees) but different topologies. Only the branches supported in all trees are displayed in a **strict consensus tree**. Branches occurring in 50% or more of the minimum length cladograms are shown on a **majority rule consensus tree**.

Consistency indices of trees are a measure for homoplasy (see **5.2.3**: CI of characters). It is 1 for a tree with no homoplasy at all and would be 0.75 if 50% of the characters have a CI of 0.5 (CI of 1 in the other 50%), or 0.5 if all characters have a CI of 0.5.

The **Bremer support** (or branch support value) is frequently used in morphology-based studies, and also in molecular investigations (Bremer 1994). Certain limitations were pointed out in a study by DeBrye (2001). The branch support value of a clade is determined by calculating the strict consensus tree of the most parsimonious trees and all trees with a defined number of additional steps until the specific branch collapses. Clades collapsing with the addition of a single step have a branch support value (BS) of 0, while all other clades have a BS of at least 1. The latter will be tested by subsequently adding more additional steps. Dichotomies collapsing with the addition of 2 steps have a BS of 1 and so on. A function to obtain the Bremer support automatically is implemented in TNT.

Bootstrapping and **jackknifing** are techniques to obtain support indices (Felsenstein 1983; Farris et al. 1996). They are widely used, especially in molecular systematics. Their value in morphology-based phylogenetic investigations is sometimes compromised due to the small number of characters. In bootstrap analyses columns of a matrix are randomly sampled to build a bootstrap data set of the same size. A large number of resampled matrices are built and analyzed (>100). Majority consensus trees are used to summarize how often a particular branch is found when the resampled matrices are analyzed. In jackknifing resampled data sets are created via the deletion of characters (usually e^{-1}). The summary of the support values is as in bootstrapping. Branches with bootstrap or jackknife values below 80% are generally considered as weakly supported.

[Hennig (1950, 1966); Mitchener & Sokal (1957); Sneath & Sokal (1973); Mayr (1975); Eldredge & Crawcraft (1980); Watrous & Wheeler (1981); Maddison et al. (1984); Farris (1988, 1989); Felsenstein (1983); Swofford (1998); de Pinna (1991); Forey et al. (1992); Sudhaus & Rehfeld (1992); Minelli (1993); Nixon & Carpenter (1993); Bremer (1994); Goloboff (1995, 2005); Meier (1997); Richter & Meier (1994); Brower & Schawaroch (1996); Farris et al. (1996); Bininda-Emonds et al. (1998); Swoffort (1998); Prendini (2001); Wägele (2001); DeBry (2001); Goloboff et al. (2003, 2008); Maddison & Maddison (2003, 2005); Bergsten (2005); Komarek & Beutel (2007); Beutel et al. (2008a, 2011); Schuh & Brower (2009); Holland et al. (2010); Niehuis et al. (2012); Blanke et al. (2013)]

6 The orders of Hexapoda

The position of Hexapoda. The systematic relationships between hexapods and the other extant groups of Euarthropoda (Chelicerata, Crustacea, Myriapoda) have been discussed intensively in the last ten years. This issue plays a crucial role with respect to the monophyly of Hexapoda, the reconstruction of the groundplan, and the interpretation of the early evolutionary history. The oldest definite hexapod fossils are described from the famous Early Devonian Rhynie Chert of Scotland (ca. 408 Ma; Grimaldi 2009). However, recent molecular studies suggest that the group is actually much older and probably originated in the Cambrian (542–488 Ma) (N. Wahlberg, pers. comm.). Considering this long evolutionary history before the rise of the terrestrial crown group Hexapoda, it is surprising that up to now no primarily aquatic members of the stem group were identified. †*Devonohexapodus bocksbergiensis* from the Hunsrück Slate in West Germany was interpreted as a stem group hexapod by Haas et al. (2003). However, the species was synonymized with †*Wingertshellicus backesi* by Kühl & Rust (2009) and it was clearly demonstrated by these authors that the suggested phylogenetic affinity with Hexapoda was based on misinterpretations. †*Wingertshellicus* likely belongs to the stem group of the entire Euarthropoda (Kühl & Rust 2009).

Apparently, the fossil record does not help to elucidate the position of Hexapoda. It was a long accepted hypothesis that insects in the widest sense form a monophyletic group Tracheata (=Antennata or Atelocerata) together with Myriapoda. This implied that the crucial change from aquatic to terrestrial habitats was accomplished by a common ancestor of both lineages in the Silurian (443–416 Ma). Consequently, almost all arguments suggesting a clade Tracheata were adaptations to the terrestrial life style, such as for instance the tracheal system (extremely variable in myriapods, absent or vestigial in some basal hexapods), the Malpighian tubules (absent in Collembola), the loss of the second antenna (locomotor organ of planktonic crustacean nauplius larvae), the loss of the ventral food rim (only suitable for food uptake in the aquatic environment), and the sperm transmission via a spermatophore (also in some crustacean groups). Analyses of different molecular data sets published in the last decade unequivocally refute Tracheata and support a monophyletic lineage Pancrustacea or Tetraconata comprising Hexapoda and crustaceans (e.g., Giribet et al. 2001; Regier et al. 2010; Meusemann et al. 2010; von Reumont et al. 2011). The name Tetraconata refers to a four-partite crystalline cone of the ommatidia of the compound eyes, one of few presumptive morphological autapomorphies of the group (Richter 2002). However, a similar condition does also occur in scutigeromorph myriapods (Müller et al. 2003).

Whereas hexapods were placed as the sistergroup of a clade Crustacea in some earlier studies (e.g., Giribet et al. 2002), the recent phylogenomic investigations place them as a subordinate group of Tetraconata (Regier et al. 2010; Meusemann et al. 2010; von Reumont et al. 2012). However, their precise position within this lineage is not unequivocally clarified yet. Interestingly, a detailed investigation of brain structures by Fanenbruck et al. (2004) suggests a clade comprising Hexapods, the "higher crustaceans" (Malacostraca), and the mysterious cave-dwelling Remipedia. Fanenbruck et al. (2004) identified the presence of lateral neuropils of the 1st antenna (antennula), olfactory lobes with glomeruli, and a medulla terminalis with hemiellipsoid bodies as evolutionary novelties and potential synapomorphies of the three groups. Partly congruent with this hypothesis, Regier et al. (2010) obtained a clade comprising Hexapoda, Remipedia, and the miniaturized Cephalocarida, but closer affinities with branchiopod crustaceans were suggested by Meusemann et al. (2010). As pointed out above, the precise placement of hexapods within Pancrustacea remains to be clarified.

Independent of the issue of the closest relatives of Hexapoda, the well founded Tetraconata- or Pancrustacea-hypothesis suggests an evolutionary scenario completely different from that implied by the Tracheata-concept. Apparently ancestral hexapods have acquired terrestrial habits later than the myriapods, probably in the early Devonian, and have acquired all features linked with this change of habitat independently (see below).

The monophyly of Hexapoda. The monophyly of Hexapoda (Fig. 6.A: 1) was rarely challenged and this is not surprising considering the uniformity of the group with respect to the general body organization and the appendages. The most conspicuous presumptive autapomorphy is the subdivision of the postcephalic body into a three-segmented thorax and a primarily 11-segmented abdomen. The functional background of this – a far-reaching division of labor – is arguably one factor which has contributed to the unparalleled diversification of the group. The locomotor functions and the associated structures (legs, wings, strongly developed muscles) are concentrated in the thoracic segments, whereas the abdomen contains the major part of the digestive tract, the excretory organs (Malpighian tubules), a strongly developed fat body (storage and mobilization of energy-rich-substances), and the genital organs.

Based on analyses of the entire mitochondrial genome, Nardi et al. (2003) postulated the non-monophyly of Hexapoda, with Collembola placed among some crustacean taxa. However, this study is problematic in several ways. The taxon sampling is insufficient as crucial groups like Protura and Diplura are missing. Not only Collembola, but also a louse and the honey bee were placed outside of Hexapoda, which is highly implausible. The mitochondrial genome apparently evolves too fast to reconstruct splitting events that took place in the Devonian or even earlier. Considering the unequivocal support of Hexapoda in recent studies based on single genes (e.g., Kjer et al. 2006) and phylogenomic data (Regier et al. 2010; Meusemann et al. 2010) and the morphological evidence (e.g., Klass 2009; Grimaldi 2010), we consider the monophyly of Hexapoda as sufficiently supported. It should be emphasized in this context that the status as a subordinate group of Tetraconata adds an entire series of new potential autapomorphies to the traditional ones (see below).

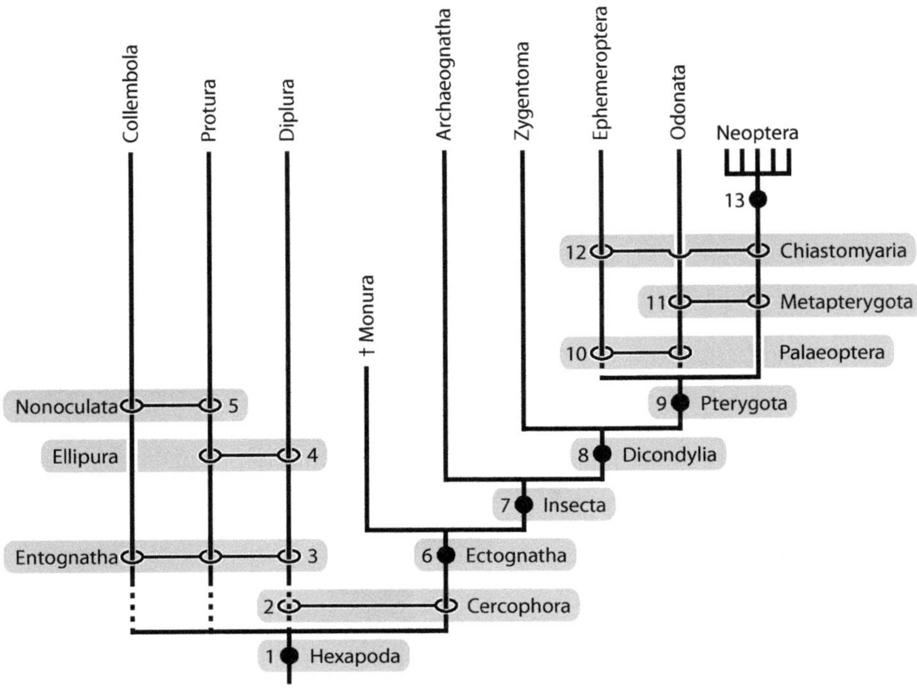

Fig. 6.A: Cladogram showing basal branching pattern in Hexapoda, with selected potential apomorphies. **1**, postcephalic body divided into 3-segmented thorax and 11-segmented abdomen, 6-segmented legs (patella reduced), tracheal system with lateral spiracles and tracheoles directly supplying organs with oxygen (groundplan?), Malpighian tubules (absent in Collembola), 2nd antenna absent, fused 2nd maxillae forming true labium, ventral food rim absent, postcephalic gnathocoxae absent, expopodites reduced (styli?), no primary aquatic larva, loss of mandibular palp, midgut gland absent; **2**, cerci, paired claws, microtubuli pattern of sperm axonoeme $9 + 9 \times 2 + 2$, loss of the temporal organ; **3**, entognathous mouthparts, compound eyes distinctly reduced or absent; **4**, advanced entognathous condition (paired mouthparts in separate pouches), linea ventralis, abdominal spiracles missing (hexapod groundplan?); **5**, eyes absent; **6**, flagellar antenna, Johnston's organ (?), tentorial bridge (?), large leg-like maxillary palp (groundplan), pretarsus largely reduced, ovipositor, terminal filament; **7**, subdivision of tarsus, elongated gonapophyses; **8**, secondary mandibular joint, anterior tentorial arms with plate-like connection, tracheal system with anastomoses, closed amniotic cavity; **9**, wings and associated structures; **10**, shortened, bristle-like antennae, loss of antennal circulatory organs (adults), dentisetae of lacinia, larval galea and lacinia fused; **11**, no subimago, secondary mandibular joint present as ball-and-socket joint, tracheae of wings and legs connected with spiracle of following segment, terminal filament absent, abdominal spiracles with closing muscles; **12**, flight apparatus with indirect flight muscles, internal insemination with postabdominal penis; **13**, ability to flex wings.

Autapomorphies of Hexapoda

- Postcephalic body divided into a 3-segmented thorax and an 11-segmented abdomen
- Abdominal appendages modified or reduced, segment X always without appendages

This character is obviously correlated with the specific tagmosis of Hexapoda.

- 6-segmented legs (patella reduced)

Additional potential autapomorphies implied by the Pancrustacea concept

- tracheal system with lateral spiracles and tracheoles directly supplying organs with oxygen (weakly developed or absent in apterygote hexapods)
- Malpighian tubules attached at the midgut-hindgut border (absent in Collembola)
- 2nd antenna absent
- fused 2nd maxillae form a true labium (proximal elements unpaired)
- ventral food rim absent, postcephalic gnathocoxae completely reduced
- expopodites distinctly or completely reduced (styli as they occur in apterygote lineages may represent vestigial exopodites)
- sperm transfer via spermatophore
- no primary aquatic larva
- loss of the mandibular palp
- reduction of the midgut gland
- reduction of the nephridial organs

[Giribet et al. (2001); Richter (2002); Haas et al. (2003); Müller et al. (2003); Nardi et al. (2003); Fanenbruck et al. (2004); Kjer et al. (2006); Klass (2009); Kühl & Rust (2009); Grimaldi (2010); Meusemann et al. (2010); Regier et al. (2010); von Reumont et al. (2012)]

The entognathous orders. The interrelationships of the three entognathous orders are still discussed controversially. Hennig (1969) suggested a clade **Entognatha** based on entognathism (mandibles and first maxillae sunk in gnathal pouches) and some reductional features such as the complete or partial reduction of the compound eyes and the Malpighian tubules (Fig. 6.A: 3). It is apparent that these apomorphies could be partly related to living in soil or similar substrates, and partly to miniaturization. It was pointed out in recent morphological studies that the entognathous condition differs distinctly in Diplura on one hand, and in Protura and Collembola on the other (Koch 2001). The specific condition found in the latter two orders is an additional argument that they form a clade **Ellipura** (Fig. 6.A: 4), which is also characterized by the absence of abdominal spiracles (also missing in Campodeidae). A monophyletic group **Cercophora** Willmann, 2003 (= Euentomata Imms, 1936; = Insecta Kukalová-Peck 1991), i.e. a sister group relationship between Diplura and Ectognatha, implies the non-monophyly of Entognatha (Fig. 6.A: 2). It is supported by the following features:

- Appendages of ultimate trunk segment differentiated as cerci (Cercophora)
- Paired claws
- Microtubuli pattern 9 + 9 × 2 + 2 in the sperm axoneme (additional outer ring of nine microtubules)
- Loss of the temporal organ (Tömösvary organ, pseudoculus)

In contrast to this concept, most recent molecular studies (e.g., Meusemann et al. 2010) support a clade **Nonoculata** including the eyeless Diplura and Protura (Fig. 6.A: 5). The complete loss of light sense organs is a potential synapomorphy of both groups. However, well-developed eyes were apparently present in the extinct †*Testajapyx*, a genus possibly related to extant Japygidae (Diplura) (s. below).

[Imms (1936); Hennig (1969); Kraus & Kraus (1984); Kukalová-Peck (1991); Friedrich & Tautz (1995); Koch (2001); Beutel & Gorb (2001); Giribet et al. (2001); Richter (2002); Nardi et al. (2003); Willmann (2003); Bitsch & Bitsch (2004); Fanenbruck et al. (2004); Meusemann et al. (2010); Regier et al. (2010)]

6.1 Collembola (Greek *kolla* = glue, *embolon* = peg, English common name: springtails)

Diversity and distribution. Collembola were already described by naturalists of the 17th century. With almost 8,000 known species they are by far the most diverse order of the apterygote (=primarily wingless) insects. The distribution is worldwide, including arctic and Antarctic areas, and also high mountain regions. Many endemic species are known, especially from caves, islands and Antarctica.

Autapomorphies
- Jumping capacity
- Six abdominal segments (Figs 6.1.1, 6.1.2)
- Abdominal ganglia concentrated into single nerve mass, fused to metathoracic ganglia
- Genital opening at abdominal segment V
- Ventral tube (Figs 6.1.1, 6.1.2)
- Retincalum (Fig. 6.1.2)
- Furculum (groundplan) (Figs 6.1.1, 6.1.2)
- Malpighian tubules absent (Fig. 6.1.2)

Taxonomy. The group is subdivided into four suborders, the Poduromorpha, Entomobryomorpha, Neelipleona, and Symphypleona. Poduromorpha currently comprise 11 families (e.g., Hypogastruridae, Onychiuridae, Neanuridae), Entomobryomorpha also 11 (e.g., Isotomidae, Entomobryidae, Tomoceridae), Neelipleona only a single family (Neelidae), and Symphypleona nine (e.g., Sminthuridae, Bourletiellidae).

Fig. 6.1.1: Collembola (Isotomidae), SEM micrograph, lateral view. Abbr.: furc – furculum, ret – retinaculum, tita – tibiotarsus, vetu – ventral tube. Scale bar: 500 μm. Courtesy H. Pohl.

Diagnosis. Usually small or very small and wingless (Fig. 6.1.1). Antennae 4-segmented. Pronotum often reduced, particularly in Entomobryomorpha. Abdomen short, 6-segmented, bearing ventral tube at its base and usually a small retinaculum and a strongly developed furculum. Usually with jumping capacity.

Morphology. Total length of most species 1–3 mm, but 9 mm in *Tetrodontophora bielanensis* (Onychiuridae) and 17 mm in *Holacanthella duospinosa* (Neanuridae). The body is either moderately elongate (Figs 6.1.1, 6.1.2) or very compact and almost globular (Sminthuridae). The cuticle is weakly sclerotized. The color is often whitish but strong pigmentation occurs (hypodermal cells), sometimes resulting in conspicuous color patterns. Tubercles or cuticular scales are present (Fig. 6.1.3) or absent. Many different types of setae occur.

The head (Figs 6.1.4) is usually orthognathous or subprognathous but can be prognathous in more specialized forms (e.g., *Neanura*). It is compact or moderately elongate, and the apical region bearing the preoral opening is cone-like in forms with sucking-piercing mouthparts (e.g., Neanuridae). The neck region is usually exposed. Compound eyes are distinctly reduced (maximum number of ommatidia eight) or absent; in addition to isolated ommatidia, up to six ocelli can occur (laterally, medially and dorsally), but they are always subepidermal, devoid of a cornea, and usually unpigmented. Temporal organs (Tömösvary organs, postantennal organs, pseudoculus) are usually present (especially in species living in soil), but absent in some genera of Hypogastruridae, Entomobryidae, Tomoceridae, and Sminthuridae. The homology of head sutures is problematic. A clypeofrontal transverse ridge is present at least in

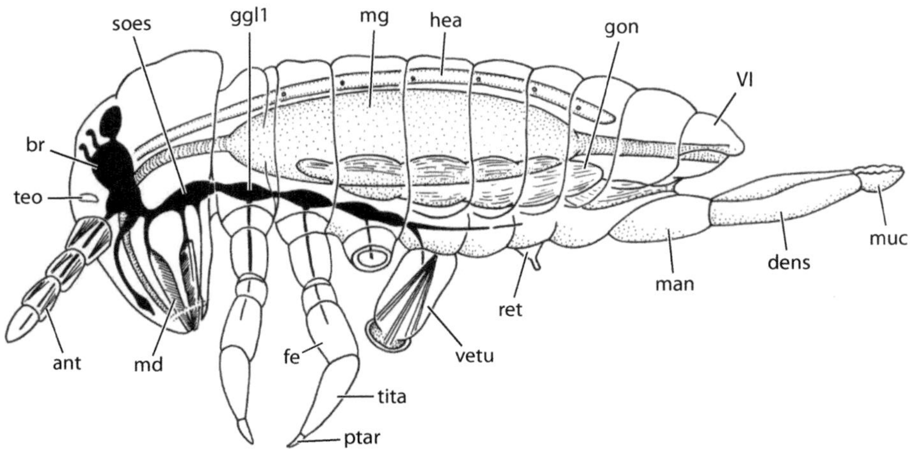

Fig. 6.1.2: Collembola, habitus and anatomy, lateral view. Abbr.: ant – antenna, br – brain, dens – dens of furculum, fe – femur, ggl1 – prothoracic ganglion, gon – internal genital organs, hea – heart, md – mandibles, mg – midgut, man – manubrium, muc – mucro, ptar – pretarsus, ret – retinaculum, soes – suboesophageal ganglion, teo – temporal organ (Tömösvary organ), tita – tibiotarsus, vetu – ventral tube, VI – segment VI. Redrawn from Dunger (2005), after Weber (1949).

some cases. The antennae are primarily 4-segmented, but antennomeres can be secondarily subdivided; all segments except the apical one are equipped with muscles; as in the other entognathous orders Johnston's organ is missing. Mandibles and maxillae are enclosed by extensive head duplicatures (plicae orales). A ventrome-

Fig. 6.1.3: Collembola, cuticular surface structure, SEM micrograph. Scale bar: 2 μm.

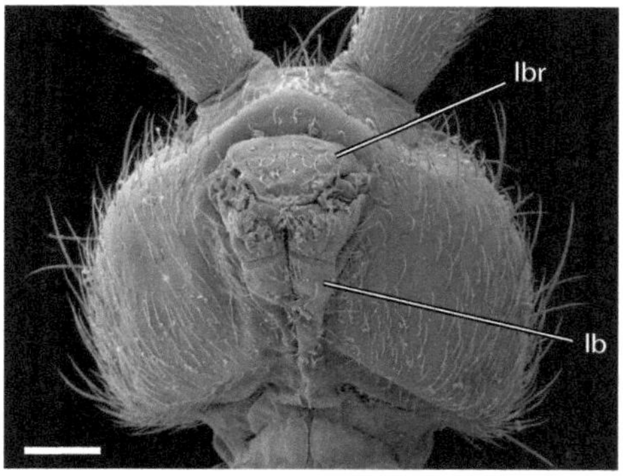

Fig. 6.1.4: Collembola (Isotomidae), SEM micrograph, head ventral view. Abbr.: lb – labium, lbr – labrum. Scale bar: 100 μm. Courtesy H. Pohl.

dian furrow, the linea ventralis, extends from the labium over the thoracic sternites towards the ventral tube; in some species, the linea ventralis additionally forms an elaborate system of branching grooves on the head. The mandibles and maxillae on each side are inserted in separate pouches. Most collembolans have biting mandibles with a molar part and incisor teeth (groundplan), but the mandibles can be elongated, acuminate and devoid of a molar region in species with a cone-shaped apical head region (e.g., *Neanura*), or even completely reduced. The maxillary palp is membranous and 1-segmented; the mala (maxillary head, capitulum) often bears a toothed outer lobe and fimbriate inner lobes, the latter specialized for sweeping small food particles (e.g., fungal spores) towards the mouth opening. Paired distal parts of the labium border the mouth cone ventrally and bear remnants of the labial palps (unsegmented setiferous lobes). Labial nephridia and salivary glands of the labial segment jointly open via the median cleft between the paired distal halves of the labium into the preoral chamber or linea ventralis, respectively; additional salivary glands are usually located in the genal region and open into the gnathal pouches (enlarged in some piercing-sucking species). The head endoskeleton comprises paired cuticular rods that are likely homologous to the tentorial apodemes in insects, and an elaborate system of interconnected ligamentous (=collagenous) structures ("pseudotentorium") held in position by muscles. A pair of spiracles is sometimes present in the cervical region (Actaletidae, most Symphypleona), but they are completely missing in all poduromorphan and most entomobryomorphan taxa.

The legs and other thoracic elements are weakly sclerotized. The pronotum is well-developed in Poduromorpha, but reduced in Entomobryomorpha and Symphypleona (Fig. 6.1.1). The sternites and pleurites are often very indistinct or not recognizable, whereas the tergites are usually more strongly sclerotized than the surrounding areas. The prolegs are shorter than the middle and hindlegs; the coxa and trochanter

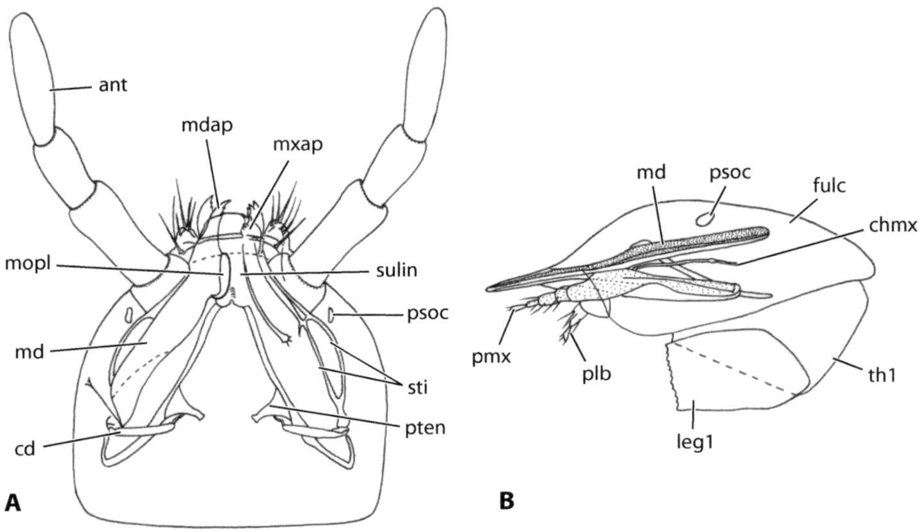

Fig. 6.1.5: Collembola, head, dorsal view and lateral view (transparent cuticle). A, ventral view; B, lateral view. Abbr.: ant – antenna, cd – cardo, chmx – channel of maxillary gland, fulc – fulcrum, leg1 – fore leg, md – mandibles, mdap – mandibular apex, mopl – molar plate, mxap – maxillary apex, plb – palpus labialis, pmx – palpus maxillaris, psoc – pseudoculus (temporal organ), pten – posterior tentorium, sti – stipes, sulin – superlinguae, th1 – prothorax. Redrawn from Naumann et al. (1991).

and the femur are often subdivided; the tibiotarsal joint is usually reduced. Distally, the elongate tibiotarsus bears the pretarsus; it is equipped with an unpaired claw, which is often serrate, and usually bears an empodium and specialized, club-shaped setae; these structures together form a cleaning apparatus. Spiracles are missing on the thorax and abdomen.

The abdomen is composed of only six segments (Figs 6.1.1, 6.1.2). Segments V and VI are usually reduced in size and tend to fuse. The segmental borders are hardly recognizable in the Symphypleona. Segment IV is often enlarged in species with jumping capacity. The weakly sclerotized tergites are extended to the lateral regions of the segments. The appendages on the ventral side of the abdomen are the most conspicuous autapomorphies of Collembola. Segment I bears the unpaired ventral tube, which is composed of a basal plate, a tube-like element with valves, and paired extrusible apical vesicles, which can be strongly elongated by haemolymph pressure in some taxa; the ventral tube is operated by 14 pairs of muscles. Segment III bears the small retinaculum, which holds the furculum in its resting position with its paired distal rami. The furculum of segment IV is composed of a basal manubrium and paired distal structures, the dens, which may bear apical mucrones. External male and female genitalia are missing. The genital opening lies on the ventral side of segment V. Segment VI bears the anal lobes and the anus.

Malpighian tubules are absent (Fig. 6.1.2). The tracheal system is absent or strongly reduced (single pair of cervical spiracles). The testes and ovaries are very large in relation to the body size; they reach the mesothorax anteriorly.

Biology. Collembola can be extremely abundant. Up to 200,000 individuals can occur per m^2, with the approximate number of species in the populations ranging between 50 and 100 under favorable conditions. In contrast to the other apterygote lineages, springtails occur in many different habitats. This includes soil, decaying vegetation, tide pools, fresh water habitats, caves and even tree canopies. Larger species tend to be more exposed, whereas small unpigmented and blind forms with short legs and reduced jumping apparatus live in small cavities in soil.

Food substrates used by collembolans are spores or hyphae of fungi, bacteria, algae, decaying leaves and wood, detritus, feces of invertebrates, and pollen. Carnivorous habits are rare and very few species are predacious.

Reproduction and development. The reproductive behavior varies strongly within the group. Males of some species deposit their spermatophores randomly, whereas complicated mating rituals are known from others. In some species females occasionally consume the spermatophore.

A total cleavage until the eight cell stadium and a subsequent superficial cleavage pattern is described for the large species *Tetrodontophora bielanensis*. Springtails molt up to 50 times, but maturity is reached after stage 6–14.

Fossil record. The Lower Devonian collembolan †*Rhyniella praecursor* (Rhynie Chert, Scotland) is one of the oldest known members of Hexapoda. It already shows the typical features of the group. After a large gap in the fossil record more recent species are described from Cretaceous deposits. Collembolans are quite common in Cenozoic amber (Engel & Grimaldi 2005).

Economic importance. Large species can cause damage by feeding on young leaves and negative impact on fungiculture is also described. Due to their high population densities collembolans can play a very important role in transforming organic matter into humus. They are also indicators of negative side effects of pesticides.

[Weber (1949); Schaller (1970); Naumann et al. (1991); Hopkin (1997); Dunger (2005); Engel & Grimaldi (2005)]

6.2 Protura (Greek *protos* = first, primitive, *oura* = tail, English common name: coneheads)

Diversity and distribution. Protura, which were discovered in 1907, are one of the smaller groups of hexapods with about 750 described species. Considering the very small size, the cryptic habits, and limited collecting efforts it appears likely that the real diversity is distinctly higher. Protura are distributed in all zoogeographic regions.

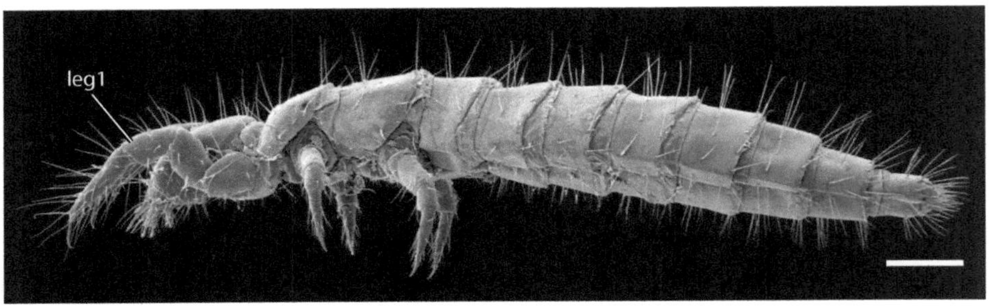

Fig. 6.2.1: Protura (Acerentomidae), SEM micrograph, lateral view. Abbr.: leg1 – fore leg. Scale bar: 100 µm. Courtesy H. Pohl.

Autapomorphies
- Complete loss of antennae (Figs 6.2.1, 6.2.2)
- Tetrapod walking, forelegs mainly functioning as tactile organs (Fig. 6.2.1.)
- Complete loss of compound eyes and ocelli (also potential synapomorphies with Diplura, see above) (Figs 6.2.1, 6.2.2)
- Paired defensive glands opening at the posterior tergal margin of abdominal segment VIII
- Meso- and metathoracic spiracles open on terga (Fig. 6.2.2A)
- Genital chamber housing extrusible genitalia in both sexes

Taxonomy. Three superfamilies are distinguished, Acerentomoidea (Hesperentomidae, Protentomidae, Acerellidae, Acerentomidae, Nipponentomidae, Berberentulidae), Eosentomoidea (Eosentomidae, Antelientomidae) (Fig. 6.2.2A), and Sinentomoidea (Sinentomidae, Fujientomidae). The latter comprise only five species. They occur in China, Korea and Japan (e.g., *Sinentomon erythranum*).

Diagnosis. Very small and slender, blind, and usually unpigmented (Figs 6.2.1–6.2.3). Antennae missing and functionally replaced by the enlarged forelegs as tactile organs. Without caudal abdominal appendages.

Morphology. Total length usually 0.8–1.3 mm, largest species 2.6 mm (*Nipponentomon kamui*). The body is slender and subparallel, and gradually narrowing towards the abdominal apex (Figs 6.2.1–6.2.3).

The prognathous head (Fig. 6.2.4) is oval, elongated, laterally rounded and moderately tapering anteriorly. Posteriorly it is not or only very slightly retracted into the prothorax. The foramen occipitale is large. Compound eyes, ocelli and antennae are absent (Figs 6.2.1–6.2.4). Temporal organs are present and superficially resemble eyes. The frontal and coronal sutures (ecdysial lines) are missing. A clypeofrontal transverse ridge separates the clypeolabrum from the rest of the head capsule; it sometimes forms an elongate, acuminate rostrum (e.g., *Acerentomon*). Like in Collembola the mouthparts are enclosed by duplicatures, with the mandibles and maxillae

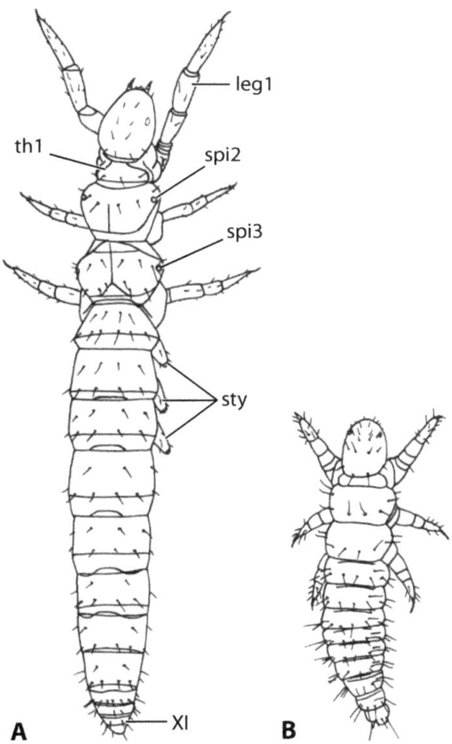

Fig. 6.2.2: Protura, dorsal view. A, *Eosentomon transitorium* (Eosentomidae); B, *Nosekiella danica* (Acerentomidae), 1st stage ("prelarva"). Abbr.: leg1 – fore leg, spi2/3, meso-, metathoracic spiracles, th1 – prothorax, sty – styli, XI – segment XI. Redrawn from Dunger (2005), after Berlese (1909) and Tuxen (1964).

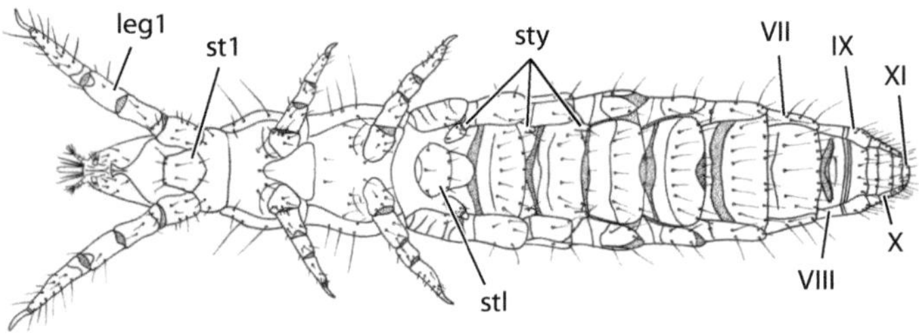

Fig. 6.2.3: *Acerentomon* sp. (Acerentomidae), ventral view.
Abbr.: leg1 – fore leg, st1 – prosternum, stl – abdominal sternum I, sty – styli, VII–XI – segments VII–XI. Redrawn from Westheide & Rieger (2007) (original B. Balkenhol, Osnabrück).

in separate pouches. A linea ventralis is also present; it reaches the cervical membrane posteriorly. The mandibles are primarily equipped with biting teeth (e.g. *Eosentomon*, *Sinentomon*), but transformed into stylets in most species (Fig. 6.2.4). The galea is primarily weakly sclerotized and short, and equipped with finger-like processes; the maxillary palp is composed of four or fewer segments; apically, the distal palpomere bears a bundle of sensorial bristles (Fig. 6.2.4). The distal parts of the labium, which bear the short labial palps, are confined to the anteroven-

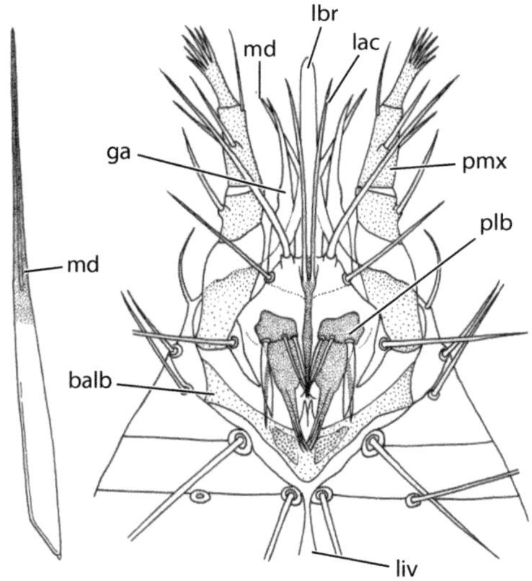

Fig. 6.2.4: *Acerentomon* sp. (Acerentomidae), head and mandible, ventral view. Abbr.: balb –base of labium, ga – galea, lac – lacinia, lbr – labrum, liv – linea ventralis, md – mandible, plb – palpus labialis, pmx – palpus maxillaris. Redrawn from Dunger (2005), after Berlese (1909).

tral head region. The cuticular endoskeleton comprises ridges at the hind margin of the head capsule and an X-shaped element ("fulcro-tentorium"), which is likely homologous with the tentorium of insects. A dorsomedian postoccipital apodeme is the area of insertion of a paired thoraco-cephalic muscle. The deutocerebrum is reduced (loss of antennae).

Cervical sclerites appear to be present in some cases. As the head is partly shifted onto the dorsal side of the prothorax, the pronotum is very small (Figs 6.2.1–6.2.3). The largely membranous dorsal and lateral prothoracic regions guarantee a high movability of the tactile prolegs. In contrast to these areas, the sternal sclerites are well-developed. The mesothorax is slightly smaller than the metathorax but otherwise similar; the tergites of both segments are distinctly developed. The middle and hindlegs function as locomotor organs (functional quadrupedalism); all legs are 6-segmented; the short, ring-shaped pretarsus apically bears an unpaired claw, a ventral empodium, and a dorsal sensorial appendage. Spiracles are usually missing, but present on the meso- und metatergum of Eosentomoidea and Sinentomidae; the originating tracheae extend through the entire body.

The abdomen is composed of 12 or 11 segments and a segment-like telson (homology uncertain) (Figs 6.2.2, 6.2.3). The terga are distinctly developed. The posterior segments decrease in size. The terminal segments IX–XII are much smaller than segments I–VIII. Vestigial 1- or 2-segmented appendages are ventrolaterally inserted on segments I–II (Fig. 6.2.3). Paired defense glands are present at the hind margin of segment VIII. Cerci are absent. The similar male and female genitalia ("squama genitalis") are enclosed in a genital chamber that opens between segments XI and XII (telson?). The male genital apparatus is composed of a basal bipartite periphallus and a retractile distal phallus; the latter is composed of an unpaired, bipartite proximal element ("stipes") and 2-segmented, paired distally inserted styli. The proximal element of the female squama genitalis is a dorsally open semicircular structure (perigynium). The females also bear paired styli, which are usually equipped with an elongate and acuminate, or short and knob-shaped acrostylus (absent in Eosentomidae but present in Sinentomidae).

Six Malpighian papillae are present. Large paired defensive glands are located in the segments V (VI)–VIII (IX). The testes and sac-shaped neo-panoistic ovaries are very large, filling out large areas of the abdomen and parts of the posterior thorax.

Biology. The biology of proturans is very insufficiently known. They live in moist substrates, in humus, decaying wood, leave litter, or moss. Very high numbers of individuals can occur in suitable habitats. The stylet-like paired mouthparts are used for piercing fungal hyphae, especially of mykorrhiza (observed in *Eosentomon* and *Acerentomon*).

Reproduction and development. One observation suggests that copulation takes place. The males do not form spermatophores. Some species are facultatively parthenogenetic. The full number of body segments is reached at the end of the postembryonic development, which lasts 3–5 months. The abdomen of the first (prelarva) and second stage is composed of nine segments (Fig. 6.2.2B), ten are present in the third, and 12 (or 11 + telson) in the 'maturus junior'. Usually the sexually mature fifth stage follows. An additional preimaginal stage occurs in males of Acerentomoidea. In contrast to Collembola and Diplura molting of mature adults is unknown.

Fossil record. Fossils are unknown.

Economic importance. Economically irrelevant as far as presently known.

[Berlese (1909); Tuxen (1964); Yin (1965); Hennig (1969); Janetschek (1970); François et al. (1992); Dunger (2005); Westheide & Rieger (2007)]

6.3 Diplura (Greek *diplos* = double, *oura* = tail, English common name: two-pronged bristletails)

Diversity and Distribution. Diplura comprise approximately 1,000 described species. They have a world-wide distribution but are absent from extreme habitats such as the arctic and Antarctic regions, high mountain areas above 3,500 m, and deserts. *Atlasjapyx atlas* (Gigasjapyginae), the longest dipluran species, occurs only in Sichuan (China).

Autapomorphies
- Complete absence of eyes (also in Protura) (Figs 6.3.1–6.3.3)
- Mandibles devoid of molar part (also in Protura)
- Prognathous head (Figs 6.3.1, 6.3.3)
- Ten-segmented abdomen (Fig. 6.3.2)
- Gonoporus in both sexes on a retractile genital papilla between abdominal segments VIII and IX

Taxonomy. The group is traditionally subdivided into the suborders Rhabdura and Dicellurata (each ca. 500 spp.), based on the shape of their cerci (filiform and segmented versus unsegmented and pincer-shaped). The former comprises Projapygoidea (Projapygidae, Anajapygidae, Octostigmatidae) and Campodeoidea (Procampodeidae, Campodeidae) (Figs 6.3.1A, 6.3.2A), the latter Japygoidea (Japygidae, Dinjapygidae, Evalljapygidae, Heterojapygidae, Parajapygidae) (Figs 6.3.1B, 6.3.2.B). A more recent classification currently supported by molecular studies recognizes Campodeina (= Campodeoidea) and Japygina (= Projapygoidea + Japygoidea) as the major clades.

Fig. 6.3.1: Diplura, SEM micrographs, lateral view. A, Campodeidae. B. Japygidae. Abbr.: cer – cercus. Scale bar: 500 µm. Courtesy H. Pohl.

Diagnosis. Wingless and eyeless entognathous insects. Ten-segmented abdomen with styli, coxal vesicles and terminal cerci. Usually small, slender, weakly sclerotized and more or less unpigmented, but occasionally very large (see below).

Morphology. The total length usually ranges between 4–12 mm, but *Atlasjapyx atlas* measures 58 mm. The elongate and slender body is parallel-sided or subparallel. It is adapted to moving efficiently in narrow crevices, especially in the short-legged Japygoidea.

The prognathous head (Figs 6.3.1, 6.3.3) is dorsoventrally flattened. Eyes and temporal organs are absent (Fig. 6.3.3). A dorsomedian longitudinal suture (coronal suture) and a dorsal transverse suture are present or absent. The antennae are composed of 13–70 segments, long and slender (Campodeoidea), shortened and less slender (Projapygoidea), or retractile in a telescope-like manner (Japygoidea). The plica orales are less extensive compared to proturans and collembolans and a linea ventralis is not present; the biting paired mouthparts are enclosed in a joint pouch on either side (Koch 2001). The labium is fully exposed and forms the ventral closure of the head (Fig. 6.3.3). The primary mandibular joint is reduced; the elongate mandibles are interconnected by a transverse tendon; they bear terminal teeth and an unsclerotized prostheca is present in Rhabdura. The maxillary palp comprises 1–3

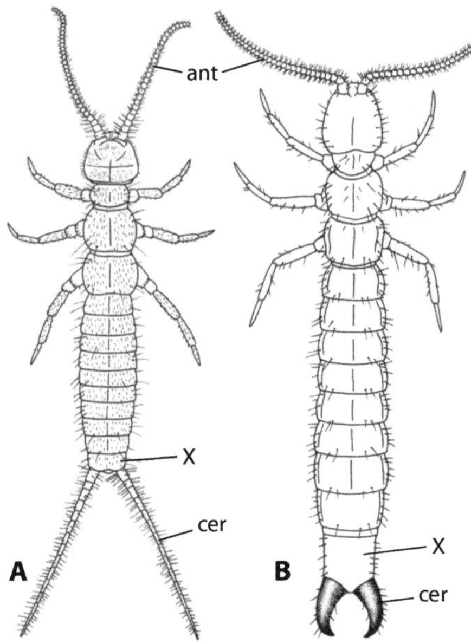

Fig. 6.3.2: Diplura, dorsal view. A, *Campodea staphylinus* (Campodeidae); B, *Japyx* sp. (Japygidae). Abbr.: ant – antenna, cer – cerci, X – segment X. Redrawn from Dunger (2005), after Eisenbeis & Wichard (1985).

segments. The labium is divided into an extensive postmentum and a prementum, which bears the more or less strongly modified palps (strongly reduced in Campodeidae). Anterior tentorial arms and short posterior tentorial apodemes are present, and also a mesodermal, ligamentous endoskeleton. Lateral cervical sclerites occur.

The prothorax is distinctly smaller than the meso- and metathorax, which are similar in size and structure. The tergites are usually medially divided by a suture or zone of weakness. Strongly developed muscles originate on Y-shaped sternal apodemes (sometimes reduced to a median branch in the prothorax, e.g., Campodeidae). The legs are shortened in Japygidae; the tarsus is undivided; the very small pretarsus bears a pair of large claws and sometimes a medial dactylus (indistinct in some Campodeidae, Procampodeidae, Projapygidae, Heterojapygidae, and Dinjapygidae). The number of thoracic spiracles varies from two pairs (e.g., Projapygidae, Parajapygidae) to three (Campodeidae) or even four (e.g., Octostigmatidae, Japygidae).

The abdomen is 10-segmented (Figs 6.3.1, 6.3.2). The coxal plates are fully fused with the sternites (coxosternites). Large, pubescent subcoxal appendages (probably modified eversible vesicles) are present on segment I of Campodeidae and Projapygoidea, and stylet-like styli on segments II–VII; the remaining subgroups have styli (peg-like in Japygoidea) on abdominal segments I–VII. Coxal vesicles with retractor muscles are usually present on segments I–VII, but are restricted to segments II

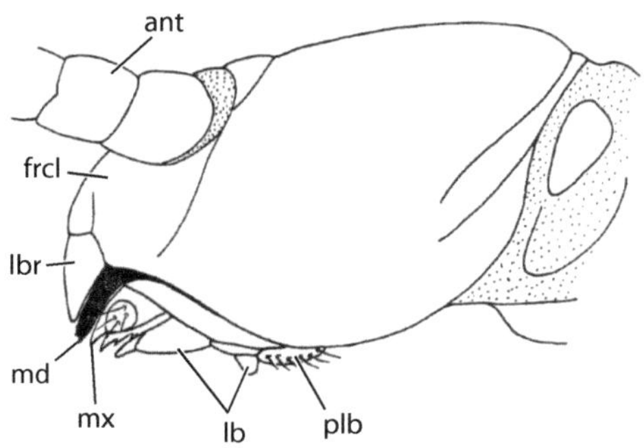

Fig. 6.3.3: *Campodea* sp. (Campodeidae), head, lateral view. Abbr.: ant – antenna, frcl – frontoclypeus, lb – labium, lbr – labrum, md – mandible, mx –maxilla, plb – palpus labialis. Redrawn from Francois (1970).

and III in Parajapygidae, absent (or transformed into subcoxal appendages, respectively) on segment I in Projapygoidea and Campodeoidea, and entirely missing on all abdominal segments in Projapygidae. The cerci are inserted on the terminal segment X; they are filiform and multisegmented in Campodeoidea, short, annulated and associated with silk glands in Projapygoidea, and unsegmented, stout and pincer-shaped in Japygoidea (Fig. 6.3.2). Abdominal spiracles are restricted to segments I–VII in Projapygoidea and Japygoidea; they are entirely absent in Campodeoidea.

The tracheal system is well-developed (Xué et al. 1994). The Malpighian papillae are similar to those of proturans. They are absent in the genus *Japyx*.

Biology. The slender diplurans are specialized on narrow spaces with constant microclimatic conditions. Japygid species may occur as deep as 1 m in soil. Some species produce a web of silk-like secretions. Campodeidae are omnivorous. They usually feed on different kinds of organic matter and microorganisms, but sometimes also on carrion. They are very rarely predacious. Japygidae are almost exclusively predators but fungal hyphae were also identified in gut contents. They usually catch their prey with the specialized cerci. Projapygids were observed to use their cercal silk secretions to capture prey.

Reproduction and development. Males deposit ca. 200 short-stalked spermatophores per week. Direct contact with females was not observed yet. The released spermatozoa fertilize the eggs in the atrium of the female genital opening. Eggs are deposited in litter or under stones by females of Campodeoidea. Brood care is unknown in this group, whereas females protecting the eggs in self-made cavities in soil occur in Japygidae. The cleavage is superficial after the 6th division. The first instar is enclosed by an embryonic membrane and morphologically not fully developed and immo-

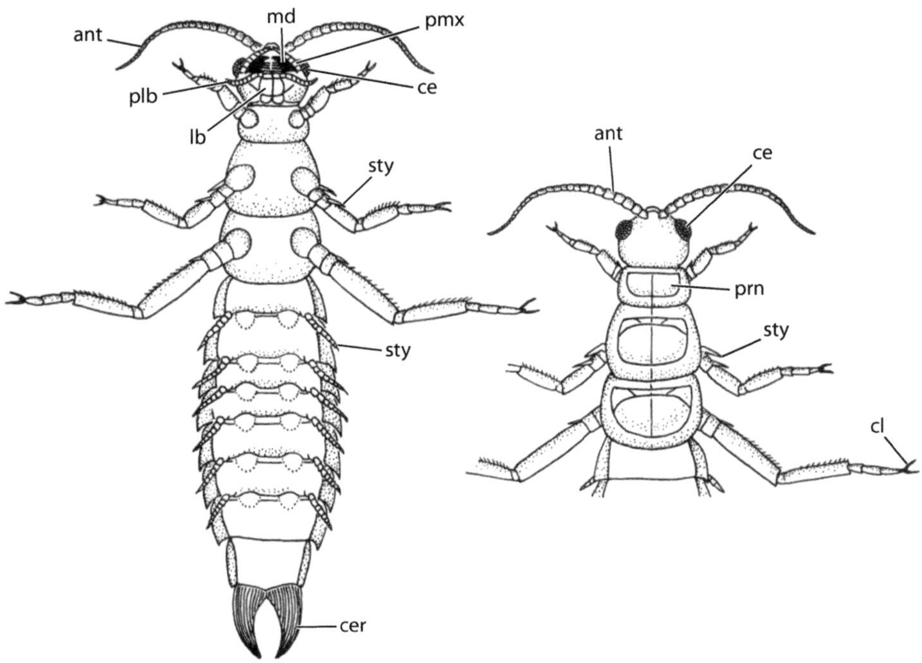

Fig. 6.3.4: †*Testajapyx thomasi* Kukalová-Peck 1987, Carboniferous (Mazon Creek, Illinois). Abbr.: ant – antenna, ce – compound eye, cer – cerci, cl – claw, lb – labium, md – mandibles, plb – labial palp, pmx – palpus maxillaris, prn – pronotum, sty – styli. Redrawn from Kukalová-Peck (1987).

bile. Molting continues after reaching maturity (stage 8–11 in Campodeoidea), about 15 times per year in Campodeoidea, and less frequently in Japygoidea.

Fossil record. †*Testajapyx thomasi* (Fig. 6.3.4) from the Upper Carboniferous (Mazon Creek) (Kukaová-Peck 1987) is a controversial fossil (Grimaldi & Engel 2005). Unlike most extant diplurans (and other entognathous hexapods) it is very large (47.5 mm), and it possessed well-developed compound eyes and largely exposed mouthparts. The unsegmented grasping cerci suggest affinities with extant Japygidae. Few fossils definitely belonging to Diplura are known. They are from Lower Cretaceous deposits and Tertiary amber (Grimaldi & Engel 2005).

Economic importance. The herbivorous species *Octostigma herbivora* is listed as potential agricultural pest.

[Francois (1970); Eisenbeis & Wichard (1985); Kukalová-Peck (1987); Xué et al. (1994); Koch (2001); Dunger (2005); Grimaldi & Engel (2005); Pass & Szucsich (2011)]

Protapteron indicum. A highly unusual wingless species from the Malabar Coast of India – *Protapteron indicum* (Fig. 6.A.1) – was described by the Russian entomologist Schepotieff (1909). The specimens were allegedly collected on the husks of coconuts. The description published in the Zoologische Jahrbücher für Anatomie (Jena) was exceptionally precise and contained numerous anatomical details including renderings of microtome sections. Schepotieff (1909) interpreted *Protapteron* as a connecting link between Protura and Diplura. Judging from the specific entognathous condition and the presence of a linea ventralis, the species would be closer to Ellipura. The habitus illustrations suggest close affinities to Protura. However, very thin multisegmented antennae are present. Four pairs of 2-segmented appendages are present on the anterior abdominal segments. Cerci are lacking. The presence of 10 structures resembling ocelli on the head and on the ventral side of the prothorax is a unique feature in Hexapoda.

The author never made material from the type series (allegedly collected in 1908) available to any other entomologist and no further specimens were collected or identified in collections. *Protapteron* is probably one of the most spectacular fakes in the history of Entomology (e.g., Tuxen [1931]).

[Schepotieff (1909); Tuxen (1931)]

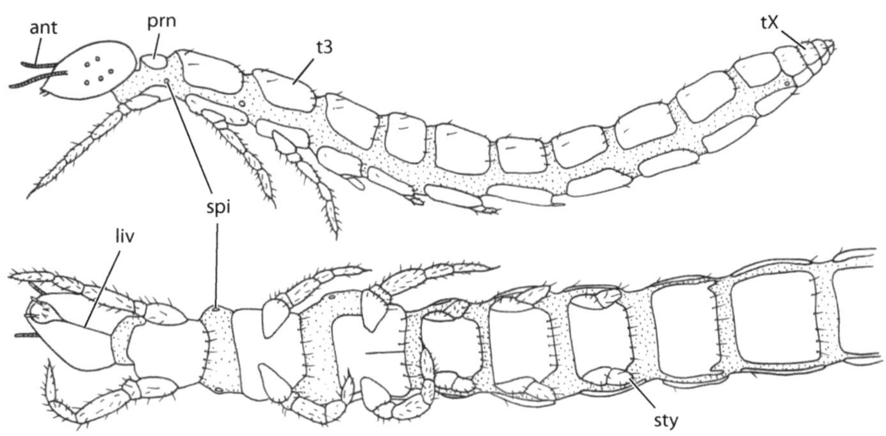

Fig. 6.A.1: *Protapteron indicum* (Schepotieff 1909). Abbr.: ant – antenna, liv – linea ventralis, prn – pronotum, spi – spiracle, sty – stylus, t3 – metanotum, tX – abdominal tergite X. Redrawn from Schepotieff (1909).

Ectognatha, Insecta and Dicondylia

A strongly supported and undisputed hexapod subgroup is Insecta (Hexapoda excl. Collembola, Protura and Diplura), also known as Ectognatha (e.g., Hennig 1969). In

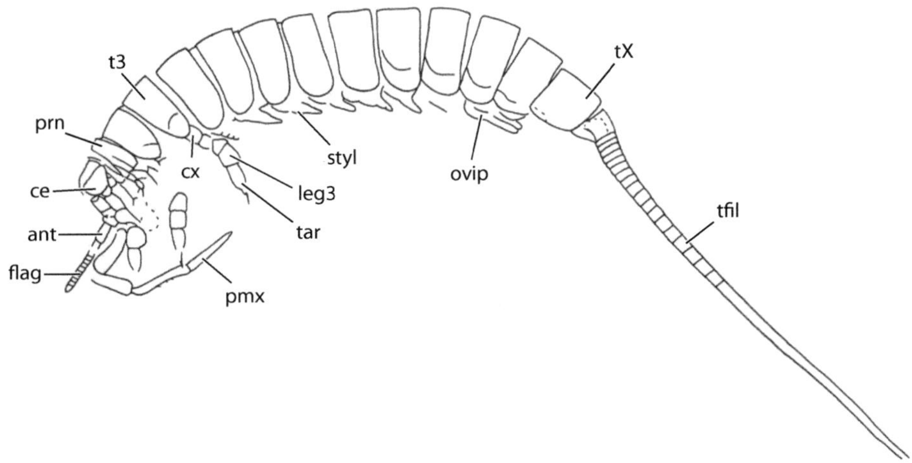

Fig. 6.A.2: †*Dasyleptus brogniarti* (†Monura), Lower Permian. Abbr.: ant – antenna, ce – compound eye, cx – coxa, flag – antennal flagellum, leg3 – hindleg, ovip – ovipositor, pmx – palpus maxillaris, prn – pronotum, styl – styli, t3 – metanotum, tX – abdominal tergite X, tar – tarsus, tfil – terminal filament (paracercus?). Redrawn from Willmann (2005a) (after Rasnitsyn 1999).

the following we consistently use Ectognatha for a more inclusive group also containing the extinct †Monura (Fig. 6.A: 6), whereas Insecta refers to a clade comprising only Dicondylia and Archaeognatha (Fig. 6.A: 7) unambiguously documented in the fossil record since the Devonian (mandible of †*Rhyniognatha hirsti* [Rhynie Chert, Scotland]), but it is likely that they originated distinctly earlier (Engel & Grimaldi 2004; Wheat & Wahlberg 2013).

The most convincing autapomorphies of Ectognatha are modifications of the antenna and the external female genital apparatus (autapomorphies see below). A plesiomorphic feature compared to the entognathous lineages is the complete exposure of the mouthparts (ectognathous condition). Within Insecta Archaeognatha are the extant group with the maximum number of preserved plesiomorphies (e.g., monocondylic mandibles, lack of a distinctly developed gonangulum, almost complete series of abdominal styli and coxal vesicles). †Monura, a group known from Carboniferous and Permian deposits, is very likely the sistergroup of Insecta. Plesiomorphies preserved in this extinct lineage (with the corresponding apomorphies in Insecta) are the undivided tarsi and the short ovipositor. The interpretation of the absence of cerci (Rasnitsyn 1999) is ambiguous as long as the systematic position of Diplura (cerci well-developed) is unclear. The 7-segmented, leg-like maxillary palp is very similar to the condition found in Archaeognatha. This suggests that this may be an additional apomorphic groundplan feature of Insecta, with secondarily shortened maxillary palps in Dicondylia.

Autapomorphies of Ectognatha (e.g., Bitsch & Bitsch 2000, 2004; Klass 2009)
- flagellar antenna (large scapus and pedicellus, multisegmented flagellum)
- large leg-like maxillary palp (?)
- pretarsus largely reduced, base of claw in contact with distal tarsomere
- ovipositor formed by appendages of abdominal segments VIII and IX
- multisegmented terminal filament

Autapomorphies of Insecta (=Ectognatha excl. †Monura)
- subdivision of the tarsus (3 or 5 tarsomeres)
- ovipositor with elongated gonapophyses

The following features belong either to the groundplan of Ectognatha or Insecta (not verifiable in monuran fossils):
- posterior tentorial bridge (=corpotentorium) connecting the posterior tentorial arms
- Johnston's organ in the pedicellus
- antennal vessels separated from cephalic aorta
- labial glands open on anterior labial base (salivarium between labium and hypopharynx)
- two pairs of thoracic spiracles (?) (up to four in Diplura)
- amniotic cavity formed by folds of the serosa enclosing the ventral side
- base of sperm flagellum with 1–3 accessory bodies, derived from the centriole adjunct (absent in entognathous hexapods) (Jamieson 1987)

Dicondylia. Dicondylia comprise the primarily wingless Zygentoma (silverfish) and the megadiverse Pterygota (Fig. 6.A: 8). The group is well-supported by morphological features and also confirmed by analyses of molecular data sets including transcriptomes. The most important dicondylian apomorphy is the second (anterior) mandibular joint, with a socket on the mandible and a condyle on the head capsule (Bitsch & Bitsch 2000). This articulation is still a gliding device with a relatively high intrinsic movability in Zygentoma and larvae of Ephemeroptera, whereas an anterior ball-and-socket joint is characteristic for Odonata and Neoptera (Staniczek 2000). The mandible of the Devonian *Rhyniognatha hirsti* (see above) displays a typical dicondylic articulation. This shows that Dicondylia or even Pterygota (Engel & Grimaldi 2004) were already present in this period.

Autapomorphies of Dicondylia (e.g., Bitsch & Bitsch 2004; Klass 2009)
- secondary (anterior) mandibular joint (gliding device in the groundplan)
- anterior tentorial arms with plate-like connection, fused with (posterior) tentorial bridge in Maindroniidae (Zygentoma) and Pterygota
- tracheal system with regular longitudinal and transverse connecting branches (anastomoses)

- egg with micropyle for fertilization
- amniotic cavity closed

The presence of a distinctly developed gonangulum is questionable as a dicondylian autapomorphy (e.g., Klass 2009).

The monophyly of Zygentoma was discussed with respect to Lepidotrichidae. This family is known from Baltic amber, and the extant species *Tricholepidion gertschi* was discovered in California about 50 years ago (Wygodzinsky 1961). *Tricholepidion* has preserved some ancestral features which are not found in the remaining Zygentoma and Pterygota, such as the presence of a well-developed ligamentous head endoskeleton and the presence of a transverse mandibular apodeme (both also present in Archaeognatha). Other plesiomorphies of *Tricholepidion* are the presence of ocelli, well-developed compound eyes, and the presence of styli and coxal vesicles on the pregenital abdominal segments. These features suggest a possible sistergroup relationship between Lepidotrichidae and the remaining Dicondylia (with paraphyletic Zygentoma). However, recent analyses of transcriptomes unambiguously support a clade Zygentoma including Lepidotrichidae (see also **VI.5. Zygentoma**).

[Schepotieff (1909); Wygodzinsky (1961); Hennig (1969); Jamieson (1987); Rasnitsyn (1999); Bitsch & Bitsch (2000, 2004); Staniczek (2000); Engel & Grimaldi (2004); Willmann (2005a); Klass (2009); Wheat & Wahlberg (2013)]

6.4 Archaeognatha (=Microcoryphia; Greek *archaeos* = ancient, *gnatha* = jaw, English common name: jumping bristletails)

Diversity and distribution. Archaeognatha are a small order with approximately 500 described species. The distribution is world-wide, with a broad spectrum of habitats, including mountains up to 5,000 m, polar regions, arctic tundra, sea and lake shores, and tropical rain forests.

Taxonomy. The order is subdivided into the extinct †Triassomachiloidea and the extant Machiloidea (crown group Archaeognatha), the latter comprising the families Machilidae (42 genera, ca. 310 spp.) and Meinertellidae (19 genera, ca. 170 species).

Diagnosis. Medium sized wingless insects with a fusiform body (Fig. 6.4.1). Compound eyes enlarged. Maxillary palps 7-segmented, large and leg-like. Abdominal apex bearing paired multisegmented cerci and an unpaired long terminal filament.

Autapomorphies

- Compound eyes enlarged (Figs 6.4.1, 6.4.2A)
- Jumping capacity involving specialized thoracic torsion muscles

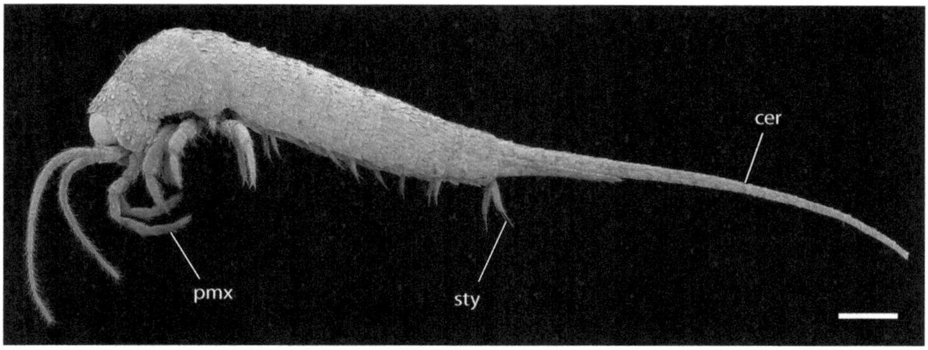

Fig. 6.4.1: *Lepismachilis y-signata* (Machilidae), SEM micrograph, lateral view. Abbr.: cer – cerci, pmx – palpus maxillaris, sty – stylus. Scale bar: 500 µm. Courtesy H. Pohl.

- Maxillary palps very large and 7-segmented (possibly a groundplan feature of Ectognatha and Insecta as a similar condition is found in the extinct †Monura)? (Figs 6.4.1, 6.4.2A)
- Styli of meso- and metacoxa present (possibly plesiomorphic)?
- Meso- and metatrochantin absent
- Spiracle of abdominal segment I reduced

Morphology. The maximum size is 25 mm but most species range between 10 mm and 15 mm (terminal filament and cercus excluded). The body is fusiform and the thorax is distinctly convex dorsally. The cuticle is mostly thin and flexible, even though some regions are strongly sclerotized (e.g., mandibular articulation); it is characterized by a vestiture of scales and bears a multitude of sensory setae. Glands associated with the integument are also present. Pigments stored in epidermal cells can result in distinctive color patterns, in addition to patterns resulting from iridescent or colored scales (white, light yellow, brown, black).

The head (Fig. 6.4.2A) is hypognathous and moderately retracted into the prothorax posteriorly. It is mainly characterized by the very large, dorsally adjacent compound eyes; three well-developed ocelli are also present. In the posterior head region an occipital phragma is developed, probably homologous to the postoccipital sulcus or ridge of pterygote insects. The frontoclypeal transverse strengthening ridge is absent; the border between the two areas is indicated by the position of the median ocellus. The clypeus is moderately protruding and anteriorly connected with the movable labrum. The antennal articulation areas are approximated due to the large size of the compound eyes. The flagellate antennae are composed of a large scapus with a pair of intrinsic muscles, a moderately sized pedicellus containing a chordotonal organ (Johnston's organ), and a very long, filiform and multisegmented flagellum; the flagellomeres except for the proximal ones appear subdivided by joint-

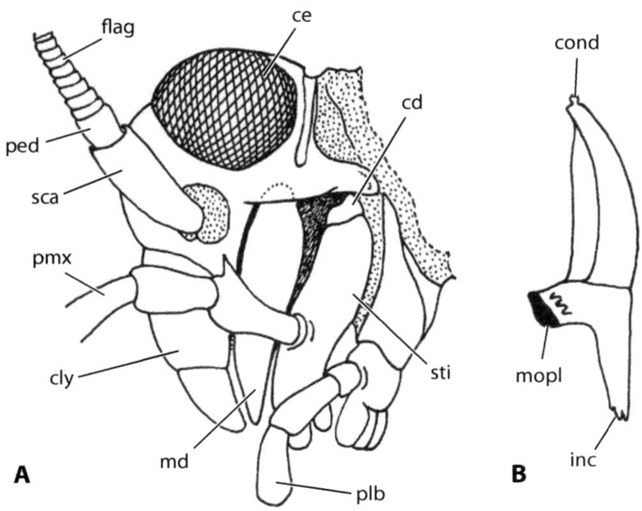

Fig. 6.4.2: Machilidae. A, *Neomachilis maoricus*, head, lateral view, ectognathous mouthparts; B, *Machilis* sp., mandible. Abbr.: cd – cardo, ce – compound eye, cly – clypeus, cond – condyle (primary mandibular joint), flag – flagellum, inc – incisivi, md – mandible, mopl – molar plate, ped – pedicellus, plb – palpus labialis, pmx – palpus maxillaris, sca – scapus, sti – stipes. Redrawn from Dunger (2005), after Snodgrass (1935).

lets. The elongate mandibles articulate with the head capsule with a single posterior joint (primary mandibular joint) (Fig. 6.4.2B); the name Archaeognatha (archaeos = ancient, gnatha = jaw) refers to the absence of the secondary mandibular joint; a molar plate is present mesally, about midlenght between the base and several small apical teeth. The main body of the maxilla is formed by a small nearly rectangular cardo and a large, mesally open stipes; the galea is rounded and unsclerotised; the sclerotized lacinia bears two lateral and one mesal tooth, and a brush of stiff setae; the palp is very large, conspicuous and leg-like; it is composed of seven palpomeres. The labium is composed of an unpaired postmentum and a paired prementum with anteriorly bilobed glossae and paraglossae, and moderately long 3-segmented palps (Fig. 6.4.2A). The hypopharynx is well-developed, with paired superlinguae and an unpaired lingua; these elements are posteriorly connected by a suspensorium. Between the base of the superlingua and the prelabium lies the opening of the salivary channel (salivarium), which is connected with the labial glands. The head endoskeleton is in a transitional stage; it comprises true tentorial elements, the fulcrum, and ligamentous parts (endosterna); the tentorium is a hollow structure; it comprises separated posterior and anterior plates, the latter bearing a long posterior intrusion suspended at the posterodorsal head capsule by muscular fibrillae; the paired elements of the posterior tentorium are connected by a transverse bar (tentorial bridge) and a metatentorial plate (Sturm & Machida 2001: fig. 8.4). An important plesiomorphic muscular feature is the presence of very strong tentorial muscles of the mandibles. Three glands are associated with the oral cavity, the paired mandibular glands, the paired acinous glands, and the labial glands (or labial kidneys).

The thorax is characterized by the broad paranotal lobes of the simple, undivided tergites, which cover the pleural areas of each of the three segments. The prothoracic paranotal lobe reaches the compound eyes anteriorly. The large mesotergite and the moderately sized metatergite are strongly arched (Fig. 6.4.1). The meso- and metathorax are rigidly united, whereas the prothorax can be bent downwards in the sagittal plane. The pleura are composed of two largely concentric supracoxal arcs (katapleuron, anapleuron). The sternal regions are not distinctly subdivided and differ scarcely in the three segments. The coxae of the well-developed legs are large but relatively slender; a unique feature of Archaeognatha is the presence of a stylus on the meso- and metacoxae, or only on the latter; in contrast to the entognathous hexapods, the tarsus is subdivided into three or rarely two tarsomeres, and the pretarsus, which bears double claws, is partly reduced; groups of specialized tarsal adhesive hairs (scopulae) are present in few genera (e.g., *Meinertellus*). The thoracic locomotor system is characterized by the torsion muscles (looped muscles), which are part of the jumping mechanism (see below); in Machilidae they are usually also present in most abdominal segments. The meso- and metathorax each bear a pair of spiracles.

The abdomen is composed of eleven segments. The tergites also bear lateral paranotal lobes. On the ventral side the small sternites are flanked by large coxites, which bear styli equipped with muscles on segments II–IX, and one or two pairs of eversible coxal vesicles usually on segments I–VII (Fig. 6.4.3). Segment X bears a long and multisegmented terminal filament and shorter multisegmented cerci (Fig. 6.4.1). The male and female gonopods insert on coxites VIII and IX. Tergite XI and sternites X and XI are reduced. The ovipositor is composed of paired gonocoxae VIII and IX, a gonostylus IX (inserted posteriorly on gonocoxite IX), and annulated gonapophyses VIII and IX which are connected with each other; a gonangulum as it occurs in Zygentoma and pterygote groups with an orthopteroid ovipositor, is indistinctly developed or missing. The male genital apparatus comprises the penis (aedeagus and phallobase) inserted on coxite IX and segmented parameres inserted on segment IX and sometimes also on VIII.

Transverse anastomoses of the tracheal system are largely or completely missing. The largely unmodified digestive tract is equipped with six midgut caeca; a proventriculus is not developed. Approximately 20 slender Malpighian tubules insert at the midgut-hindgut border. The heart with eleven paired ostia (*Machilis*) extends from the mesothorax to abdominal segment X. Segment IX bears the genital openings. The ovarioles are panoistic. Spermathecae are missing. The testes are composed of three or four follicles. Two vasa deferentia are present on both sides; they are connected by transverse ducts and a median reservoir is also present.

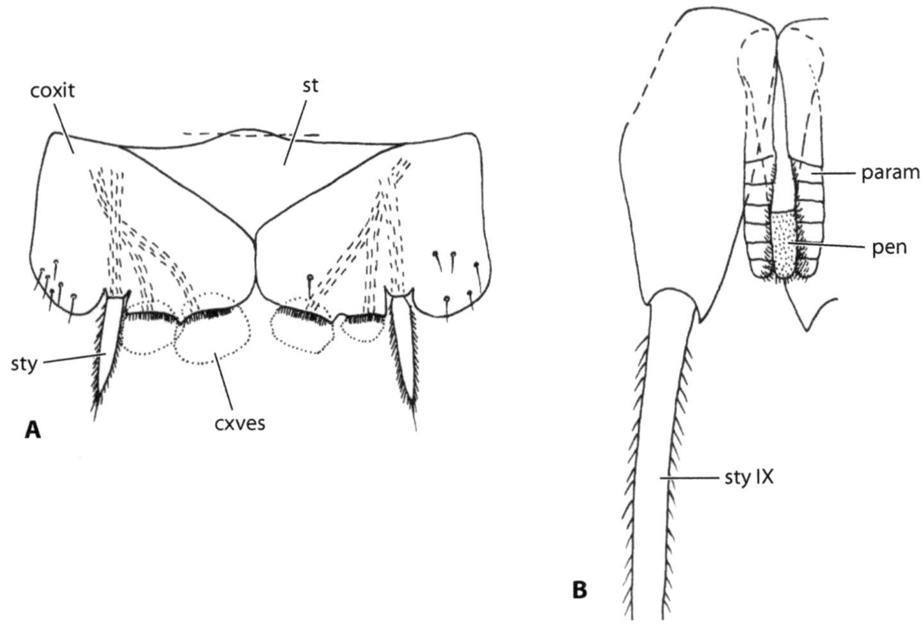

Fig. 6.4.3: *Machilis* sp. (Machilidae), abdominal appendages. A, segment V, ventral view; B, coxite IX of male with genital appendages. Abbr.: cxves– coxal vesicles, param – paramere, pen – penis, st – sternite V, sty – stylus, sty IX – gonostylus IX. Redrawn from Dunger (2005), after Sturm & Bach de Roca (1993).

Biology. Archaeognatha are ground oriented and generally require a high level of humidity. Only few species occur in more arid regions. Water can be absorbed by the coxal vesicles. The food spectrum mainly comprises algae, lichens and moss; carnivorous habits possibly occur as a rare exception but this needs confirmation. Archaeognathans move very swiftly over the substrate, but interrupted by frequent pauses. When disturbed they jump over distances of 10–20 cm and 10 cm high; before the take-off both ends of the body are abruptly bent downwards, using the looped muscles; the jump is also supported by the maxillary palps, forelegs, cerci and terminal filament. The strongly developed maxillary palps are not involved in the normal locomotion; they are only exceptionally used when the insects climb steep and smooth walls.

Reproduction and development. Sperm transfer is usually indirect, but with contact between the males and females and specific patterns of mating behavior. The males usually deposit sperm droplets on carrier threads, but apparently directly on the ovipositor in *Petrobius*. The females deposit 15–30 eggs in small cavities in the soil substrate using their ovipositor. The cleavage is total initially, but then superficial. The embryonic development is completed in ca. 7–12 months. The serosal or amnioserosal fold and cavity (apparently homologous to these formations in Dicondylia) are formed in the blastokinesis, a distinctive phenomenon in the ontogenesis of Insecta. The postembryonic development is accomplished in 3–12 months. The first stage hatches without scales and macrochaetae (large bristle-like setae); the forward projecting laciniae function as oviruptors. Scales appear after the second molt, coxal styli and gonapophyses in a later stage. Maturity is reached in the 8th or 9th stage, but molting continues after that. The maximum age is three years in *Machilis* (Dunger 2005).

Fossil record. The oldest insect fossils from the Devonian resemble archaeognathans, but the record of extinct taxa which can be unambiguously assigned to this order is sparse. †*Triassomachilis uralensis* (†Triassomachilidae), the oldest Mesozoic representative of the group, displays the typical thoracic hump but lacks the enlarged compound eyes and the enlarged maxillary palps. The other fossil archaeognathans are preserved in Cretaceous or Tertiary amber (Grimaldi & Engel 2005).

Economic importance. Archaeognatha are economically irrelevant.

[Snodgrass (1935); Sturm & Machida (2001); Dunger (2005); Grimaldi & Engel (2005)]

6.5 Zygentoma (Greek *zyg* = bridge, *entoma* = insect, English common names: silverfish and firebrats)

Diversity and Distribution. With ca. 510 described species (L. Mendes, pers. comm.) the diversity of Zygentoma is comparable to that of Archaeognatha. The distribution is world-wide, with a preference for tropical and subtropical regions.

Taxonomy. The order is subdivided into the four the families Maindroniidae (3 spp. in Chile, Afghanistan, Arabia and Sudan), Lepismatidae (more than 250 spp., world-wide) (Fig. 6.5.1), Nicoletiidae (more than 150 spp.), Protrinemuridae (10 spp.) (Mendes 2002), and Lepidotrichidae. Lepidotrichidae comprise only the single extant species *Tricholepidion gertschi* (Northern California) and its systematic position is disputed (see **Insecta [=Ectognatha] and Dicondylia**).

Diagnosis. Often characterized by the silvery glitter of cuticular scales (silverfish). Very similar to Archaeognatha in size and shape and some other features such as the presence of three long terminal appendages. Zygentoma differ from Archaeognatha by the presence of the secondary mandibular joint, the reduced or absent compound eyes and ocelli (not in Lepidotrichidae) (Figs 6.5.1, 6.5.2), the distinctly flattened body with horizontally oriented and flattened coxae, and the reduced number or absence of coxal vesicles and styli (not in Lepidotrichidae).

Fig. 6.5.1: *Lepisma saccharina* (Lepismatidae), SEM micrograph, lateral view. Abbr.: cer – cerci, cxstyl – coxal stylus, tfil – terminal filament. Scale bar: 500 µm. Courtesy H. Pohl.

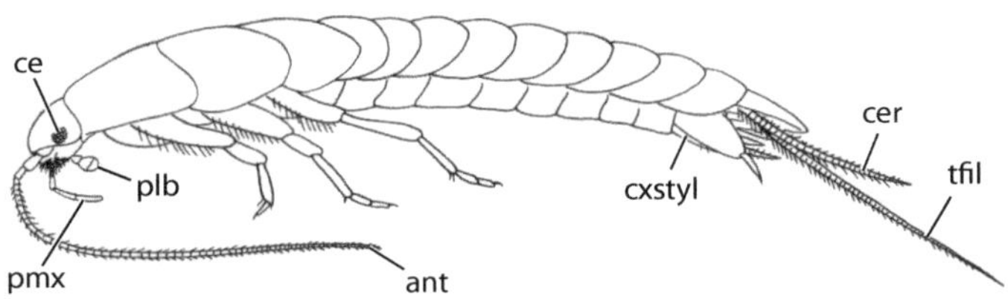

Fig. 6.5.2: *Lepisma saccharina* (Lepismatidae), lateral view. Abbr.: ant – antenna, ce – compound eye, cer – cerci, cxstyl – coxal stylus, plb – palpus labialis, pmx – palpus maxillaris, tfil – terminal filament. Redrawn from Dunger (2005) (after Handschin 1929).

Autapomorphies. The monophyly of the order including Lepidotrichidae is disputed with respect to *Tricholepidion* (see above).

- Four-segmented labial palps
- Hypopharyngeal superlinguae absent (?) (usually also missing in Pterygota)
- Sperm conjugation (?) (not in all groups, the specific condition varies)
- Specific articulation of the cerci

Morphology. Usually medium sized (ca. 12–15 mm), but ranging between 1.4 mm and 20 mm (without terminal appendages). Like in Archaeognatha the body is usually fusiform and tapering posteriorly, but more sub-parallel body shapes also occur. The thorax is more flattened (Fig. 6.5.2). The cuticle is generally thin and not strongly sclerotized. Zygentomans are fragile insects and easily loose body appendages. Scales are present (Lepismatidae, some Nicoletiidae) or absent (e.g., Lepidotrichidae). Brown or

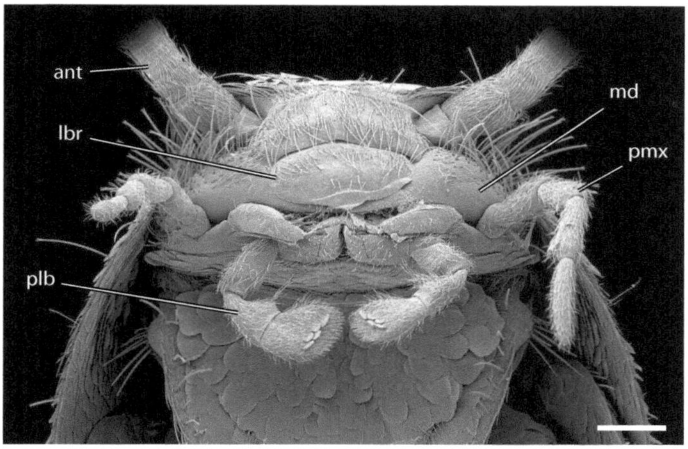

Fig. 6.5.3: *Lepisma saccharina* (Lepismatidae), SEM micrograph, head frontal view. Abbr.: ant – antenna, lbr – labrum, md – mandible, plb – palpus labialis, pmx – palpus maxillaris. Scale bar: 250 µm. Courtesy H. Pohl.

blackish hypodermal pigmentation occurs but many species are largely or completely unpigmented.

The orthognathous head is slightly retracted into the prothorax. Ocelli are present in Lepidotrichidae; they are greatly reduced in Lepismatidae (Figs 6.5.1, 6.5.2), where they form a median frontal organ, and absent in the other groups; the compound eyes are normally developed (Lepidotrichidae), composed of 12 ommatidia (Lepismatidae), or completely reduced; all species of Nicoletiidae are eyeless. A complete postoccipital ridge is present. The coronal and frontal sutures form the typical Y-shaped dorsal ecdysial lines. A distinct transverse strengthening ridge is present between the antennae. The labrum is free. The flagellate antenna is inserted above the mandibular articulations and anterad the compound eye (if present); it is similar to the archaeognathan antenna, but the jointlets are missing. In contrast to Archaeognatha a secondary (anterior) mandibular articulation is present; it is not a ball-and-socket joint as in most pterygote insects but rather a gliding device, with a considerable flexibility at the mandibular base; mesally, distad the secondary joint, a tuberculate molar area and a group of apically bifid stiff setae are present; the mesal mandibular margin bears a brush of long setae (Chaudonneret [1950]: *Thermobia domestica*). The maxillae represent the typical orthopteroid type with a transverse cardo and a large stipes bearing the endite lobes and palp; the galea is apically rounded and mainly sensorial; the lacinia is sclerotized and apically pointed; its mesal edge bears stout hook-like setae distally, and straight, apically bifid stiff setae more proximally; the moderately long palp is 6-segmented (or 5-segmented with a distinct, segment-like palpifer) (Chaudonneret [1950]: *Thermobia domestica*). The labium is composed of a postmentum which is much wider than long, and a complex prementum which is ventrally divided by a longitudinal sagittal furrow; in contrast to Archaeognatha the glossae and paraglossae are undivided; the latter is not distinctly separated from the anterolateral prementum; the palp is 4-segmented, with a small basal segment, and two flattened distal palpomeres; the apical segment is usually broadened and equipped with specialized sensilla. In contrast to Archaeognatha, the hypopharynx lacks superlinguae. The anterior and posterior tentorial arms are connected; a large plate-like central element is sometimes present. The mandibular musculature is similar to that of Archaeognatha despite of the modified mandibular articulation; however, the transverse mandibular tendon (ligamentous head endoskeleton) is only retained in *Tricholepidion*. In contrast to Staniczek (2000), a hypopharyngo-mandibular muscle is developed in this genus. A pair of labial kidneys or two pairs of salivary glands are present.

The simple thoracic tergites usually bear paranotal lobes like in Archaeognatha (Fig. 6.5.2). Three small sclerites are present in the pleural region; they bear lobe-like posterior extensions in Lepismatidae. The coxae are large, flattened and posteriorly directed, almost parallel to the longitudinal body axis; they do not bear styli; the maximum number of tarsomeres is five; the pretarsus bears two lateral claws and one or several middle claws. The thorax as a whole is distinctly flattened, and looped thoracic muscles (characteristic for Archaeognatha) are absent.

The 11-segmented abdomen is usually more distinctly separated from the thorax than in Archaeognatha. The tergites are less convex. The large fused sternites are separated from the laterosternites in Lepidotrichidae and some Nicoletiidae. Coxal vesicles are present on segments II–VII in Lepidotrichidae, in contrast to all other groups; styli are usually present on segments II–IX (Fig. 6.5.2), but sometimes only on segments VII–IX or entirely lacking. The terminal filament and cerci are similar to those of Archaeognatha but lack scales. The ovipositor is elongate like in Archaeognatha; a triangular gonangulum connecting the gonocoxae VIII and IX and tergite IX is distinctly developed.

The foregut is generally characterized by a long ingluvies, and a well-developed proventriculus occurs in Lepismatidae. Like in Pterygota well-developed transverse anastomoses of the tracheal system are present. The ovaries are composed of 2–7 (Lepidotrichidae) panoistic ovarioles; a spermatheca and accessory glands open into the genital atrium. The testes usually contain 2–7 follicles but a higher number is present in Lepidotrichidae.

Biology. Zygentoma mostly prefer humid environments but some species also occur in drier regions. They generally prefer higher temperatures but avoid exposure to sunlight. The group is characterized by a tendency to live in cryptic habitats, linked with a trend to reduce the light sense organs. Nicoletiidae and Protrinemuridae are generally eyeless, and species of these families live in caves or in crevices, or are inquilines of ants or termites (Nicoletiidae partim). The synanthropous lepismatid species

Lepisma saccharina is specialized on moist places such as bathrooms or kitchens, and *Thermobia domestica* frequently occurs in bakeries. Zygentoma have a broad food spectrum including fungi, algae and other plants, but also cereals, paste, paper, starch in clothes, rayon fabrics, dried meats, and carrion of silverfish.

Reproduction and development. Zygentoma are usually bisexual but parthenogenesis apparently occurs, at least in some species of Nicoletiidae. The sperm transfer is external. Males of *Lepisma* deposit signal threads and a bottle-shaped spermatophore on the soil substrate after having made contact with the female. The light-colored eggs deposited by females of *Lepisma* become darker with the progressive development; eggs of a whitish-grey color are deposited by females of *Nicoletia*. In contrast to Archaeognatha the entire cleavage is superficial. The amnioserosal fold and amniotic cavity are well-developed.

Fossil record. Possible zygentoman fragments are reported from the Devonian and a species from the Late Carboniferous (Czech Republic) was tentatively assigned to the order (*Carbotiplura kukalovae*). The fossil record of the group is almost entirely restricted to resins of the Cretaceous and Tertiary (Grimaldi & Engel 2005).

Economic importance. The economic impact of Zygentoma is limited. Some species of Nicoletiidae can cause minor damage in agriculture, and the synanthropic species *Lepisma saccharina* can be a nuisance.

[Handschin (1929); Chaudonneret (1948; 1950); Staniczek (2000); Sturm & Machida (2001); Mendes (2002); Dunger (2005); Grimaldi & Engel (2005)]

Pterygota

There is little doubt that the acquisition of wings was the most important single innovation in the evolution of Hexapoda (Fig. 6.A: 9). This complex derived feature is a convincing autapomorphy of Pterygota which comprise ca. 99% of all known hexapod species. Apparently the new escape mechanism and the qualitatively improved dispersal capacity have triggered a tremendous radiation in the second half of the Palaeozoic. This is not only suggested by the extreme discrepancy in species numbers (ca. 11,500 described species in the apterygote orders versus ca. 915,000 in Pterygota) but also by the very low diversity of the secondary wingless orders and groups with a strong tendency to reduced the wings (Grylloblattodea: 29 spp., Mantophasmatodea: 18 spp., Zoraptera: ca. 40 spp.).

The ancestor to all winged insects likely diverged from the common ancestor of Zygentoma in the late Devonian, ca. 384 Ma (Wheat & Wahlberg 2013). It was suggested that wings evolved during a period with an exceptionally high oxygen level

(Late Devonian to Late Carboniferous, ca. 375–250 Ma), which also favoured gigantism in different lineages (e.g., species of Odonata and †Palaeodictyoptera with a wing span of nearly 60 cm). However, confirmation by independent (molecular) data is needed (Wheat & Wahlberg 2013).

Grimaldi & Engel (2005) stated that "how, when and why insect wings originated is one of the most perplexing conundrums in evolution" (see also Wheat & Wahlberg [2013]). Two main competing scenarios are presently discussed, the **paranotal lobes concept** and the **exite or gill concept** (s. Grimaldi & Engel [2005] for a detailed discussion).

Snodgrass (1935), Hennig (1969), Hamilton (1971) and others assumed that thoracic paranotal lobes (paranota, paratergites) similar to those found in extant Zygentoma were precursors of wings. This hypothesis is tentatively supported by some Palaeozoic fossils (e.g., Palaeodictyoptera, Eugeropteridae; e.g., Grimaldi & Engel 2005; Klass 2009) which displayed pterothoracic wings *and* similar but smaller prothoracic winglets resembling moderately extended prothoracic paranotal lobes with a recognizable venation. Hasenfuss (2002) outlined an "evolutionary pathway to insect flight starting from lepismatid organization". Several preconditions for flight were pointed out in this study: the movability of lateral paranotal lobes attained by a membranous strip between the tergite and paratergite, the utilization of pre-existing muscles (present in the groundplan of Insecta), and the integration of subcoxal sclerites (present in Zygentoma) in the pleural wing joint as a pivot. Dropping from plants and gradually improved gliding ("wings as parachutes") were considered as intermediate stages eventually leading to active flight (Hasenfuss 2002).

In the alternative **exite or gill concept** (e.g., Woodworth 1906; Wigglesworth 1973; Kukalová-Peck 1983) it is assumed that wings evolved from external appendages of the leg base (exites, e.g., thoracic styli) serially homologous to lateral abdominal gills as they occur in larvae of Ephemeroptera. This was tentatively supported by developmental evidence (Averov & Cohen 1997). It was suggested that exite-homologue precursors of wings were used for sailing ("skimming") on the water surface (e.g., Marden 2003). Various criticisms of the exite or gill concept (e.g., Rasnitsyn 2003) are summarized in Grimaldi & Engel (2005).

Another crucial evolutionary novelty of Pterygota is the internal fertilization with a postabdominal copulatory organ (aedeagus). It is apparent that this is more efficient than external transfer of a spermatophore (apterygote lineages, groundplan of Hexapoda) and it is likely that this also contributed to the evolutionary success of Pterygota. Whether this doubtlessly derived feature is an autapomorphy of Pterygota is an unresolvable question. A completely different mechanism with a secondary copulatory organ at the base of the abdomen evolved in dragonflies and damselflies. Two scenarios are equally parsiomonious: the unique configuration of Odonata may have evolved directly from a mode without copulation or may represent a strongly modified version of an ancestral pterygote condition with internal fertilization with a postabdominal aedeagus (Dallai et al. 2013).

Another potential autapomorphy of Pterygota is the presence of an epicuticular wax layer. This is apparently linked with the necessity to limit water loss via the body surface. Considering the selective pressure resulting from the increased exposure of winged insects, it cannot be excluded that this feature evolved several times independently. Additional autapomorphies of Pterygota are listed in the following:

Autapomorphies (e.g., Kristensen 1991; Hasenfuss 2002; Klass 2009)
- meso- and metathoracic wings with specific venation and articulation
- extensive sternal and pleural sclerotization
- pleuron stiffened by distinctly developed pleural ridge
- delimitation of the meso- and metascutellum
- subalare present as delimited element above pleuron
- ligamentous endoskeleton completely or largely reduced
- anterior and posterior tentorium fused (pre- and metatentorium)
- loss of the paired circumoesophageal vessels
- legs with ligamentous diaphragm separating the haemolymph currents
- abdominal coxal vesicles completely reduced
- lateral parts of abdominal segment XI reduced
- epicuticular wax layer (?)
- internal fertilization (?)

[Woodworth (1906); Snodgrass (1935); Hennig (1969); Hamilton (1971); Wigglesworth (1973); Kukalová-Peck (1983); Kristensen (1991); Averof & Cohen (1997); Hasenfuss (2002); Marden (2003); Rasnitsyn (2003); Grimaldi & Engel (2005); Klass (2009); Dallai et al. (2013); Wheat & Wahlberg (2013)]

The basal splitting events in Pterygota – the "Palaeoptera-Problem"

The interrelationships of the three major pterygote lineages – Odonata, Ephemeroptera and Neoptera – are one of the most persistent challenges in systematic entomology. In the traditional Palaeoptera concept (Fig. 6.A: 10) Ephemeroptera (mayflies) are the sistergroup of Odonata (dragonflies and damselflies) (e.g., Hennig 1969; Kukalová-Peck 1991). As a conspicuous plesiomorphic feature both groups share the inability to flex the wings over the abdomen. Potential synapomorphies suggesting a monophyletic origin include the aquatic larvae (naiads), shortened, bristle like antennae, characters of the wing venation and articulation, and the configuration of the maxilla (e.g., Hennig 1969; Bechly et al. 2001; Haas & Kukalová-Peck 2001; Willkommen & Hörnschemeyer, 2007; Kukalová-Peck 2008). A sister-group relationship between Ephemeroptera and Neoptera (Chiastomyaria concept; Fig. 6.A: 12) is suggested by the mode of direct sperm transfer involving a postabdominal intromittent organ (aedeagus, penis), the dominant role of indirect flight muscles, and analyses of rRNA sequences (Boudreaux 1979; Kjer 2004; Simon et al. 2009). A sistergroup relationship between Odonata and Neoptera (Metapterygota concept; Fig. 6.A: 11) was mainly suggested by characters of the mandibles and their muscles (Staniczek 2001) and derived features of the respiratory system (Kristensen 1991). It was also supported in molecular studies (Ogden & Whiting 2003; Terry & Whiting 2005) and an additional argument was the absence of an immature winged subimago, which is preserved in Ephemeroptera. The Metapterygota concept was widely accepted for some time (e.g., Beutel & Gorb 2001, 2006), but the Palaeoptera hypothesis gained strong support in recent studies. It was shown by Blanke et al. (2012a) that Staniczek's interpretation of mandibular features (Staniczek 2001) suffers from a serious lack of information on head structures and associated muscles of Odonata. Moreover, it was demonstrated by Blanke et al. (2012b) that serious artifacts can result from 'concerted convergence', i.e. linked character transformations within functional complexes such as for instance the mandibular articulation and musculature.

The "Palaeoptera" problem should still be considered as an unsolved issue. However, a sistergroup relationship between Ephemeroptera and Odonata (Palaeoptera) appears as the most plausible option presently.

Potential synapomorphies of Ephemeroptera and Odonata (Palaeoptera)
(e.g., Hennig 1969; Kukalová-Peck 1991; Blanke et al. 2012a, b)
- Shortened, bristle-like antennae (probably not in the groundplan of both groups) (?)
- Pedicellus longer than scapus
- Loss of antennal circulatory organs in adults
- Lacinia with single invisivus (Blanke et al. 2012a)
- Dentisetae (extremely enlarged subapical bristles of the lacinia) (Blanke et al. 2012a)
- Parts of the archaedictyon (network of crossveins) rearranged to form long straight lines between longitudinal veins
- Aquatic larvae (polarity?)
- Galea and lacinia fused in larvae

Potential synapomorphies of Ephemeroptera and Neoptera (Chiastomyaria)
(Boudreaux 1979)
- Indirect flight musculature plays a dominant role in flight
- Internal fertilization with a postabdominal aedeagus

Potential synapomorphies of Odonata and Neoptera (Metapterygota)
(Kristensen 1991; Staniczek 2001; Wheeler et al. 2001)
- Subimago missing (no molt of an immature winged stage)
- Secondary mandibular joint modified as ball-and-socket joint (mandible moves in one plane)

- Tentoriomandibular muscles partly reduced, mandible mainly moved by cephalic adductor and abductor (?)
- Terminal filament always reduced in adults (present in few plecopteran larvae)
- Tracheae of wings and legs connected with spiracle of following segment
- Occlusor muscles directly inserted on abdominal spiracles

[Hennig (1969); Boudreaux (1979); Kristensen (1991); Kukalová-Peck (1991, 2008); Bechly et al. (2001); Beutel & Gorb (2001, 2006); Bechly et al. (2001); Haas & Kukalová-Peck (2001); Staniczek (2001); Ogden & Whiting (2003); Kjer (2004); Terry & Whiting (2005); Willkommen & Hörnschemeyer (2007); Simon et al. (2009); Blanke et al. (2012a, b)]

6.6 Ephemeroptera (Greek *ephemeros* = short-lived, *pteron* = wing, refers to the brief lifespan of adults, English common name: mayflies)

Diversity and Distribution. Ephemeroptera are a medium sized insect order with over 3,000 extant species (Barber-James et al. 2008). The group has a worldwide distribution, with the exception of Antarctica and some Oceanic islands. Most species occur in the tropical zone. Mayflies reach the highest generic diversity in the Neotropics, while the Palearctic has the lowest generic diversity but a high species diversity (Barber-James et al. 2008). Endemism is common and even occurs on the family level. Rallidentidae and Siphlaenigmatidae are restricted to New Zealand, and Dipteromimidae to Japan. According to Barber-James et al. (2007) numerous unknown genera and species await description, especially in tropical areas.

Autapomorphies
- Mouthparts of adults vestigial (Fig. 6.6.3)
- Compound eyes of males enlarged (Fig. 6.6.3)
- Palmén's organ (extension of head tracheae)
- Males with additional joint between tibia and proximal tarsomere
- Midgut filled with air
- Hindwing reduced in size (Fig. 6.6.1)
- Anal region of wings reduced
- Naiads with lateral abdominal tracheal gills (Fig. 6.6.2)

Taxonomy. The precise position of Ephemeroptera within Pterygota is a long disputed matter. They were either placed as sistergroup of Odonata (Palaeoptera), as sistergroup of a clade comprising Neoptera and Odonata (Metapterygota), or as sistergroup of Neoptera (Chiastomyaria) (see **The basal splitting events in Pterygota – the "Palaeoptera problem"**). The intraordinal phylogeny and classification are still disputed. McCafferty (1991) subdivided the group into four suborders, Carapacea

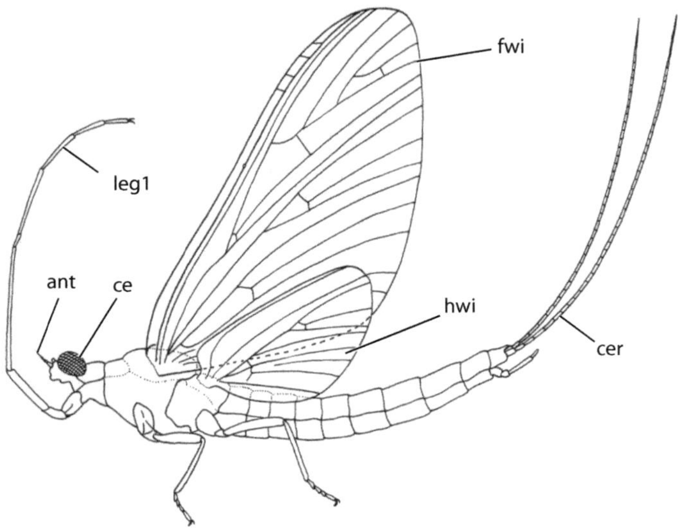

Fig. 6.6.1: *Rhithrogena* sp. (Heptageniidae), habitus. Abbr.: ant – antenna, ce – compound eye, cer – cerci, fwi – fare wing, hwi – hind wing, leg1 – fore leg. Redrawn from Bauernfeind (2005) (from Mizzaro 1983).

(Baetiscidae, Prosopistomatidae), Furcatergalia (four infraorders including Pannota, e.g., Leptophlebiidae, Behningiidae, Potamanthidae, Euthyplociidae, Ephemeridae), Setisura (Arthropleidae, Heptageniidae) and Pisciforma (possibly paraphyletic; e.g., Baetidae, Ameletopsidae, Oniscigastridae, Rallidentidae, Siphlonuridae). In a recent study based on morphological characters and molecular data (12S, 16S, 18S, 28S, H3 genes) Ogden et al. (2009) suggested a sistergroup relationship between the Asian genus *Siphluriscus* (Siphluriscidae) and the remaining Ephemeroptera. Carapacea, Furcatergalia, Fossoria, Panota, Caenoidea and Ephemerelloidea were supported as clades (Ogden et al. 2009: figs 5, 7). Soldán (2003) recognized 37 families with a total of 376 genera. Forty two families (with more than 400 genera) are listed in Barber-James et al. (2008).

Diagnosis. Adult mayflies are characterized by large, often sexually dimorphic compound eyes, reduced mouthparts, two pairs of unequally sized membranous wings (hindwings smaller or rarely absent), and three long terminal appendages (terminal filament and cerci). The aquatic immatures (naiads) possess lateral tracheal gills (Fig. 6.6.2).

Morphology of adults. Mayflies are usually medium sized insects. The body size ranges between 2 mm and 50 mm. The short-lived adults are slender, very fragile, and

Fig. 6.6.2: Ephemeroptera, naiads, SEM micrographs. A, Heptageniidae, lateral view, courtesy H. Pohl; B, *Ephemera* sp. (Ephemeridae), dorsal view. Abbr.: cer – cerci, tgil – tracheal gills. Scale bars: 1 mm.

unable to fold back the wings (Fig. 6.6.1). The cuticle is thin and weakly sclerotized except for articulations and muscle attachment areas.

The small head is relatively well sclerotized. It is triangular in dorsal view. Its main function is bearing the large compound eyes (Fig. 6.6.3), which are always well-developed and sometimes turban-shaped or completely subdivided in males; three ocelli are present. The antennae are very short and composed of scapus, pedicellus, and a short, bristle-like flagellum. The adult mouthparts are strongly reduced and not suitable for processing food (no food uptake in the adult stage). Vestigial mandibles are present in the groundplan, but most ephemeropteran adults only possess greatly reduced maxillae and a labium. The suboesophageal complex is strongly reduced in correlation with the reduced mouthparts. The Palmén's organ is an enlarged, bulla-shaped extension of the head tracheae.

The prothorax is small and composed of only few skeletal elements (Fig. 6.6.1). The pronotum is transverse and rectangular. Single pleural plates are present or absent. The sternum is usually triangular and shield-like; the well-developed profurcae originate at its caudal margin. The mesothorax is the largest segment and dorsally strongly convex. Posteriorly it is firmly connected with the metathorax; both segments form a functional unit. The large mesotergum is subdivided into a pre-

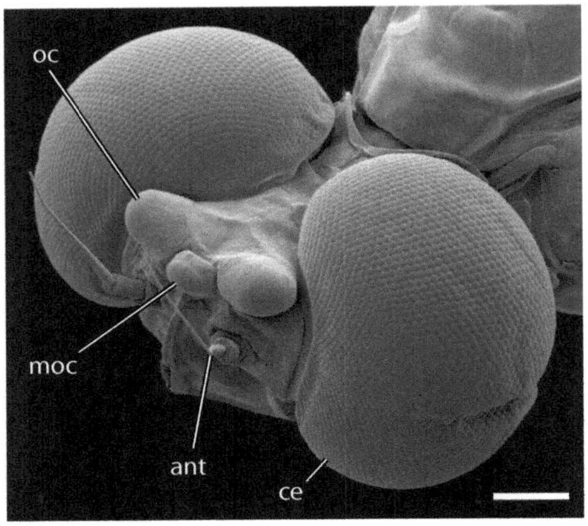

Fig. 6.6.3: *Baetis* sp. (Baetidae), male head, anterolateral view, SEM micrograph. Abbr.: ant – antenna, ce – compound eye, moc – median ocellus, oc –ocellus. Scale bar: 200 µm. Courtesy H. Pohl.

scutum, scutum, scutellum and postnotum; the scutum is often subdivided into oval fields which serve as attachment areas for the dorsoventral indirect flight muscles. The episterna reach onto the ventral side of the segment; they are fused with the epimera. The ventral side of the segment is covered by a sternum and poststernum; the paired mesofurcae originate at the posterior mesosternal margin; they are smaller than their prothoracic counterparts. The metathorax is distinctly reduced in size. Dorsally scutum, scutellum and postnotum can be distinguished, laterally the episternum and epimeron, and ventrally an undivided sternum. The metafurca is present or absent. The wings are held in an upright position at rest (Fig. 6.6.1) and cannot be folded back over the abdomen (palaeopteran condition); the triangular forewings are always larger than the hindwings; coupling mechanisms connecting both pairs are not present; the anterior wing margin is strengthened by a costal bridge and reinforced anterior longitudinal veins; the venation is reticulate, with numerous cross veins; the anal region is reduced; the hindwings can be largely or completely reduced (Caenidae, many baetids). The thoracic musculature is similar to what is found in most neopteran insects, with strongly developed indirect flight muscles. The legs of adults are usually less strongly developed than in naiads and different degrees of reduction occur; two claws articulate with the pretarsus, and one of them is often modified as an attachment device (claw pad); the forelegs of males are usually elongated and a specialized additional tibiotarsal joint is present; they are used for grasping the females during copulation.

The abdomen is composed of ten distinct segments, with well-developed tergites I–X. The spiracles I–VIII are inserted on the lateral tergal areas of the respective seg-

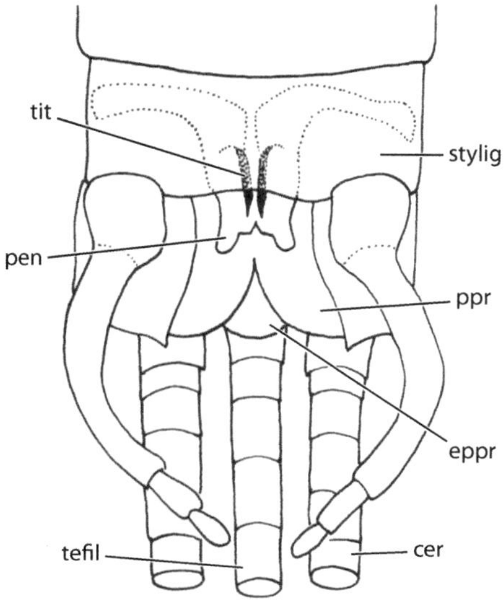

Fig. 6.6.4: *Ephemera* sp. (Ephemeridae), male postabdomen, ventral view. Abbr.: cer –cerci, eppr – epiproct (segment XI), pen – penis, ppr – paraproct (segment XI), stylig – styliger, tefil – terminal filament, tit – titilator. Redrawn from Bauernfeind (2005) (from Landa 1969).

ments. Sternite X is not developed. A vestigial segment XI is represented by a ventral sclerite (paraproct) at the base of the cerci. The cerci and the terminal filament are usually multisegmented and elongate, but the latter is absent in some groups. The paired genital ducts of females open behind sternite VII, which is rarely elongated and resembling an ovipositor, sometimes also involving sternite VIII (Leptophlebiidae); in some groups it covers a genital vestibulum as a subgenital plate (e.g., Heptageniidae, Siphlonuridae) (Bauernfeind 2005). In males a caudal extension of sternite IX forms an articulated styliger-plate (forceps base, gonocoxite) (Fig. 6.6.4); it bears the segmented gonostyli, which together function like a clasping apparatus during copulation (forceps). The paired penis is movably connected with segment IX by the more strongly sclerotized penis arms and often reinforced by species specific sclerotizations; the penis lobes are often equipped with thorns, fields of bristles or titillators, which may functionas as accessory attachment devices (Bauernfeind 2005).

The midgut is disconnected from the fore- and hindgut and filled with air. The number of Malpighian tubules ranges between 40 and 160. The testes contain numerous follicles; the terminal part of the vasa deferentia is strongly extended, thus forming the vesiculae seminales which contain the mature sperm. The ovarioles are panoistic or meroistic.

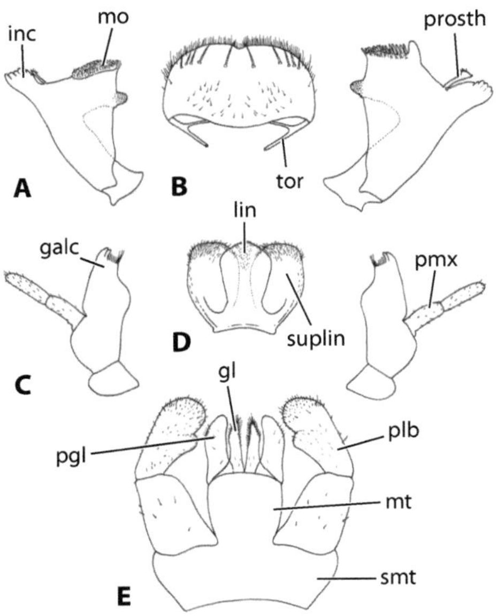

Fig. 6.6.5: *Baetis* sp. (Baetidae), naiad, mouthparts. A, mandible; B, labrum; C, maxilla; D, hypopharynx; E, labium. Abbr.: galc – galealacinia, gl – glossa, inc – mandibular incisivi, lin – lingua, mo – mola, mt – mentum, pgl – paraglossa, plb – palpus labialis, pmx – palpus maxillaris, prosth – mandibular prostheca, smt – submentum, suplin – superlinguae, tor – tormae. Redrawn from Bauernfeind (2005) (from Grandi 1960).

Morphology, immature stages (naiads). As the aquatic immature stages of Ephemeroptera (Fig. 6.6.2) possess specific larval structures and adaptations, they could be addressed as larvae (instead of nymphs) despite of the hemimetabolous development. They are also referred to as naiads, a term which is used in the following.

The head is triangular or transversely ellipsoid in frontal view. The compound eyes are usually well-developed and larger in males. The labrum is free. The multisegmented antennae are usually relatively short but less strongly reduced than in the adults. The mouthparts are well-developed (Fig. 6.6.5); they are basically of the orthopteroid (biting) type, but modified in many different ways depending on the feeding habits (Fig. 6.6.6); they can be modified as burrowing, scraping or filtering devices. The mandibles (Fig. 6.6.5) bear an articulated mesal process (prostheca). The endite lobes of the maxilla are fused (galeolacinia). The hypopharyngeal superlinguae are usually well-developed.

In some groups with flattened naiads (Heptageniidae, Behningiidae), the prothorax can be relatively large and broadened (Fig. 6.6.2A). Like in the adults the mesothorax is larger than the metathorax and both segments form a functional unit. In

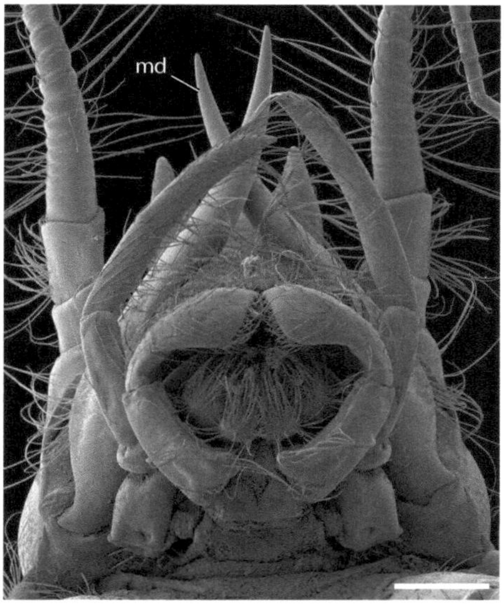

Fig. 6.6.6: *Ephemera* sp. (Ephemeridae), naiad, head ventral view, SEM micrograph, see schematic drawings in Fig. 6.6.5 for details. Abbr.: md – mandible. Scale bar: 200 µm.

some specialized families the mesothorax is distinctly extended posteriorly and in the extreme case it forms a carapax-like duplicature which covers not only the metathorax but also the abdominal segments I–VI (Baetiscidae). The thoracic legs as a whole are more strongly developed than in the adults, but the tarsus is 1-segmented and only a single claw is present; they can be modified in many different ways depending on the lifestyle (see below); in some groups the forelegs bear fringes of hairs as a filtering device (e.g., Oligoneuriidae); in burrowing naiads they are modified as shovel-like structures (e.g., Ephemeridae); tube-like gills inserting on the coxal bases are only rarely present (e.g., *Baetodes*). The external wing buds appear in a later immature stage; they appear black before the molt to the winged subimago; the buds of the forewings are largely fused in the Pannota (see Taxonomy).

The 11-segmented abdomen of the naiads is flattened or cylindrical depending on the life style. Posteriorly directed, plate-like structures can be present on the posterolateral margins of flattened forms, and thorn-like outgrowths of tergites or fringes of hairs on the pleural or sternal regions also occur. The main characteristic is the presence of seven pairs of gills inserted on the dorsal side; the gills can be tube-like and densely pubescent (burrowing forms) but are usually divided longitudinally; in many cases they are composed of a flattened anterior lamella and a posterior tube-like element. Like in the adults the abdominal apex bears long cerci and usually also a terminal filament.

The tracheal system is closed, without functional spiracles. The digestive tract is normally developed; caeca are absent.

Biology. The naiads are generally aquatic. They occur in a very broad variety of habitats including brackish water, and usually stay under rocks, among submerged vegetation or in the sediment. The life habits are very diverse; five different types can be distinguished; burrowers (e.g., Ephemeridae) use gravelly substrate, mud, or rarely wood (e.g., *Asthenopus*); naiads of the dorsoventrally flattened type stay on the surface of stones in stronger current or in the narrow spaces below them; free swimming naiads (e.g., *Baetis*) are characterized by a strongly developed abdominal musculature and dense fringes of hairs on their terminal appendages; naiads of the crawling (e.g., Ephemerellidae) and wriggling type (Leptophlebiidae) mainly use submerged plants, leave litter or move among gravel; specialized species living on drifting sand in large rivers are characterized by strongly elongated legs or dense pads of microtrichia. The food spectrum of the naiads is very broad; it includes algae, diatoms, fungi, and detritus (filter feeders, scrapers), and few species are predaceous (e.g., Behningiidae).

Reproduction and development. On average 150–200 large, round or oval eggs are deposited by the females. The duration of the embryonic development varies greatly and obligatory or facultative egg diapause or delayed hatching occur. As in the apterygote orders numerous molts take place after hatching (20–40 times), and in contrast to all other pterygote insects an additional final ecdysis of a fully winged preimaginal stage (subimago) takes place. The duration of the development depends on the species and the temperature of the habitat; it may last only 16 days or several years and sometimes includes a diapause; some species are multivoltine. The freshly hatched first instars do not possess gills and also differ in other features from the following stages. The last stage molts on the water surface or very close to the edge of the aquatic habitat. The fully winged subimago is characterized by less bright colors, shorter body appendages, and a cuticle with a non-wettable, dense vestiture of microtrichia. After a short period ranging between only few minutes and two days the adults emerge. Synchronized hatching is typical and related to the very short time span of the mature mayflies; it can lead to mass emergence, sometimes with tens of millions of individuals (see below). Copulation takes place in flight. Usually the adult males form swarms (mating flight) and move rhythmically in vertical direction. Females entering the agglomeration of males are grasped with the forelegs and the forceps embrace the female abdomen; the terminal appendages of the males are bent forward horizontally between the wings of the females, thus stabilizing the flight. The formation of swarms by females is the exception. Oviposition takes place immediately after mating; the eggs are usually deposited on the surface of stagnant or more frequently running water bodies, where they sink to the bottom. The tip of the abdomen may be inserted into the water or packages of eggs are "air dropped" (Grimaldi & Engel 2005).

In some groups the abdomen is simply ruptured upon contact with the water surface (*Caenis*). Some mayflies carefully deposit their eggs on suitable substrate such as for instance stones under water.

Fossil record. Some fossils from the Upper Carboniferous (e.g., *Lithoneura* [from Mazon Creek]) may belong to Ephemeroptera or are close relatives (Grimaldi & Engel 2005). The earliest definitive mayflies are known from early Permian deposits (ca. 290 Ma). The group probably reached its highest diversity in the Mesozoic (Brittain & Sartori 2003).

Economic importance. In some regions of Africa and New Guinea mayflies are used as human food. Many species play an important role in aquatic ecosystems, especially as food of predaceous fish species (e.g., trout). In cases of mass emergence extreme numbers of mayflies can cause traffic problems. Wood boring naiads of some tropical species can cause a limited economic damage.

[Illies et al. (1968); Landa (1969); Mizzaro (1983); McCafferty (1991); Brittain & Sartori (2003); Soldán (2003); Bauernfeind (2005); Grimaldi & Engel (2005); Barber-James et al. (2008); Ogden et al. (2009)]

6.7 **Odonata** (Greek *odontos* = tooth, refers to mandibular teeth, English common names: damselflies, dragonflies)

Diversity and Distribution. Like Ephemeroptera, Odonata are a medium sized insect order. Approximately 5,600 extant and 600 fossil species are described. The distribution is worldwide, with the exception of the circumpolar regions. The highest number of species is presently recorded from the Oriental region, but the diversity is also high in Africa, and the number of described South American species is rapidly increasing.

Autapomorphies
- Head unusually large (Fig. 6.7.1)
- Cervical region strongly constricted, with specialized sensorial pads
- Antennae short, with bristle-like flagellum (Fig. 6.7.1A)
- Pterothoracic skewness
- Legs forming a prey catching basket, not used as locomotor organs
- Wings moved by indirect-direct flight system
- Abdomen long and slender, stabilizing flight
- Secondary copulatory organ at the abdominal base of males
- Labium of naiads transformed into prehensile prey catching apparatus (labial mask) (Fig. 6.7.2)

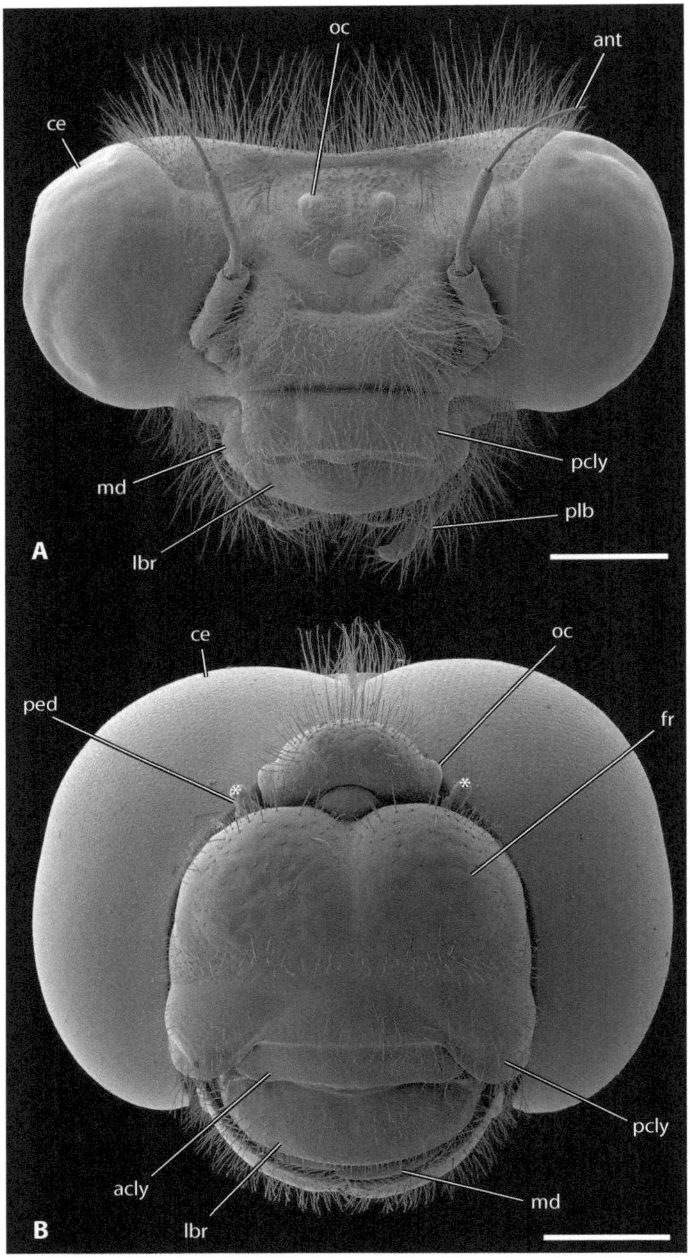

Fig. 6.7.1: Odonata, adult head, frontal view, SEM micrographs. A, *Calopteryx splendens* (Zygoptera, Calopterygidae). B, *Sympetrum sanguinea* (Anisoptera, Libellulidae). Abbr.: acly – anteclypeus, ant – antenna, ce – compound eye, fr – frons, lbr – labrum, md – mandible, oc – ocellus, pcly – postclypeus, ped – pedicellus, plb – palpus labialis, * – flagellum lacking. Scale bar: 1 mm.

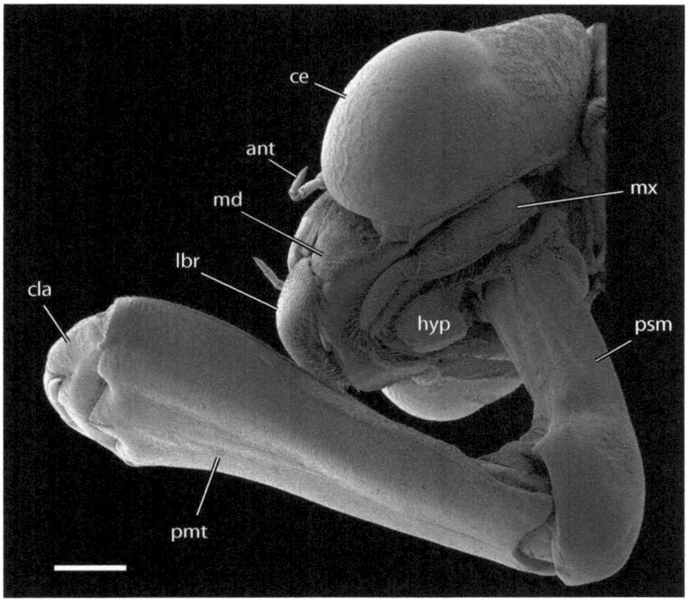

Fig. 6.7.2: *Aeshna cyanea* (Anisoptera, Aeshnidae), head of naiad, ventrolateral view, SEM micrograph. Abbr.: ant – antenna, ce – compound eye, cla – claps, hyp – hypopharynx, lbr – labrum, md – mandible, mx – maxilla, pmt – prementum, psm – postmentum. Scale bar: 1 mm.

Taxonomy. Odonata are a basal lineage of Pterygota. Like Ephemeroptera they lack the ability to fold back the wings over the abdomen. The corresponding apomorphic condition, foldable wings, characterizes the Neoptera, which comprise all other pterygote orders. The group is subdivided into two monophyletic suborders Zygoptera (damselflies; e.g., Calopterygidae, Lestidae, Coenagrionidae) and Epiprocta. The former are characterized by a laterally strongly expanded head, naiads with three long caudal gills, and almost identical fore- and hindwings, the latter by greatly enlarged compound eyes, the internal rectal gills of the naiads (often also used as a jet-propulsion system), and differently shaped fore- and hindwings. The last feature is reflected by the name Anisoptera. Epiprocta are subdivided into the small infraorder Epiophlebioptera (2 spp., Japan, Himalaya) and the comparably large group Anisoptera (dragonflies: e.g., Aeshnidae, Gomphidae, Cordulegastridae, Libellulidae).

Diagnosis. Medium sized or very large, colorful, very attractive insects. Wings stiff and usually transparent, with a reticulate venation. Abdomen elongate and usually cylindrical. Males with secondary copulatory apparatus at the base of the abdomen. Larvae (naiads) almost exclusively aquatic. Labium transformed into a prey catching apparatus.

Morphology of adults. Odonata, especially Anisoptera, are large insects. The maximum length of extant species is found among Zygoptera (*Megaloprepus caerulatus*, length 150 mm, wingspan 200 mm). The body length usually ranges between 30 and 90 mm. Extinct species from Upper Carboniferous deposits belong to the largest know insects (ca. 370 mm body length and up to 750 mm wingspan). The cuticle is usually characterized by conspicuous color patterns with different pigments and pigment bodies in the epidermis or in deeper layers of tissue.

The large orthognathous head is mainly characterized by very large hemispherical, eucone compound eyes (Fig. 6.7.1). Approximately 10,000 ommatidia per eye are present in Zygoptera and ca. 30,000 in Anisoptera. The foramen occipitale is strongly narrowed and the head is freely movable; together with fields of sensilla in the cervical region the head functions as a sense organ for centrifugal forces. The shape of the head capsule is transverse in Zygoptera (Fig. 6.7.1A) and posteriorly concave in Anisoptera. Three well-developed ocelli are present (Fig. 6.7.1). The vertex is usually not distinctly separated from the frons but subdivided by a transverse interantennal ridge between the eyes. The clypeus is large and subdivided into a postclypeus and anteclypeus. The frontoclypeal (epistomal) transverse ridge is distinct. The transverse labrum is movably connected with the anterior clypeal margin (Fig. 6.7.1). The antennae are short; the flagellum is bristle-like and composed of only 3–6 segments. The mouthparts are well-developed and of the orthopteroid (biting) type. The secondary (anterior) mandibular articulation is a ball-and-socket joint; consequently the axis of movement of the mandible is restricted to a transverse level; the apical part of the mandible is divided into three acute teeth and additional teeth are present on the broad basal molar area. The maxilla is composed of cardo, stipes (divided into basi- and mediostipes), lacinia and a one segmented palp; a galea is lacking; the lacinia is hook-shaped and sclerotized, with one acute apical tooth and spines and finer setae (proximally) along its mesal edge; the palp is apically pointed and densely covered with minute hairs. The broad labium is composed of a postmentum and prementum, the latter bearing a median lobe, possibly representing fused glossae and paraglossae, and paired lateral 1-segmented palps; the palps are equipped with two appendages, the fixed mesal terminal hook and a movable hook, which is not equipped with muscles. The tentorium is complete, with a bridge and posterior, dorsal and anterior arms. The mandibles are almost exclusively moved by the very large cranial adductor and the less strongly developed abductor. Salivary glands are present but small.

The prothorax is small and movably connected with the pterothorax. Females of Zygoptera possess species specific pronotal devices fitting with the male abdominal clasping apparatus. The pterothoracic pleural sclerites are well-developed and like the entire segments posteriorly slanting (thoracic skewness); consequently the coxal articulations are shifted anteriorly and the wing articulation posteriorly. The mesepisterna are greatly enlarged anterodorsally and adjacent with each other anterad the wing base, thus forming distinctly developed “shoulders”. Two pairs of wings (Fig. 6.7.3) with a reticulate venation with numerous transverse veins are always

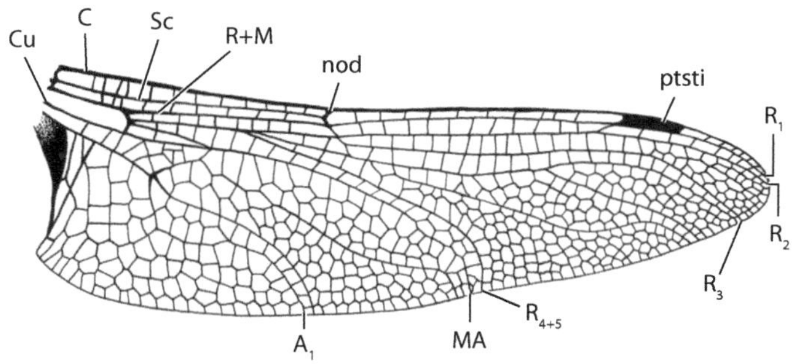

Fig. 6.7.3: Hindwing, *Aeshna* sp. (Anisoptera, Aeshnidae). Abbr.: A_1 – 1st anal vein, C – costa, Cu – cubitus, ptsti – pterostigma, MA – media anterior, nod – nodus, $R_{1-3/4+5}$ – branches of radius, R+M – radius + media, Sc – subcosta. Redrawn from Xylander & Günther (2005), original Conci & Nielsen (1956), modified.

present; the articulatory sclerites which are usually present in Neoptera are missing with the exception of pterale 1; like in Ephemeroptera the wings cannot be folded back over the abdomen ("Palaeoptera"); they are held in an upright (Zygoptera, Epiophlebioptera) or more or less horizontal position (Anisoptera) at rest; the main longitudinal veins are basally extended, forming two plate-like structures, which articulate directly with the tergites; the nodus subdivides the anterior wing margin at about half-length; the pigmented pterostigma at the anterolateral wing margin is a thickened area enclosed by veins and filled with haemolymph; it has likely an aerodynamic effect and probably also functions as a visual signal; the fore- and hindwings are moved independently and are never connected. In contrast to Ephemeroptera and Neoptera the wing movements are almost exclusively triggered by strongly developed direct flight muscles, especially large muscles attached to the basalare and subalare. The legs form a prey catching basket; they are equipped with thorns and suitable for grasping but not used for walking; the protibial spurs are often broadened and part of a cleaning device; the tarsi are 3-segmented; apically they bear double claws; attachment devices are absent.

The 11-segmented abdomen (Fig. 6.7.4) is long, slender and usually cylindrical; it stabilizes the body during flight. In Anisoptera constrictions occur and the abdomen is flattened in some groups (e.g., some Libellulidae). The large tergites enclose the abdominal segments dorsally and also laterally. The sternites are small. A pair of unsegmented and usually short cerci is inserted at the hind margin of segment X. During copulation they are used by males for grasping the females, together with the paraprocts or the epiproct, the large valves of segment XI. A secondary copulatory apparatus is present at the abdominal base of males (Fig. 6.7.4B). It is formed by sternite II and parts of sternite III; its substructures are the lamina anterior, two pairs of lateral clasping organs (hamuli anteriores and posteriores), the median ligula, and

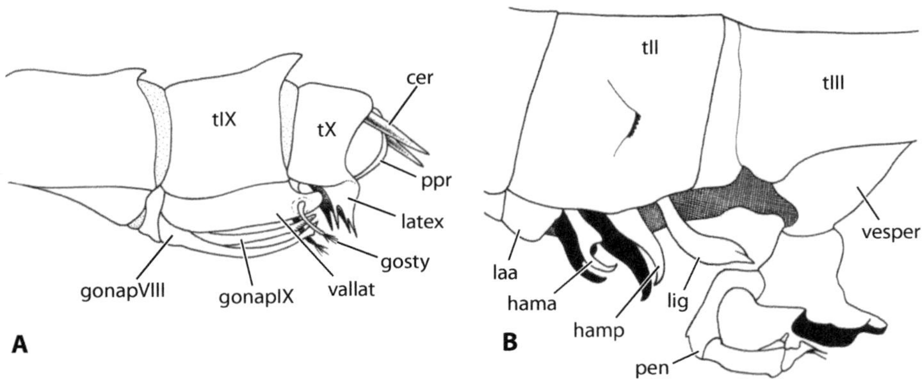

Fig. 6.7.4: Abdominal structures of Anisoptera, lateral view. A, female postabdomen of *Tetracanthagyna* sp. (Aeshnidae); B, male anterior abdomen of *Ophiogomphus bison* (Gomphidae), with secondary copulatory apparatus. Abbr.: cer – cerci, gonapVIII/IX – gonapophyses VIII/IX, gosty – gonostylus, hama – hamulus anterior, hamp – hamulus posterior, laa – lamina anterior, lateX – laterotergite X, lig – ligula, pen – penis, ppr – paraproct, tII/III/IX/X – tergites II/III/IX/X, vallat – valvula lateralis (IX), vesper – vesica sperminalis. Redrawn from Xylander & Günther (2005), originals Tuxen (1907) and Needham & Westfall (1955).

the vesicula seminalis; the intromittent organ (functional penis) can be formed by different elements such as the ligula (Zygoptera) or an outgrowth of the vesicula seminalis (Anisoptera). The male genital opening lies at the abdominal sternite X, that of females posterad sternum VIII. In Zygoptera and some groups of Anisoptera (e.g., Aeshnidae) an ovipositor (Fig. 6.7.4A) is present composed of gonapophyses VIII and IX and gonocoxae IX, which often bear gonostyli; the ovipositor is strongly reduced in most anisopterans in correlation with epiphytic egg deposition.

The digestive tract is straight; an ingluvies and a proventriculus are developed posterior to the long oesophagus; between three and six rectal papillae are present. The number of Malpighian tubules varies between 50 and 200; they are arranged in five or six groups. The main elements of the well-developed tracheal system are three longitudinal stems. The circulatory system is represented by the heart with eight pairs of ostia, the cephalic aorta, and two pairs of dorsal ampullae at the wing bases. The testes comprise numerous vesicles; they extend from abdominal segment IV to VIII in Anisoptera and between segments VII–IX in Zygoptera. The elongate ovaries are composed of numerous panoistic ovarioles; the short paired oviducts unite as a short egg duct; this is continuous with the vagina, which is dorsally extended as a bursa copulatrix, also referred to as spermatheca (see St. Quentin & Beier 1968). Paired accessory glands are usually present in females.

Morphology, immature stages (naiads). Like in Ephemeroptera the immature stages of Odonata possess specific larval structures (Fig. 6.7.2). They are also addressed as naiads in the following.

The head is less movable than in the adults. The compound eyes are smaller, and the antennae longer. The labium is strongly modified, forming a large, folded, prehensile prey-catching device, the labial mask (autapomorphy) (Fig. 6.7.2). Distally it bears movable palps with a mesally extended proximal segment and a distal hook-like second palpomere.

In contrast to the adults, the thoracic segments are not skewed. The legs are well-developed. The thoracic spiracles are closed like those of the abdomen.

The abdomen is always shorter than in adults. The caudal valves are developed as elongate, leave-like tracheal gills in Zygoptera and some tropical species possess lateral segmented tracheal gills, which are not homologous to the similar organs of ephemeropteran naiads. Together with the epiproct and two paraprocts the apically pointed cerci form a quinquepartite anal pyramid in Anisoptera and Epiophlebioptera, which may function as a defensive structure. Gas exchange in the anisopteran naiads takes place in a rectal chamber lined with gill pads; a powerful forward thrust using a jet-propulsion mechanism is achieved by rapid ejection of water from this chamber.

The salivary glands are absent and the proventriculus is much larger than in the adults. The genital structures are formed in a late stage of the postembryonic development.

Biology. Adult Odonata are commonly known and very attractive insects and also popular objects of ethologists ("bird watcher's insects"). They are also intensively studied under aspects of biomechanics and physiology and other fields of science. They are often observed using methods applied in ornithology, and they were the first insects for which territorial behavior was described. Likewise, sperm displacement during copulation was observed for the first time in dragonflies. This had a strong impact on the understanding of evolutionary processes and reproductive behavior well beyond Odonata. Dragonflies have also been extensively studied regarding thermoregulation and the biomechanics of their flight. Their ecological significance as mosquito catchers in Malaria regions or their function as indicators of water quality are only a glimpse of the topics stressed in the last decades.

In contrast to most other insects Odonata are active during daytime and exposed, and largely rely on the visual system in their orientation. Typically they are the largest insects at the waterside, facing only birds and larger spiders as predators. Zygoptera usually rest on horizontal branches or rocks at the edge of water bodies and occasionally undertake short flights for catching prey, chasing away males of the same species, or mating (percher behavior). In contrast, anisopteran species fly continuously and usually cling to reeds or other vertical items during the short resting intervals (flier behavior). Whereas damselflies fly moderately well, the larger dragonflies are characterized by a powerful maneuvered flight, moving both pairs of wings and the wings of either side independently by using the direct flight muscles. Prey is identified visually, grasped with the catching basked formed by the legs, and devoured in the air by most dragonflies. Many zygopteran species also catch their prey in the air but perch for feeding. Smaller damselflies pick small arthropods from stems. Tropical pseudostigmatine species hover at spider webs and pick out suitable prey (Grimaldi & Engel 2005).

The naiads are usually aquatic or at least semiaquatic. In the typical case they are ambush or sit-and-wait predators. For catching prey they use the folded prehensile labial mask. They usually stay at the bottom of their aquatic habitats, in mud, among submerged parts of water plants or under stones or submerged wood. Naiads of some tropical odonatans can stay for longer periods in moist environments outside of the water and naiads of some species of *Megalagrion* (Hawaii) and *Antipodophlebia* (Australia) are terrestrial.

Reproduction and development. The reproductive behavior of Odonata differs strongly from that of all other insects. Three different stages can be distinguished. Conspecific males and females recognize each other first by the flight behavior, then by the coloration and overall body shape (Grimaldi & Engel 2005). After species specific courtship behavior the males grasp the females with the legs at first (precopulatory tandem). In the next step, the male bends the abdomen thus bringing the secondary copulatory apparatus in contact with the male genital opening at segment IX and filling it with sperm (intramale sperm translocation). Then females are grasped with the clasping apparatus of the abdominal apex, either in the neck region (Epiprocta, Pfau 2002) or at the prothorax (Zygoptera) (copulatory tandem). In some species sticky secretions are released by the females. Then both males and females bend their abdomen and form the characteristic tandem-linkage ("wheel position", copulation wheel). The female genital opening between sternites VIII and IX is brought into contact with the secondary copulatory organ and the spermatheca is filled with sperm. The copulation is achieved during flight in some species but usually the tandem rests on suitable substrate (e.g., reed or twigs). Males and females can copulate several times per day. Males of many species defend their territory against other males. They attempt to fertilize as many females as possible and to prevent copulation with other males. They are also able to remove sperm transferred to the female genital tract in earlier copulations.

The egg laying behavior can be similar to that of Ephemeroptera (e.g., on the water surface) but endophytic oviposition in tissue of plants or in submerged logs also occurs, or eggs are deposited in mud at the edge of the water body. The hatching pronymphs live from nutrients contained in the eggs. In the following postembryonic development the nymphs molt 9–14 times. The final molt usually takes place at dusk or dawn, with the final instar clinging to reed or other suitable objects outside of the water. The hatching adult is still unpigmented. During a time span between few hours and two days the coloration develops and the insects become able to fly. Usually a combination of haemolymph pressure and gravitational pull is involved in the shaping of the fully functional body of the adults.

Fossil record. Odonata have a rich fossil record. The oldest representatives of the stem group (†Protodonata, "Griffenflies") are known from Upper Carboniferous deposits (Xylander & Günther 2005). The largest known insect of all times, †*Meganeuropsis perimiana* (Early Permian), is assigned to †Protodonata (Grimaldi & Engel 2005). †Protanisoptera is documented from the Lower to the Upper Permian (Grimaldi & Engel 2005). They possibly form the sistergroup of extant Odonata. The earliest definite Zygoptera are †Triassolestidae from Triassic deposits. The earliest fossil representatives of Anisoptera are also from the Triassic. Epiophlebiidae are not recorded as fossils (Grimaldi & Engel 2005; see also Bechly [1996] and Rehn [2003]).

Economic importance. The economic importance is limited. The naiads may occasionally cause damage in fish ponds. As pointed out above, odonatans can play a positive role in the control of *Anopheles* and other vectors of diseases.

[Tuxen (1907); Needham & Westfall (1955); Conci & Nielsen (1956); St. Quentin & Beier (1968); Bechly (1996); Lohmann (1996); Corbet (1999); Pfau (2002); Rehn (2003); Grimaldi & Engel (2005); Xylander & Günther (2005); Bybee et al. (2008)]

Neoptera (Pterygota excl. Ephemeroptera and Odonata)

The monophyly of Neoptera, i.e. all pterygote orders except for Ephemeroptera and Odonata, is almost generally accepted (e.g., Kristensen 1975; Klass 2009; Fig. 6.A: 13). It was challenged in recent studies based on primary sequence data of complete mitochondrial genomes (Lin et al. 2010; Zhang et al. 2010), with a clade Ephemeroptera + Plecoptera as sister group to all remaining pterygote lineages. However, this very likely an artifact underlining the limited suitability of mitochondrial genes in high level insect systematics.

The most important apomorphy supporting Neoptera is the ability to fold back the wings over the abdomen, linked with a complex of modifications at the wing base. This is apparently a key innovation in pterygote insects. The ability of the adult winged insects to hide in narrow crevices or loose substrates has certainly contributed to the success of the megadiverse group.

Autapomorphies of Neoptera
- Ability to flex the wings flat or roof-like over the abdomen
- Median plate of wing base subdivided (Figs 1.3.6.1, 1.3.6.2)
- Flexion lines within wing base (Fig. 1.3.6.2)

- Pleural muscle inserted on 3rd axillary sclerite functions as wing flexor
- Longitudinal veins detached from axillary sclerites (Fig. 1.3.6.2)
- Third valvulae (gonoplacs) form a sheath for the 1st and 2nd valvulae (groundplan)
- Arolium present (groundplan?)

[Kristensen (1975); Klass (2009); Lin et al. (2010); Zhang et al. (2010)]

Polyneoptera

Polyneoptera (=Lower Neoptera) comprise the hemimetabolous neopteran orders Mantophasmatodea, Grylloblattodea, Dermaptera, Embioptera, Phasmatodea, Orthoptera, Plecoptera, Zoraptera, Blattodea (incl. Isoptera) and Mantodea (Fig. 6.B: 1). These groups are mostly characterized by plesiomorphic features. The monophyly is disputed but was consistently supported in recent studies based on morphology (e.g., Yoshizawa 2011) and molecular data (e.g., Ishiwata et al. 2011). Despite of considerable progress in the last years, the phylogenetic relationships of the orders are still partly obscure. The phylogenetic backbone of the lineage and especially the placement of the successful Orthoptera (grasshoppers, etc.) and Dictyoptera (roaches, termites and mantids) are insufficiently clarified (Fig. 6.B). A breakthrough is the unambiguous placement of Zoraptera (ground lice) in Polyneoptera, but the precise position of this enigmatic small order is still unclear. Characters of the eggs and spermatozoa tentatively suggest a clade comprising Zoraptera, Embioptera and Phasmatodea (Dallai et al. 2012a, b) but this needs further confirmation. A monophylum Eukinolabia which includes the latter two groups appears to be well-supported by various independent character systems (e.g., Friedemann et al. 2012; Fig. 6.B: 3). This implies the non-monophyly of Orthopterida (Orthoptera + Phasmatodea), a grouping suggested for instance by Willmann (2005a). A well-supported supraordinal unit is Dictyoptera comprising Mantodea and Blattodea incl. Isoptera (Fig. 6.B: 2). The placement of Isoptera as a subordinate group of roaches (sister group of the genus *Cryptocercus*) is strongly supported by different data sets (e.g., Lo et al. 2000; Klass & Meier 2006). This renders the traditional order "Blattaria" paraphyletic and implies a subordinal rank of Isoptera. A recently revealed well-supported clade is Xenonomia (Fig. 6.B: 5). It includes the very small orders Grylloblattodea (=Notoptera, ice crawlers) and Mantophasmatodea (heelwalkers, "gladiators"), which were discovered only recently (Klass et al. 2002). The placement of Plecoptera (stoneflies) is still problematic. A basal position as the sistergroup of the remaining polyneopteran orders (incl. Zoraptera) is a serious option. Fully developed, pad-like tarsal euplantulae and a fan-like folding pattern of the anal field are potential apomorphies of Polyneoptera excl. Plecoptera. However, the latter feature does also occur in basal plecopteran representatives. The terms Paurometabola or Pliconeoptera were used for a unit comprising the remaining Polyneoptera without Zoraptera (see e.g., Hennig 1969). As polyneopteran affinities of

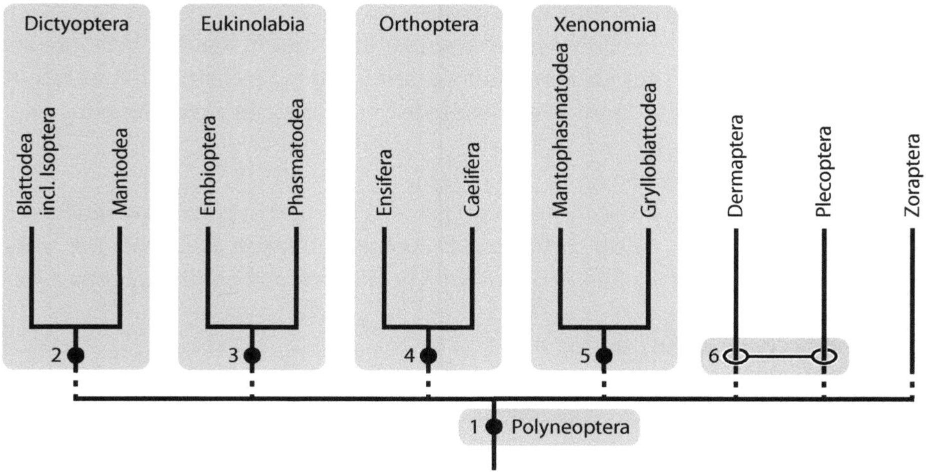

Fig. 6.B: Polyneopteran relationships, with selected potential apomorphies. **1**, enlarged anal field of hindwings, tarsal euplantulae, heart with excurrent ostia, tegmina (groundplan?); **2**, accessory anterior tentorial arms medially connected, membranous mandibular postmola, cubitus of forewing arcuate, specific armature of proventriculus, opening muscles of abdominal spiracles inserted on paratergites, female genital atrium, elongated female sternite VII forms subgenital lobe with movable terminal lobe, joint between gonangulum and paratergite IX modified as hinge, egg cases (oothecae); **3**, premental origin of M. tentorioparaglossalis, M. tentorioscapalis medialis, eggs with operculum and micropylar plate, prothoracic epimeron reduced; **4**, pronotum saddle-like, hindlegs modified as jumping legs; **5**, complete loss of wings, modifications of labium and hypopharynx, unique construction of antennal heart including membranous connection with brain.

the ground lice appear to be settled, the implied concepts are obsolete. A sister group relationship between Plecoptera and Dermaptera was suggested in several molecular studies (e.g., Ishiwata et al. 2011), but this is insufficiently supported (Fig. 6.B: 6). A possible placement of stoneflies as the sister taxon of the entire remaining neopteran lineages was discussed by Hennig (1969) (see also Kristensen 1995), implying the non-monophyly of Polyneoptera. The loss of the transverse stipital muscle and a terrestrial larval development were considered as potential apomorphies of Neoptera excl. Plecoptera.

Autapomorphies and supraordinal polyneopteran groups
Polyneoptera

- enlarged anal field of the hindwings
  The large anal field is secondarily missing in Zoraptera, Embioptera and termites (wings absent or strongly simplified), and the wings are completely missing in Grylloblattodea and Mantophasmatodea.

- euplantulae (tarsal attachment pads)
  The euplantulae are well-developed pad-like structures in roaches, orthopterans and other groups. They are less distinctly developed in Plecoptera (Beutel & Gorb 2006) and greatly reduced (Grylloblattodea) or absent in several groups (e.g., Isoptera, Zoraptera).
- heart with excurrent ostia
  A heart with both incurrent and excurrent ostia is found in polyneopteran insects with the exception of the miniaturized Zoraptera (Pass et al. 2006). The situation in Mantophasmatodea is unknown. The position and number of ostia varies between the groups.
- modifications of the wing joint
  Yoshizawa (2011) proposed four modifications of the wing joint as potential synapomorphies of the polyneopteran lineages. However two of them imply reversal in Blattodea and one is highly homoplasious.
- Tegmina (?)
  This feature is highly questionable as an autapomorphy of the entire Polyneoptera. The typical condition is only found in Mantodea, roaches, Orthoptera and Phasmatodea.

Dictyoptera. Dictyoptera are a medium sized polyneopteran lineage. Its two monophyletic subgroups, the Mantodea and Blattodea (including Isoptera), differ strongly in their morphology and biology. Nevertheless, the monophyletic origin is widely recognized and well-supported by morphological and molecular data (Fig. 6.B: 2).

Autapomorphies
- Accessory anterior tentorial arms medially connected ("perforated corpotentorium")
- Membranous mandibular postmola
- Cubitus in the forewing arcuate
- Specific armature of the proventriculus
- Opening muscles of abdominal spiracles inserted on paratergites
- Female genital atrium
- Elongated female sternite VII forms subgenital lobe with movable terminal lobe
- Joint between gonangulum and paratergite IX modified as a hinge
- Egg cases (oothecae), formed with secretions from accessory glands
- Modifications of the male genital apparatus

The presence of Dictyoptera s.str. (crown group) is documented by Jurassic fossils, whereas roach-like members of the stem group were already abundant in the Carboniferous.

That termites (Isoptera) are highly specialized xylophagous and eusocial roaches was one of the most important insights in systematic entomology in the last decades (see below). The more inclusive clade will be referred to as Blattodea in the following and the traditional order (i.e. excl. Isoptera) as “Blattaria”.

Xenonomia (Grylloblattodea + Manophasmatodea)
- Complete loss of wings
- Modifications of the labium and hypopharynx (Wipfler et al. 2011)
- Unique construction of the antennal heart including a membranous connection with the brain (Pass et al. 2006)
- DNA sequence data (Ishiwata et al. 2011) and neuropeptides (Gäde & Šimek 2010)

Eukinolabia (Embioptera + Phasmatodea)
- Premental origin of M. tentorioparaglossalis
- Presence of M. tentorioscapalis medialis
- Eggs with an operculum and a micropylar plate
- Prothoracic epimeron reduced
- Molecular data (Ishiwata et al. 2011)

[Hennig (1969); Kristensen (1995); Lo et al. (2000); Klass et al. (2002); Willmann (2005a); Beutel & Gorb (2006); Klass & Meier (2006); Pass et al. (2006); Gäde & Šimek (2010); Ishiwata et al. (2011); Wipfler et al. (2011); Yoshizawa (2011); Dallai et al. (2012a, b); Friedemann et al. (2012)]

6.8 **Plecoptera** (Greek *pleco* = folded, *pteron* = wing, English common name: stoneflies)

Diversity and Distribution. About 3,500 species are presently known. Plecopterans occur on all continents except Antarctica. They are almost completely missing on oceanic islands. The immature stages (naiads) of most species develop in cool mountain streams. The diversity of the group is high in temperate regions with suitable aquatic habitats, but many species of the family Perlidae from tropical countries have been described in recent years. The maximum height reached by a plecopteran species is ca. 5,600 m in the Himalayas.

Autapomorphies
- Naiads aquatic (primarily specialized on well oxygenated streams) (possibly groundplan of Pterygota?) (Fig. 6.8.1)
- Paired internal genital organs anteriorly connected, forming a closed loop
- Males with two pairs of loop-shaped seminal vesicles stacked upon each other
- Naiads with specialized intersegmental muscles in the thorax and abdomen

Fig. 6.8.1: Plecoptera, (Chloroperlidae), naiad, lateral view, SEM micrograph. Abbr.: ant – antenna, cer – cercus, plb – palpus labialis, pmx – palpus maxillaris, sti – stipes, wanl – wing anlagen. Scale bar: 1 mm. Courtesy H. Pohl.

Taxonomy. The monophyly of the group was disputed but appears to be well founded (Zwick 1980, 2000). Plecoptera are divided into the two suborders Antarctoperlaria and Arctoperlaria. The former group is restricted to South America, Australia, and New Zealand. It comprises two superfamilies, Eusthenioidea (Eustheniidae and Diamphipnoidae), and Gripopterygoidea (Austroperlidae and Gripopterygidae). An autapomorphy of this Gondwanan lineage is the presence of floriform chlorid cells. The suborder Arctoperlaria is characterized by a complex of features related to the habit of causing substrate vibrations before mating. Arctoperlaria is also subdivided into two subgroups. Euholognatha contains the family Scopuridae and its sistergroup Nemouroidea. The latter comprises five families, the strictly Holarctic Taeniopterygidae and Capniidae, Leuctridae which occur in cool, temperate and tropical regions, the very species rich Nemouridae, and Notonemouridae occurring on the southern hemisphere including Madagascar, Tasmania and New Zealand. The eight species of the Japanese-Korean Scopuridae are completely wingless. Systellognatha comprise the families Peltoperlidae, Styloperlidae, Pteronarcyidae, Perlidae, Perlodidae, and Chloroperlidae. The adults of this lineage are usually non-feeding (Zwick 2005).

Diagnosis. Slender, cylindrical or slightly flattened insects, usually with inconspicuous coloration. Almost always with a pair of segmented cerci and two similar and well-developed pairs of wings, usually folded back flat in the resting position. Terminal filament absent in adults and usually also in naiads. Naiads aquatic, usually developing in cool streams.

Morphology of adults. Most plecopteran species are medium sized insects between 8 and 20 mm. The minimum size is ca. 3.5 mm (e.g., *Tasmanocerca*, *Capnopsis*) and the maximum ca. 40 mm (e.g., *Pteronarcys*, *Diamphipnoa*). The body is slender and

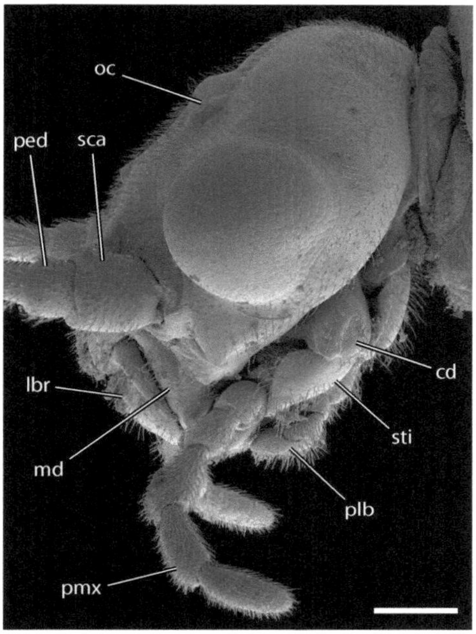

Fig. 6.8.2: Plecoptera (Nemouridae), adult head, lateral view, SEM micrograph. Abbr.: cd – cardo, lbr – labrum, md – mandible, oc – ocellus, ped – pedicellus, plb – palpus labialis, pmx – palpus maxillaris, sca – scapus, sti – stipes. Scale bar: 200 μm.

cylindrical or moderately flattened. The coloration is usually brown or yellowish. The cuticle is moderately or weakly sclerotized, smooth or covered by a vestiture of fine setae.

The head is prognathous or subprognathous but lacks a gula or hypostomal bridge (Fig. 6.8.2). Posteroventrally it is closed by the labium. The compound eyes are well-developed. Three ocelli are present in the groundplan but the anterior unpaired ocellus or all three can be absent. The frontal and coronal sutures are distinct in the groundplan. The frontoclypeal transverse strengthening ridge is usually present but absent in Perloidea and Peltoperlidae. The free labrum is trapezoid or rectangular with rounded edges, and sometimes strongly transverse; tormae are present posterolaterally; the anterior edge is often sinuate. The antennae are long (usually ca. 1/3–1/2 of the body length), multisegmented and filiform; the scapus is distinctly larger than the other antennomeres and inserted in a wide articulatory membrane on the anterolateral head region. The mouthparts are of the orthopteroid, i.e. unmodified biting type. The mandibles are primarily well-developed (reduced in adults of Systellognatha), stout, and more or less triangular; in the distal region they bear several teeth; a brush of microtrichia (penicillum) is present proximad a tuberculate molar area in the middle region of the mesal edge; short setae are irregularly arranged or in rows on different regions of the mandibular surface. The cardo is indistinctly divided into a basicardo and disticardo; the stipes bears the endite lobes and the 5-segmented palps; the lacinia bears apical and subapical teeth and mesally directed spines. The labium is composed of a large submentum, a short mentum, and a prementum bearing the glossae and paraglossae and 3-segmented palps. The hypopharynx is triangular, rectangular or rounded and bears posteriorly directed microtrichia on its surface; laterally and dorsally it is reinforced by a suspensorium, which is divided into several small sclerotized elements; anterior to them lies the opening of the maxillary glands. The ducts of the salivary glands open into the funnel-shaped salivarium. The tentorium is well-developed and complete; the attachment sites of the dorsal arms are visible anterolaterad the paired ocelli. Muscle attachment areas are also more or less distinctly recognizable externally. Paired laterocervicalia are embedded in the cervical membrane (Zwick 1980, 2005).

The prothorax is relatively large and movably connected with the pterothorax. The pronotum is flat and shield-like and usually divided by a thin median line (more distinct in naiads), which is not a zone of weakness; its marginal area is sometimes delimited by a fine line and bent upwards; it is connected with the pleuron by a membranous area. The pleural sclerotized elements are small in some groups; in the typical case two arcuate sclerites are present, the anepisternum and the trochantin; posteriorly they are adjacent along the vertical pleural ridge and separated by the paracoxal suture. The basisternum and furcasternum are sometimes present as separate elements, but usually fused, thus forming a single extensive prosternal sclerite, which reaches the laterocervicalia anteriorly and the spina posteriorly. The prothoracic spina is present, but the spinasternum is greatly reduced or missing. The furcal arms arise between the procoxae or slightly posterad. A ventral cervicale, a presternite, a postfurcasternum, and a postcoxal bridge occur in some groups. The two pterothoracic segments are very similar. The terga are usually divided by a median line like the pronotum; they are composed of a flat antecosta, a short acrotergite, a prescutum, a large scutum (often subdivided), and the distinctly convex scutellum; laterally a complex connection is formed between the scutum and the wing base, involving articulatory processes and three axillary sclerites; the basalare and subalare are usually well-developed. The pleural region is subdivided by the pleural ridge into the episternum and epimeron, the former subdivided into the anepisternum and katepisternum. The anterior part of the sternal region is usually covered by a broad sclerite formed by the basisternum and preepisternum. The posterior furcasternum lies between the widely separated furcal invagination sites. A spinasternum is present in the mesothorax but absent in the metathorax, where the spina is missing (Zwick 1980). The furcae are connected with the pleural arm by a pleurosternal muscle. The two pairs of thoracic spiracles are well-developed and normally placed in the intersegmental regions. The membranous wings (Fig. 6.8.3) are usually more or less well-developed, but completely absent in some taxa (e.g., Scopuridae, *Baikaloperla*); primarily they are projecting beyond the abdominal apex but quite often shortened;

in the resting position they are folded at the base (autapomorphy of Neoptera) and either held horizontally or slightly curved following the contour of the abdominal segments; the forewings are relatively slender, whereas the hindwings are characterized by an enlarged fan-shaped anal field with several folding lines (sometimes reduced); coupling mechanisms connecting the meso- and metathoracic wings are absent; a rich venation is probably part of the groundplan, with many transverse veins and a reticular pattern, but reductions occur in different lineages; the media (M) originates from the wing base in the forewings, but from the radial sector (radius posterior) in the hindwing. The legs are largely unmodified; the short coxae are distinctly separated from each other medially and divided into a basicoxa and meron; the trochanter is small and triangular; the long femora are more or less strongly flattened; the elongate tibiae fit into a ventral femoral rim; apically they bear a pair of spurs; the tarsus is composed of three tarsomeres; apically it bears well-developed, simple paired claws; an arolium is almost always present; the tarsal euplantulae vary considerably; they can be present as soft pads or as a simple longitudinal bulge, and they are sometimes absent. The thoracic muscle system is generally well-developed and complex, with the likely exception of the few wingless forms (Zwick 1980, 2005).

The abdomen is weakly sclerotized and cylindrical or moderately flattened. It is composed of ten distinctly developed segments and vestiges of segment XI (Zwick 1980). The segments are usually composed of simple plate-like tergites and sternites and a connecting pleural membrane. Tergite and sternite X usually form a ring-shaped structure; sternite X is only present as a narrow connection in Nemouridae. Fusions occur also in other segments (especially in naiads). Tergite I is closely connected with the metathorax and shorter than the following tergites; sternite I is usually fused with the metasternum and rarely present as an identifiable separate structure (e.g., in Pteronarcyidae, Austroperlidae). Sternite X is usually well-developed but reduced in females and naiads of some groups. Segment XI is represented by an unpaired dorsal epiproct (supra-anal lobe) and the ventral paraprocts (sub-anal lobes). The cerci insert at the upper edge of the paraprocts; they are multisegmented in the groundplan but often shortened and 1-segmented in some groups, with an apical tubercle representing the fused and vestigial distal cercomeres. The filament-like terminal appendage of some Gripopterygoidea is formed by the base of the epiproct (Zwick 1980, 2005). Spiracles are present on segments II-VIII. The female genital opening lies behind sternite VIII, which is only slightly modified as a subgenital plate. The primary ovipositor is reduced; neoformations functionally replacing it occur in some groups. The male genital opening lies behind segment IX, which can be distinctly modified. The primary aedeagus is reduced. The terminal region of the internal genital organs is covered by a cuticle and can be extruded by haemolymph pressure as a simple secondary copulatory organ with two gonopores (Zwick 2005).

The digestive tract is characterized by an unusually long oesophagus, whereas the midgut and hindgut are short; the proventriculus is vestigial or completely reduced; midgut caeca are usually absent, but 2–6 occur in Perloidea. Between

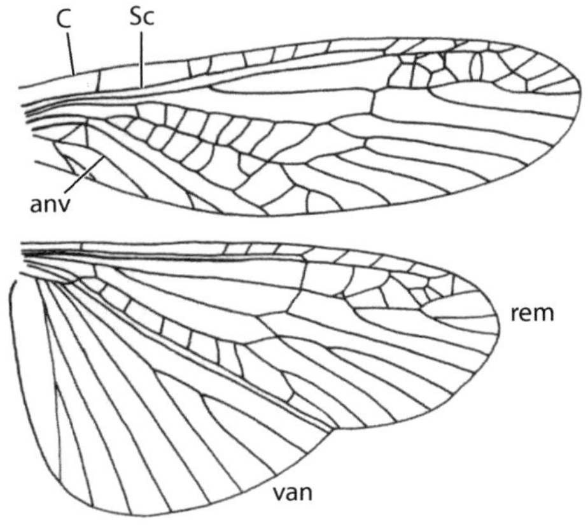

Fig. 6.8.3: Wings, *Perlodes microcephalus* (Perlodidae). Abbr.: anv – anal veins, C – costa, rem – remigium (costal field), Sc – subcosta, van – vannus (anal field). Redrawn from Zwick (2005) (from Aubert 1959).

twenty and hundred thin Malpighian tubules insert at the midgut-hindgut border. The paired gonads are anteriorly connected in both sexes (see above: autapomorphies). The ovaries are composed of many panoistic ovarioles, which are either arranged separately or arise from a common duct. A spermatheca is usually present, whereas a bursa copulatrix is missing in most groups (Theischinger 1991). Two pairs of loop-shaped seminal vesicles are present in males, on both sides placed upon each other (see above: autapomorphies). The testes are elongate strands of follicles in the groundplan, connected in each segment to the vas deferens by a short duct. The internal male genital organs vary strongly within the group in correlation with modifications of the terminal segments and the formation of secondary copulatory organs (see above) (Zwick 2005).

Morphology, immature stages (naiads). The immature stages (Figs 6.8.1, 6.8.4) are similar to the adults in most features like in other hemimetabolous lineages. However, they display some specific larval characteristics which are mostly correlated with the aquatic habits. Like in Odonata and Ephemeroptera they are adressed as naiads in the following. The body is cylindrical or more or less strongly flattened, depending of the preferred microhabitat. Gills are absent in the 1st instar naiads hatching from the eggs; in the following stages gills may be present on different body parts, for instance on the cervical membrane, the ventral side of the thorax, laterally on the abdominal segments (similar to the condition in ephemeropteran naiads), or on the apex of the abdomen. Fringes of setae may be present on the lateral segmental margins, on the legs, on the cerci, or medially on the dorsal side of the abdomen; the setae can be club-shaped or spatulate or modified in other ways. The unsclerotized ventral

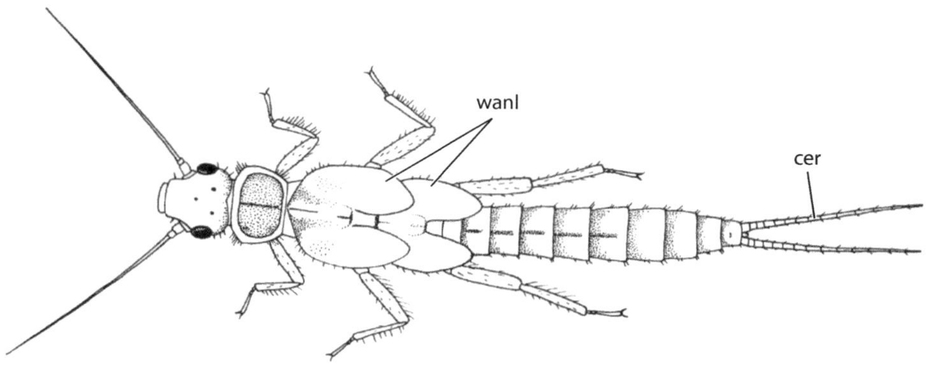

Fig. 6.8.4: *Xanthoperla apicalis* (Chloroperlidae), nymph, final stage. Abbr.: cer – cercus, wanl – wing anlagen. Redrawn from Zwick (2005) (from Aubert 1959).

sides of the postcephalic segments, intersegmental regions, and the bases of gills may be densely covered by chloride cells. The number of antennomeres and cercomeres increases during the immature development. The ocelli appear and the number of ommatidia increases. The wing anlagen develop in the last stages as rigid sheaths. A median terminal filamentous gill occurs in Austroperlidae; whether it is homologous with the terminal filament of Archaeognatha, Zygentoma and Ephemeroptera is unclear.

Biology. The adults are almost always found close to the aquatic habitat of the naiads, on the riparian vegetation, rocks, logs or trees, sometimes as assemblages of numerous individuals ready for reproduction. Only one wingless species is apparently permanently submerged (Lake Tahoe benthic stonefly). The flying abilities are generally modest in smaller species but larger species are often capable of prolonged and swift flight (e.g., Perlidae). The fore- and hindwings are not connected by coupling mechanisms, but are moved synchronously. Many species are nocturnal or crepuscular. The adults often appear inactive in the daylight hours, but males of Nemouridae, Chloroperlidae and Isoperlinae are very agile in search of a mating partner. Consumption of food plays a minor role in species hatching as adult females with developed eggs and distinct reductions of mandibles occur in large species of Systellognatha. Adults of other plecopterans use pollen, algae, moss particles, fungi or detritus as food substrates. Several species of Nemouroidea have to feed before mating.

The main part of the life cycle is the immature development, which almost always takes place in aquatic habitats, usually in cold and well oxygenated lakes and rivers. Naiads of very few species are terrestrial but occur only in very moist habitats (Grimaldi & Engel 2005). In the typical case gas exchange takes place on the body surface in the naiads hatching from the eggs and also in the following stages, but functional gills are usually present on different body parts after the first instar. The feeding habits of naiads are highly variable and several functional types can be distinguished (e.g., scrapers, ambush predators). Many species are predators of other aquatic arthropods, but others rely on decaying leaves, on layers of algae on stones, or on fine detritus particles in the cavity system of the river bed. The naiads are very intolerant of water pollution. Consequently their presence is an indicator of very good water quality.

Reproduction and development. A characteristic mating behavior is often displayed before copulation. Males of some species produce rhythmic substrate vibrations using their abdomen, which are perceived by females with specific sensorial organs in the legs. Hundreds or even thousands of eggs are produced. Initially they are carried as balls on the female abdomen and later deposited in the water. The entire developmental cycle can take up to five years. The first instar naiads hatch from the eggs after 2–3 weeks (Nemouroidea) or several months. Indirect development involving a diapause, apparently mainly during an unfavorable season, is not a rare exception. The number of immature ecdyses is usually high, varying between 10 and 17 (P. Zwick, pers. comm.). The pharate adult leaves the water and molts in the riparian zone.

Fossil record. The affinities of certain Carboniferous and Lower Permian fossils assigned to Plecoptera are questionable. Several Lower Permian lineages such as for instance †Lemmatophoridae, †Probnidae and †Liomopteridae belong to the paraphyletic stem group of the order, traditionally addressed as “Protoperlaria” (Grimaldi & Engel 2005). †Lemmatophoridae had terrestrial nymphs, and 5-segmented tarsi were preserved in this family and also in †Liomopteridae. Numerous fossils are known from Mesozoic deposits including representatives of extant families.

Economic importance. The naiads of Plecoptera play a certain role as prey of fresh water fish, especially of trouts and their relatives (Salmonidae). They are also indicators of good water quality. Historically, the very large adults of Pteronarcyidae were used as food by native Americans in California.

[Aubert (1959); Zwick (1980, 2000, 2005); Theischinger (1991); Grimaldi & Engel (2005)]

6.9 Dermaptera (Greek *derma* = skin, *pteron* = wing, English common name: earwigs)

Diversity and distribution. About 2,000 species are described. The group has a worldwide distribution. It reaches the highest diversity in the tropical and subtropical regions. Only 20 species occur in Western Europe.

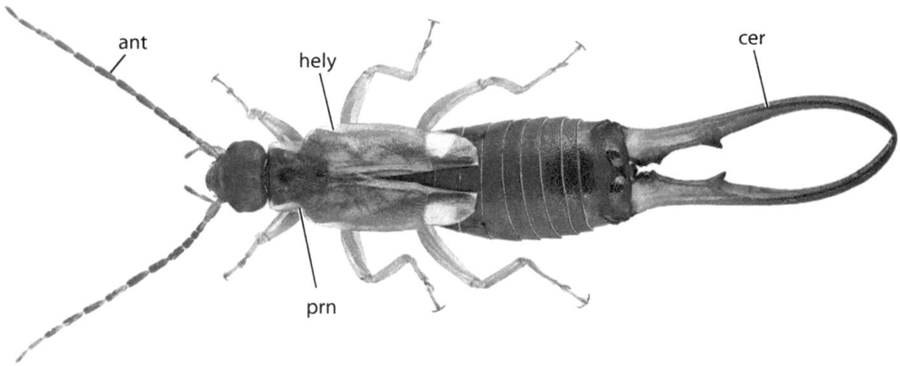

Fig. 6.9.1: *Forficula auricularia* (Forficulidae), male, dorsal view. Abbr.: ant – antenna, cer – cerci, hely – hemelytra, prn – pronotum. Photo courtesy H. Pohl.

Autapomorphies
- Forewings modified as strongly shortened and sclerotized tegmina (Fig. 6.9.1)
- Hindwings fan-shaped and strongly folded (Fig. 6.9.3)
- Cerci of adults 1-segmented, strongly sclerotized and curved (males)
- Maternal brood care (also in Embioptera)
- External genital organ of males with paired distal elements

Taxonomy. Extant dermapterans are divided into the two very small suborders Arixeniina and Hemimerina, and the large suborder Forficulina, which comprises more than 99% of the known species. Arixeniina occur in Indonesia, Malaysia and the Philippines and comprise two genera with a total of five species, all of them epizoic on bats. Hemimerina comprise two genera and 11 species, which are restricted to the Afrotropics and occur only on Pouched rats (Cricetomyinae). Whether the nine families assigned to Forficulina (e.g., Karschiellidae, Pygidicranidae, Forficulidae, Apachyidae) form a monophylum is questionable. Anisolabidae (349 spp. [monophyly uncertain]), Forficulidae (490 spp.) and Spongiphoridae (512 spp.) are the largest families. The last two families form a clade Eudermaptera together with Chelisodidae (96 spp.).

Diagnosis. Distinctly flattened and slender insects, usually with a subparallel body. The forewings are transformed into strongly shortened, evenly sclerotized tegmina. The adult cerci are transformed into 1-segmented sclerotized structures resembling a pair of forceps.

Morphology. Dermapterans are dorsoventrally flattened and usually slender and parallel-sided or subparallel. Most species are medium-sized insects with a body length between 10 and 15 mm. The smallest earwig *Eugerax poecilium* (Forficulidae) meas-

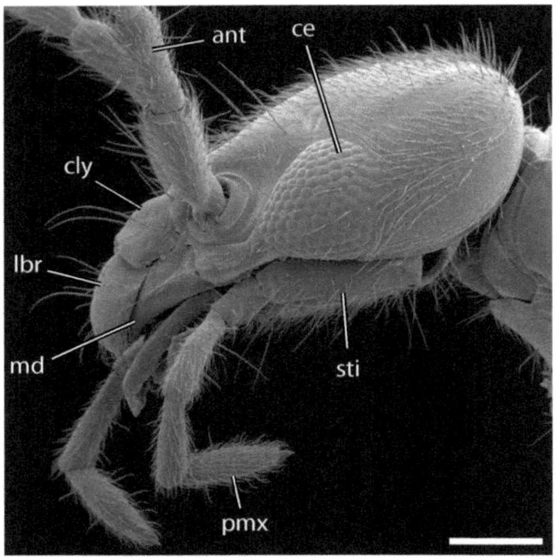

Fig. 6.9.2: *Labia minor* (Spongiphoridae), adult head, lateral view, SEM micrograph. Abbr.: ant – antenna, ce – compound eye, cly – clypeus, lbr – labrum, md – mandible, pmx – palpus maxillaris, sti – stipes. Scale bar: 200 µm. Courtesy H. Pohl.

ures ca. 3 mm. *Labidura herculeana*, (Labiduridae) the Giant earwig of St. Helena, is the largest species with 85 mm. The coloration is almost always yellowish, brownish or black. The cuticle is usually smooth, i.e. without strong sculpture, spines or denticles, and moderately sclerotized.

The head is distinctly prognathous, broad and flattened (Figs 6.9.1, 6.9.2). In dorsal view it appears roughly triangular with rounded edges or heart-shaped and is freely movable against the prothorax. The foramen occipitale is narrowed by a gula; a complete postoccipital ridge originating from the posterior tentorial grooves encloses it laterally and dorsally. A curved lateral strengthening ridge was referred to as occipital ridge by Günther & Herter (1974) but it was pointed out by the authors that it is no segmental border and not homologous to "occipital sutures" occurring in other polyneopteran lineages. The compound eyes are usually well-developed (groundplan) (Fig. 6.9.2) but can be partly (Arixeniina) or completely reduced (Hemimerina and cave dwelling species); ocelli are always absent. The coronal and frontal are usually distinct. The frontoclypeal strengthening ridge is present. The clypeus is subdivided into a transparent anteclypeus and a sclerotized postclypeus. The ovoid labrum is free. The filiform antennae insert at the anterolateral edges of the head capsule. The mouthparts are of the orthopteroid type and largely unmodified (Fig. 6.9.2). The stout mandibles bear two or three pointed apical teeth connected by sharp edges, and a distinctly developed molar part is present basally. The maxillae are elongated; the stipes is subdivided into a basistipes and a mediostipes; the slender lacinia usually bears

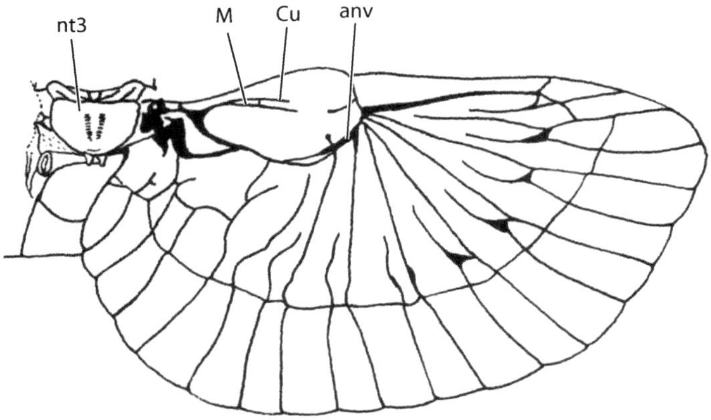

Fig. 6.9.3: Metanotum and unfolded hindwing, *Labidura* sp. (Labiduridae). Abbr.: anv – anal vein, Cu – cubitus, M – media, nt3 – metanotum. Redrawn from Haas (2005), from Albouy & Gaussanel (1990).

two apical teeth and the galea is elongate and cylindrical; the palp is 5-segmented. The postlabium is divided into the submentum and mentum; the prementum bears fused glossae and paraglossae and a 3-segmented palp. The hypopharynx is present as a large tongue-like lobe; it is strengthened by several dorsal sclerites, possibly representing the suspensorium; its ventral side is usually sclerotized and plate-like; the membranous apical part is flanked by superlinguae; they are usually small, but large and lobe-like in *Hemimerus*. The tentorium is strongly developed, with a massive, plate-like corpotentorium, short posterior arms, relatively slender dorsal arms, and strongly developed anterior arms, which are broadened and shovel-like in Forficulinae. Paired and unpaired cervical sclerites are embedded in the cervical membrane.

The prothorax is freely movable against the pterothorax. The pronotum is disc-shaped (Fig. 6.9.1), usually more or less rectangular with rounded edges; its variable length to width ratio is used for identification; a narrow acrotergite is recognizable anteriorly; the lateral edges overlap with the pleuron and the posterior margin with the dorsal side of the mesothorax. The propleuron is divided into an episternum (with a separate preepisternum) and epimeron by an oblique pleural ridge (almost horizontal in *Hemimerus*); the proepimeron usually covers the first spiracle (not in *Hemimerus*); a minute postpleural sclerite is present or absent. The protrochantin is large and triangular. The basisternum is a longitudinal-rectangular plate anterior to the procoxae; a presternum is usually present but in most cases connected with the prosternum; the profurcae arise between the coxae; the median sclerite posterior to them is the sternellum; the spinasternum is present but very small. The meso- and metathorax form a compact unit. The tergal regions are composed of the usual elements, i.e. the prescutum, scutum and scutellum. The mesopostnotum is missing.

The metanotum is similar to the mesonotum but distinctly larger. The phragmata are reduced in flightless forms, even if wings are present. The pleural ridges separating the episterna form the epimera are almost parallel to the tergites. A preepisternum is absent. The metepimeron is distinctly elongated. The mesosternite (=basisternum) is an undivided plate-like structure overlapping with the metasternite posteriorly; at its posterior end it is strongly inflected and bears the furcae. The metasternite is divided into the large anterior basisternum and the smaller posterior sternellum (=furcasternum). The metafurcal arms are connected with the metapleural ridges by muscles. A spinasternum is not present. The venation of the sclerotized and distinctly shortened forewings is reduced (tegmina) (Fig. 6.9.1); the remigium of the hindwings (Fig. 6.9.3) is strongly narrowed and modified, whereas the anal field is enlarged and fan-shaped and characterized by a complicated and unique folding pattern, fan-wise and subsequently transversely (autapomorphy); about 40% of all species are flightless and reduction of tegmina and hindwings does occur within very few genera (e.g., *Anisolabis*, *Euborellia*). The slender walking-type legs are largely unmodified. The short coxae are wider than long and inserted laterally; a basal furrow separates a narrow, ring-shaped basicoxite from the main part; two femoral keels are probably present in the groundplan, but the femora are rounded in more derived taxa; the middle section of the femora is slightly widened; tibial and femoral spurs are absent; the 3-segmented tarsi are usually equipped with brushes of adhesive microtrichia; tarsomere 2 is elongated in Chelisochidae and heart-shaped in Forficulidae; plantulae occur in Hemimerina; their homology with true euplantulae is not fully clarified; the paired claws are well-developed. An arolium occurs only in few (non-related) genera (Haas 2005).

The abdomen (Fig. 6.9.1) is characterized by a high intrinsic movability. Tergite I is firmly connected with the metanotum, but a separating line is still recognizable; sternite I is completely reduced. Tergites and sternites overlap with the corresponding elements of the following segments; this results in a characteristic zig-zag pattern in lateral view. Eight tergites are visible in females and ten in males and immatures. The genital opening lies behind sternite IX. In males sternite IX forms a more or less elongated subgenital plate; it covers the male genital chamber on the ventral side and reaches the base of the cerci posteriorly. The male genital appendages differ distinctly from those of other polyneopteran insects; the proximal portion is unpaired but the distal part paired; the paired ejaculatory ducts enter the paired lateral genital lobes ("penes"), which bear a sclerotized more or less elongated virga distally. The subgenital plate of females (sternite VII) usually covers the three pairs of valvulae of the ovipositor completely; the ovipositor is elongate, thin and curved in some Hemimerina, partly exposed, and apparently not suitable for oviposition in more solid substrates. The cerci are sclerotized and 1-segmented (rarely composed of several cercomeres in nymphs); they are simple and straight in females and immature stages, but highly variable in males, often curved and equipped with teeth or ridges; in the typical case they form a grasping device resembling pincers.

A weakly developed proventriculus and small caeca are present in the anterior region of the digestive tract; six rectal papillae are present. The ovarioles are meroistic-polytrophic. The paired testes are divided into two compartments.

Biology. Most species are nocturnal. The slender body is suitable for penetrating narrow crevices such as fissures in soil or wood or narrow spaces under bark or stones. The feeding habits vary strongly within the group. Saprophagy is arguably a groundplan feature, but there are also herbivorous, predacious and omnivorous species. Many species are apparently not specialized in their diet and can rely on different food sources. The specialized cerci are used as defensive organs and also during copulation. About 40% of all species are secondarily flightless. Eyeless species dwelling in caves are known from Hawaii and South Africa. The species of Arixeniina and Hemimerina are epizoic (i.e. not parasitic) on mammals. Species of *Arixenia* are specialized on skin folds and the gular pouch of Malaysian hairless bulldog bats. *Araeomerus* (Hemimerina) is found in nests of Hamster rats (*Beamys*), and *Hemimerus* on African Giant pouched rats (*Cricetomys*).

Reproduction and development. A complex mating behavior and maternal brood care are characteristic for the group. The females usually deposit up to 50 eggs and sometimes several batches per fertilization. The eggs are placed in clusters in crevices or small cavities in debris or soil and licked by the females to avoid infection with fungi (*Forficula*). They are actively defended with the cerci and relocated if necessary. The postembryonic development comprises 4–6 nymphal stages. The early stages feed on food regurgitated by the females and on their exuviae. They are protected by the females until after the second molt. The newborn nymphs of the viviparous Arixeniina and Hemimerina are ca. 4 mm long. In both taxa, the embryos are supplied by a pseudoplacenta.

Fossil record. Whether the Permian †Protelytroptera belong to the dermapteran stem group (e.g., Haas & Kukalová-Peck 2001) is not entirely clear. The oldest fossils definitely belonging to the order are tegmina form the Late Triassic-Early Jurassic (Grimaldi & Engel 2005). More recent fossils traditionally assigned to the paraphyletic "Archidermaptera" are known from the Late Jurassic and Early Cretaceous (e.g., †Protodiplatyidae). The Jurassic-Cretaceous †Eodermaptera (†Bellodermatidae, †Semenoviolidae, †Turanodermatidae) are the sistergroup of the dermapteran crown group, i.e. Neodermaptera, both forming a clade Pandermaptera (Grimaldi & Engel 2005; Zhao et al. 2010). Fossil dermapterans are very similar to the extant species in their general appearance but have maintained some plesiomorphies such as the presence of three ocelli, 5-segmented tarsi, and cerci composed of several cercomeres (Haas 2005).

Economic importance. The economic importance of Dermaptera is very limited. Some may play a certain role as predators of harmful insects such as aphids and stem boring insects. Other species may cause some damage to cultivated plants by feeding on flowers and other soft plant tissue.

[Günther & Herter (1974); Albouy & Coussanel (1990); Haas & Kukalová-Peck (2001); Grimaldi & Engel (2005); Haas (2005); Jarvis et al. (2005); Zhao et al. (2010)]

Eukinolabia (Embioptera + Phasmatodea)

6.10 **Embioptera** (Embiodea, Embiidina, Greek *embios* = lively, *pteron* = wing, English common name: webspinners)

Diversity and Distribution. Embioptera is one of the small insect orders. The group comprises ca. 360 species. They occur in tropical and subtropical regions and in some warmer areas of the temperate zones, and also marginally in semixeric areas. In tropical rain forests they can reach an altitude of ca. 3,500 m.

Autapomorphies
- Tarsomere 1 of forelegs strongly enlarged (Figs 6.10.1, 6.10.2), silk glands present in protarsomeres 1 and 2 (up to 200 small gland subunits)
- Hairy soles of tarsomeres (also common in Dermaptera)
- Wing dimorphism, females wingless, wings of males (Fig. 6.10.1) simplified (or also reduced)
- Female brood care (also in Dermaptera)

Taxonomy. The order is presently subdivided into 13 extant families, the Clothodidae, Australembiidae, Anisembiidae, Andesembiidae, Archembiidae, Notoligotomidae, Embiidae, Ptilocerembiidae, Scelembiidae, Teratembiidae, Oligotomidae, Embonychidae, and Paedembiidae. This classification was recently suggested based on analyses of 96 morphological characters and 5 genes (Miller et al. 2012). The monophyly of several families (e.g., Embiidae) is still questionable.

Diagnosis. Elongate, cylindrical insects of medium size. Always wingless as females. Males also wingless or with simplified dehiscent wings (Fig. 6.10.1). Protarsomere 1 strongly enlarged.

Morphology. The total size range is 3–25 mm but most species are between 8 and 12 mm long. The body is elongate, parallel-sided and cylindrical, adapted to moving efficiently forwards and backwards in galleries (Fig. 6.10.1). The cuticle is thin and smooth and the coloration usually dark brown.

Fig. 6.10.1: *Oligotoma nigra* (Oligotimidae), habitus of winged male, lateral view, SEM micrograph. Scale bar: 1 mm. Courtesy H. Pohl.

The elongate, ovoid head is prognathous and ventrally closed by an elongate gula. The compound eyes are oval or kidney-shaped, small in females, but larger and distinctly convex in males. Ocelli are usually absent but vestiges are present in some cases (e.g., *Antipaluria*, some species of *Clothoda*). The clypeus is separated from the frons by a distinct but medially obliterated transverse strengthening ridge (epistomal ridge); it is divided into a transparent anteclypeus and a sclerotized posterior postclypeus. The transverse and evenly rounded labrum is movably connected to the anterior clypeal margin; the tormae are well-developed; the complete set of labral muscles is present (M. labroepipharyngalis, M. frontolabralis, M. frontoepipharyngalis). The filiform or moniliform antennae are inserted anterolaterally in front of the eyes; they are composed of 13–43 segments and are rarely longer than the head and thorax combined; they are moved by three tentorioscapal muscles. The mandibles are often sexually dimorphic, equipped with a mola in females and transformed into elongate clasping devices in males. The ventral mouthparts are largely unmodified. The maxillae are sexually dimorphic but less distinctly than the mandibles; they are more compact in females (observed in *Embia*; Rähle 1970); the lacinia and galea are normally developed; the palps are 5-segmented. The labium is composed of a plate-like, roughly quadrangular postmentum (designated as submentum by Rähle [1970], only well-developed in adults) and a well-developed prementum with paraglossae and small glossae; the palps are 3-segmented; like in Phasmatodea an additional dorsal flexor of the paraglossae is present. The hypopharynx is large, subdivided into an anterior and posterior part, and axe-shaped (observed in *Embia*; Rähle 1970). The salivarium and the paired salivary glands are well-developed. The latter reach the mesothorax posteriorly. The tentorium is complete and strongly developed; the anterior arms are widened anteriorly; the corpotentorium is rectangular.

Two laterocervicalia connect the head with the prothoracic pleura. The prothorax is usually slightly narrower than the pterothoracic segments. The meso- and

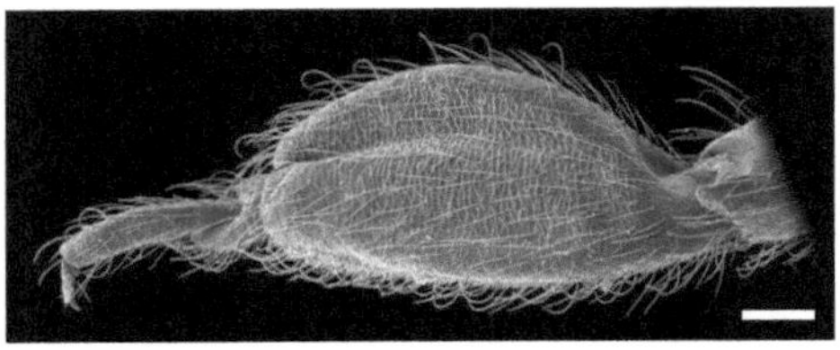

Fig. 6.10.2: *Oligotoma nigra* (Oligotimidae), distal part of fore leg, with enlarged basal tarsomere, SEM micrograph. Scale bar: 100 µm. Courtesy H. Pohl.

metathorax are similar in their overall structure and distinctly simplified in flightless individuals. The legs (Fig. 6.10.1) are relatively short and the coxae widely separated; the femora and tibiae of the forelegs and middle legs are largely unmodified; the metafemora are often thickened and equipped with a strong musculature; the basal protarsomere is greatly enlarged; voluminous silk glands are present in protarsomeres 1 and 2 of nymphs and adults; tarsomeres 1 and 2 are densely covered with adhesive hairs on their ventral side and equipped with one or two euplantulae; tarsomere 3 bears paired claws with a basal seta; the pretarsal arolium and pulvilli are absent. The dehiscent wings of alate males (Fig. 6.10.1) are narrowed at their base and the anal field is greatly reduced, with only one longitudinal vein; the hindwings are slightly smaller but otherwise very similar to the membranous forewings; the venation is simplified and the wing margin is set with microtrichia; a specialized feature is the presence of sinuses in the wings, which collapse when the haemolymph is withdrawn into the body.

The cylindrical, parallel-sided abdomen is composed of ten distinct segments. Segment XI is vestigial, with greatly reduced epiproct and paraprocts. The cerci are usually 2-segmented but sometimes 1-segmented in males. Segments I–VIII are similar in size and shape. The terminal segments of males (Fig. 6.10.3B, C) are transformed into accessory copulatory structures. They are usually distinctly asymmetric but symmetrical in the groundplan. Tergite X is often divided into two halves with specific caudal modifications. The left cercus usually bears one or several mesally directed projections on its basal part. Sternite IX forms a large, often asymmetric subgenital plate (hypandrium). Gonostyli are absent. A strongly simplified penis is only rarely present. The ductus ejaculatorius usually opens posterad the hypandrium. The subgenital plate is formed by sternite VIII in females (Fig. 6.10.3A). An ovipositor is not developed. The epiproct is an unsclerotized lobe hidden by tergite X, which is often pointed. The triangular sclerotized paraprocts flank the anus.

A toothed proventriculus is present; the midgut lacks caeca. The number of Malpighian tubules varies between 14 and 30. In addition to the eleven lateral ostia of the dorsal vessel unpaired ventromedian openings are present in the pterothoracic segments and abdominal segment I. The paired testes are composed of five serially

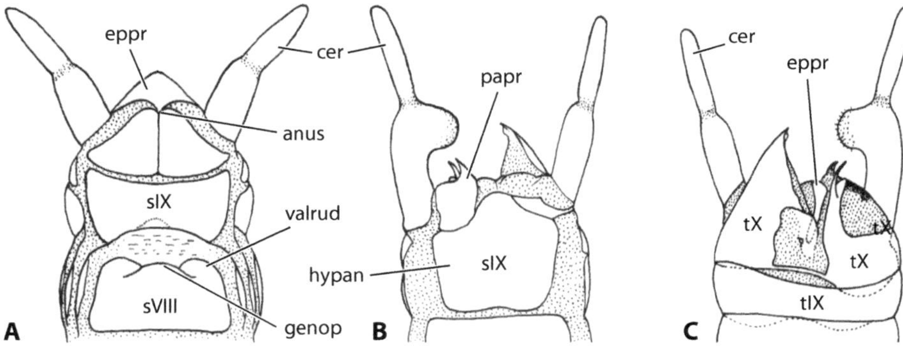

Fig. 6.10.3: Terminal abdominal segments. A, female, ventral; B, male, ventral; C, male dorsal. Abbr.: cer – cercus, eppr – epiproct, genop – genital opening, hypan – hypandrium, papr – paraproct, sVIII/IX – sternum VIII/IX, tIX/X – tergum IX/X, valrud – vestigial valvula. Redrawn from Günther (2005) (from Tuxen 1970).

arranged vesicles. Tube-like accessory glands open into the proximal ductus ejaculatorius. Five pairs of panoistic ovarioles are present in females.

Biology. The gregarious webspinners occur in soil, leave litter, under stones, in rotting wood, in shrubs, and under bark of trees. They produce connected systems of galleries using silk secreted by glands in the protarsomeres 1 and 2. The cylindrical body size and the collapsible wings (of males) allow them to move efficiently forwards and backwards in the tunnels, which protect them from predators and parasites. Silk is spun by all nymphal instars and adults. It is extruded through specialized hollow setae. Females share a nest with their offspring (gymnopaedium) and take close care of the young nymphs. Connected nests can form an extensive tunnel system with colonies comprising many individuals. Nymphs stay permanently in the galleries, whereas young mature males and females occasionally leave the system. Alate males fly over short distances. They are attracted by light. Wingless males usually mate with females in nearby colonies. Females and nymphs mainly feed on decaying plant material such as rotting bark or leaves, but also on fresh parts of mosses, lichen, algae and microorganisms. The short-lived males do not consume food.

Reproduction and development. Webspinners are primarily bisexual but thelytokous parthenogenesis occurs. Males search for a gallery containing a mature female. They perforate the silk layer and perform a mating ritual, which includes trilling with the antennae, before grasping the head of the female with the elongated mandibles. The apical part of the male abdomen is locked with the female terminal segments before the spermatophore is transmitted. The ovoid or urn-shaped eggs (ca. 80–220) are deposited in the galleries, either singly or in groups. In some taxa they are covered with a web by the females. The female brood care includes licking and protecting the eggs and sometimes feeding the young nymphs with chewed up material. Four nymphal molts take place after hatching. In some species the nymphs move away from the females very early to build their own nest.

Fossil record. Fossil webspinners are known from Baltic and Dominican amber. Permian fossils assigned to Embioptera are questionable.

Economic importance. The economic impact is negligible. One species damages orchids in greeneries.

[Kaltenbach (1968); Rähle (1970); Tuxen (1970); Ross (1966, 1970, 2000, 2006, 2007); Günther (2005); Engel & Grimaldi (2006); Szumik et al. (2008); Miller et al. (2012)]

6.11 Phasmatodea (Latin *phasma* = apparition, English common names: stick and leaf insects)

Diversity and distribution. About 3,030 known species are known. They are mainly distributed in the tropical and subtropical regions. Only 17 species occur in the Mediterranean area. The group reaches its highest diversity in Southeast Asia (ca. 1,500 spp.), followed by South America and Australia.

Autapomorphies
- Frons bulging anterolaterally
- Ventral apodemes of the posterior tentorial arms (= trabeculae tentorii)
- Tube-shaped defensive glands opening at the anterolateral edges of the prothorax
- Midgut with pear-shaped appendage
- Trochanter and femur rigidly connected
- Male sternite X modified as vomer
- Midgut with pear- or cone-shaped glands with tubular appendages
- Sexual dimorphism with larger and less slender females

In earlier phylogenetic concepts Phasmatodea were considered as a subgroup of Orthoptera or until recently as the sistergroup of this lineage (e.g., Willmann 2005a). Recent studies based on morphological and molecular data show that they are very likely the sistergroup of Embioptera.

Taxonomy. Timematodea with the only genus *Timema* (21 wingless species in Mexico and California) is the sistergroup of the entire remaining Phasmatodea (Euphasmatodea) (Fig. 6.11.1A). The next branch is Agathemeridae, also with a single genus (*Agathemera*, eight wingless species in South America). All species of Euphasmatodea excl. Agathemeridae (ca. 3,000) were placed in ca. 400 genera of the family Neophasmat-

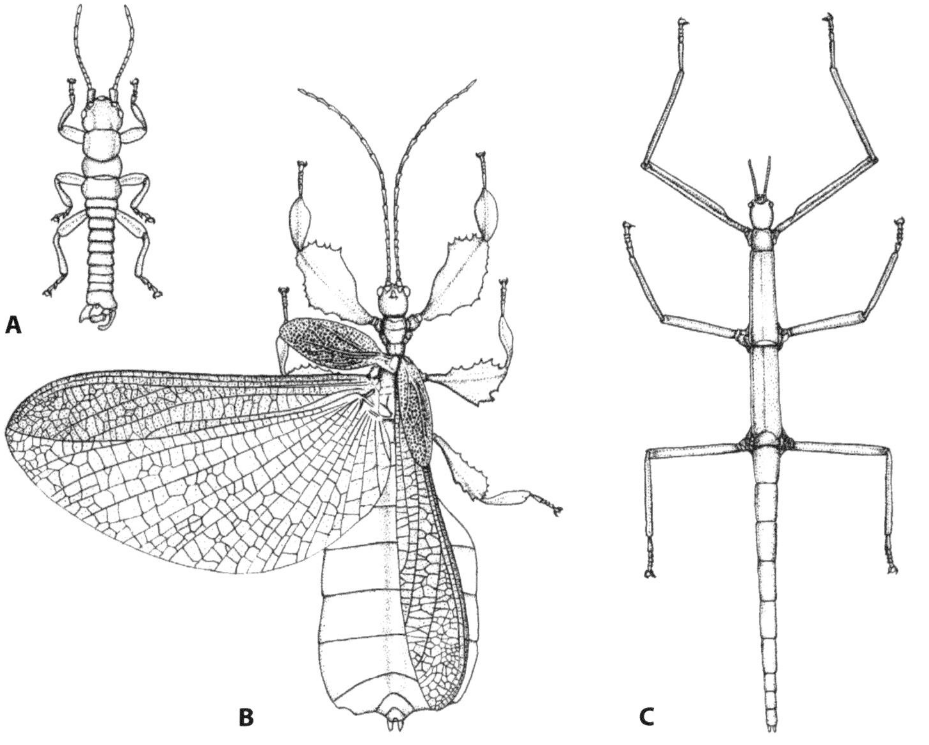

Fig. 6.11.1: Phasmatodea, habitus. A, *Timema podura* (Timematidae), male; B, *Phyllium bioculatum* (Phylliidae); C, *Bacillus rossius* (Bacillidae). Redrawn from Bradler (2005).

idae in a recent classification (Bradler 2005, 2009), but alternative systems exist with more than ten families and numerous subfamilies. The relationships within this large lineage are still largely unclear. Most currently recognized subfamilies (e.g. Bacillinae, Eurycanthinae, Phasmatinae, Tropidoderinae, Xeroderinae) and tribes are likely not monophyletic (Bradler 2005, 2009).

Diagnosis. Large or very large, with a rather compact cylindrical body (Timematodea, Agathemeridae) or with conspicuous camouflage mechanisms, either mimicking branches, twigs (stick insects) or leaves (leaf insects, Phylliinae) (Fig. 6.11.1).

Morphology. The body of Phasmatodea is more or less slender and elongated. Males of *Timema cristinae* belong to the smallest representatives of the order with 11.6 mm. With over 35 cm females of *Phobaeticus chani* are longer than any other known extant insect species. The basal Timematodea and Agathemeridae (both wingless) are cylinder-shaped and largely unmodified (Fig. 6.11.1A), whereas the other members of the group either display a stick-like (Fig. 6.11.1C) or a leaf-like habitus (Fig. 6.11.1B) .

The head is almost always prognathous and strongly sclerotized. The foramen occipitale is usually narrowed by a gula (not in the groundplan). The compound eyes are relatively small. Three ocelli are often present but only two in *Phasma*. Ocelli are lacking in all flightless forms, and also in many winged species. A coronal suture and frontal sutures are present in the groundplan. The frons is bulging anterolaterally (potential autapomorphy). The free labrum is medially incised, thus appearing bilobed; two extrinsic and one intrinsic labral muscles are present. An antennifer is present in Timematodea but absent in Euphasmatodea (autapomorphy). The antennae are almost always filiform; they are slightly pectinate in males of some species and an antennal stridulatory organ occurs in Phylliinae; the flagellum is usually composed of numerous segments (up to 100) but only one flagellomere is present in *Paragongylopus*. The mouthparts are orthopteroid. The stout mandibles bear a well-developed, strongly sclerotized mola; prominent incisivi and molar teeth are only preserved in Timematodea (plesiomorphy). The maxilla is largely unmodified; the lacinia usually bears three apical teeth (two in Timematodea, 5–6 in *Phyllium*); an appendage of the galea, the galeolobulus, is present in Euphasmatodea (autapomorphy); the palp is always 5-segmented. The labial postmentum is subdivided into a posterior submentum and a seemingly tripartite mentum; the prementum bears the 3-segmented palps and large paraglossae, which laterally and anteriorly enfold the smaller glossae. The semimembranous hypopharynx lies above the mentum and prementum. The salivary ducts open separately in Euphasmatodea (autapomorphy) but not in Timematodea (groundplan). The tentorium is complete; the dorsal arms are slightly thinner than the other elements (*Phyllium*); the posterior arms bear ventral apodemes, the trabeculae tentorii (autapomorphy). Bipartite paired laterocervicalia are embedded in the cervical membrane; they articulate with the prosternum.

The thoracic segments are about equally sized in the groundplan (Fig. 6.11.1A). The prothorax is never elongated. The ducts of the defensive glands open at the anterolateral edges of the pronotum (autapomorphy). The prothoracic anapleurites and coxopleurites are usually fused but not in Timematodea and Aschiphasmatinae. A profurca is only present in Timematodea (plesiomorphy). The pterothoracic segments are almost always distinctly elongated in Euphasmatodea (Fig. 6.11.1C), with conspicuously enlarged episterna (not in Timematodea and Agathemeridae). Wings are absent in the basal lineages and in more than 50% of all species (Fig. 6.11.1A, C); they are more often well-developed in males than in females; brachypterism is rather common aside from the complete loss of wings; the forewings (if present) are leathery tegmina; they are distinctly narrower than the large hindwings and often shorter; in many cases they are only represented by small scale-like structures; the costal field of the hindwing also shows an increased degree of sclerotization; in the resting position it covers the large membranous anal field; the venation is uniform; the costal vein forms the anterior margin of the hindwings but is missing in the forewing; the subcosta is always unbranched; the radius is usually more strongly developed than the other veins, and the media is the only branched vein in the forewings (exception:

Heteropteryx with a branched radius sector); only one anal vein is present in the forewing but an entire series with a fan-shaped arrangement in the hindwing. The femora and trochanters of the largely unmodified walking legs are rigidly connected; the tarsi are usually 5-segmented and the basal tarsomere can be strongly elongated; tarsomeres 1–3 are fused in Timematodea (autapomorphy); large euplantulae are present on the ventral side of tarsomeres 1–4; a large arolium is inserted between the claws; the euplantulae and arolium of Timematodea and Agathemeridae are very similar to the corresponding structures of Mantophasmatodea, i.e. unusually large and densely covered with acanthae.

The abdomen is 11-segmented (Fig. 6.11.1). Tergum I is free in Timematodea (groundplan) but fused with the metanotum in Euphasmatodea, thus forming a „segmentum medianum" (euphasmatodean autapomorphy); sternite I is always firmly connected with the metathorax. Segment IX of males is divided into an anterior and a posterior portion. The posterior part (subgenital plate, poculum) covers the aedeagus. Segment X forms a characteristic element of the male genital apparatus, the curved vomer (secondarily reduced in some groups), which clasps the female sternum VII during copulation. In females sternite VIII is often enlarged and forms a subgenital plate which covers the short orthopteroid ovipositor. Three pairs of valves are present in the groundplan; the gonostyli IX (gonoplacs) are often reduced; all three valves are missing in Agathemeridae (autapomorphy); a distinct gonangulum connecting the gonobases (gonocoxae) VIII and IX is usually present; secondary ovipositors occur in some groups. Segment XI bears the short unsegmented cerci; it is composed of the dorsal epiproct (supraanal plate) and the lateral paraprocts, which are fused and form an unpaired lobe in Agathemeridae (autapomorphy).

The abdominal musculature is distinctly modified. The longitudinal muscles appear shortened in elongated species. Their origin is shifted posteriorly in the segments. The dorsoventral muscles are subdivided into isolated fibers. A large ingluvies is present but the proventriculus is distinctly (Timematodea) or completely reduced; the caeca of the relatively long midgut are vestigial or completely reduced; the posterior midgut bears pear- or cone-shaped glands, each of them with tube-like appendages (four in *Timema*, 130 in *Phyllium*). Numerous Malpighian tubules are present (Bradler 2005). The paired testes are composed of numerous follicles enclosed by a peritoneal membrane; they usually appear as uniform simple tubes (not in Timematodea). The ejaculatory duct bears at least six tube-like accessory glands. In females different glands and a paired or unpaired receptaculum seminis can be dorsally attached to a bursa copulatrix, which is often present above the short unpaired part of the oviduct (oviductus communis). The number of panoistic ovarioles per ovary varies between seven and 103.

Biology. Phasmatodeans are exclusively phytophagous and usually nocturnal. Most species are characterized by conspicuous camouflage mechanisms, which have evolved within the groups in two distinctly different ways (see above) (Fig. 6.11.1B, C).

Some species are able to change their color by shifting pigment granules in the epidermis.

Phasmatodeans stay in the foliage, on branches, on stems, or among herbaceous plants. In the typical case rhythmical swaying or rocking movements of the body enhance the camouflage effects. Additional protection against predators is provided by secretions of the defensive glands. Aposematic color patterns occur in species with very effective defensive substances. The appearance of phasmatodeans can vary considerably within species. Horn-like structures or thorns can be present on different body parts of some individuals but missing in others. Morphs with and without green coloration occur in some species. Geographic variation of morphological features is also common.

Despite camouflage mechanisms and defensive glands phasmatodeans are the preferred prey of certain bird species. Early nymphal stages are also preyed upon by ants and spiders. Important parasites belong to Tachinidae (Diptera) and Chrysididae (Hymenoptera).

Reproduction and development. Phasmatodeans usually reproduce sexually but facultative parthenogenesis is relatively common. Obligatory parthenogenesis is only known of few species. Specific mating behavior is unknown. Males of some species produce sounds by knocking on the substrate with their abdomen to attract females. During copulation the abdomen of the males forms a bent of ca. 180°. The vomer and corresponding parts of the female genital apparatus form a clasping mechanism. The eggs are characterized by a conspicuous lid-like structure, the operculum. In many cases an additional knob-like capitulum is present, which facilitates the transport and dispersal of the eggs by ants (myrmecochory). The eggs are deposited in leaf litter or on the ground, or attached singly or in batches to suitable substrates with secretions. The total number of eggs varies between 100 and 1,300, depending on the species. There are 4–8 nymphal stages (Bradler 2005).

Fossil record. Presumptive phasmatodeans from Permian deposits (e.g., Rasnitsyn & Quicke 2002) are questionable (Grimaldi & Engel 2005). Mesozoic fossil species likely belong to the stem group of Phasmatodea (Willmann 2003). Tertiary phasmatodean fossils are rare.

Economic importance. Only few phasmatodeans are relevant as pests in agriculture or forestry. Outbreaks of some Australian species can result in defoliation of *Eucalyptus* forests (e.g., *Podacanthus wilkinsoni*, *Didymuria violescens*), and *Graeffea coccophaga* and *G. crouani* can cause serious damage in coconut plantations in the South Pacific (Bradler 2005).

[Beier (1968b); Rasnitsyn & Quicke (2002); Willmann (2003, 2005a); Bradler (2005, 2009); Grimaldi & Engel (2005); Friedemann et al. (2012)]

6.12 Orthoptera (Greek *ortho* = straight, *pteron* = wing, English common names: katydids = bush-crickets, crickets, grasshoppers, locusts)

Diversity and distribution. Orthoptera is by far the most species rich polyneopteran lineage. It comprises about 22,500 described species and has a worldwide distribution, with the highest diversity in tropical and subtropical regions, occurring from lowland up to subalpine zones, from rain-forests to semi-deserts, and also in caves. Some species are synanthropic.

Taxonomy. Orthoptera is divided into the two suborders Ensifera and Caelifera. The former comprises six extant superfamilies (Tettigonioidea, Stenopelmatoidea, Rhaphidophoroidea, Hagloidea, Grylloidea, Schizodactyloidea) and the latter eight (Acridoidea, Trigonopterygoidea, Pneumoroidea, Pyrgomorphoidea, Tanaoceroidea, Eumastacoidea, Tetrigoidea, Tridactyloidea). Caelifera are strongly supported by apomorphies (e.g., shortened antennae, strongly modified ovipositor) whereas Ensifera are mainly characterized by plesiomorphic features, but are consistently supported as a monophylum in phylogenetic studies. A characteristic which may have evolved twice within this suborder is the protibial tympanal organ.

Diagnosis. Usually medium sized or large insects with a saddle-shaped pronotum and forewings modified as leathery tegmina. Hindlegs usually elongated and modified for jumping. Orthopterans are often able to produce stridulatory sounds.

Autapomorphies
- Pronotum saddle-shaped (strongly elongated in Proscopiidae) (Fig. 6.12.1)
- Distinctly developed precostal field of forewings
- Tarsi with four segments or less (Fig. 6.12.1)
- Metathoracic spiracles shifted anteriorly, enlarged, with two originating tracheae
- Hindlegs modified as jumping legs, femora enlarged (groundplan)
- Wing pads twisted in later nymphal instars

Morphology. Orthoptera are moderately small to very large insects, ranging in body length from 2 to 200 mm. Most species are moderately elongated and slightly compressed laterally, but different modifications occur, such as for instance a branch-like or very stout body. Wings are usually present but reductions took place in different lineages. Color patterns mainly comprising brown, green, red, black and yellow are frequent. The setation is inconspicuous but a great variety of thorns, tubercles, grooves and ridges can occur on the head and thorax.

The large head (Fig. 6.12.1) is almost always orthognathous or hypognathous, only in rare cases subprognathous; in frontal view it appears ovoid or nearly circular (e.g., *Gryllus*). The posterior part is slightly retracted into the prothorax. The com-

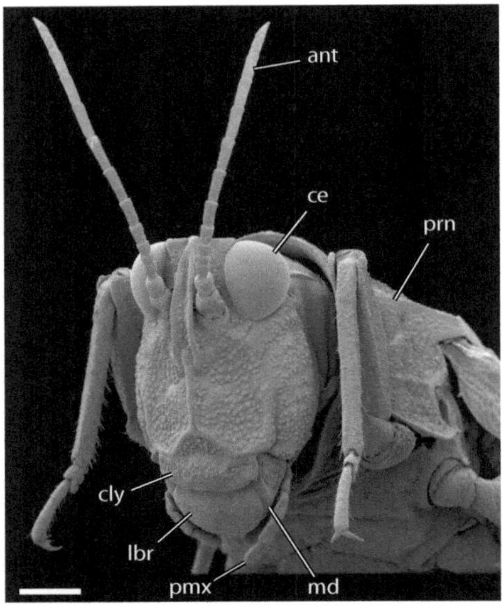

Fig. 6.12.1: *Tetrix* sp. (Tetrigidae), head, anterolateral view, SEM micrograph. Abbr.: ant – antenna, ce – compound eye, cly – clypeus, lbr – labrum, md – mandible, pmx – palpus maxillaris, prn – pronotum. Scale bar: 500 µm. Courtesy H. Pohl.

pound eyes are usually well-developed and integrated in the contour of the head capsule; they are primarily large but distinctly reduced in size in Myrmecophilidae (e.g., *Myrmecophilus*) and few Stenopelmatidae (e.g., *Stenopelmatus*), and completely missing in few Rhaphidophoridae living in rodent nests (e.g., *Typhloceuthophilus*); three ocelli are usually present and arranged in a triangle on the frontal region. The coronal and frontal sutures are primarily present. The postoccipital ridge is present and the genal ridge is usually also well-developed. A distinct occipital ridge surrounds the foramen occipitale in *Locusta*. The clypeus is usually separated from the frons by the transverse clypeofrontal strengthening ridge; a short incision ("clypeal suture") is usually present at the lateral clypeal margin, incompletely dividing the sclerite into a postclypeal and anteclypeal region. The antennal foramina lie usually on the upper frontal region between the compound eyes, sometimes strongly approximated, rarely between the compound eyes and the mandibular articulation. The labrum is almost always free and often large and covering the distal parts of the mandibles in frontal view; on its largely membranous epipharyngeal side it is variously equipped with sclerotizations and fields of dorsally directed microtrichia. The antennae are usually filiform and primarily long and multisegmented; they are distinctly shortened in Caelifera and usually slightly flattened (7–30 antennomeres); club- or comb-shaped antennae are rare exceptions. The mouthparts are of a very plesiomorphic ("orthopteroid") type. The large biting mandibles bear strongly sclerotized teeth (incisor area) and are primarily equipped with a well-developed molar area; in predacious ensiferans the mandibles are more elongate and the mola is reduced in size. The ventral mouthparts (maxillae and labium) are largely unmodified and equipped with the typical elements. The proximal labium comprising the submentum and mentum forms the posterior closure of the head. The tentorium is well-developed and complete. Cervical sclerites are present.

The prothorax is mainly characterized by the extension of the paranota, resulting in the saddle-shaped pronotum (Fig. 6.12.1). Anteriorly the pronotum overlaps with the vertex and posteriorly with the wing bases; usually it is subdivided by several transverse furrows; the paranota can be closed ventrally in burrowing forms, and the pronotum is strongly extended posteriorly in Tetrigidae, thus covering the hindwings and the abdomen; various spines or other surface modifications occur. The pterothoracic segments form a compact functional unit. The metathorax is larger than the mesothorax in species with normally developed flight organs. Both segments show a largely unmodified pattern of tergal, pleural and sternal elements. Meso- and metathoracic spiracles are present; the latter are enlarged, shifted anteriorly, and divided, giving rise to two tracheal branches. The wings are primarily well-developed but reductions occur in different lineages (e.g., Myrmecophilidae and Rhaphidophoridae); at rest they are either held horizontally on top of the abdomen or in a roof-like position; the forewings are leathery tegmina and comparatively narrow; the anal field of the hindwings is large and folded in a fan-like manner as in several other polyneopteran lineages; stridulatory organs are usually present at the basal part of the tegmina of ensiferans (asymmetric in Tettigoniidae), and therefore missing in wingless taxa; in most caeliferans the hind femora with the inner pars stridens interact with the radial vein on the external side of the tegmina, but other stridulatory and sound-producing mechanisms occur (e.g., vibrational signals in Tetrigidae). The fore- and middle legs are normally sized walking legs; a protibial tympanal organ is usually present in Ensifera (Fig. 6.12.2B, D) and the forelegs of Gryllotalpidae are modified as stout burrowing organs; the hindlegs are modified as jumping legs, with a strongly enlarged femur and strongly developed femoral musculature; reversal occurs in few lineages (e.g., Gryllotalpidae); in few taxa the hind tibiae are also enlarged and adapted for swimming (e.g., *Paulinia* [Pauliniidae]); four or three tarsomeres are present in Ensifera (Fig. 6.12.1) and three or rarely two in Caelifera; one-segmented tarsi are a rare exception; well-developed euplantulae are present on the ventral side of the tarsomeres; a distal arolium is present in Caelifera but missing in Ensifera.

Segment I of the 11-segmented abdomen is closely attached to the metathorax; in most caeliferans it bears a large and complex tympanal organ. Elements of stridulatory organs are occasionally present on the basal segments. Tergum XI (epiproct) is sometimes fused with tergum X. Sternite VIII forms the subgenital plate in females (Fig. 6.12.3) and sternite IX in males (Fig. 6.12.4), where it is often elongated. The cerci at the posterior margin of tergite X are 1-segmented and short and cone-shaped in

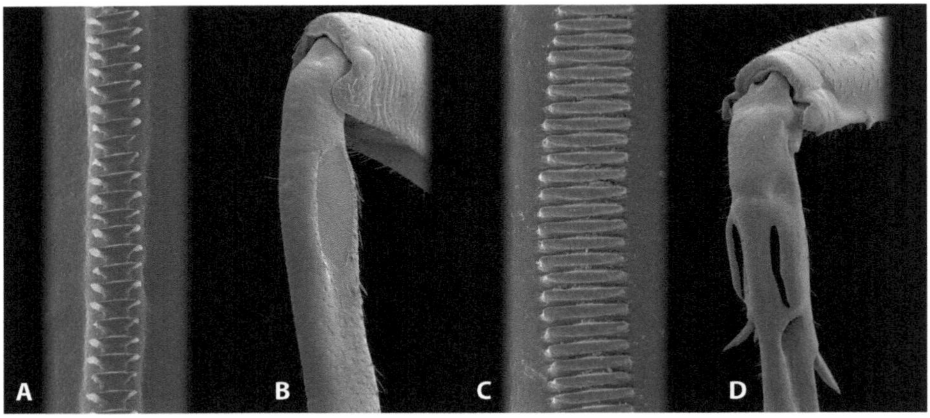

Fig. 6.12.2: Stridulatory and tympanal organs of Ensifera, SEM micrographs. *Gryllus campestris* (Gryllidae), A, stridulatory file; B, protibial tympanal organ; *Tettigonia viridissima* (Tettigoniidae), C, stridulatory file; D, protibial tympanal organ. Courtesy H. Pohl.

females (Fig. 6.12.3); in males they are often involved in the copulation as clasping organs. In male ensiferans the subgenital plate is fused with the gonocoxites and usually bears the styli (reduced in Gryllidae) (Fig. 6.12.4); the simple copulatory organ is unsclerotized and formed by 1–3 pairs of phallus lobes which enclose the opening of the endophallus (phallotrema); the dorsal phallus lobes usually bear sclerotized structures, which are sometimes hook-shaped (titillators, epiphallus) (Fig. 6.12.4). Styli are absent in Caelifera; the well-developed copulatory organ (phallic complex) lies in the genital chamber above the subgenital plate; it is composed of an epiphallus and the phallus, the latter comprising the ectophallic cingulum, the caudal aedeagus, and the endophallic plates. Female ensiferans are equipped with an elongate, typical orthopteroid ovipositor (Ensifera=sword bearers) (Fig. 6.12.3A); the ovipositor is strongly shortened and modified in Caelifera (Fig. 6.12.3B), apparently an adaptation for depositing the eggs in soil; it is composed of well-developed ventral and lateral valves (gonapophyses) and reduced mesal valves (Groll & Günther 2005).

A proventriculus is present in Ensifera and midgut caeca are present in both suborders; a varying number of ampullae is present at the midgut-hindgut border where numerous thin Malpighian tubules originate. The paired testes vary strongly in shape; the distal part of the paired vasa deferentia sometimes forms a spiral-shaped epididymis in Ensifera (Gryllacrididae). Male accessory glands can fill out large parts of the abdomen. The ovaries are composed of a varying number of panoistic ovarioles; a very short unpaired oviduct is present or absent. The receptaculum seminis is often unpaired and asymmetric and connected with the vagina by a long duct. Different accessory glands occur in females (colleterial glands in Ensifera, pseudocolleterial glands in Caelifera) (Groll & Günther 2005).

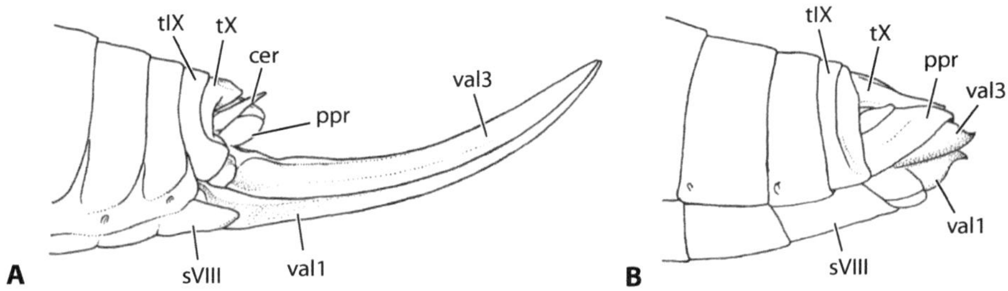

Fig. 6.12.3: Orthoptera, female postabdomen. A, *Pholidoptera* sp. (Tettigoniidae); B, *Chorthippus* sp. (Acridiidae); Abbr.: cer – cerci, ppr – paraproct, sVIII – sternum VIII, tIX/X – tergum IX/X, val 1/3 – 1st/3rd valve. Redrawn from Groll & Günther (2005).

Biology. Orthoptera prefer warm conditions and are basically terrestrial insects living on or among plants. Species more or less strongly adapted to aquatic habitats occur in different lineages (e.g., Tetrigidae, Tridactylidae) and the South-American Pauliniidae are semi-aquatic. Other specialized orthopterans dig in soil (Gryllotalpidae) or live in caves (e.g., Rhaphidophoridae). Stenopelmatids use saliva and soil or sand particles to build tubes. Myrmecophilids (ant crickets) are inquilines in ant colonies (Grimaldi & Engel 2005).

Most species of Ensifera are nocturnal and many predacious forms occur in this lineage. Carnivorous species, however, usually complement their diet by consuming different plant parts such as for instance fruit, seeds or flowers. Prey may include insects, snails and even small vertebrates (e.g., lizards), but ensiferans also consume carrion. In contrast to the typical ensiferan pattern Caelifera are mainly diurnal and strictly phytophagous.

Camouflage mechanisms are common and different somatolytic forms of homochromy, mimesis and mimicry occur. This includes color patterns, different surface modifications, and modifications of the body shape, resulting in a leave- or branch-like habitus (e.g., Pseudophyllinae, Proscopiidae), superficially similar to phenomena occurring in Phasmatodea. Some specialized species are mimics of ants or wasps and others blend in with lichens or bark. Defense mechanisms (mainly in Caelifera) include mechanical ones, like striking with the hindlegs or displaying conspicuous color patterns on the hindwings, and also chemical defense, like regurgitation of digestive fluids and secretion of defensive substances by prothoracic, tracheal or abdominal glands. Shedding the hindlegs at preformed zones of weakness is also common.

The acoustic communication systems have apparently played a crucial role in the evolution of orthopterans. Different tympanal and stridulatory organs have evolved independently within the group (Fig. 6.12.2). Ensiferans usually stridulate by rubbing the tegmina one upon the other (tegmino-tegminal stridulation), whereas most caeliferans use femoral ridges as pars stridens and the radial veins of the tegmina as plec-

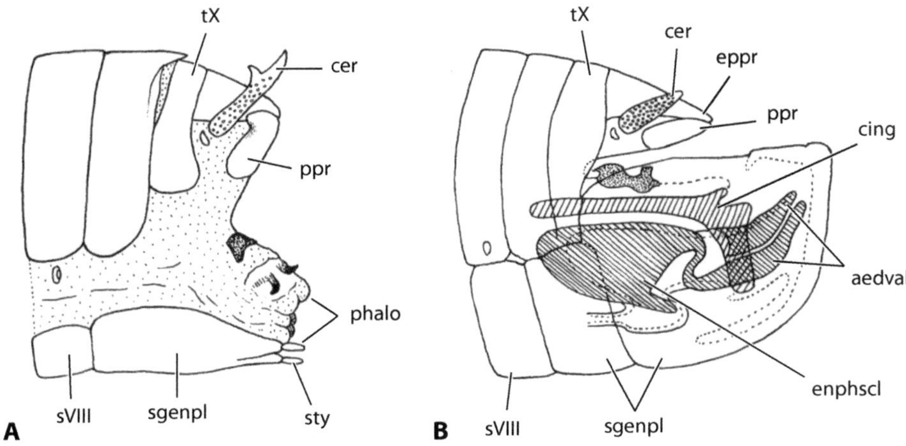

Fig. 6.12.4: Orthoptera, male postabdomen. A, *Pholidoptera* sp. (Tettigoniidae); B, *Chorthippus* sp. (Acridiidae); Abbr.: aedval – aedeagus valves, cer – cerci, cing – cingulum, enphscl – endophallic sclerite, eppr – epiproct, phalo – phallic lobes, ppr – paraproct, sVIII – sternum VIII, sgenpl – subgenital plate, sty – styli, tX – tergum X. Redrawn from Groll & Günther (2005) after Tuxen (1970).

trum (femuro-tegminal stridulation). The acoustic signals are used for intraspecific communication in different contexts. The sounds, which are generally species specific, are mainly or exclusively produced by males. The main purpose is to attract females of the same species and a side effect is to prevent mating between closely related species. The species specific sounds are also used to hold populations together or to defend territories.

Reproduction and development. Orthopterans are usually bisexual but different forms of parthenogenesis occur. In courtship acoustic communication (see above) plays an important role. The males produce a spermatophore, which can be structurally very complex. It is transmitted into the female genital opening by the male copulatory organ. A nuptial gift (spermatophylax) attached by male ensiferans to the spermatophore is consumed by the females. The eggs are deposited in or on plant tissue with the orthopteroid ovipositor or more often in soil, especially by females of Caelifera (ovipositor shortened and strongly modified). A worm-shaped larva without fully movable legs hatches from caeliferan eggs, but immediately begins to molt. Otherwise, the postembryonic development follows the typical hemimetabolous pattern, with nymphs very similar to the adults except for the smaller size and the absence of fully developed wings and genital organs. The number of molts normally varies between four and 11.

Fossil record. Among the polyneopteran lineages Orthoptera are exceptionally well represented in the fossil record, surpassed only by Blattodea and their extinct relatives. According to Grimaldi & Engel (2005) Paleozoic orthopterans (paraphyletic †Oedischioids) likely belong to the stem lineage of the entire order. The earliest representative of Ensifera was described by Béthoux et al. (2002) from the Late Permian of France (†Oedischiidae). The earliest definite representatives of Caelifera where found in Triassic deposits of Asia, Australia and Europe. Early Permian forms (e.g., †Permoraphidiidae, †Permelcanidae) were disputed with respect to their subordinal placement. They were assigned to Caelifera by Béthoux & Nel (2002).

Economic importance. Larger species are traditionally used as seasonally available protein-rich food in different cultures. Locusts are already mentioned in very old sources (e.g., the Bible) as the cause for disastrous famines. The most harmful species belong to the caeliferan family Acridiidae. Approximately ten species are known to cause damage to vegetation and crops on a large scale. Swarms of *Schistocerca gregaria* can comprise approximately 50 billion of individuals and cover thousands of square miles (Grimaldi & Engel 2005).

[Yuasa (1920); Uvarov (1966, 1977); Tuxen (1970); Beier (1972); Chapman & Joern (1990); Gangwere et al. (1997); Gwynne (2001); Béthoux & Nel (2002); Béthoux et al. (2002); Grimaldi & Engel (2005); Groll & Günther (2005)]

6.13 Zoraptera (Greek *zor* = pure, *aptera* = unwinged, English common names: ground lice, angel insects)

Diversity and distribution. The group introduced by Silvestri in 1913 belongs to the smallest insect orders. It comprises only 39 extant and nine fossil species. They occur in tropical regions with the exception of Australia, but are occasionally also found in subtropical or even temperate areas. They were apparently introduced in northeastern areas of the USA. Several new species were recently discovered in Malaysia (Mashimo Y., pers. comm.). The real diversity of the inconspicuous insects (Fig. 6.13.1) may be distinctly higher than what is presently known.

Taxonomy. All known extant species are placed in the family Zorotypidae. The systematic placement was long disputed. Zoraptera were considered as sistergroup of Acercaria by Hennig (1969) (=Paraneoptera). Recent studies clearly support a placement in a clade Polyneoptera.

Diagnosis. Small dimorphic insects, either wingless, eyeless and unpigmented, or winged, with compound eyes and ocelli and a distinct pigmentation. Tarsi 2-segmented and cerci 1-segmented.

Fig. 6.13.1: *Zorotypus weidneri* (Zorotypidae), SEM micrographs. A, head of blind, wingless form, lateral view; B, male of alate form (distal parts of wings removed), lateral view. Abbr.: ant – antenna, ce – compound eye, cer – cercus, fwi – forewing, hwi – hindwing, md – mandible, pmx – palpus maxillaris, sVIII – abdominal sternite VIII. Scale bar: 250 μm. Courtesy H. Pohl.

Autapomorphies

- Polymorphism
- Wing venation strongly simplified (in alate specimens)
- Tarsi 2-segmented
- Formation of small colonies with different morphs

Morphology. Small, inconspicuous insects with a total length of 1.5–2.5 mm (Fig. 6.13.1). The body is subcylindrical. The coloration of alate specimens is brownish, whereas the wingless exemplars are largely unpigmented. The cuticle bears a vestiture of setae of different length.

The head (Fig. 6.13.2) is orthognathous. Its posterior part is exposed. It is about as broad as long and triangular in frontal view. The compound eyes are completely reduced (wingless forms) (Fig. 6.13.2A) or moderately sized and composed of large ommatidia (Fig. 6.13.1B). The ocelli are arranged in a triangle and widely separated from each other (Fig. 6.13.2B) or completely reduced (wingless forms). The frontal and coronal sutures are absent. The clypeal region is divided into an anterior transparent anteclypeus and a posterior postclypeus, which is strongly narrowed between the antennal insertion areas and posteriorly fused with the frontal region. The semicircular labrum is articulated with the anterior clypeal margin and retractile. The large antennal foramina are separated from each other by less than their diameter (Fig. 6.13.2). The filiform antennae are 9-segmented and about as long as the head and thorax combined; the scapus and the distal flagellomeres are elongated; the flagellomeres bear a dense vestiture of setae of different lengths. The biting mandibles bear

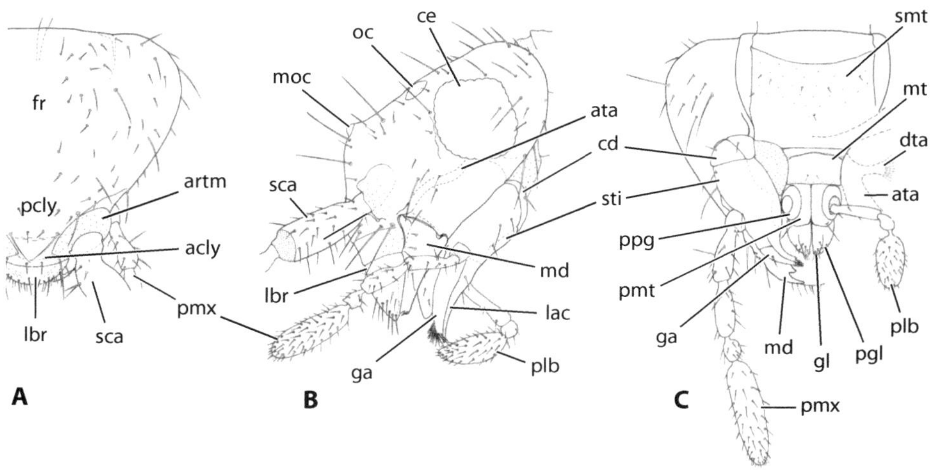

Fig. 6.13.2: *Zorotypus hubbardi* (Zorotypidae), head. A, wingless form, dorsal view; B, alate form, lateral view; C, wingless form, ventral view. Abbr.: acly – anteclypeus, artm – articulary membrane, ata – anterior tentorial arm, cd – cardo, ce – compound eye, dta – dorsal tentorial arm, fr – frons, ga – galea, gl – glossa, lac – lacinia, lbr – labrum, md – mandible, moc – median ocellus, mt – mentum, oc – ocellus, pcly – postclypeal area, pgl – paraglossa, plb – palpus labialis, pmt – prementum, pmx – palpus maxillaris, ppg – palpiger, sca – scapus, smt – submentum, sti – stipes. From Beutel & Weide (2005), modified.

five distal teeth and a proximal prominent molar region; a brush like prostheca is present on the left mandible. The lacinia of the largely unmodified maxilla is slender and devoid of any mesally directed spines. The semimembranous galea bears mesally directed spines on its apical part. The labium is composed of a very short submentum, a quadrate mentum, and a large prementum which is medially divided by a ridge and bears the 3-segmented palp and glossae and paraglossae. The large distal maxillary and labial palpomeres bear a dense vestiture of short setae. The foramen occipitale is narrowed by a weakly sclerotized, transverse gular area (in *Zorotypus hubbardi*; Beutel & Weide 2005). The tentorium is complete. The bridge is broad and the flattened anterior arms are the most extensive parts. The head articulates with the anterior pair of lateral cervical sclerites.

The thorax (Fig. 6.13.3) is less strongly sclerotized and simplified in the wingless forms. All morphs have an extensive cervical membrane reinforced by unpaired dorsal and ventral cervicalia and two pairs of lateral cervical sclerites. The prothorax is slightly smaller than the head. The large and saddle-shaped pronotum appears approximately rectangular in dorsal view, with rounded anterior and posterior corners. Two pleurites are present; the posterior one bears the pleural arm which is connected to the profurca by a very short muscle; the homology of the anterior pleurite is unclear. The slender, slightly curved protrochantin articulates with the procoxa.

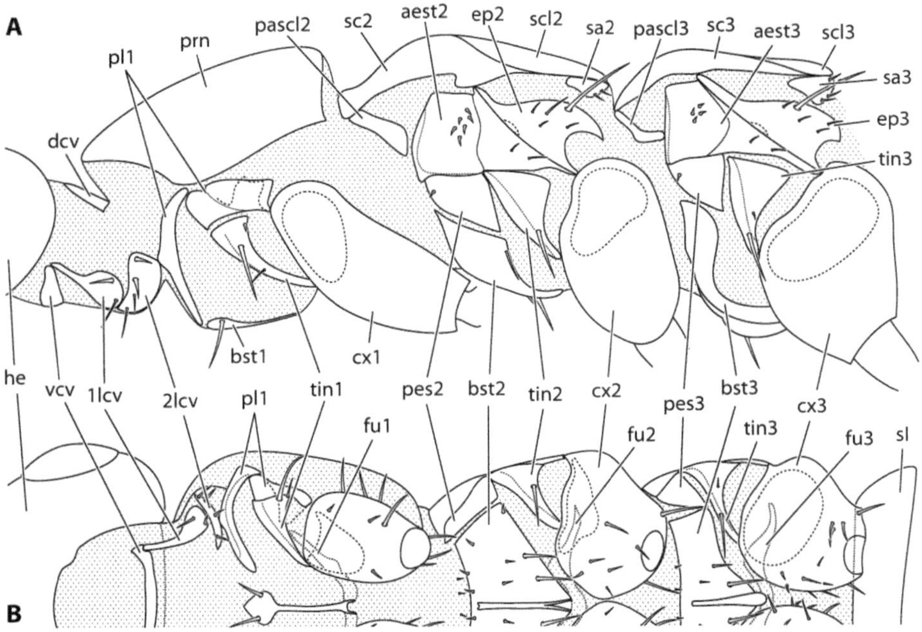

Fig. 6.13.3: *Zorotypus weidneri* (Zorotypidae), winged morph, thoracic skeleton. A, lateral view; B, ventral view. Abbr.: 1/2lcv – 1st/2nd lateral cervical sclerite, aest2/3 – mes-/metanepisternum, bst1/2/3 – pro-/meso-/metathoracic basisternum, cx1/2/3 – pro-/meso-/metacoxa, dcv – dorsal cervical sclerite, ep2/3 – mes-/metepimeron, fu1/2/3 – pro-/meso-/metafurca, he – head, pascl2/3 – meso-/metathoracic prealar sclerite, pes2/3 – meso-/metathoracic preepisternum, pl1 – propleuron, prn – pronotum, sl – abdominal sternite I, sa2/3 – meso-/metathoracic subalare, sc2/3 – meso-/metascutum, scl2/3 – meso-/metascutellum, tin1/2/3 – pro-/meso-/metatrochantin, vcv – ventral cervical sclerite. From Friedrich & Beutel (2008), modified.

The small, spoon-shaped basisternum bears a paired, flattened profurca at its posterior end. The well-developed prospina defines the prothoracic-mesothoracic border. A distinct, spinasternal sclerotization is recognizable externally. The mesothorax is slightly smaller than the prothorax, but larger than the metathorax. In the alate forms the notum is divided into a prescutum, a large scutum, a smaller and triangular scutellum, and a postnotum. The mesophragma is well-developed. The subalare is embedded in the pleural membrane close to the lateral scutal margin. The extensive mesopleuron is formed by a large anepisternum anterad the oblique pleural ridge and a smaller, posteriorly pointed epimeron behind it. The pleural suture is indistinct dorsally. A prealar bridge is formed by the well-developed prealar sclerite interconnecting the prescutum and anepisternum. A distinct preepisternum is present between the anterior half of the anepisternum and the basisternum. The freely movable trochantin is large and almost triangular and articulates with the anterior coxal rim. The largest portion of the sternal region is the broad basisternum. The well-developed furca has a

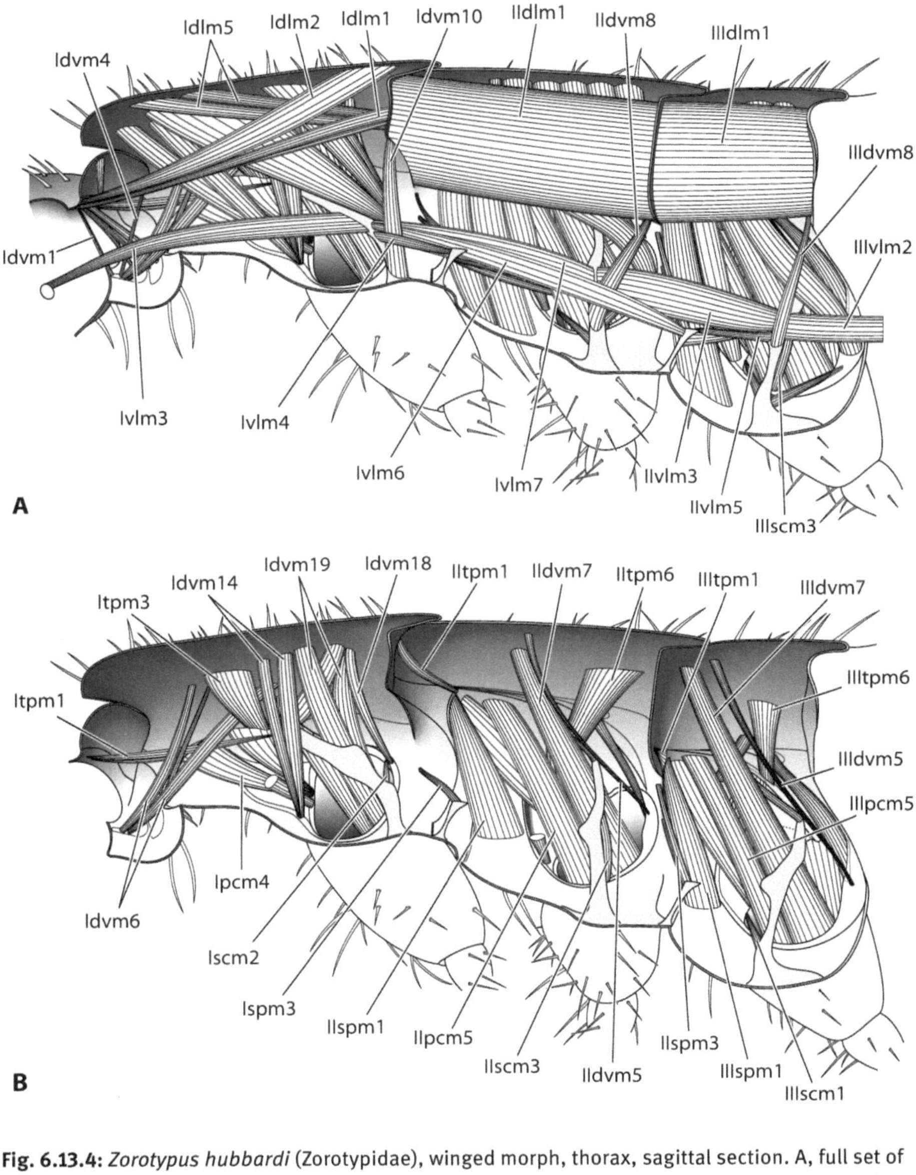

Fig. 6.13.4: *Zorotypus hubbardi* (Zorotypidae), winged morph, thorax, sagittal section. A, full set of muscles; B, mesal muscles removed. See Fig. 6.13.3 for sclerites and Table II (end of chapter 1.3) for muscle descriptions. From Friedrich & Beutel (2008), modified.

common stem, in contrast to the profurca. A mesospina is present. The metathorax is similar to the mesothorax but distinctly smaller (Fig. 6.13.3). The metaspina is absent. The ambulatory legs are slender and elongated; the large coxae are widely separated and posteriorly directed; the metafemora are more or less strongly thickened bearing stout spines interacting with a row of short spines on the tibia; the 2-segmented tarsi bear simple paired claws; an arolium or other attachment structures are absent. Both pairs of wings of the alate forms are membranous. The venation is strongly simplified, especially in the smaller hindwings. A fringe of microtrichia is present along the margin. The thoracic muscle system is well-developed (Fig. 6.13.4) and close to the neopteran groundplan condition (e.g., complex system of sternum-associated muscles; see Friedrich & Beutel 2008).

The abdomen (Fig. 6.13.1) is broadly connected with the thorax. It is composed of 11 segments, but the posterior segments are strongly reduced. The 1-segmented cerci insert laterally on tergite IX. Tergites VIII–XI of males bear specific processes which are involved in the copulation. Sternite IX and gonostyli are missing. The male genitalia differ strikingly within the group (e.g., Hünefeld 2007; Dallai et al. 2012a, b) (Fig. 6.13.5). Accessory glands are probably always present. The male intromittent organ (Fig. 6.13.5.B: introm) which transfers a large spermatophore is a membranous pouch-like structure, at least in most cases. A strongly elongated, very thin and coiled element (flagellum) within a spiral-shaped pouch (Fig. 6.13.5.B: sppo) is present in some species (see below). It is not connected with the ejaculatory duct and not involved in sperm transfer (Dallai et al. 2011, 2012a, b, 2013). In females sternite VIII forms a subgenital plate anterior to the genital opening. A spermathecal duct and spermatheca are present. The ovipositor is reduced.

The tracheal system is holopneustic. The abdominal ganglionic chain is strongly concentrated, forming two large complexes. Six Malpighian tubules are present. Five accessory glands occur in males. The paired testes are composed of two follicles and distinctly elongated in some species. The ovaries comprise 4–6 panoistic ovarioles.

Biology. Zoraptera are ground-oriented and found in rotting wood, under bark in decaying vegetation and in soil. They are subsocial, with the individual-rich colonies comprising strongly pigmented individuals with shed wings, weakly pigmented and unpigmented adults without wings and eyes, and nymphs with and without wing buds. Nymphs and adults feed on fungal hyphae and spores and are also scavengers and occasionally cannibalistic. At least in captivity zorapterans are efficient predators of Collembola. Bacteria are abundant in the gut contents, but endosymbiotic protozoans where not found so far.

Reproduction and development. In contrast to the highly uniform habitus of zorapterans, the genital organs, especially of males, vary tremendously (see above; Fig. 6.13.5). A strongly elongated and coiled male flagellum has possibly evolved several times independently (e.g., *Zorotypus barberi*, *Z. gurneyi*; Dallai et al. 2012a). As pointed out above this structure is not involved in sperm transfer but is likely used to remove or damage large spermatozoa from previous copulations in the spermathecal duct. The sperm cells of Zoraptera are large and strongly elongated. This condition is possibly related with a reproductive strategy of males of some species trying

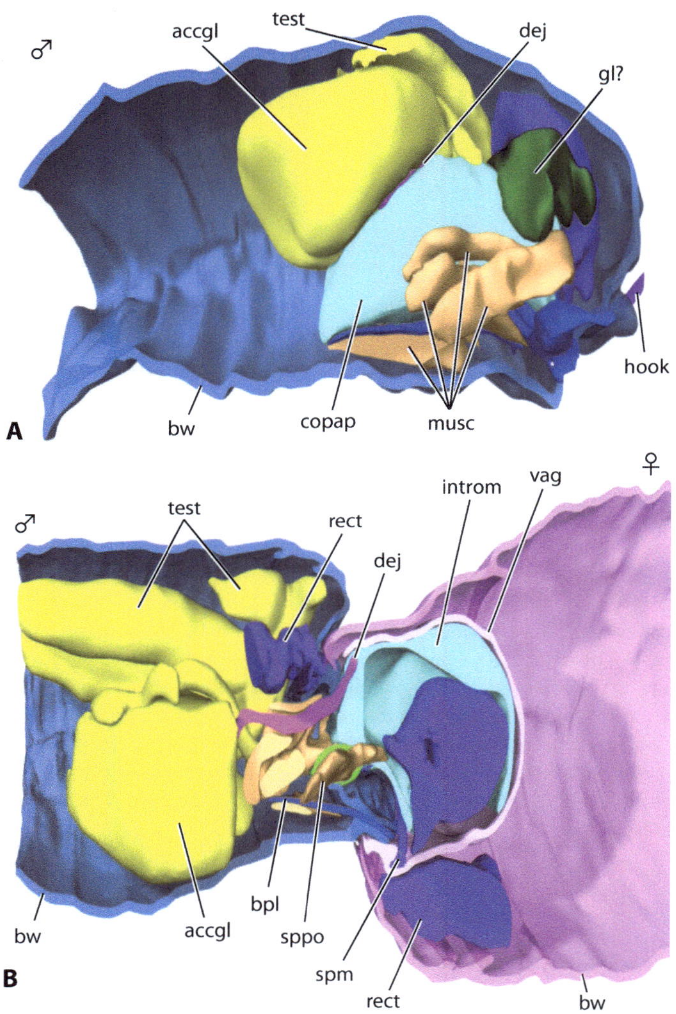

Fig. 6.13.5: Postabdomen and genitalia of *Zorotypus* species (Zorotypidae). A, *Zorotypus weidneri*, male; B, *Zorotypus caudelli*, male (with spiral-shaped flagellum) and female in copula. Abbr.: accgl – accessory gland, bpl – basal plate, bw – body wall, copap – copulatory apparatus in repose, dej – ductus ejaculatorius, gl? – possible gland, introm – male intromittent organ, musc – muscles, rect – rectum, spm – spermatophore, sppo – spiral-shaped pouch with coiled flagellum, test – testes, vag – vagina. Courtesy Y. Matsumura.

to avoid re-mating of females with other males after the first copulation. A large spermatophore is probably produced and transferred by all species. A part of the cover is likely consumed by females of some species (nutptial gift) (Dallai et al. 2013). The

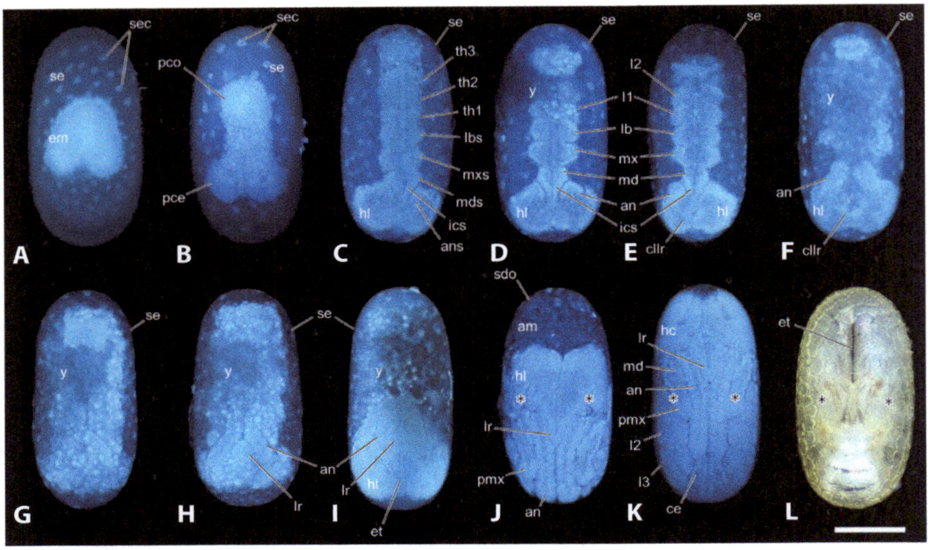

Fig. 6.13.6: *Zorotypus caudelli* (Zorotypidae), embryonic development with short germ embryo, A-I ventral view of egg. J–L dorsal view. Stages 1 (A) – 12 (L). Scale bar = 200. Abbr.: am – amnion, an – antenna, ans – antennal segment, ce – compound eye, cllr – clypeolabrum, em – embryo, et – egg tooth, hc – head capsule, hl – head lobe, ics – intercalary segment, l1–3 – pro-, meso- and metathoracic legs, lb – labium, lbs – labial segment, lr – labrum, md – mandible, mds – mandibular segment, mx – maxilla, mxs – maxillary segment, pce – protocephalon, pco – protocorm, pmx – palpus maxillaris, sdo – secondary dorsal organ, se – serosa, sec – serosal cell, syc – secondary yolk cell, th1–3 – pro-, meso- and metathorax, y – yolk. White and black arrowheads indicate cephalic and caudal end of the embryo, respectively, asterisks the position of the micropyle. Courtesy M. Machida and Y. Mashimo.

egg of *Zorotypus caudelli* was described in detail by Mashimo et al. (2011). The outer layer displays a honeycomb pattern and is composed of an endo- and exochorion, and a thin vitelline membrane. The exochorion is perforated by numerous branched aeropyles. A pair of micropyles is present at the equator on the dorsal side of the egg. A small heart-shaped embryo is formed on the dorsal side of the egg by a fusion of paired blastoderm regions with a higher cellular density (Fig. 6.13.6). The orientation of its anteroposterior axis is opposed to that of the egg. The embryo extends along the egg surface and after reaching its full length migrates into the yolk (anatrepsis). After staying and developing there for a certain period it re-appears on the surface (katatrepsis), accompanied by a reversion of its anteroposterior axis, finally taking its position on the ventral side of the egg. The definitive dorsal closure is completed there. Finally the prelarva perforates the chorion using a long egg tooth formed by the embryonic cuticle and hatches. The first nymphal stage is eyeless and the antenna comprises eight segments. The adult stage is reached after 3–5 molts.

Fossil record. Fossil zorapteran species are known from Burmese and Dominican amber.

Economic importance. Economically irrelevant.

[Hennig (1969); Engel & Grimaldi (2002); Beutel & Weide (2005); Hünefeld (2007); Friedrich & Beutel (2008); Mashimo et al. (2011); Dallai et al. (2012a, b, 2013)]

Xenonomia (Grylloblattodea + Mantophasmatodea)

6.14 **Grylloblattodea** (=Notoptera, Latin *gryllus* = cricket, *blatta* = roach, refers to the similarity with these two groups, English common names: ice crawlers, rock crawlers, icebugs) (Benjamin Wipfler)

Diversity and distribution. Grylloblattodea is one of the insect groups with a very low diversity. It is restricted to the Holarctic region. Presently 32 species are described in five genera. *Grylloblatta* occurs in northwestern North America and the other four genera in Japan, the Korean peninsula, the Russian Pacific coast, northeastern and northwestern China, and the Altai and Sayan mountain ranges. The two species occurring in China are *Galloisiana sinensis* (northeastern China, Jilin) (Fig. 6.14.1) and *Grylloblattella cheni* (border region between Xinjiang Autonomous Region of China, Republic of Mongolia, Russia and Kazakhstan). They are considered as very rare (Bai et al. 2010). Grylloblattodean species occur mostly in mountainous and high alpine regions, often at the edges of glaciers. Some species of *Galloisiana* live in dense forests and some are specialized cave dwellers. *Grylloblattina* occurs in elevations of 5–300 m above sea level.

Taxonomy. The single family Grylloblattidae comprises the genera *Grylloblatta*, *Grylloblattina*, *Grylloblattella*, *Galloisiana* and *Namkungia*.

Diagnosis. Medium sized to large, secondary flightless insects. Compound eyes small or completely reduced. Ocelli absent. Head appendages generalized. Tarsi 5-segmented. Attachment devices largely reduced. Cerci multisegmented.

Autapomorphies
- Unpaired extrusible membranous vesicle between sternite I and II
- Vestigial euplantulae
- Preference for cold temperatures
- Ground oriented

Fig. 6.14.1: *Galloisiana chinensis* (Grylloblattidae). Locality: Mt. Changbaishan, Jilin Prov., China. Courtesy S. Schoville.

Morphology. Most species range in size between 20 and 24 mm. The smallest species measures 14 mm and the largest 34 mm. The body is elongate, subparallel and moderately flattened dorsoventrally (Fig. 6.14.1). All extant species are of brownish or pale coloration and wingless.

The head (Fig. 6.14.2) is prognathous, moderately flattened, freely movable, and only very slightly retracted into the prothorax. The foramen occipitale is roughly quadrangular, slightly rounded dorsally, and very slightly narrowing towards the submentum. The head capsule is rounded laterally and posteriorly and slightly declined in relation to the longitudinal body axis; it is reinforced by an occipital ridge. The smooth cuticle is nearly transparent. The coloration is light brown in *Galloisiana*. A vestiture of some long and numerous short setae is present. Ocelli are absent. The compound eyes are small (if present); they are composed of ca. 60 pseudocone ommatidia in *Galloisiana* and *Grylloblatta*; in some Asian species they are completely reduced; a thick, sclerotized circumocular ridge completely encloses the eyes (if present). The frontal sutures and the long coronal suture are visible as very thin unsclerotized lines; the frontal sutures are straight and form a shallow angle; anterolaterally they nearly reach the antennal sockets. The frontoclypeal (epistomal) ridge is straight and distinctly visible externally. The clypeus is divided into a postclypeus and a transparent anteclypeus; its anterior margin forms a shallow angle; long setae are present along the anterior postclypeal margin. The labrum is connected with the anteclypeus by an internal membranous fold and is retractile; it narrows towards its rounded anterior margin; the basolateral tormae are posteriorly elongated; they serve as attachment areas of the lateral retractors (M. frontoepipharyngalis); the mesal retractors are also present (M. frontolabralis), whereas the intrinsic muscle is missing. The antennal foramina are large, nearly round, separated by about twice their own diameter, and enclosed by a low circumantennal strengthening ridge; a distinct anterolateral antennifer is present; the foramina are separated from the dorsal mandibular articulation

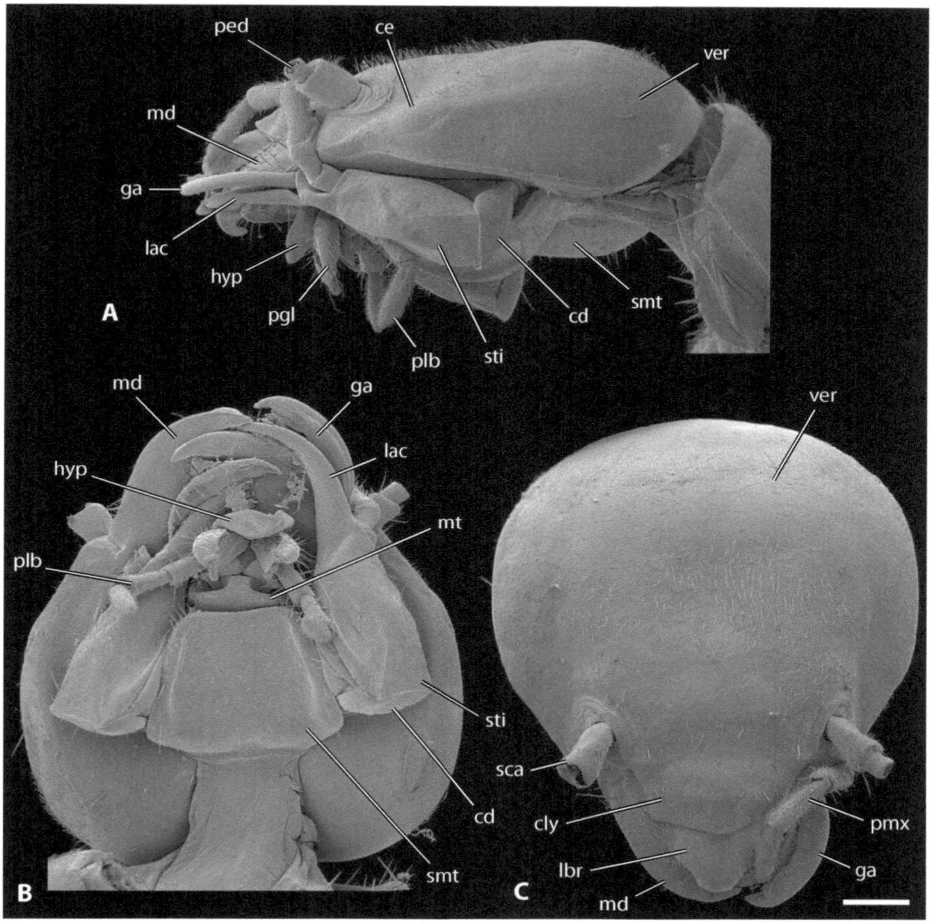

Fig. 6.14.2: *Galloisiana yuasai* (Grylloblattidae), head, SEM micrographs. A, lateral view, B, ventral view, C, dorsal view. Abbr.: cd – cardo, ce – compound eye, cly – clypeus, ga – galea, hyp – hypopharynx, lac – lacinia, lbr – labrum, md – mandible, mt – mentum, ped – pedicellus, pgl – paraglossa, plb – palpus labialis, pmx – palpus maxillaris, sca – scapus, smt – submentum, sti – stipes, ver – vertex. Scale bar: 500 μm. From Wipfler et al. (2011), modified.

by a very narrow cuticular bridge. The filiform and multisegmented antennae reach the terminal abdominal segments posteriorly; the scapus is the largest segment; the pedicellus is very short (Fig. 6.14.2A). The well-developed dicondylic mandibles are slightly asymmetric; two incisivi are present on the left mandible and three on the right in *Galloisiana*, while two are present on both in *Grylloblatta*. A distinct basal Z-shaped mesal cutting edge is present, with small teeth along the entire length, and two molar protuberances at the anterior end; a mola and prostheca are not developed (Fig. 6.14.3B). The maxillae are generalized; the transverse cardo articulates laterad

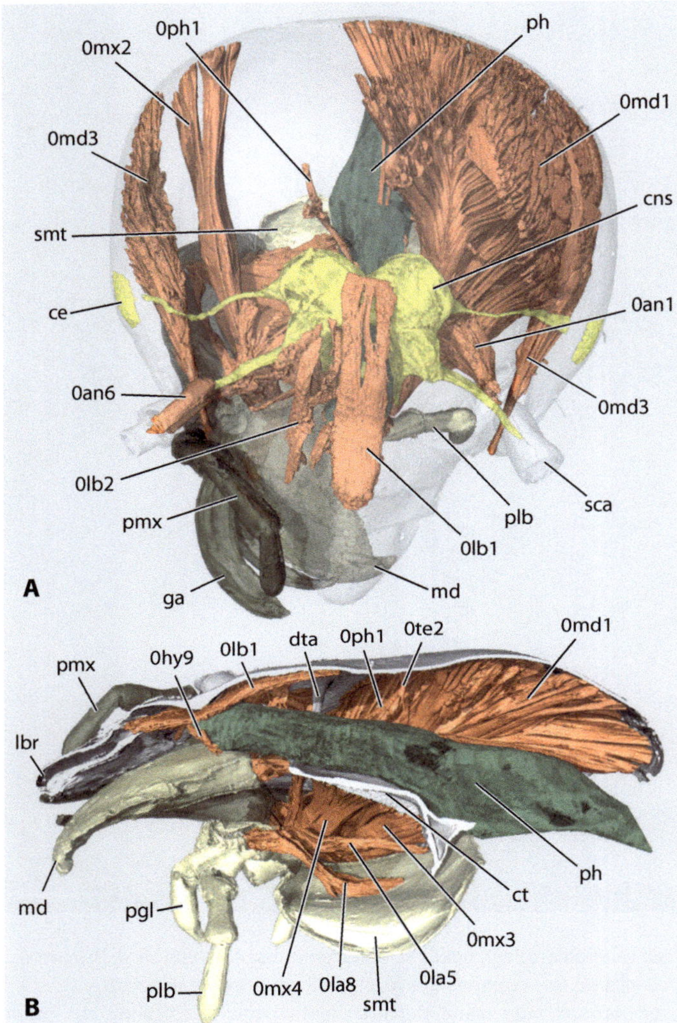

Fig. 6.14.3: *Galloisiana yuasai* (Grylloblattidae), head with muscles, 3D reconstruction. A, anterior view; B, sagittal section. Abbr.: ce – compound eye, cns – central nervous system, ct – corpotentorium, dta – dorsal tentorial arm, ga – galea, lbr – labrum, md – mandible, pgl – paraglossa, ph – pharynx, plb – palpus labialis, pmx – palpus maxillaris, sca – scapus, smt – submentum, 0an1 – M. tentorioscapalis anterior, 0an6 – M. scapopedicellaris lateralis, 0hy9 – M. transversalis buccae, 0la5 – M. tentoriopraementalis inferior, 0la8 – M. submentopraementalis, 0lb1 – M. frontolabralis, 0lb2 – M. frontoepipharyngalis, 0md1 – M. craniomandibularis internus, 0md3 – M. craniomandibularis externus, 0mx2 – M. craniolacinialis, 0mx3 – M. tentoriocardinalis, 0mx4 – M. tentoriostipitalis, 0ph1 – M. verticopharyngalis, 0te2 – M. tentoriofrontalis anterior. Courtesy B. Wipfler.

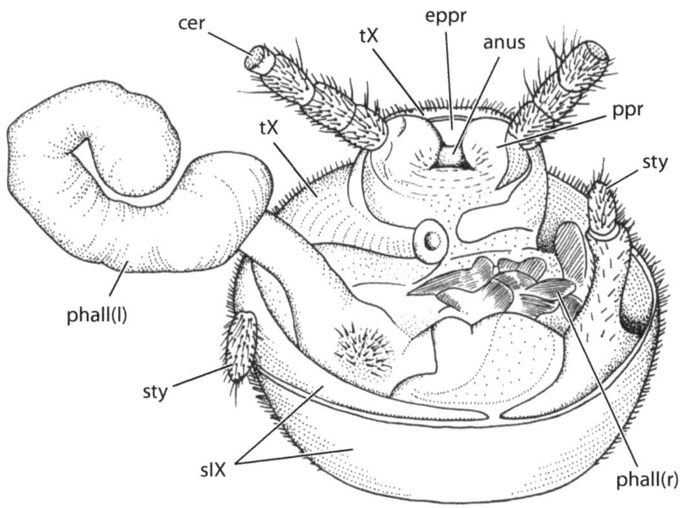

Fig. 6.14.4: *Grylloblatta campodeiformis* (Grylloblattidae), male terminal abdominal segments, ventrocaudal view. Abbr.: cer – cercus, eppr – epiproct, phall(l/r) – phallomere element left/right, ppr – paraproct, sIX – sternum IX, sty – stylus, tX – tergum X. Redrawn from Klass (2005) (after Walker 1919).

the submental base; the stipes is about three times as long as wide; it bears a 5-segmented palp, a well-developed galea, and an elongate, hook-shaped lacinia with an apodeme at its base. The labium is composed of a large trapezoidal submentum, a small mentum, and a prementum with a deep cleft at its anterior margin; the labial palp is 3-segmented and orientated posteriorly. The hypopharynx is tongue-like and reaches beyond the anterior margin of the paraglossae and glossae (Fig. 6.14.2). M. tentoriohypopharyngalis is reduced. A gula is not developed. The tentorium is complete. The short posterior arms are broad and bear strongly developed ventral apodemes (trabeculae tentorii); the massive tentorial bridge (corpotentorium) is long and very broad; a ventral ridge is present along its entire length (Fig. 6.14.3B); the anterior tentorial arms are flat and articulate with almost the entire pleurostomal sulcus; the dorsal arms arise at the conjunction of the corpotentorium and the anterior arms; they are broadly fused with the frons distinctly posterior to the antennal socket. The occipital region of the head articulates ventrolaterally with lateral cervical sclerites. The cephalic musculature is well-developed (Fig. 6.14.3).

The thorax is characterized by the complete loss of both pairs of wings and correlated modifications. The prothorax is about as long as the meso- and metathorax combined (Fig. 6.14.1). The pronotum is parallel-sided and distinctly longer than wide. The propleura is dorsally fused with the pronotum and characterized by a wing-like, internal anapleural process that serves as an attachment area for muscles. The pterothoracic segments are distinctly simplified. The tergites are not subdivided by

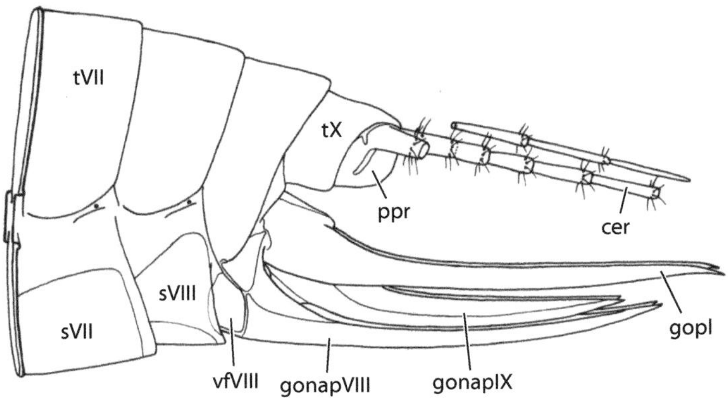

Fig. 6.14.5: *Grylloblatta campodeiformis* (Grylloblattidae), female terminal abdominal segments, lateral view, distal part of cercus detached. Abbr.: cer – cerci, gonapVIII – gonapophysis VIII (1st valve), gonapIX – gonapophysis IX (2nd valve), gopl – gonoplacs (3rd valve), ppr – paraproct, sVII/VIII – sternum VII/VIII, tVII/X – tergum VII/X, vfVIII – valvifer VIII. Redrawn from Klass (2005).

ridges and not forming phragmata; in contrast to the pronotum, the lateral tergal parts are slightly bent downwards, thus partly covering the upper pleural region. The pleura and the sternal regions are weakly sclerotized; the pterothoracic pleural ridges are very oblique; they form a vestigial pleural wing process anterodorsally. The furcal arms are widely separated. Spinae are present in all three segments. The meso- and metathoracic spiracles are placed in the intersegmental membrane anterior to the segmental border. The indirect flight muscles are reduced. The slender generalized legs bear a sparse vestiture of short setae; the large conical coxae are widely separated; the femora appear very slightly curved or at least convex along their anterior margin; the slender tibiae are very slightly widening distally; at their apex they bear a pair of short spines; the slender 5-segmented tarsi are elongated on the hindlegs; the euplantulae are vestigial; they are paired on tarsomeres 1–4 but unpaired on the apical tarsal segment; arolium, pulvilli and empodium are missing; the paired claws are short and simple.

The abdomen (Fig. 6.14.5) is composed of ten distinctly developed segments. Sternite I is partly reduced; an unpaired retractable membranous vesicle posterior to this sclerite may be part of a gland or function as a water uptake device. Tergites I–VIII and sternites II–VIII are simple transverse plates. The pleural membranes are extensive, and fairly wide membranous areas are also present between the sternites. The very small spiracles I–VIII are embedded in the pleural membranes connecting the tergites and sternites I–VIII. Segment X forms a closed ring-like sclerite in females whereas the ventral part is membranous in males. The male genital apparatus (Fig. 6.14.4) is distinctly asymmetric; a subgenital lobe is formed by sternite IX; It is divided into an unpaired anterior part and paired posterior portions, the latter equipped with a pair of 1-segmented pubescent styli; the external genitalia, i.e. the phallomeres are asymmetric; the left phallomere bears a tube-like membranous process and small sclerotizations, whereas the right one is formed by an array of several strongly sclerotized elements. The testes are composed of numerous follicles; the vasa deferentia bear a longitudinal vesicula seminalis distally; together with two pairs of tube-like accessory glands they open into the unpaired ductus ejaculatorius. The female genital apparatus (Fig. 6.14.5) displays a well-developed ovipositor comprising the valvifers of segments VIII and IX, a distinct triangular gonangulum, and three long valves. Fourteen panoistic ovarioles open into the lateral oviduct on both sides. The unpaired receptaculum seminis bears a large terminal vesicle. Accessory glands are absent in females. The filiform cerci comprise 5–10 cercomeres and are inserted posterior to segment X in both sexes; they bear trichobothria on the distal region of each segment except the apical one.

The central nervous system is characterized by relatively small optic lobes in the brain and the fusion of the pairs of ganglia of the metathorax and abdominal segment I. In the foregut an ingluvies is present and strongly expanded ventrally; it is posteriorly connected with a small proventriculus with cuticular teeth arranged in three rows and a strongly developed muscularis. The short and wide midgut bears a pair of large diverticula anteriorly. The number of long Malpighian tubules varies between 14 and 24. The heart is located between abdominal segment IX and the mesothorax; it is anteriorly continuous with the aorta cephalic; laterally it is suspended by segmental alary muscles including a pair originating in segment X; incurrent ostia are present in all segments, while excurrent ostia are missing in the posterior region. The thoracic spiracles possess a closing apparatus whereas it is absent in those of the abdomen.

Biology. A highly unusual feature of the flightless and strictly ground oriented grylloblattodeans is the preference for cold habitats (see above). By carrying out vertical migrations they stay close to their optimum of humidity and temperature (ca. 5°C in *Grylloblatta campodeiformes*). They are mostly nocturnal or cavernicolous and display a distinct trend towards a cryptic life style. Usually they are found in narrow spaces among stones or rocks, but also under snow, in moss, in soil substrate, in leave litter, or in rotting logs. Ice crawlers are primarily carnivorous and largely depend on dead insects (e.g., Collembola) or other dead or slow moving arthropods. In high alpine regions, *Grylloblatta* species search for dead insects on the edges of snowfields during the night. Injured individuals of the same species may be attacked and cannibalism occurs also in the context of copulation.

The copulation takes 0.5–4 hours. The large black eggs are deposited 10–50 days later with the long ovipositor, on the surface of stones or rocks, in soil, in rotting wood or among dropped leaves.

Development and immature stages. Eight successive stages occur during the hemimetabolous development, which takes 3-6 years from hatching of the first instar to the adult stage. The nymphs differ from the adults by their size, the lower degree of pigmentation and sclerotization, and by the number of segments of the antennae and the cerci.

Fossil record. Extinct grylloblattodean species are known from Baltic amber. The assignment of several extinct taxa to the order (e.g., †Protoblattaria, †Protoperlaria, or †Paraplecoptera) is highly questionable. As pointed out in Grimaldi & Engel (2005) Grylloblattodea assume the role of a "taxonomic wastebasket" for fossil lineages like the former "†Protorthoptera". The treatment of the order as the "stem group" of several extant and extinct insect lineages such as for instance †Protelytroptera, Plecoptera or Dermaptera (see e.g., Storozhenko 1997) is based on an unsound evolutionary and phylogenetic concept.

Economic importance. Unknown, apparently negligible.

[Walker (1919; 1931, 1933, 1938); Ando (1982); Storozhenko (1997); Grimaldi & Engel (2005); Klass (2005); Bai et al. (2010); Wipfler et al. (2011)]

6.15 **Mantophasmatodea** (the name refers to the similarity with Mantodea and Phasmatodea, English common names: heelwalkers, gladiators) (Benjamin Wipfler)

Diversity and Distribution. Eighteen species are described and they occur only in the southern part of Africa. Presently there are records for Namibia (Fig. 6.15.1), South Africa and Tanzania.

Autapomorphies

- Unique structure of antennal flagellum with 14 cylindrical basiflagellomeres (distal ones subdivided) and seven spindle-shaped distiflagellomeres (1st very long, 2nd very short)
- Distinctive equipment of sensilla of antennomeres, with apertures on 1st and 6th distiflagellum
- Arolium strongly enlarged and densely covered with spines, not in contact with the substrate during normal locomotion (size, shape and vestiture of arolium similar in basal Phasmatodea) (Figs 6.15.1, 6.15.3)
- Very large spiny euplantulae (similar in basal Phasmatodea)
- Frontal region with x-shaped apodeme
- Male genital apparatus with horseshoe-shaped, vomer-like sclerite formed by sternite X

The complete loss of the wings and simplifications of associated structures are possibly synapomorphies with Grylloblattodea.

Fig. 6.15.1: *Striatophasma naukluftense* (Austrophasmatidae) from Namibia, in copula. Photo courtesy R. Predel.

Taxonomy. The phylogeny and taxonomy of the order, which was only recently introduced (Klass et al. 2002), was investigated intensively in the last ten years (e.g., Klass et al. 2003; Damgaard et al. 2008). Recently two new genera (*Striatophasma*, *Pachyphasma*) were described from the Brandberg massiv in Namibia (Wipfler et al. 2012). The total number of species is presently 18 and they are placed in 12 genera. Phylogenetic analyses of Predel et al. (2012) revealed two monophyletic lineages: one containing all South African Austrophasmatidae (*Austrophasma*, *Hemilobophasma*, *Karoophasma*, *Lobatophasma*, *Namaquaphasma*, *Viridiphasma*; Klass et al. 2003; Eberhard et al. 2009) and the Namibian genus *Striatophasma* (Wipfler et al. 2012), and the other one all the remaining Namibian species presently placed in Mantophasmatidae.

Diagnosis. Medium sized or large wingless insects with an orthognathous head. Antennae with cylindrical basiflagellomeres and spindle-shaped distiflagellomeres. Arolium and euplantulae unusually large and densely covered with spines.

Morphology. The maximum size is ca. 32 mm in males (*Tyrannophasma gladiator*) and ca. 35 mm in females (*Striatophasma naukluftense*). The body is slender, with a parallel-sided thorax and an abdomen which is very slightly rounded laterally in males and more distinctly in females. The insects are moderately sclerotized and the predominant colors are green and brown, with or without a patchwork pattern, which may sometimes also contain grey.

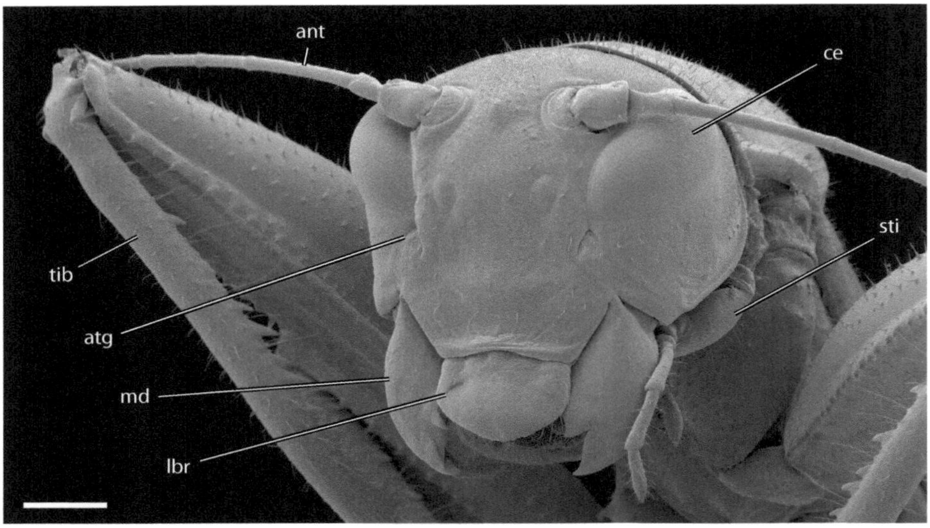

Fig. 6.15.2: *Mantophasma zephyra* (Mantophasmatidae), head, anterolateral view, SEM micrograph. Abbr.: ant – antenna, atg – anterior tentorial groove, ce – compound eye, lbr – labrum, md – mandible, sti – stipes, tib – fore tibia. Scale bar: 250 μm. Courtesy H. Pohl.

The orthognathous head (Fig. 6.15.2) is slightly retracted into the prothorax dorsally. The shape is triangular or globular in frontal view; the length ratio between the maximum width and the length of the head is approximately 1 : 1. The vestiture of short setae is sparse. The postoccipital ridge enclosing the large foramen occipitale laterally and dorsally is narrow. The more or less oval compound eyes vary in size but are always composed of numerous small ommatidia (Fig. 6.15.2); a circumocular sulcus is present; ocelli are absent. The frons is characterized by an externally visible x-shaped apodeme. The coronal and frontal sutures and the transverse frontoclypeal strengthening ridge are indistinct or absent, whereas the subgenal ridge is distinct. The clypeus is subdivided into an anterior transparent anteclypeus and a posterior postclypeal area which is continuous with the frontal region. The labrum is free and the full set of intrinsic and extrinsic muscles is present. The multisegmented, long and thin antennae are inserted between the compound eyes (Fig. 6.15.2); the comparatively large scapus articulates with a distinct ventral antennifer; the flagellum displays a characteristic pattern of cylindrical basiflagellomeres and spindle-shaped distiflagellomeres, and a specific vestiture of sensilla (Drilling & Klass 2010); a transverse muscle connecting the antennal ampullae is present. The mouthparts are generalized. The mandibles are apically pointed and equipped with a mesal cutting edge; asymmetrically arranged teeth fitting with the counterparts on the opposite side are present on the middle region of the mesal mandibular edge; a mola is not developed. The maxilla is largely unmodified, with a lobe-like, largely semimembranous galea, a sclerotized curved lacinia with an apical and a subapical tooth and mesally directed spines, and a 5-segmented palp; a distinctly developed palpifer is not present. The labium is composed of a basal submentum, a mentum, and a deeply cleft prementum bearing the glossae, paraglossae and 3-segmented palps. The hypopharynx is a semimembranous tongue-like structure posterolaterally reinforced by a curved suspensorium; its anterior surface is covered with minute microtrichia; its upper part forms the floor of a short prepharyngeal tube. The salivarium is present as a flat pocket between the posterior side of the hypopharynx and the anterior wall of the anterior labium. A gula is absent. The tentorium is complete. The bridge is elongated but lacks a "perforation", i.e. medially connected accessory anterior arms are absent; the dorsal arms are less strongly developed than the other elements and do not reach the head capsule dorsally; they are attached to the sulcus of the antennal articulatory area by a thin ligament (*Karoophasma*). The head musculature is largely unmodified. The dumbbell-shaped brain and suboesophageal ganglion appear small in relation to the head size. Two pairs of lateral cervical sclerites are present.

The thorax is parallel-sided. The metathorax is distinctly smaller than the other two segments (Fig. 6.15.1). All three tergites are simple undivided plates and overlap slightly with the tergites of the following segment. A transverse furrow is present along the anterior margin of the oval or rectangular pronotum and a less distinct furrow along the posterior margin. Internal tergal ridges and intersegmental phragmata are absent. Distinct pleural ridges separate the pleuron of all three segments into an anterior episternum and a posterior epimeron; they are strongly slanted in the pterothoracic segments. The wings are completely reduced. The legs are long and slender; the conical and fairly elongated coxae articulate with small trochanters; the profemur is slightly expanded; the tibiae of the fore- and middle legs bear thorns (Fig. 6.15.2) which play a role in the prey catching behavior (see below); the tarsi are 5-segmented; tarsomeres 1–4 bear large, bilobed, cushion-like euplantulae with a dense vestiture of short spines; the arolium is pan-shaped and very large, and also densely covered with short spines (Fig. 6.15.3); it is articulated at its base and normally raised from the ground (the English name heelwalkers refers to this condition); in exceptional situations the arolium is brought in contact with the substrate (see below).

The abdomen is composed of ten distinctly developed segments. The tergites and sternites of the pregenital segments are simple plates connected by wide pleural membranes, which bear the very small abdominal spiracles I–VIII. Tergite and sternite I are distinctly shortened and movably connected with the adjacent metathoracic elements. Segment X is largely membranous in both sexes on its ventral side; however, a transverse sclerotization is present anteroventrally in males; it articulates with the anterolateral edges of tergite X and bears a horseshoe-shaped, vomer-like process medially. The cerci are 1-segmented in both sexes, but elongated and strongly curved in males. The male genital apparatus is distinctly asymmetric; the undivided sternite IX forms a large, slightly asymmetric subgenital lobe; ventromedially it bears a small spatulate process (drumming organ) with or without a distinct dorsal sclero-

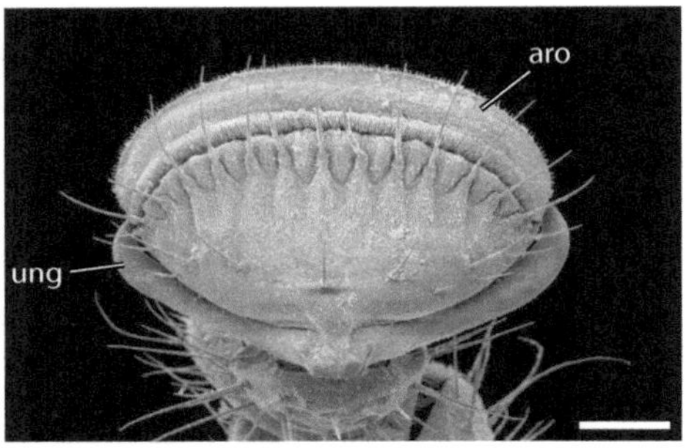

Fig. 6.15.3: *Mantophasma zephyra* (Mantophasmatidae), pretarsus, dorsal view, SEM micrograph. Abbr.: aro – arolium, ung – ungues. Scale bar: 150 µm. Courtesy H. Pohl.

tisation; the large, phallomeres are inserted posterior to it; they are largely membranous but contain also sclerotized elements, which are important for the identification on the species level; styli are always absent. The ovipositor is composed of three pairs of short valves; a gonangulum is present at its base; the lateral valves (gonoplacs) are strongly sclerotized and claw-shaped; mesally they are fused with the lightly sclerotized mesal valves; a keel along the line of fusion fits into a rim of the ventral valves (see Klass et al. [2003] for more details).

A proventriculus is present, with a well-developed musculature and 24 weakly sclerotized longitudinal bands with minute papillae; a pair of voluminous lateral caeca is present at the anterior end of the midgut. Numerous long Malpighian tubules insert at the midgut-hindgut border. Ten paired ovarioles are present in females.

Biology. Mantophasmatodea almost exclusively occur in arid and semiarid sclerophyll scrub biomes of southwestern and eastern Africa. The predacious insects are probably nocturnal and are usually found in low shrubs to which they are adapted in their coloration. Their spectrum of prey comprises various other insects but also spiders. The forelegs are the main catching devices but the middle legs can be also used depending on the size of the prey. The specialized arolia are not involved. They are only activated under specific conditions, such as for instance on smooth surfaces or when the insects carry large prey.

Reproduction and development. Mature males and females communicate using vibrations produced by beating the ground with the abdomen. Males possess a specific drumming organ (see above) whereas the females use the entire abdomen. The copula (Fig. 6.15.1) can last for up to three days. In the final stage the male is often eaten by the female. The eggs are deposited in soil. Up to 12 eggs can be glued together as packages also including soil particles. The eggs of *Mantophasma zephyra* are 2.7 mm long and 0.9 mm wide (Machida et al. 2004).

Fossil record. Specimens are known from Baltic amber and the Jurassic of China (Huang et al. 2008).

Economic importance. Mantophasmatodea are economically irrelevant.

[Klass et al. (2002, 2003, 2005); Machida et al. (2004); Baum et al. (2007); Damgaard et al. (2008); Huang et al. (2008); Eberhard et al. (2009); Drilling & Klass (2010); Predel et al. (2012); Wipfler et al. (2012)]

Dictyoptera (Mantodea + Blattodea incl. Isoptera)

6.16 Mantodea (Greek *mantis* = diviner, English common names: mantises, praying mantises) (Benjamin Wipfler)

Diversity and distribution. Mantodea comprise about 2,300 described species. They are largely absent in the temperate zones. Their main areas of distribution are tropical and subtropical regions. They reach the highest diversity in the Afrotropis.

Fig. 6.16.1: *Mantis* sp. (Mantidae) from Tunisia. Photo courtesy L. Möckel.

Autapomorphies. The monophyly of the group is undisputed and strongly supported. Most features are closely related with the specialized predacious habits.
- Head highly movable (Figs 6.16.1, 6.16.2)
- Compound eyes very large (Fig. 6.16.2)
- Pronotum subdivided by transverse supracoxal ridge
- Specialized raptorial forelegs (Figs 6.16.2, 6.16.3)
- Profemural cleaning device (indistinct in Mantoididae) (Fig. 6.16.2: fecl)
- Fusion of paraproct with paratergites X
- Copulatory organ asymmetric (also in roaches)

Taxonomy. The intraordinal phylogeny is presently not well understood. Most traditional families are para- or polyphyletic. Eight families are listed in a recent classification (Klass & Ehrmann 2005): Mantoididae (one genus with nine small neotropical species), Chaeteessidae (one genus with four small neotropical species), Metallyticidae (one genus with five unusually stout species in the Malayan region), Amorphoscelidae (15 tropical and subtropical genera of the Old World with 84 small species), Eremiaphilidae (two genera with 70 species in North Africa and Western Asia), Hymenopodidae (55 genera with 281 small or medium-sized circumtropical species), Mantidae (350 genera with 1,796 species) and Empusidae (nine genera with 51 species in Africa, Asia and Southern Europe).

The indistinct delimitation of the profemural cleaning device is likely a plesiomorphy preserved in Mantoididae. An apparent plesiomorphic condition only found in Chaeteessidae is the apical insertion of the tarsus, correlated with the absence of the fixed and elongated tibial spur. The 2nd anal vein of the hindwing is only preserved in Chaeteessidae and Metallyticidae. The basal branching pattern is still ambiguous.

Diagnosis. Relatively small (smallest species ca. 10 mm) to very large insects (Fig. 6.16.1) with more or less strongly elongated prothorax and raptorial forelegs (Fig. 6.16.2). Triangular head with very large compound eyes (Fig. 6.16.2). Prothorax almost always elongated. Forewings leathery (tegmina) or wings reduced. Males with distinctly asymmetrical copulatory organ. Females deposit oothecae. Camouflage mechanisms occur frequently but also conspicuous colorations (Metallyticidae).

Morphology. The total size range is 10–145 mm, but most species measure between 25 mm and 70 mm. Most species are elongate and slender but several groups have a relatively stout habitus (e.g., Metallyticidae).

The orthognathous head is movable in all directions. It is triangular and wider than long in frontal view and bears a frontal shield (Fig. 6.16.2). The very large compound eyes are located at the dorsolateral head regions (Fig. 6.16.2); three ocelli are always present. In some species the vertex bears paired tubercles (e.g., *Haania*) or differently shaped median processes; they can be conical (e.g., *Pygomantis*), band-shaped (e.g., *Stenophylla*), asymmetric (e.g., *Phyllocrania*), or bifurcated (e.g.,

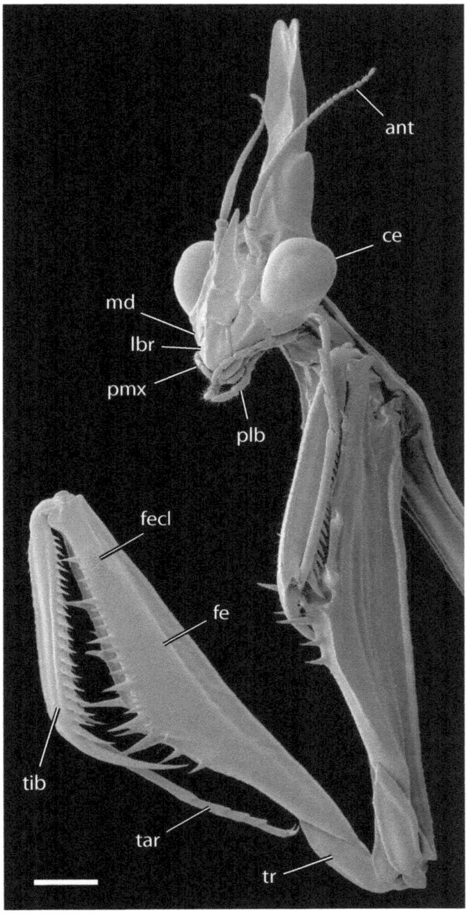

Fig. 6.16.2: *Empusa* sp. (Empusidae), head and raptorial forlegs, anterolateral view, SEM micrograph. Abbr.: ant – antenna, ce – compound eye, fe – femur, fecl – femoral cleaning field, lbr – labrum, md – mandible, plb – palpus labialis, pmx – palpus maxillaris, tar – tarsus, tib – tibia, tr – trochanter. Scale bar: 1 mm. Courtesy H. Pohl.

Sibylla). The labrum is free. The antennal articulatory areas lie in front of and laterad the ocelli, distant form the mandibular base. The thin multisegmented antennae can be as long as the entire body; they are rarely serrate or symmetrically pectinate (males of Empusidae); the number of antennomeres is usually higher in males (92 versus 86 in *Mantis*). The mouthparts are orthopteroid (Fig. 6.16.2). Apically the mandibles bear two (right side) or three (left side) acute teeth; proximally a sharp mandibular cutting edge and one or two additional teeth are present; a sclerotized mola with a grinding surface is entirely lacking, whereas an unsclerotized postmola is present.

The maxillae are largely unmodified; the lacinia lies within a fold of the galea. The labium is composed of a postmentum and a deeply cleft prementum. The tentorium is complete, with thin dorsal arm and medially connected accessory anterior arms ("perforated corpotentorium").

The prothorax is usually elongated (Figs 6.16.1, 6.16.2) and can be up to 20 times longer than wide (*Leptocola stanleyana*) and make up almost half of the total body length (*Angela*, *Schizocephala*). The movability between the pro- and mesothorax is unusually high. The pronotum is divided in an anterior and a posterior part by a transverse internal supracoxal ridge, which is often visible externally as a furrow; it often bears a narrow anterior duplicature and a lateral duplicature of varying length; in some species lateral processes are part of a camouflage mechanism. The prothoracic basisternite is fused with the praecoxite, and the furcasternite with the pronotum, thus creating a transverse bridge stabilizing the forelegs. The highly specialized raptorial forelegs (Figs 6.16.1–6.16.3) function like a subchela, with the tibia striking against the femur; they are equipped with a species specific armature of spines; a pleural process prevents overstretching of the elongate coxae during the strike; a profemural field of bristles (not clearly defined in *Mantoida*) functions as a cleaning device (e.g., for compound eyes); the tibia usually bears a fixed, elongated and apically pointed process which fits into a mesal furrow of the femur (absent in Chaeteessidae). The pterothorax is almost always shorter than the prothorax and the movability between its segments is minimal. The meso- and metathorax are very similar in their size and construction. The nota are composed of a prescutum, scutum and scutellum; small lobe-like phragmata are formed by the prescutum laterally. The basisternites are large and apparently fused with the praecoxites; the furcasternites are narrow; the mesothoracic spinasternum is fused with the metathoracic basisternum. The metathoracic spinasternum is free, but lacks a spina. The pleural sutures are oblique and dorsally obliterated in wingless species. An unpaired acoustic organ ("cyclopean ear") between the metacoxae is present or absent. The pterothoracic legs are largely unmodified; the coxae of all legs are elongated and the tarsi are almost 5-segmented (4-segmented in *Heteronutarsus*); tarsomeres 1–4 bear euplantulae; an arolium is lacking. The wings vary strongly within the group; they can be well-developed and extend beyond the abdominal apex posteriorly, but partial or complete reduction occurs frequently; the forewings are usually leathery tegmina (not in Mantoididae and Chaeteessidae), whereas the hindwings (if present) are always membranous. The meso- and metathoracic spiracles lie in the intersegmental membranes. The flight muscles are well-developed in winged forms.

The abdomen is broadly connected with the metathorax and composed of ten distinctly developed segments. Tergite I is slightly and sternite I distinctly reduced; they are separated from the metathoracic spiracle by membranous areas. The abdominal sternites are undivided except for those of the terminal region. The paratergites I–VII bear the spiracles; paratergite I is small and widely separated from the tergite, whereas paratergites II–VIII are separated from the corresponding tergites by a fold

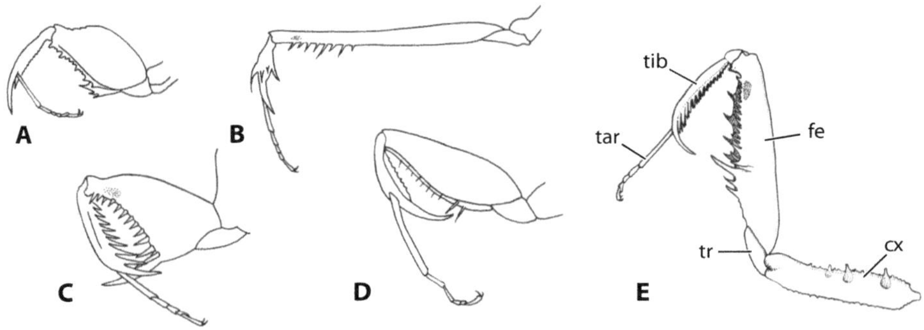

Fig. 6.16.3: Raptorial forlegs. A, *Otomantis aurita* (Hymenopodidae); B, *Thesprotia filum* (Mantidae); C, *Mantoida brunneriana* (Mantoididae); D, *Gyromantis kraussii* (Amorphoscelidae); E, *Sphodromantis viridis* (Mantidae). Abbr.: cx – coxa, fe – femur, tar – tarsus, tib – tibia, tr – trochanter. Redrawn from Klass & Ehrmann (2005) (from different original sources).

and a narrow membrane. The paraprocts of the reduced segment XI are fused with the paratergites X. Tergite X is rarely elongated. A vestigial epiproct (tergite XI) is usually recognizable posterior to it. The multisegmented cerci articulate with the hind margin of tergite X. In males the subgenital plate (sternite IX) is strongly elongated and forms an asymmetric subgenital lobe, which bears a pair of styli; the asymmetric phallomeres forming the copulatory apparatus are inserted on the ventral side posterior to the lobe and covered by it; additional elements are the right epiphallus and the left epiphallus + hypophallus; the part on the right side forms a clasping apparatus. In females the subgenital plate is formed by sternite VII; this trough-shaped structure, also referred to as subgenital lobe, distally bears a pair of flexible terminal lobes; the genital atrium contains the well-developed ovipositor in its resting position (Klass & Ehrmann 2005).

The posterior foregut comprises a very large ingluvies which reaches the abdomen posteriorly, and a complex funnel-shaped proventriculus with a strongly developed musculature and primary (6) and secondary plicae (5), each equipped with a specific armature; the midgut bears eight caeca at its anterior margin; six rectal papillae are present. Approximately 100 thin Malpighian tubules insert at the midgut-hindgut border. The testes are composed of numerous follicles. Numerous tubes of the male accessory glands open into the ductus ejaculatorius. Paired ovaries each with numerous panoistic ovarioles, paired (proximally) and unpaired oviducts, an unpaired receptaculum seminis with a terminal vesicle, and accessory glands form the internal female genital apparatus.

Biology. Mantodea are very well adapted to their habitats and camouflage mechanisms are common, for instance with a habitus resembling leaves or twigs, or resulting from specific processes or color patterns.

All species are carnivorous and most of them are ambush predators, waiting for prey in a rigid posture. Exceptions are species of Eremiaphilidae, which actively hunt prey on the ground in deserts. The food spectrum comprises other insects, spiders and small vertebrates, such as frogs, lizards, or young birds. The raptorial legs (Fig. 6.16.3) carry out extremely rapid movements (catching process ca. 60 ms). When disturbed, mantises assume a defensive posture with raised and spread raptorial legs, let themselves drop, or escape using their legs or wings. Species with a preference for bark usually switch to the other side of the tree. All species of Mantodea are solitary and usually stay in one locality and habitat.

Reproduction and development. Mantises reproduce almost exclusively sexually. Parthenogenesis is a rare exception and it is obligatory in only one species, *Brunneria borealis*. Sexual cannibalism can occur in all phases of mating, but is apparently an exception in the natural environment. The copulation can be completed without problems after the male has been decapitated by the female. The females deposit typical oothecae, which are formed using foam-like secretions of the accessory glands. They contain between 10 and 400 eggs and are attached to branches, leaves, or stones, or are buried in the ground. One female can deposit up to 20 oothecae.

Fossil record. The assignment of Paleozoic fossils to Mantodea is highly questionable. The earliest definite fossil mantises are described from Cretaceous deposits (Klass & Ehrmann 2005). Cretaceous species are conserved as impression fossils and also as inclusions in amber. The most diverse family Mantidae (ca. 75% of all extant species) radiated in the Tertiary, like termitid isopterans (Grimaldi & Engel 2005).

Economic importance. Mantises are economically irrelevant.

[Beier (1968a); Grimaldi & Engel (2005); Klass & Ehrmann (2005)]

6.17 Blattodea (Latin *blatta* = roach, English common names: roaches [for "Blattaria"] and termites [for Isoptera]) (Benjamin Wipfler)

Diversity and distribution. Blattodea are one of the largest orders of Polyneoptera with ca. 7,600 described species. Ca. 4,600 of them are roaches ("Blattaria") and ca. 3,000 termites. Roaches and termites (also known as white ants) generally prefer warm temperatures and mainly occur in tropical regions. Termites reach 40° of latitude. Some hemerophile roaches are cosmopolitan (e.g., *Periplaneta americana*, *Blatta orientalis*, *Blattella germanica*).

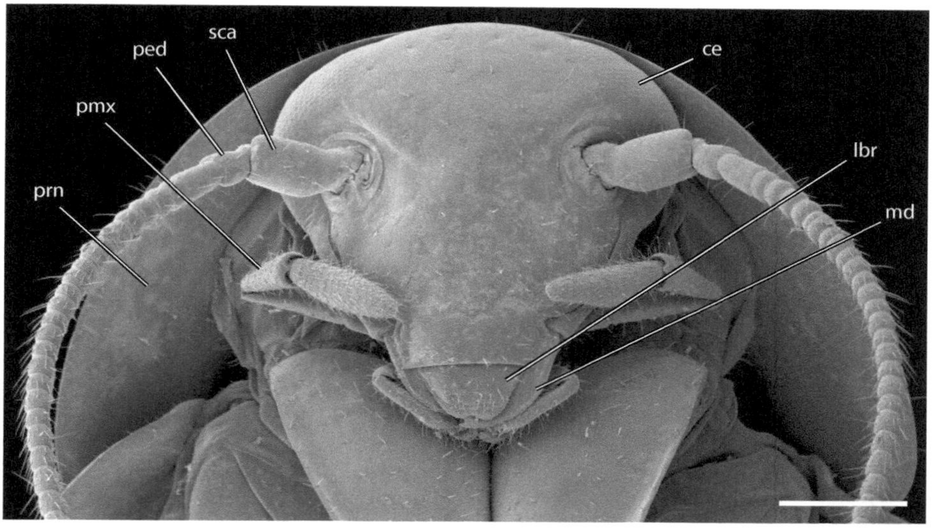

Fig. 6.17.1: *Ectobius sylvestris* (Ectobiidae), head, ventral view, SEM micrograph. Abbr.: ce – compound eye, lbr – labrum, md – mandible, ped – pedicellus, pmx – palpus maxillaris, prn – pronotum, sca – scapus. Scale bar: 500 µm. Courtesy H. Pohl.

Autapomorphies. Due to profound differences in the morphology and lifestyle of most roaches and termites apomorphic groundplan features supporting the clade Blattodea are scarce. However, its monophyly was well-supported in studies based on morphological (Klass 1995) and molecular (e.g., Lo et al. 2000) data.

- Median ocellus absent
- Specialized fat body cells with endosymbionts

Further potential autapomorphies include modifications of the head musculature and of the proventriculus. That the roach genus *Cryptocercus* (Cryptocercidae) (Fig. 6.17.3) is the sistergroup of Isoptera (Fig. 6.17.2) is well established (see above). Synapomorphies are the presence of endosymbiotic protists ("flagellates") not occurring in any other group, specific modifications of the proventriculus, moniliform antennae, monogamy, extended biparental care, allogrooming, and proctodeal trophallaxis (Klass 1995; Bohn 2005b; Grimaldi & Engel 2005).

Taxonomy. The phylogenetic relationships of the blattarian lineages are not well understood. A preliminary classification mainly based on the comprehensive works of McKittrick (1964) and Klass (1995) was recently proposed by Bohn (2005a). Eight families are distinguished but the very heterogeneous "Ectobiidae" (="Blattellidae") (five subfamilies, ca. 1,800 spp.) were clearly identified as paraphyletic. Blattidae comprise two subfamilies and ca. 650 species, Corydiidae (=Polyphagidae) five sub-

Fig. 6.17.2: *Reticulitermes lucifugus* (Rhinotermitidae), habitus lateral view, SEM micrographs. A, worker. B, soldier. Abbr.: ant – antenna, cer – cercus, lbr – labrum, md – mandible, plb – palpus labialis, pmx – palpus maxillaris. Scale bar: 500 µm. Courtesy H. Pohl.

families and ca. 230 species, and Blaberidae nine subfamilies and ca. 1,200 species. Blaberidae are subdivided into three subunits by Bohn (2005), the blaberoid-complex (Zetoborinae, Blaberinae, Panesthiinae), the panchloroid-complex (Pycnoscelinae, Diplopterinae, Panchlorinae, Oxyhaloinae), and the epilamproid-complex (Epilamprinae, Perisphaeriinae). Smaller families are the Tryonicidae (11 species in the Australian region), Lamproblattidae (eight wingless species in one genus, tropical America), and Nocticolidae (20 very small species, most of them cavernicolous or termitophiles). Cryptocercidae (eight species, one genus in northeastern Asia [*Cryptocercus relictus*] and the USA) (Fig. 6.17.3) are the sistergroup of Isoptera (see above).

Within Isoptera, Mastotermitidae (Australia, one extant species, several genera with a worldwide distribution in the Cretaceous) are clearly the sistergroup of the remaining families, which are characterized by four tarsomeres, a reduced ovipositor, and the deposition of single eggs (ootheca reduced). Isoptera excluding Mastotermitidae are subdivided into six families, the Termopsidae (pantropical, 20 spp., three subfamilies), Kalotermitidae (worldwide, 411 spp., 21 genera), Hodotermitidae (Africa, western Asia, India, 15 spp., three genera), Rhinotermitidae (worldwide, 305 spp., six subfamilies, 15 genera), Serritermitidae (Brazil, one species), and Termitidae ("higher termites", pantropical, ca. 1,900 spp., four subfamilies, ca. 240 genera) (Bohn 2005b).

Fig. 6.17.3: *Cryptocercus punctulatus* (Crypocercidae), Mountain Lake Biological Station, University of Virginia. Photo courtesy of A. Thipaksorn.

Morphology. The size of roaches varies between 2.5 mm (*Attaphila*) and ca. 10 cm (*Megaloblatta*). The body is dorsoventrally flattened and usually oval with an indistinct or obsolete angle between the pronotum and pterothorax (Figs 6.17.3, 6.17.4). The head is almost always largely or completely covered by the large and flat pronotum (Fig. 6.17.1). The body length of termites ranges between 5 mm (Nasutitermitinae) and 22 mm (*Macrotermes goliath*); physogastric queens of *Macrotermes natalensis* can even reach a length of 14 cm. In termites the small pronotum never covers the head (Figs 6.17.2, 6.17.7). All termites form colonies with castes distinctly differing in their morphology and other aspects (autapomorphy).

The head of roaches is ortho- or hypognathous and usually triangular with rounded edges in frontal view (Fig. 6.17.1). The compound eyes are usually kidney-shaped and they are reduced in some subgroups; the paired ocelli are present close to the margin of the compound eyes (groundplan) or absent. The broad labrum is free and retractable (Fig. 6.17.1: lbr). The antennae are multisegmented and filiform. The mouthparts are orthopteroid. The apical part of the mandibles is divided into three or four pointed incisive; a flat mola and a membranous postmola are present. The lacinia lies in a fold of the galea; one or two pointed teeth are present apically; the palp is 5-segmented. The labial postmentum is undivided; the prementum is deeply cleft anteromedially; glossae and paraglossae are well-developed; the labial palp is 3-segmented. A gula is absent. In addition to the medially connected accessory anterior tentorial arms paired processes of the corpotentorium are often present (osteotendons). One dorsal, two lateral and two ventral cervical sclerites are embedded in the cervical membrane of roaches. The head of termites is prognathous and nearly circular or oval in dorsal view. The more or less round compound eyes are only composed of a few hundred ommatidia and usually reduced in workers and soldiers (not in Hodotermitidae); a pair of ocelli is usually present but missing in Termopsidae. The mandibles are asymmetric in most species and sometimes strongly enlarged in soldiers (Figs 6.17.2B, 6.17.7, 6.17.8A, B). Soldiers of some Rhinotermitidae and Termitidae bear a very prominent opening of the frontal gland on their frontal region (Figs 6.17.8C). Two lateral cervical sclerites are present in most groups but rarely also three (additional unpaired ventral sclerite). The head of workers is similar to that of members of the winged caste. It is usually enlarged in soldiers (Fig. 6.17.8).

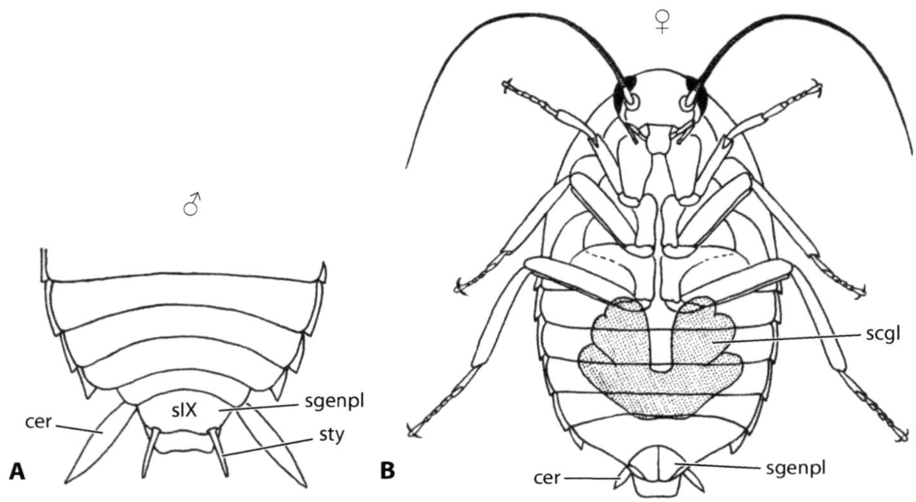

Fig. 6.17.4: Blattodea, ventral view. A, male of *Methana marginalis* (Blattidae), posterior abdominal segments; B, female of *Eurycotis floridana* (Blattidae). Abbr.: cer – cerci, sIX – sternum IX, scgl – scent gland, sgenpl – subgenital plate, sty – styli. Redrawn from Bohn (2005), after Roth (1991) and Roth & Alsop (1978).

The thorax of roaches is characterized by the large and often more or less round pronotum (Fig. 6.17.1: prn), which usually covers the head at least partly (see above). In termites the pronotum is elongate and the head always exposed. The three thoracic segments are always movably connected with each other. Like in Mantodea the meso- and metanotum are composed of a prescutum, scutum and scutellum; the prescutum forms small lobe-shaped phragmata. A small sclerite, the tegulum, is present anterior to the three axillary sclerites. The pleural ridge of all three segments is distinctly slanting posteriorly. The forewings are leathery tegmina in roaches but membranous like the hindwings in termites (Fig. 6.17.7E); the wing pairs are not connected during flight; a predetermined breaking point (humeral suture) is present at the wing base in termites, which shed their simplified wings after a short mating flight. The legs are long and slender and the large coxae are medially adjacent (Figs 6.17.1, 6.17.4B); the cone-shaped coxae are in an almost horizontal position; posteriorly a flat meron is delimited by a line continuous with the pleural suture; the trochanter is immovably connected with the femur; a preformed zone of weakness is present between both elements in roaches (rupture facilitating mechanism); the femur and tibia usually bear a vestiture of strong spines (less distinctly developed in termites); the tarsi are 5-segmented in roaches and Mastotermitidae (isopteran groundplan), but 4-segmented in the other termites; tarsomeres 1–4 are usually equipped with euplantulae in roaches, but lacking them in termites (Fig. 1.3.5.4); an arolium is present or absent (present only in alate adults of termites); reduction of attachment structures often occurs in cavernicolous forms.

The abdomen is broadly connected with the metathorax (Figs 6.17.3, 6.17.4, 6.17.7). It is composed of ten distinctly developed segments. The width decreases from segment VIII. The cerci of roaches can be composed of up to 20 segments whereas only 1-8 cercomeres are present in termites. The copulatory apparatus of male roaches is strongly asymmetric. In termites it is either completely reduced or present as a simple and symmetric penis papilla. A strongly elongated subgenital plate, a genital atrium (Fig. 6.17.5), and a short ovipositor composed of three paired valves and associated basal sclerites is present in female roaches. In termites the ovipositor is only maintained in Mastotermitidae (plesiomorphy).

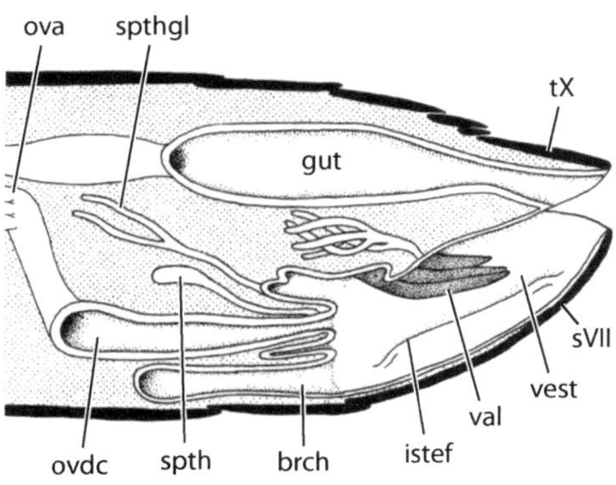

Fig. 6.17.5: Blaberidae, abdomen, sagittal section, schematized. Abbr.: brch – brood chamber, istef – intersternal fold, ova – ovar, ovdc – oviduct, sVII – sternum VII, spth – spermatheca, spthgl – spermathecal gland, tX – tergum X, val – valvulae, vest – vestibulum. Redrawn from Bohn (2005), after McKittrick (1964), modified.

Blattodea are characterized by a broad array of glands. Up to six different glands can be present in the head of roaches. Mandibular glands produce aggregation pheromones in Blaberidae (*Blaberus*) and probably also in termites. The paired salivary glands located on both sides of the ingluvies are composed of loosely arranged acini and a reservoir. Laterocervical glands occur in the neck region of nymphs and adults

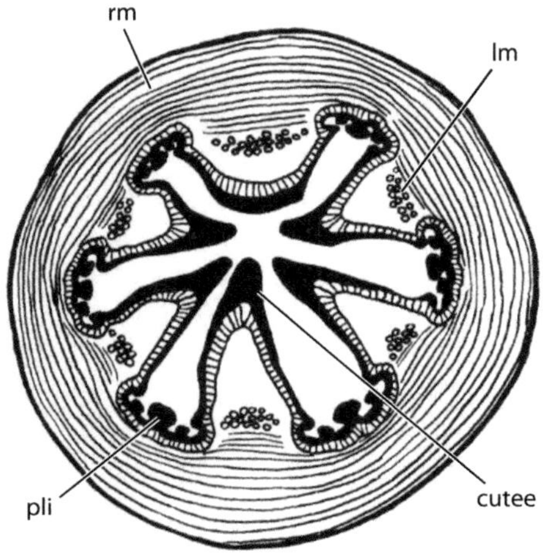

Fig. 6.17.6: Blattodea, proventriculus, cross section. Abbr.: cutee – cuticular teeth, lm – longitudinal muscle, pli – plica, rm – ring muscle. Redrawn from Bohn (2005), after Beier (1974).

of Blattidae. Different glands also occur in the tergal, pleural, or sternal abdominal regions in different lineages. Pleural scent glands occur in Corydiidae, and sternal glands in nymphs and adults of Blattidae (Bohn 2005) and in Isoptera. A characteristic feature of Isoptera is the presence of glands in the distal parts of the legs. Frontal glands occur in Rhinotermitidae and Termitidae (often enlarged in soldiers) (Bohn 2005). Six longitudinal folds are present in the oesophagus of roaches; a very large and dilatable ingluvies reaches deep into the abdomen (often asymmetric in termites); a well-developed proventriculus is present at the posterior end of the foregut, with primary, secondary and tertiary plicae (in some groups), cuticular teeth and a strongly developed ring musculature (Fig. 6.17.6); the valvula cardiaca separate the foregut from the relatively short midgut, which bears caeca (usually 6–10, absent in some smaller roaches and most groups of termites) at its anterior end. Up to 100 thin Malpighian tubules insert at the border region between the midgut and hindgut in roaches (only four in termites). In Isoptera and *Cryptocercus* (sistergroup) the posterior hindgut is extended as a pouch or fermenting chamber containing endosymbiotic flagellates (Bohn 2005). The fat body is strongly developed and contains bacteriocytes with endosymbiotic bacteria (Bohn 2005). The heart extends between the prothorax (roaches) or mesothorax (termites) and abdominal segment IX; twelve or eleven (termites) segmental alary muscles are present and 12 or 11 (termites) paired ostia; six lateral vessels (two in the thorax, four in the abdomen) are present in roaches but absent in Isoptera; accessory pulsatile organs are present at the bases of the antennae and at the wings bases of roaches (Bohn 2005). The paired testes are composed of numerous follicles (up to 30 in roaches, 8–50 in termites). Accessory glands are

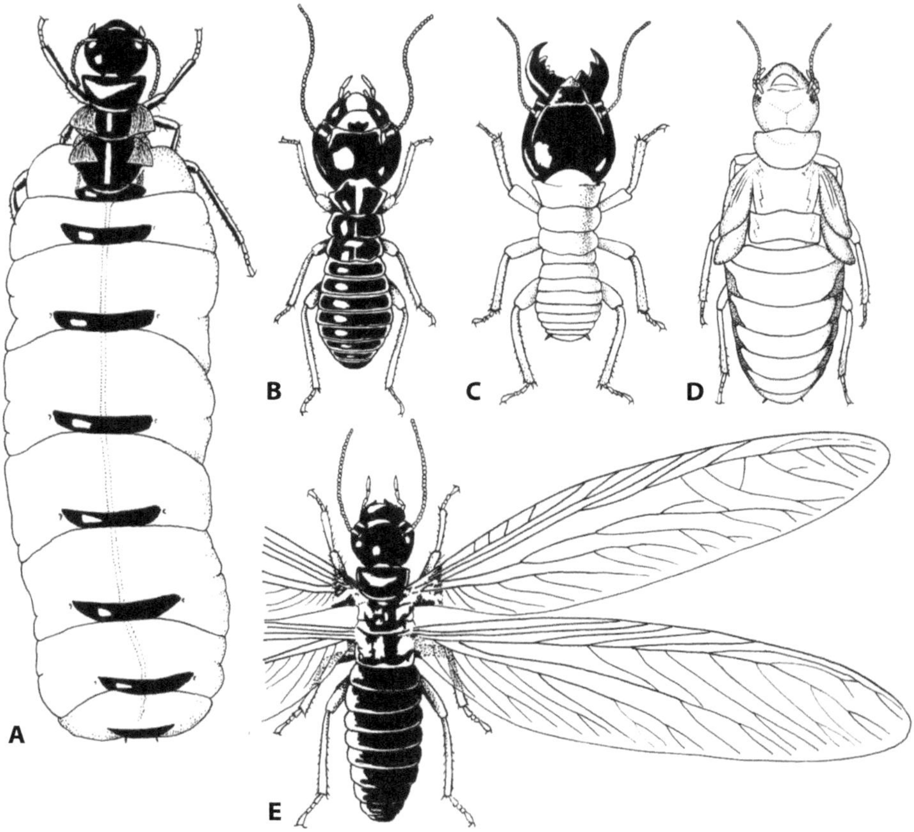

Fig. 6.17.7: Isoptera, dorsal view, castes of the harvester termite *Hodotermes mossambicus* (Hodotermitidae), dorsal view. A, physogastric queen; B, worker; C, soldier; D, replacement sexual morph; E, winged reproductive morph. Redrawn from Bohn (2005), after Coatan (1952).

present in roaches and basal isopteran lineages. Between eight and 20 panoistic ovarioles are usually present in roaches, whereas the number in termites varies between seven (*Kalotermes*) and several thousand (*Macrotermes*).

Biology. Roaches are usually nocturnal and lead a cryptic life, often covering under stones, bark or logs. Most species live in tropical rain forests. They usually feed on decomposing organic materials and are important saprotrophs. Likewise, termites play an essential role in the recycling of plant materials and are strongly involved in the formation of soil (pedogenesis) in tropical areas. They have the potential to produce immense amounts of hydrogen and arguably belong to the most efficient "bioreactors".

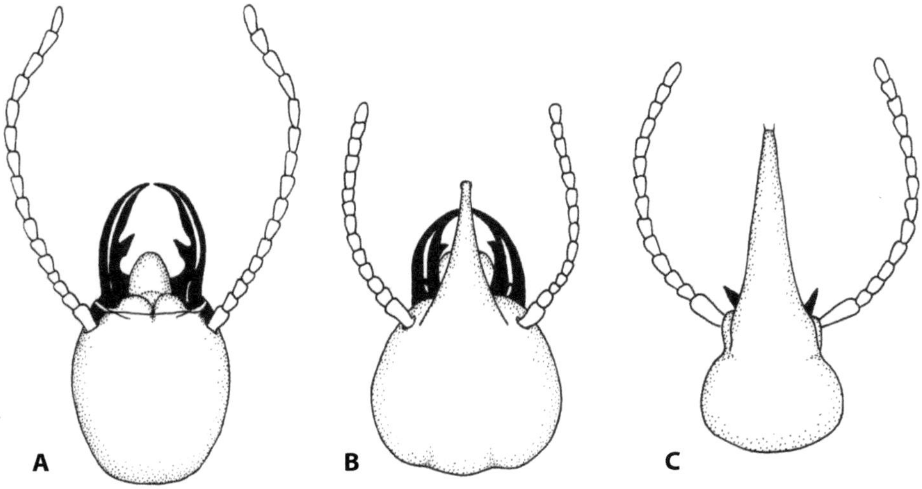

Fig. 6.17.8: Isoptera (Termitidae), heads of different species. A, *Drepanotermes* sp.; B, *Amitermes* sp.; C, *Angularitermes* sp., with vestigial mandibles. Redrawn from Bohn (2005) (after Watson & Gay [1991] and Weesner [1969]).

Xylophagous habits have evolved several times independently in roaches from less specialized phytophagous or detritophagous habits. Termites feed on cellulose in the form of plant fibers (wood, leaves, seeds etc.). Some species breed fungi on predigested wood.

Roaches are usually nocturnal and primarily solitary. However, there is a broad spectrum covering simple aggregations (e.g., *Blattella*), biparental brood care (e.g., *Salganea*) and subsociality (*Cryptocercus*). Aside from some groups of the holometabolous Apocrita (Hymenoptera), Isoptera are the only insect lineage with highly developed social systems (Figs 6.17.7, 6.17.8). They form complex colonies ("states") ranging in individual numbers from several hundred (*Reticulitermes*) to several millions (*Macrotermes*). The colonies are composed of different castes: immature workers and soldiers, nymphs with reduced wings (pseudergates) and fully winged adults (Figs 6.17.7, 6.17.8). Workers and pseudergates can transfer to other castes after molting. Females producing eggs after the mating flight are called queens. They are often physogastric and largely immobilized. Pheromones play a very important role in the communication systems of the colonies.

Reproduction and development. Pheromones play an important role in the reproduction of roaches and complex mating behavior occurs in some groups. The eggs are primarily deposited in robust oothecae but different degrees of reduction of the egg case occur. Females of Blaberidae are ovoviviparous and the protective cover of the ootheca is largely reduced. Their eggs develop in an "uterus" after leaving the oviduct. Obligatory parthenogenesis is described for *Pycnoscelus surinamensis* and *Phyllodromica subaptera*.

In termites new colonies are founded after a mating flight of winged adult males and females, or an established colony divides by budding or social fragmentation (sociotomy). Oothecae are only formed by females of *Mastotermes*. A queen of *Odontotermes* can deposit up to ca. 86,000 eggs per day.

Fossil record. Roach-like insects were very abundant in the Carboniferous. However, their systematic affinities are ambiguous and their taxonomy confusing (Grimaldi & Engel 2005). Three currently recognized Paleozoic dictyopteran families were replaced by three other families in the early Mesozoic. Fossils of extant "blattarian" families occur in the fossil record in the Early Cretaceous. *Cryptocercus* has no fossil record. Termites are the only major group of insects without an extinct family. The oldest fossils are wings and wing fragments from the Early Cretaceous (Grimaldi & Engel 2005; Bohn 2005).

Economic importance. About 1% of the roaches are associated with human habitations and some of them are important household pests. The most important species in this context, commonly referred to as cockroaches, are the cosmopolitan *Blattella germanica*, *Blatta orientalis*, *Periplaneta americana*, and the Asian *Blattella asahinai* (also reported from the southern USA since 1986). Cockroaches feed on human and pet food and a wide range of stored products. They also transport bacteria on their body surface and thus play a certain role as unspecific vectors of diseases, especially in hospitals. The control is greatly impeded by the impregnable oothecae. In general cockroaches belong to the hardiest known insects.

Termites are mainly known for the extreme damage they cause to wooden constructions (ca. 1.5 billion $ each year in the southwestern USA) and also to crops (e.g., peanuts) and plantation forests. Their economic and ecological impact is hardly paralleled by any other group of insects. Termites and their microorganisms are also potentially efficient producers of biogas (hydrogen).

[Coatan (1952); McKittrick (1964); Weesner (1969); Beier (1974); Roth & Alsop (1978); Roth (1991); Watson & Gay (1991); Klass (1995), Lo et al. (2000); Bohn (2005); Grimaldi & Engel (2005); Djernaes et al. (2012)]

Acercaria (Thysanoptera, Psocodea, Hemiptera)

Acercaria is a well-established clade comprising three distinct phylogenetic units, the Psocodea ("Psocoptera" and Phthiraptera), the Thysanoptera, and the megadiverse Hemiptera (Sternorrhyncha, Auchenorrhyncha, Coleorrhyncha and Heteroptera) (Fig. 6.C.1: 1). Paraneoptera is sometimes used as a synonym (e.g., Grimaldi &

Engel 2005) but as this group as defined by Hennig (1969) also included Zoraptera (now well established as a polyneopteran order), Acercaria is more appropriate. A tendency towards liquid feeding with piercing sucking mouthparts is characteristic for Acercaria. However, this derived condition has evolved at least two times independently. Feeding on solid particles with biting mandibles is maintained in Psocodea excl. Rhynchophthirina and Anoplura.

Psocodea are confirmed as a clade by morphological and molecular data (Fig. 6.C.1: 10). In contrast, the “Psocoptera” (bark lice) are not monophyletic. Liposcelididae, traditionally assigned to the psocopteran suborder Troctomorpha, is the sister group of Phthiraptera (true lice) (e.g., Yoshizawa & Johnson 2010). Phthiraptera is not supported in all studies based on molecular data but considering many apomorphic morphological features shared by its subgroups and the ectoparasitic habits its monophyly can be considered as very likely. Amblycera is the sistergroup of the remaining three lineages of true lice and the Rhynchophthirina the sistergroup of the specialized blood-sucking Anoplura.

Thysanoptera are strongly supported as a clade, especially but not only by autapomorphies linked with miniaturization (e.g., asymmetric mouthparts, strongly modified wings with fringes of long microtrichia). In contrast, the position of the order is still disputed. A possible clade Micracercaria (Thysanoptera + Psocodea) is suggested by the presence of a cibarial muscle with an unpaired tendon and biflagellate spermatozoa (Fig. 6.C.1: 2). Other morphological features such as for instance stylet-like mandibles suggest a clade Condylognatha (Thysanoptera + Hemiptera) (Fig. 6.C.1: 3). Considering the results of recent molecular studies (e.g., Cryan & Urban 2012) this concept appears more likely, but the placement of Thysanoptera should still be treated as an unsolved question.

With nearly 100,000 described species Hemiptera is by far the largest group of Acercaria and non-holometabolan insects (Fig. 6.C.1: 4). The high diversity is probably closely linked with phytophagy and the enormous radiation of the angiosperm plants in the Cretaceous. Hemiptera are unambiguously supported as a clade and comprise the four traditional orders Auchenorrhyncha, Sternorrhyncha, Coleorrhyncha and Heteroptera (e.g., Strümpel 2005). The monophyly of Auchenorrhyncha (Cicadina) was not confirmed in all molecular studies but appears likely considering recent molecular (Cryan & Urban 2012; Song et al. 2012) and morphological investigations (Fig. 6.C.1: 9).

The interrelationships of the four hemipteran subgroups are still insufficiently understood. A sister group relationship between Sternorrhyncha and the remaining Hemiptera (= Euhemiptera, Strümpel 2005a) was suggested by Cryan & Urban (2012) but this needs further support (Fig. 6.C.1: 5). What appears well established is a clade Heteropterodea (or Prosorrhyncha) comprising the very small relict group Coleorrhyncha and the highly successful Heteroptera (Fig. 6.C.1: 6). However, a close relationship between Coleorrhyncha and Auchenorrhyncha cannot be ruled out completely at present (Fig. 6.C.1: 7). The monophyletic origin of three of the orders is well-supported

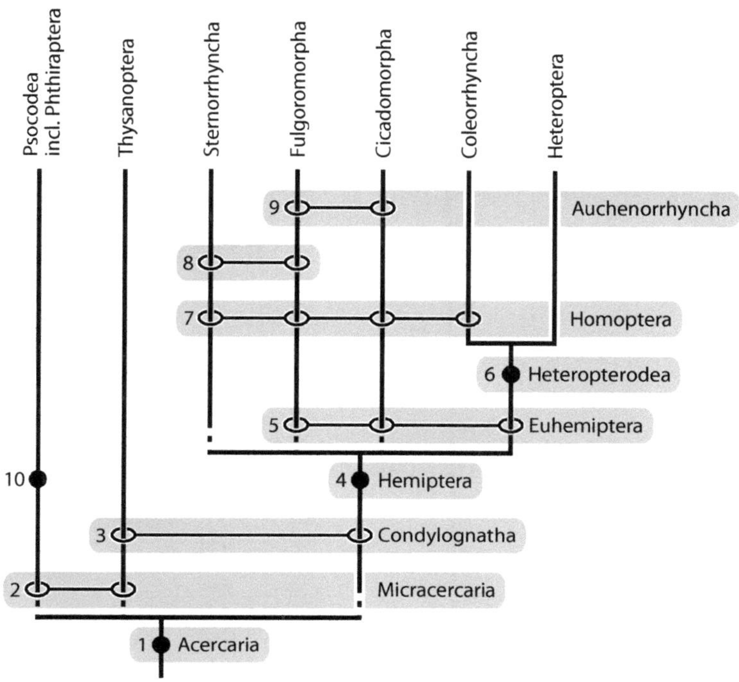

Fig. 6.C.1: Acercarian relationships, with selected potential apomorphies. **1**, cerci absent, postclypeus and cibarial dilator enlarged, lacinia detached from stipes and stylet-like, 3-segmented tarsi, single abdominal ganglionic complex, four Malpighian tubules, polytrophic ovarioles; **2**, enlarged cibarial dilator with unpaired tendon, biflagellar spermatozoa; **3**, dorsal shift of anterior tentorial pits, labrum narrowed, mandibles stylet-like (?), maxillary palps absent; **4**, four-segmented labial rostrum, labial endite lobes and palps absent, mandibles stylet-like (?), buccal pump, brain and suboesophageal ganglion form compact unit; **6**, four antennomeres, anterior ocellus reduced, head with long and thin sensorial setae, forewings horizontally placed on abdomen, tegula of forewing absent, metathoracic scent gland system; **10**, antenna with rupture-facilitating device (groundplan), cibarial water-vapor uptake apparatus, cibarial pistil-and-mortar apparatus (groundplan), cardo not present as a separate sclerite.

(Sternorrhyncha, Coleorrhyncha, Heteroptera) but is disputed in the case of Auchenorrhyncha (see e.g., Cryan & Urban 2012) (Fig. 6.C.1: 8). Different morphological or molecular data sets suggest monophyletic Auchenorrhyncha (Fig. 6.C.1: 9; see below for potential autapomorphies), or as alternatives Heteropterodea (Coleorrhyncha + Heteroptera) as sistergroup of Fulgoromorpha or Cicadomorpha, respectively.

Autapomorphies and supraordinal acercarian groups

Acercaria

- Cerci reduced
- Postclypeus and cibarial dilator muscle enlarged
- Lacinia detached from stipes and stylet-like
- Tarsi with three (groundplan) or less segments
- One single abdominal ganglionic complex
- Only four Malpighian tubules
- Polytrophic ovarioles (?)

Psocodea

- Antenna with rupture-facilitating device (groundplan)
- Cibarial water-vapor uptake apparatus
- Cibarial pistil-and-mortar apparatus (groundplan)
- Cardo not present as a separate sclerite

Liposcelididae + Phthiraptera

- Complete loss of wings
- Head and postcephalic body dorsoventrally flattened
- Compound eyes reduced
- Meso- and metanotum fused

Condylognatha (Thysanoptera + Hemiptera) (?)

- Dorsal shift of anterior tentorial pits
- Labrum narrowed
- Mandibles stylet-like (?)
- Maxillary palps absent
- Distal median plate of forewing positioned next to 2nd axillary sclerite and articulating with a convex hinge

Micracercaria (Thysanoptera + Psocodea) (?)

- Enlarged cibarial dilator with unpaired tendon
- Biflagellar spermatozoa

Hemiptera

- Four-segmented labial rostrum (Fig. 6.C.2)
- Labial endite lobes and palps absent (Fig. 6.C.2)
- Mandibles stylet-like (?) (Fig. 6.C.2)
- Buccal pump formed by parts of the cibarium and anterior pharynx
- Brain and the suboesophageal ganglion form a compact unit with a narrow passage for the foregut

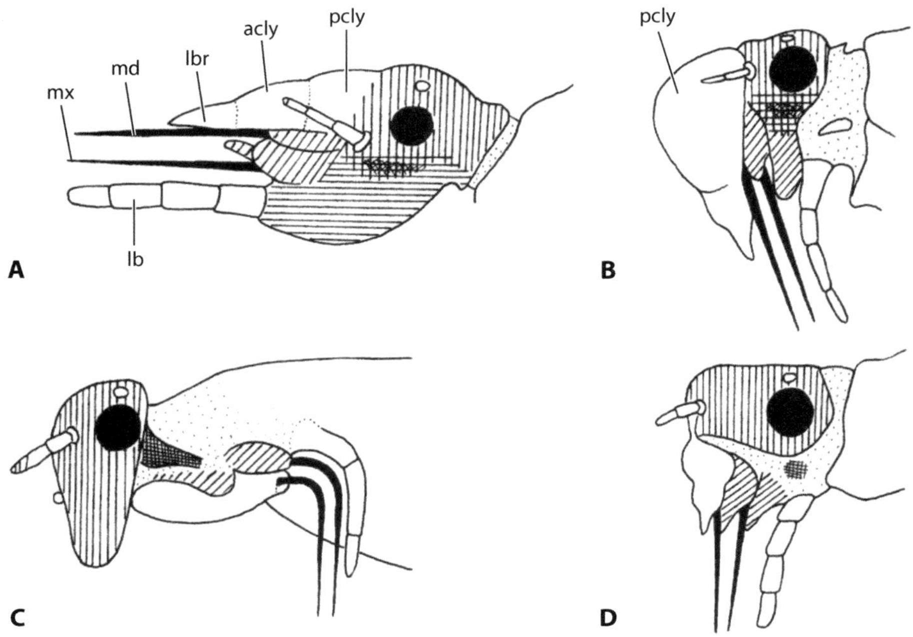

Fig. 6.C.2: Hemipteran heads, lateral view, schematized. A, Heteroptera; B, Cicadoidea; C, Psylloidae; D, Aphidina. Abbr.: acly – anteclypeus, lb – labium, lbr – labrum, md – mandible, mx – maxilla, pcly – postclypeus, horizontal lines – gula, vertical lines – epicranium (main part of head capsule), oblique lines – laminae, white – clypeolabrum, black – compound eyes and stylets. Redrawn from Strümpel (2005), after Weber (1930).

Heteropterodea (=Prosorrhyncha) (Coleorrhyncha + Heteroptera)

- Maximum number of four antennomeres
- Anterior ocellus reduced
- Head with long and thin sensorial setae (cephalic "trichobothria")
- Forewings horizontally placed on abdomen in resting position
- Tegula of forewing absent
- Metathoracic scent gland system

[Weber (1930); Parrish (1967); Hennig (1969); Lyal (1985); Strümpel (2005); Willmann (2005a); Grimaldi & Engel (2005); Yoshizawa & Saigusa (2001); Yoshizawa & Johnson (2010); Cryan & Urban (2012); Song et al. (2012)]

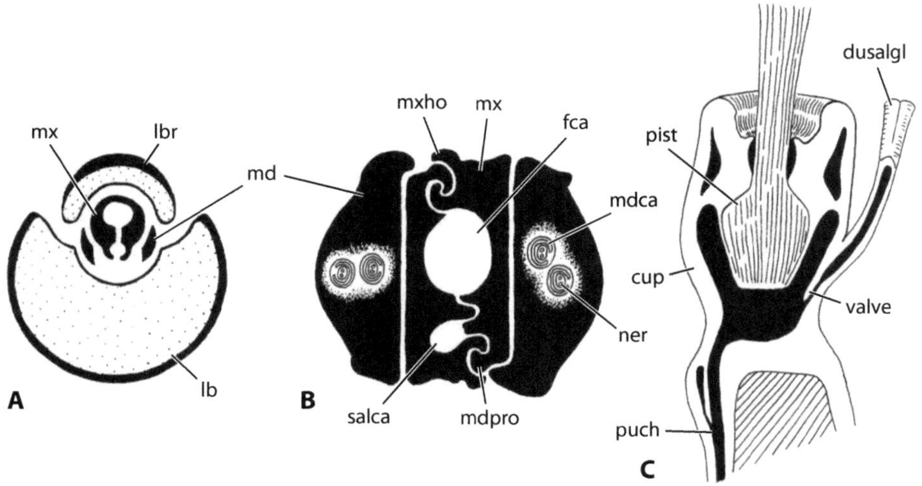

Fig. 6.C.3: Hemipteran mouthparts and associated structures. A, labial rostrum and stylets, cross section, schematized; B, mandibular and maxillary stylets, cross section; C, Salivary pump of a true bug (Pentatomidae). Abbr.: cup – cupula, dusalgl – ductus of the salivary gland, fca – food canal, lb – labium, lbr – labrum, md – mandibular stylet, mdca – mandibular canal, mdpro – mandibular process, mx – maxillary stylet, mxho – maxillary hook, ner – nerve, pist – pistill, puch – pumping chamber, salca – salivary canal. Redrawn from Strümpel (2005), after Weber (1930) and Parrish (1967).

Psocodea (Psocoptera + Phthiraptera)

6.18 **Psocoptera** (Greek *psōchos* = dust, *pteron* = wing, English common names: barklice, booklice, barkflies)

Diversity and Distribution. With about 5,500 described species Psocoptera belong to the medium-sized insect orders. The known diversity is possibly only a fraction of the real species number. The distribution is worldwide. Most species occur in the tropical zones, but specialized forms also occur in the arctic region.

Autapomorphies. The monophyly of Psocoptera was never considered as strongly supported. Barklice are mostly characterized by plesiomorphic features compared to their specialized sistergroup Phthiraptera. Few potential autapomorphies of Psocoptera were suggested by Seeger (1979):

- Chorion extremely thin, unsculptured and lacking micropyles and aeropyles
- Legs of embryos bent antero-mediad

Today there is strong evidence that the group is paraphyletic. A sistergroup relationship between the family Liposcelididae and the ectoparasitic Phthiraptera is sup-

Fig. 6.18.1: *Lepinotus* sp. (Stenopsocidae), habitus of brachypterous specimen, lateral view, SEM micrograph. Abbr.: ant – antenna, ce – compound eye, fwi – forewing, lbr – labrum, md – mandible, pcly – postclypeus, plb – palpus labialis, pmx – palpus maxillaris, prn – pronotum. Scale bar: 250 µm. Courtesy H. Pohl.

ported by morphological (Lyal 1985) and molecular evidence (Yoshizawa & Johnson 2003). Liposcelididae are characterized by a prognathous head like all phthirapterans, and the wings are reduced or short with a vestigial venation. Moreover, some liposcelidid species are found in nests of mammals and birds where they feed on debris (Grimaldi & Engel 2005).

Taxonomy. The group is currently classified into the three suborders Trogiomorpha (see Yoshizawa et al. 2006), Troctomorpha and Psocomorpha. Trogiomorpha (ca. 300 spp.) comprise the three infraorders, Prionoglaridetae, Psyllipsocetae and Atropetae and five families (e.g., Trogiidae, Psyllipsocidae). Troctomorpha (ca. 350 spp.) are divided into the two infraorders Amphientometae and Nanopsocetae, together comprising eight families (e.g., Troctopsocidae, Liposcelididae). Troctomorpha is almost certainly paraphyletic, containing the ectoparasitic Phthiraptera as a subordinate lineage. Psocomorpha, by far the most species-rich suborder, is classified into four (Lienhard & Smithers 2002) or five (Yoshizawa & Johnson 2010) infraorders and 24 families (e.g., Stenopsocidae, Psocidae).

Diagnosis. Small or very small insects. Often polymorphic. Fully winged, brachypterous or completely wingless. Head large in relation to the body size (Fig. 6.18.1). Antennae usually long and thin. With biting mandibles, a chisel-like lacinia detached from the stipes, and a specialized cibarial food processing apparatus. Wings membranous if present and usually held in a roof-like position (not in Liposcelididae). With different coupling mechanisms during flight and at rest.

Morphology of adults. The length of the small insects ranges between 0.5 mm and 10 mm (wing span 25 mm), with the majority of species measuring between 1 mm and 5 mm. The cuticle is usually thin; areas with a stronger degree of sclerotization (head, thoracic tergites and sternites, legs) usually display a specific microsculpture. Pigments of the cuticle usually results in a brownish or grey coloration. Polymorphism occurs frequently, with morphs with well-developed wings and eyes, and others with reduced or lacking wings and more or less strongly reduced light sense organs.

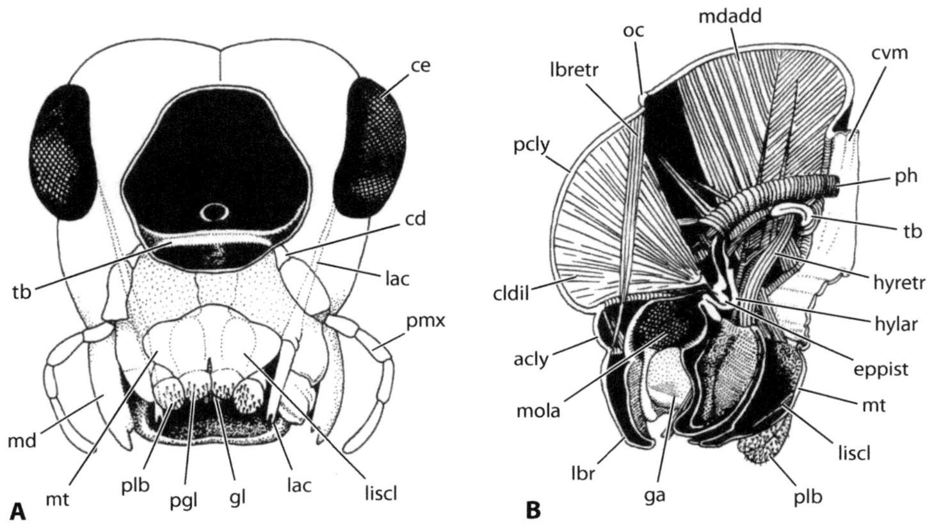

Fig. 6.18.2: *Ectopsocus briggsi* (Psocidae), head. A, posterior view; B, sagittal section. Abbr.: acly – anteclypeus, cd – cardo, ce – compound eye, cvm – cervical membrane, cldil – cibarial dilator with unpaired tendon, eppist – epipharyngeal pistill, ga – galea, gl – glossa, hylar – hypopharyngeal larynx (sclerotized concavity), hyretr – hypopharyngeal retractor, lac – lacinia, lbr – labrum, lbretr – labral retractor, liscl – lingual sclerite, md – mandible, mdadd – mandibular adductor, mt – mentum, oc – ocellus, pcly – postclypeus, pgl – paraglossa, ph – pharynx, plb – palpus labialis, pmx – palpus maxillaris, tb – tentorial bridge. Redrawn from Günther (2005), after Weber (1936).

The large, hemispherical head (Figs 6.18.1–6.18.3) is fully exposed posteriorly, with a high movability in the neck region. It is usually orthognathous, but prognathous in Liposcelididae. The frons is small compared to the large and conspicuous clypeus, which is subdivided into a narrow anteclypeus and a large and strongly convex postclypeus (Figs 6.18.2B, 6.18.3). The eucone compound eyes are usually well-developed and convex but can be reduced to few ommatidia in some cases (*Liposcelis*); three ocelli are present in the groundplan but missing in apterous forms. The transverse frontoclypeal strengthening ridge is very distinctly developed. The frontal and coronal sutures are primarily present. The labrum is free and well-developed (Figs 6.18.1, 6.18.2B); the labral nodes (nodi labrales) are paired sclerotizations of the anterior margin, which is bent inwards anterad the mandibular apices; short setae are present between the nodes; the distal epipharynx forms the roof of the anterior preoral chamber, the corrosium; its middle section, the roof of the molarium, is characterized by the molar brush, which is composed of long microtrichia and narrowed between the mandibular molae; the proximal epipharyngeal section bears a cylindrical, apically thickened sclerotization, the pistill. The antennae are filiform and usually long; the number of antennomeres ranges between 11 and 50. The mandibles are asymmetrical and stout, with a well-developed grinding mola and 1–3 teeth; a small brush of posteriorly directed microtrichia is present proximad the mola. The maxilla lacks a separate cardo; it is likely fused with the stipes; the chisel-like lacinia is detached from the stipes and sunk into an internal epidermal pouch; the galea is large and bears a very small grinding area; the palp is 4-segmented (Fig. 6.18.2A). The labium is composed of an unsclerotized postmentum and a fairly large prementum, which bears the membranous paraglossae, the very small glossae, and the small, 2- or 1-segmented palps; the palps and the paraglossae are both densely set with short sensorial setae; the glossae together form a spinning-tube (Fig. 6.18.2A). The ventral side of the hypopharynx forms the roof of the salivarium where the ducts of the salivary glands and silk glands open; its dorsal surface is characterized by a deep, cup-shaped, thick-walled concavity, the larynx; the anterior wall of the larynx is connected with the oval lingual sclerites on the posterior hypopharyngeal side by thin chitinous tubes (Fig. 6.18.2B). The tentorium is complete, with a well-developed bridge (corpotentorium), which is continuous with the strongly developed transverse posterior arms; the corpotentorium is connected with the anterior cervical sclerites by chitinous ligaments; the anterior arms together form a V-shaped structure; the dorsal arms are thin. The head musculature is characterized by a very strongly developed cibarial dilator (M. clypeopalatalis), which is attached to the cibarial roof by an unpaired tendon (like in Phthiraptera and Thysanoptera) (Fig. 6.18.2B). The wide cervical membrane connecting the head and prothorax is reinforced by an elongate anterior pair of cervical sclerites and a smaller posterior pair.

The prothorax is distinctly reduced in size in the alate forms, with the exception of Lepidopsocidae (Fig. 6.18.3). The short pronotum is often covered by the anteriorly protruding, strongly convex mesonotum. The propleuron is strongly reduced. On the ventral side a narrow sclerotized element is formed by the fused basisternum and furcasternum; it is the area of origin of the profurcal arms. The pterothoracic segments are connected with the prothorax by a membranous zone and together form a compact functional unit (fused in Lepidopsocidae) (Fig. 6.18.3). The meso- and metanotum are subdivided into a small anteromedian prescutum, a large scutum, and a smaller scutellum. In the alate forms three axillary sclerites are connected with three small processes of the lateral notal margin; together with the basalare and subalare and elements of the wing base they form the wing articulation. The postnotum is not directly connected with the wings. The mesepisterna and -epimera are large plates separated by the pleural ridge; the former is subdivided into a ventral katepisternum and a dorsal anepisternum by the anapleural suture (Fig. 6.18.3). The anterior thoracic spiracle is embedded in the membrane anterad the anepisternum. The basisternum and trochantin are fused with the katepisternum. In most other groups the metathorax is distinctly smaller than the mesothorax, in correlation with the smaller size of the hindwings. The metepisternum is short and the anapleural suture is missing; like in the mesothorax the epimeron is dorsally connected with the postnotum. The ventral side of the pterothorax is formed by a narrow, elongate sclerite extending between the mesothoracic katepisternum and the metathoracic epimeron; it is only interrupted by the meso- and metafurcal invagination sites; a spinasternum at the anterior mesothoracic margin is present or absent (Psocidae). Wings are primarily present but reductions occur frequently. Usually they are held in a roof-like position, rarely flat over the abdomen (Liposcelididae). The costa of the hindwing is held between the forewing and an apophysis of the mesothoracic spiracle (pterostigmatic locking device). The forewings are always larger and usually reach the abdominal apex posteriorly; the venation of the hindwings is similar to that of the forewings but somewhat simplified; the wing membrane is often smooth but scales or other surface modifications occur in some groups; the venation is characterized by a shortened subcosta, a pterostigma enclosed by R_1, and a nodulus formed by a basal fusion of the postcubitus and the anal vein (forewing); the hind margin of the forewings is bent downwards and interlocked with the conversely bent anterior margin of the hindwings during flight; this connection is often stabilized by hook-shaped structures on the nodulus. The flight musculature is well-developed in both pterothoracic segments of alate specimens. The three pairs of legs are similar and largely unmodified; the hindlegs are slightly longer and the hindfemora are thickened in species of *Liposcelis*, which are able to jump over small distances (ca. 1 cm); the coxae are club-shaped; a specialized organ, possibly with stridulatory function, is often present on the metacoxae (Pearman's organ); the trochanter is slightly curved and immovably connected with the femur; the long, cylindrical tibia often bears longitudinal rows of ctenobothria, thorns or setae and two spurs on its apex; the tarsus is 2- or 3-segmented and the first tarsomere is often equipped with ctenobothria; paired simple or toothed claws with a basal seta articulated with the apical tarsal segment; pulvilli are present and thread-like or membranous and distally extended as suction pads; an empodium is not developed.

The 11-segemented abdomen is barrel-shaped or rarely almost spherical. The terminal segment appears trilobite and forms the anal cone with the dorsal epiproct and the lateral paraprocts (Fig. 6.18.4); the paraprocts usually bear a sensorial field with specifically arranged trichobothria; it is usually more extensive in males; minute thorn- or hook-like structures can also occur on the epi- and paraprocts. Segments I–VII in females and I–VIII in males are largely unsclerotized, except for weakly developed transverse sclerites of segments I and II. In males (Fig. 6.18.4B) sternite IX forms a subgenital plate (hypandrium); its pattern of sclerotization is often species specific

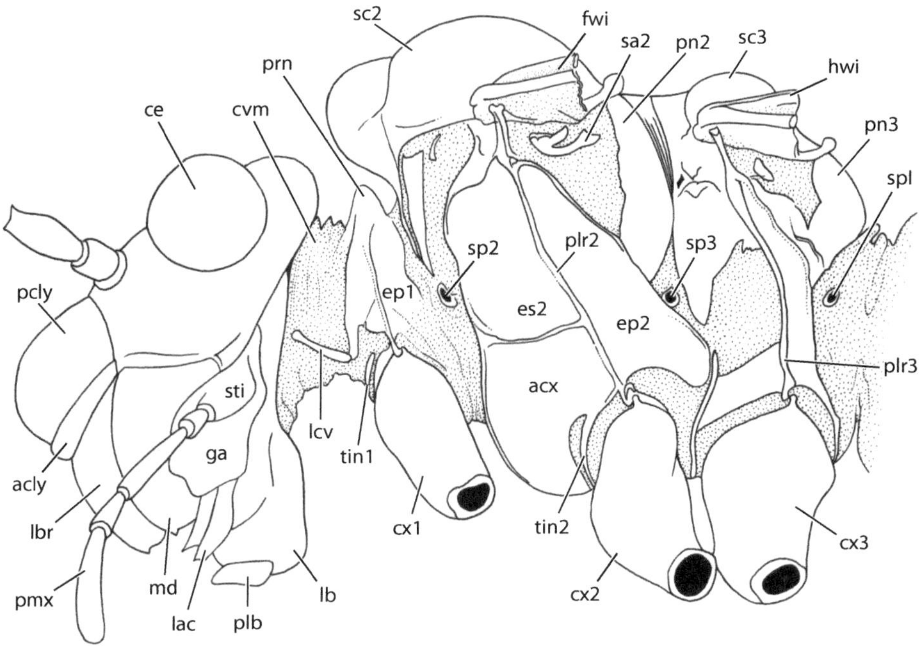

Fig. 6.18.3: *Stenopsocus* sp. (Stenopsocidae), head and thorax, lateral view. Abbr.: acly – anteclypeus, acx – antecoxa, ce – compound eye, cvm – cervical membrane, cx1/2/3 – pro-, meso-, metacoxa, ep1/2 – pro-, mesepimeron, es2 – mesepisternum, fwi – forewing, ga – galea, hwi – hindwing, lac – lacinia, lb – labium, lbr – labrum, lcv – lateral cervical sclerite, md – mandible, pcly – postclypeus, plb – palpus labialis, plr2/3 – meso-, metathoracic pleural ridge, pmx – palpus maxillaris, pn2/3 – meso-, metathoracic postnotum, prn –pronotum, sa2 – mesothoracic subalare, sc2/3 – meso-/metascutum, sp2/3/I – meso-, metathoracic spiracles, 1st abdominal spiracle, sti – stipes, tin1/2 – pro-, mesotrochantin. Redrawn from Jeannel (1951).

and complicated. Sclerotized hooks (harpagons) occurring in *Lachesilla* are possible vestiges of gonopods. The phallus (Fig. 6.18.4C) is composed of the paired parameres, which are usually connected with an unpaired basal plate (rarely with paired apodemes); the highly variable aedeagus lies between the parameres; the wall of the endophallus can be equipped with numerous sclerites or bristles. Accessory copulatory structures (e.g., hooks) can be present on the epiproct, paraprocts or tergite X. In females (Fig. 6.18.4A) the subgenital plate is formed by sternite VIII (hypogynium). The ovipositor is largely unmodified and formed by three paired valves in the groundplan; the ventral valves (segment VIII) are usually slender and elongate; the dorsal valves (segment IX) are also fairly long in most cases, but broader than the ventral ones; the external valves (3rd valves, segment IX) are short, lobe-shaped, and densely covered with setae; different degrees of reduction of the ovipositor occur; the gonan-

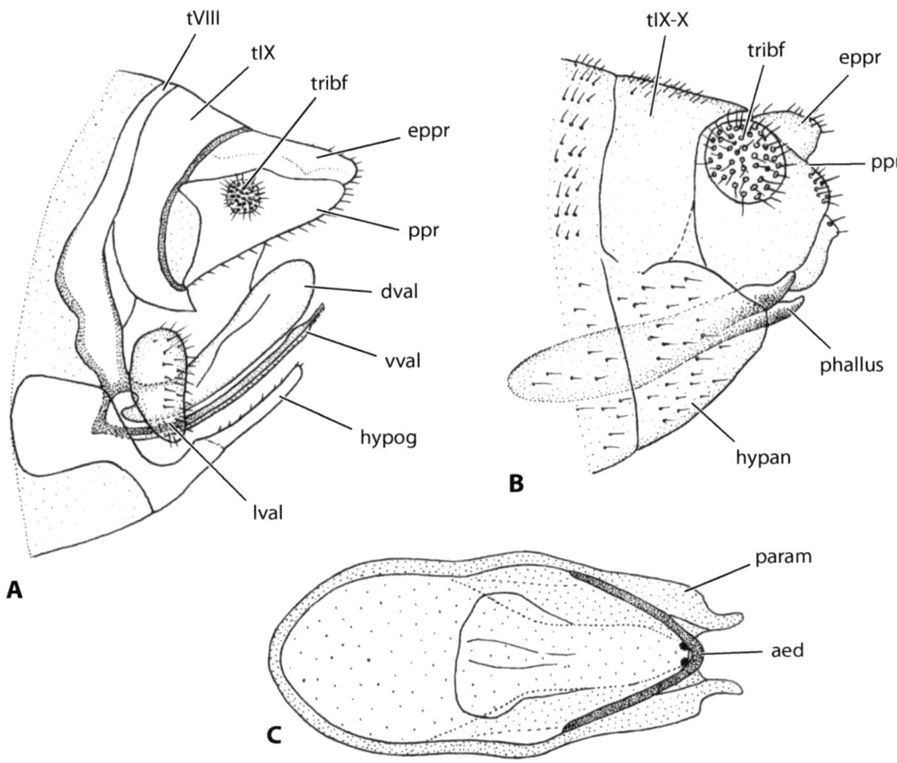

Fig. 6.18.4: Terminal abdominal segments of Psocoptera (Mesopsocidae). A, *Metylophorus* sp., female, lateral view; B, *Mesopsocus* sp., male, lateral view; C, *Mesopsocus* laticeps, phallus. Abbr.: aed – aedeagus, dval – dorsal valve (gonapophysis IX), eppr – epiproct, hypan – hypandrium, hypog – hypogynium (sternum VIII), lval – lateral (3rd) valve, param – paramere, ppr – paraproct, tVIII/IX/X – tergum VIII/IX/X, tribf – field of trichobothria, vval – ventral valve (gonapophysis VIII). Redrawn from Günther (2005), A, B after Badonell (1943), modified, C after Günther (1974).

gulum is fused with tergite IX. The opening of the spermatheca (spermaporus) lies often medially within a sclerotized dorsal plate of the genital chamber.

The postcephalic ganglionic chain is composed of three thoracic ganglia and a single abdominal ganglionic complex (autapomorphy of Acercaria). The anterior digestive tract is characterized by the highly specialized cibarial pumping chamber with the epipharyngeal pistill and the hypopharyngeal larynx; the two pairs of tube-shaped labial glands reach the mid-gut region posteriorly; the ventral pair produces saliva and the dorsal pair secretions which form silk threads; the long and thin oesophagus reaches the abdomen posteriorly; ingluvies and proventriculus are not developed; the unusually voluminous midgut forms a loop; caeca are absent; the short hindgut widens posteriorly as a rectal ampulla with four rectal papillae.

Four Malpighian tubules (autapomorphy of Acercaria) insert at the midgut-hindgut border. The short heart reaches only the middle region of the abdomen anteriorly and lacks ostia and wing muscles. The paired testes are spherical to tube-shaped and undivided, or subdivided into three follicles; the vasa deferentia are connected with a vesicular seminalis where the material for the spermatophore is produced; the ductus ejaculatorius is short. The ovaries are composed of 3–5 polytrophic ovarioles; the short oviduct leads to the wide genital chamber (vagina), which opens at the hind margin of sternite VIII (hypogynium); the receptaculum seminis opens dorsally in the genital chamber; it is connected with two hemispherical glands in Trogiomorpha.

Biology. Barklice prefer warm temperatures and dark microhabitats. Some of them are cryptic and live among leaf litter, under decaying bark, or in compost, but most species are found on the surface of bark or on branches, logs, or dead and living leaves. More specialized species occur on timber, in galleries of wood boring insects, on rocks, in caves, in nests of termites, hymenopterans, mammals or birds, in straw or hay in stables, on walls of new buildings, and also in moist human habitations (e.g., *Psyllipsocus ramburii*, *Liposcelis* sp.). Only few species are gregarious. Mechanical and acoustical signals can play a role in intraspecific communication (e.g., Pearman's organ). Psocopterans of the temperate zones fly only rarely, but mass flight over longer distances can occur under specific conditions. Camouflage mechanisms in adults are mainly based on color patterns of the wings but also of other body regions. A similar effect is achieved in nymphs with specialized gland-hairs covered with minute pieces of algae, bark, lichen, or feces. Many species produce protective webs using secretions of their labial silk glands. They can be very extensive in some tropical species (Archipsocidae). The food spectrum includes fungi, algae, lichen, films of yeasts, bacteria, glue, and organic detritus. Mold plays an important role in species specialized on human habitations.

Reproduction and development. Psocopterans are usually bisexual but different forms of parthenogenesis occur. Males are unknown in some species and occur only very rarely or only in some populations in others (thelytoky, facultative parthenogenesis). Elaborate mating behavior occurs within the group. Some species produce sounds by drumming with the abdomen on the substrate and precopulatory behavior of males may include vibrating with the wings, crawling across the female, or even standing on the head (New 1987). Males transmit a spermatophore, which only hardens within the receptaculum seminis in Trogiomorpha and Troctomorpha. Females almost generally deposit eggs. Vivipary was only observed in one tropical species. Single eggs are deposited by females of some groups and some species cover them with a blackish anal fluid sometimes containing undigested green algae. The more common pattern is oviposition on foliage or in minute crevices on bark in small groups (6–8 eggs) which are covered by a silk web. The generation cycle is generally short but the duration of the embryonic and postembryonic development varies strongly depending on the external conditions. The first instar hatches using eggbursters. The nymphs resemble small wingless adults and typically molt six times (only five times in wingless morphs). Before molting the nymphs spin a web on the surface of the substrate for better attachment and a second web as cover. Many species of the temperate regions are univoltine and hibernate as egg or nymph. The total life span rarely exceeds a few months.

Fossil record. The Permian †Permopsocida (e.g., Rasnitsyn & Quicke 2002) are apparently not monophyletic (Grimaldi & Engel 2005) and it is likely that they do not form a clade with the extant lineages of Psocoptera. They possess cerci, 4- or 5-segmented tarsi, and similar fore- and hindwings with a comparatively unmodified venation (e.g., long subcosta, indistinct pterostigma, media with four branches). Therefore even their assignment to Acercaria s.str. (excluding the stem group) is highly questionable. The oldest psocopteran fossils (wings) presumably related to extant subgroups are known from the Late Jurassic of Karatau. The first definite members of extant families are from Cretaceous ambers (e.g., Lebanon, Myanmar, New Jersey, Siberia) (Grimaldi & Engel 2005).

Economic importance. Psocoptera can be a nuisance but the economic importance is usually very limited. Damage can be caused to stored products and sometimes in natural history collections, especially if specimens (e.g., mounted mammals or birds, pinned insects) are kept in a relatively moist environment. Barklice can be efficiently controlled by strongly reducing the humidity.

[Weber (1936); Badonell (1943); Jeannel (1951); Weidner (1972); Seeger (1979); Lyal (1985); New (1987); Lienhard & Smithers (2002); Rasnitsyn & Quicke (2002); Yoshizawa & Johnson (2003); Grimaldi & Engel (2005); Günther (1974, 2005); Yoshizawa et al. (2006)]

6.19 Phthiraptera (Greek *phteir* = louse, *apteros* = unwinged, English common name: true lice)

Diversity and Distribution. Phthiraptera comprise about 5,000 described species. The monophagous, oligophagous or rarely polyphagous ectoparasites have a worldwide distribution including 66 species specialized on birds of the Antarctic and subantarctic region. Only 18% of mammal species are known as hosts and lice are entirely missing on monotremes, cetaceans, bats, and pangolins. Only one species of Amblycera occurs on marsupials. Whereas mammals are likely the ancestral hosts of Rhynchophthirina and Anoplura (Fig. 6.19.1), Amblycera and Ischnocera (Fig. 6.19.2) are probably primarily specialized on birds. Representatives of these two phthirapteran suborders are known form nearly all bird species. All families of Aves are associated with specific lineages of these ectoparasitic groups. On South American tinamous (Tinamiformes) at least 18 genera and ca. 200 species occur. Among mammals, the African rock hyrax is the host of three genera and 17 species of Amblycera and Ischnocera, and of six species of Anoplura. Rhynchophthirina (3 spp.) occur only on the African and Asian elephant, on the warthog, and on the bushpig.

Fig. 6.19.1: *Pediculus humanus*, (Anoplura, Pediculidae), habitus, anterolateral view, SEM micrograph. Abbr.: ant – antenna, sti – stigma. Scale bar: 500 µm. Courtesy H. Pohl.

Autapomorphies

- Ectoparasitic life style
- Complete loss of wings (Figs 6.19.1, 6.19.2)
- Antennae with only 3–5 segments (Figs 6.19.1, 6.19.2)
- Compound eyes with two or less ommatidia (Figs 6.19.1, 6.19.2)
- Ocelli absent
- Dorsal tentorial arms absent

Taxonomy. Phthiraptera comprise four suborders (Fig. 6.19.3), Amblycera (ca. 1,360 species, five superfamilies), Ischnocera (ca. 3,080 species, five superfamilies), Rhynchophthirina (three species, one family), and Anoplura (ca. 540 species, 5 superfamilies) (Mey 2005). The "Mallophaga", i.e. the non-anopluran Phthiraptera, are paraphyletic. Rhynchophthirina is the sistergroup of Anoplura (true lice in the strict sense), which are characterized by numerous autapomorphies, such as for instance the internalized mouthparts, the labrum transformed into a tunnel-shaped probos-

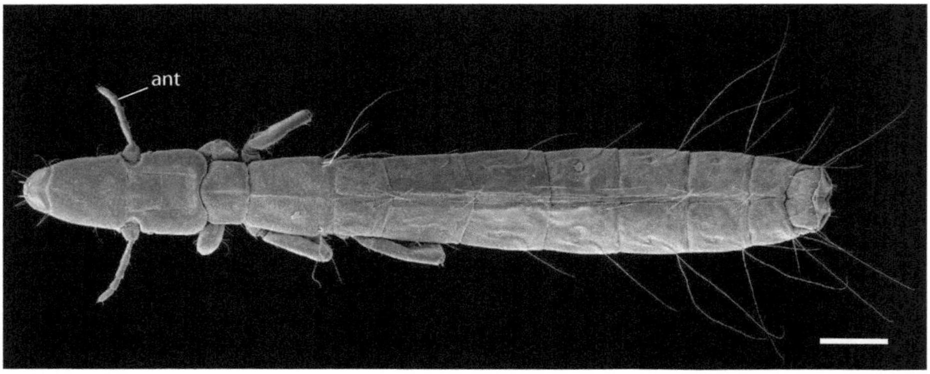

Fig. 6.19.2: *Columbicola columbae*, (Ischnocera, Philoteridae), habitus, dorsal view, SEM micrograph. Abbr.: ant – antenna. Scale bar: 200 µm. Courtesy H. Pohl.

cis, and piercing stylets formed by the labium and hypopharynx. A different sucking-piercing apparatus has evolved in Rhynchophthirina. A sistergroup relationship between Amblycera and the remaining Phthiraptera is suggested by a series of plesiomorphic features preserved in the former group, such as maxillary palps with several segments, a complete tentorium, pterothoracic segments not fused with each other, and free tarsi with two tarsomeres and two equally sized claws.

Diagnosis. Small, secondarily wingless ectoparasites of birds and mammals (Figs 6.19.1–6.19.3). Head prognathous. Antennae short, composed of 3–5 segments. Eyes distinctly reduced or absent. Legs with specialized clasping apparatus (Fig. 6.19.4).

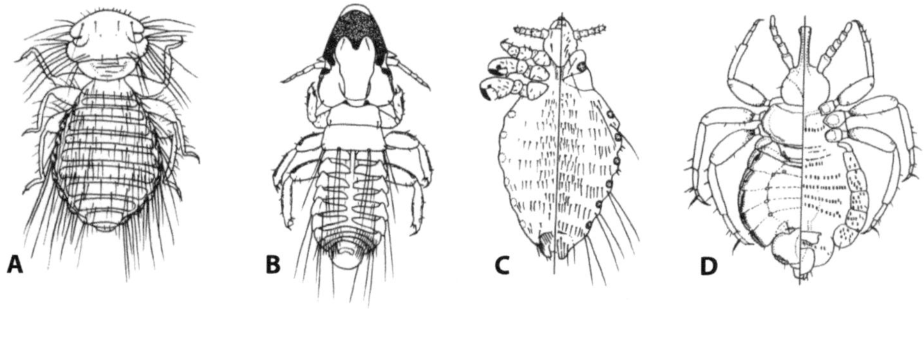

Fig. 6.19.3: Phthirapteran subgroups. A, Amblycera; B, Ischnocera; C, Anoplura; D, Rhynchophtirina (A and B on mammals and birds; C and D on mammals, combined dorsal and ventral view). Redrawn from Mey (2005) and Mauersberger & Mey (1993).

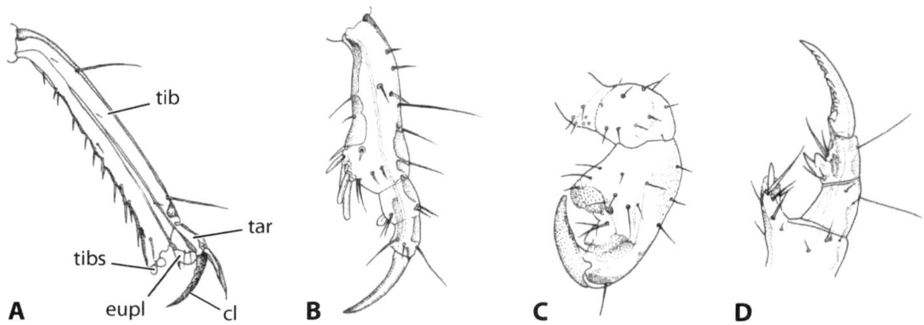

Fig. 6.19.4: Legs, from left to right, A, *Struthioliperus struthionis* (Ischnocera, Philopteridae), hindleg; B, *Trichodectes melis* (Ischnocera, Trichodectidae), hindleg; C, *Linognathus tibialis* (Anoplura, Linognathidae), hindleg; D, *Pediculus humanus capitis* (Anoplura, Pediculidae), foreleg. Abbr.: cl – claw, eupl – euplantulae, tar – tarsus, tib – tibia, tibs – tibial spur. Redrawn from Mey (2005), from v. Kéler (1957) and Mey (1998).

Morphology. The length varies between 0.35 and 11.8 mm (*Laemobothrion maximum*, Amblycera), with the majority of species measuring 2–4 mm. All species are completely wingless and the body is usually dorsoventrally flattened (Figs 6.19.1, 6.19.2). An inconspicuous brownish coloration is characteristic for the group, but bird parasites can be black or white depending on the color of the host's plumage.

The head is prognathous in all subgroups (Figs 6.19.1, 6.19.5), but the mandibles are ventrally oriented in Ischnocera. The head capsule is large, posteriorly broad and usually roughly triangular in Amblycera and Ischnocera, but distinctly reduced in size and more or less evenly rounded laterally in the two other suborders. Hook- or horn-shaped cuticular surface structures are often present, especially in Amblycera and Ischnocera, apparently serving as accessory attachment devices. A frontoclypeal transverse strengthening ridge is absent or present, but always interrupted medially. The postclypeal region is extensive but not as convex as in Psocoptera; in some ischnocerans it is subdivided by transverse and longitudinal stripes of thin cuticle; an inflected anteclypeus is present in the groundplan of Phthiraptera, but more or less strongly reduced in Anoplura; in Ischnocera it bears a median feather- or hair-groove; a strongly elongated rostrum is formed in Rhynchophthirina by an elongation of the clypeus, the genal regions and the postmentum. The frontal and coronal sutures are absent. The head capsule is strengthened by a continuous complex of internal ridges in Amblycera and Ischnocera (phthirapteran groundplan) (Tröster 1990); paired dorsal ridges (lorae) are present on the posterodorsal head region; they are anteriorly continuous with the peristomal ridges. A ligamentous obteraculum of the postoccipital region is usually present but missing in Amblycera; it stabilizes the connection between the head and thorax, thus impeding the removal of the parasite by the host. The maximum number of ommatidia is two, but the compound eyes are completely reduced in many species; ocelli are always absent. The labrum is primarily

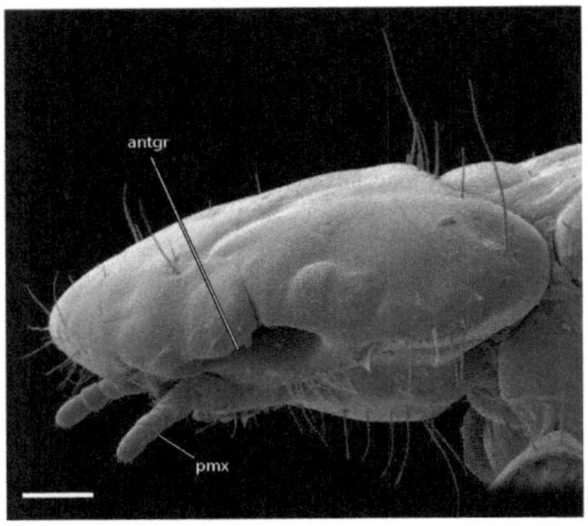

Fig. 6.19.5: *Piagetiella caputincisum*, (Amblycera, Menopodidae), head, lateral view, SEM micrograph. Abbr.: antgr – antennal groove, pmx – palpus maxillaris. Scale bar: 100 µm. Courtesy K. Friedemann.

present but largely membranous; in Ischnocera it is shifted posteriorly on the ventral side of the head capsule; its free lateral areas are bent inwards and overlapping on the ventral side in Anoplura, thus forming the characteristic tunnel-shaped proboscis (haustellum); the labrum is reduced in Rhynchophthirina, together with the maxillae, labium and hypopharynx. The short antennae are 3–5 segmented (Figs 6.19.1, 6.19.2) and quite variable in shape; in male ischnocerans they form clasping organs used during copulation; antennomere 3 is club-shaped in Amblycera, which possess a deep antennal groove on the ventrolateral head region (Fig. 6.19.5); an apical membranous field with sensilla is present in Anoplura. The mouthparts are more or less exposed in Amblycera and Ischnocera (groundplan), and shifted posteriorly in correlation with the inflected anteclypeus; they are largely invaginated in Rhynchophthirina and Anoplura. The axis of the mandibular movements is vertical in Ischnocera but horizontal in the other groups; largely unmodified mandibles with two incisivi, similar to those of Psocoptera, are present in Amblycera and Ischnocera; they serve as cutting or clasping devices; in contrast to psocopteran mandibles the basal molar projections lack grinding surfaces; the mandibles of Rhynchophthirina are small and inserted on the anterior part of the rostrum; the teeth are directed outwards; the mandibles of Anoplura are either vestigial or the dorsal part forms a beak-like piercing device. The maxillae are strongly modified; cardo and stipes are more or less completely fused with the head capsule and postmentum in the groundplan of Phthiraptera; the laciniae are reduced to very small sclerotized rods and lie within a pouch like in Psocoptera; the palp is composed of 2–4 segments in Amblycera but reduced in the other groups; the galea is large and sack-.shaped in Amblycera and Ischnocera and its mesal side bears an armature of minute hooks and bristles; it lies within the preoral cavity in anoplurans and is part of the guiding device for the piercing stylets formed by the hypopharynx (dorsal stylet) and prementum (ventral stylet). The postmentum is fused with the maxillae (see above) and its hind margin is only defined by the position of the posterior tentorial groove; the prementum is free in the groundplan; glossae and paraglossae are vestigial or absent; the palps are 1-segmented in the groundplan; the prementum forms the ventral stylet in Anoplura. The hypopharynx forms the dorsal stylet and contains the salivary channel. The tentorium is nearly complete in Amblycera but the dorsal arm is missing (phthirapteran groundplan); the posterior arms arise from the ventral wall of the head capsule at the hind margin of the postmentum; they separate the postoccipital and hypostomal ridges, which are more or less strongly reduced in Anoplura; the tentorial bridge is present in Amblycera and Ischnocera; the anterior arms are attached to the lateral parts of the transverse frontoclypeal strengthening ridge; they are not connected with the posterior arms in Ischnocera and Anoplura. Occipital apophyses are usually present in Phthiraptera but absent in Amblycera. A cibarial food-processing mechanism similar to that of Psocoptera is present in Amblycera and Ischnocera (epipharyngeal pistill and hypopharyngeal larynx), but largely or completely reduced in Rhynchophthirina and Anoplura. The head musculature is distinctly modified in correlation with modifications of the food uptake apparatus. The cerebrum is bent posteriorly as a result of the flattened head capsule; like the suboesophageal complex it lies in the posterior region of the head capsule. In contrast to Psocoptera only one pair of lateral sclerites is embedded in the cervical membrane of Phthiraptera; paired dorsal cervicalia occur in Anoplura but only an unpaired sclerite is present in Ischnocera.

The thorax is mainly characterized by the complete absence of wings (Figs 6.19.1–6.19.3). The three segments are distinctly separated in some Amblycera, but completely fused in Anoplura and Rhynchophthirina. The prothorax is well-developed, with a shield-like pronotum. The mesothorax is distinctly reduced, especially on its dorsal side; it is usually fused with the metathorax (mesometathorax) but connected with it by an intersegmental membrane in Amblycera. The three thoracic pleura are distinctly developed and form the pleurocoxal articulations ventrally. Sternites are usually present. The three pairs of legs (Fig. 6.19.4) are usually similar in their general configuration; the proleg is usually the shortest and the hindlegs slightly longer than the middle leg; the procoxae are strongly enlarged in Amblycera, whereas the metacoxae of Ischnocera are slightly larger than the others; trochanter and femur are immovably attached to each other; the tibiae are slender in Rhynchophthirina and some other groups, but often short and stout, and equipped with processes which form the proximal part of a clasping apparatus in representatives of Ischnocera; the tarsi are 2-segmented; in most amblycerans they bear two equal, extensible claws and the tarsomeres are movably connected; in Gyropoidea and all non-amblyceran groups they are largely fused with each other and with the tibia, thus forming a tibiotarsus; the unequal claws of ischnoceran bird parasites are not extensible; the large claw with its basal processes forms a clasping apparatus together with the distal part of the tibia (pes harpageus); the smaller claw is missing in Trichodectidae; the forelegs of Rhynchophthirina bear only a single claw, and single claws are present on all pairs of legs of Gliricolidae and Gyropidae; a clasping apparatus formed by the claw and a distinct mesal process of the distal tibia is characteristic for anopluran species. The thoracic musculature is strongly modified in correlation with loss of the flight function; the dorsal longitudinal muscles are completely absent in the mesothorax; a characteristic feature is the presence of thoracic transverse muscles.

The abdomen is short and broad and laterally rounded (e.g., Rhynchophthirina) or more or less elongated (Figs 6.19.1–6.19.3). It is composed of ten (Amblycera, nymphs of some ischnoceran species) or nine segments and characterized by a group-specific pattern of tergites, pleurites and sternites; the tergites and pleurites are often fused (tergopleurites) or reduced in some groups. Segment I is usually distinctly developed in Amblycera and Anoplura, but completely reduced or fused with the metathorax or segment II in the other groups. Segment II is also vestigial in Heptapsogasteridae (Ischnocera). Spiracles are present on segment III–VIII. The terminal segment is likely a product of fusion of X and XI; it bears the anal opening which is usually enclosed by a dense ring of bristles in Amblycera. A genital chamber is formed by an invagination between segments VIII and IX in females and between IX and X in males; a subgenital plate is formed by sternite VIII (or VII and VIII) in females and IX in males, whereas sternites IX (females) and X (males) form the roof of the genital chamber, respectively; a conspicuous armature of bristles at the opening of the female genital chamber is always present but varies in the phthirapteran subgroups. A pair of gonapophyses is sometimes present (e.g., Anoplura) but reduced in most cases. The male genital opening lies on the ventral side of the terminal abdominal region in Amblycera, but on the dorsal side or the abdominal apex in the other groups. Pseudostyli are only present in Trichodectoidea. The male genital apparatus, which is usually large and complex, is composed of a phallobase with an elongate basal plate, paired parameres (or epimeres) articulated or fused with it, and an endophallus, which is folded in its resting position, and everted during copulation; the endophallus usually bears specific cuticular appendages, the phallomeres.

The long oesophagus is extended as an ingluvies in the posterior thoracic region; the intima of the posterior foregut often bears an armature of minute chitinous teeth; this region is connected with a caecum in Ischnocera; food substrate is stored and predigested in this voluminous appendage; a pair of salivary glands associated with nephrocytes is present on both sides of the ingluvies and an additional third unpaired gland in Anoplura (Pawlowsky gland). The testes are composed of two or three (most amblycerans) follicles; they are connected with the vesiculae seminales by 2–3 pairs of short vasa efferentia and one pair of long vasa deferentia. The ductus ejaculatorius is usually unpaired and partly differentiated as a sperm pump (Mey 2005). The ovaries are composed of five polytrophic ovarioles; the glandular walls of the paired oviducts produce secretions for the attachment of the eggs (nits); the unpaired "uterus" opens with the gonoporus into the genital chamber; a simple or forked, balloon-shaped spermatheca is connected with the vagina by the ductus spermathecae, which forms a cup-shaped, sclerotized spermatheca in some amblycerans and ischnocerans.

Biology. Phthiraptera survive only a short time when separated from their host. The association is closer than in any other ectoparasitic insects. Lice spend their entire life cycle on the bird or mammal species they are specialized on and the host specificity and co-speciation is more distinct than in other parasitic groups (Grimaldi & Engel 2005). Suitable hosts are mainly detected using thermosensitive and olfactory sensilla on the short antennae. The ectoparasites are often transmitted during body contact (e.g., copulation) but phoresy does also occur (e.g., with hippoboscid or muscid flies). The number of individuals on a host varies strongly and depends on the size of the host species. Usually it ranges between less than ten and several thousand.

Species of Amblycera and Ischnocera (Fig. 6.19.3A, B) are usually scavengers feeding on particles of skin or feathers, secretions, small quantities of blood, or eggs or nymphs of anoplurans. All species of Rhynchophthirina and Anoplura (Fig. 6.19.3C, D) are specialized on sucking blood with strongly modified mouthparts (see above). In anoplurans the strongly specialized labrum is everted and brought into contact with the host's skin. It functions as a guiding device for the labial and hypopharyngeal stylets. When a blood vessel is perforated blood is sucked into the anterior digestive tract involving the modified semicylindrical mandibles, the maxillae, and strongly developed cibarial and pharyngeal muscles. Saliva is pumped into the perforation using the hypopharyngeal salivary channel.

Phthirapterans are usually found on a single host species (see above) and sometimes even specific body regions, such as for instance the pubic hair of humans or the bill pouch of pelicans. Many bird species are parasitized by several ischnoceran and amblyceran species, whereas only one anopluran species occurs on most mammals. *Rattus turkestanicus* is an exception as the host of six species of Anoplura.

Phthirapterans are mainly controlled by their hosts. Birds use dust baths to eliminate the ectoparasites and intensive grooming behavior occurs in mammals. Some bird parasites live in body parts inaccessible to preening such as for instance the interior of feather shafts.

Reproduction and development. The reproduction of Phthiraptera is usually bisexual but parthenogenesis has evolved independently in several lineages. Ovovivipary is only known from one species (*Meinertzhageniella lata*). Spermatophores are only formed by males of few species. The copulation takes only few seconds in some species but up to two days in some ischnoceran bird parasites (Mey 2005). The copulation posture is suprafeminal in Amblycera (males above females) and subfeminal in the other groups. The comparatively large eggs (nits) are usually cemented to shafts of hairs or feather barbules (Fig. 6.19.6) using specialized secretions of accessory glands. Only females of *Pediculus humanus humanus* deposit the nits on textile fibers. The eggs are equipped with a well-developed operculum and the chorion is often strongly sculptured. They are stalked in few species such as for instance *Haematomyzus elephantis* (Rhynchophthirina). The number is usually low (often ca. 30) but up to 300 nits are deposited by females of *Pediculus humanus capitis*. The first nymphal

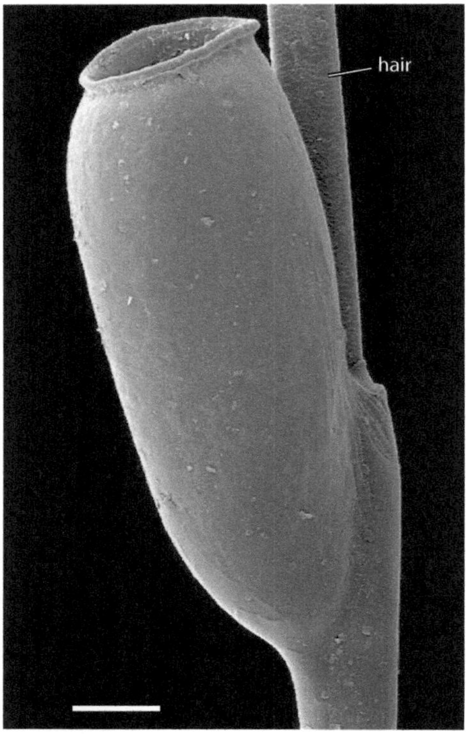

Fig. 6.19.6: *Pediculus humanus*, (Anoplura, Pediculidae), empty egg shell after hatching, SEM micrograph. Scale bar: 100 µm. Courtesy H. Pohl.

instar hatches after 5–18 days using eggbursters on the head. Three molts follow until the adult stage is reached. The nymphs are similar to the adults but smaller and less pigmented. The entire life cycle lasts only between 13 and 43 days. The maximum lifespan is ca. 100 days.

Fossil record. Due to the close association with the hosts the fossil record of Phthiraptera is extremely scarce. The oldest record is strands of hair with nits embedded in Baltic amber. The only body fossil is an amblyceran species found in Eocene oil shale at Messel (Germany).

Economic importance. Phthiraptera can have a significant negative impact on poultry, livestock and humans. Several anopluran species are vectors of important human diseases. The human body louse *Pediculus humanus* is the vector of three epidemics, the trench fever (caused by *Bartonella quintana*), relapsing fever (*Borrelia recurrentis*), and notably epidemic typhus (*Rickettsia prowazekii*), which killed millions of people in past centuries (Krenn & Aspöck 2012).

[v. Kéler (1957, 1969); Tröster (1990); Mauersberger & Mey (1993); Mey (1998, 2005); Grimaldi & Engel (2005); Krenn & Aspöck (2012)]

6.20 Thysanoptera (Greek *thusanos* = tassel, pteron = wing, English common names: thrips, fringe wings)

Diversity and Distribution. Approximately 5,822 species are known at present (see: http://anic.ento.csiro.au/thrips/identifying_thrips/classification.html). The distribution is worldwide including arctic regions. Specialized species occur in high mountain areas up to ca. 5,000 m (*Himalthrips janetscheki* in Nepal) (z. Strassen & Göllner-Scheiding 2005).

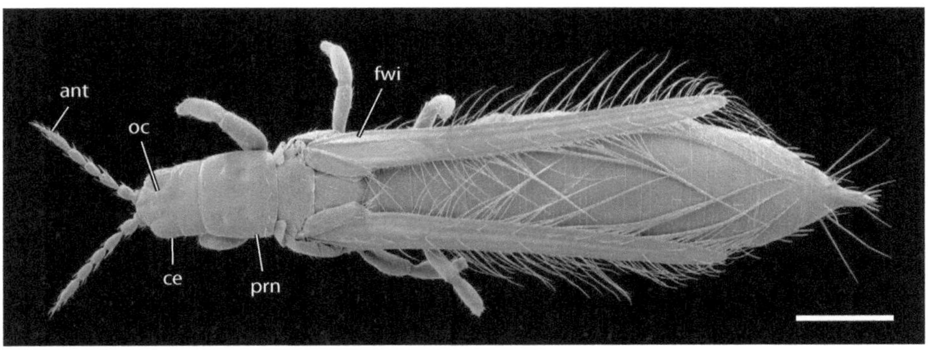

Fig. 6.20.1: *Frankliniella* sp., (Thripidae), habitus dorsal view, SEM micrograph. Abbr.: ant – antenna, ce – compound eye, fwi – forewing, oc – ocelli, prn – pronotum. Scale bar: 200 μm. Courtesy H. Pohl.

Autapomorphies
- Miniaturization
- Mouthcone (rostrum) composed of labrum, maxillae and labium
- Right mandible largely reduced (Fig. 6.20.2, 6.20.3)
- Arolium modified as eversible pretarsal bladder, claws very small 6.20.4)
- Strap-shaped wings with dense fringes of fine hairs ("cilia") (Fig. 6.20.1)
- Only two pairs of abdominal spiracles
- Development with two larval stages without external wing buds and two or three quiescent instars (6.20.5)

Taxonomy. Thysanoptera are subdivided into two suborders, the Terebrantia (eight families, e.g., Aeolothripidae [190 spp], Thripidae [1,970 spp.]) and the Tubulifera (one family, Phlaeothripidae [3,500 spp.!], alternatively 3–9 families in other classifications). A clade Tubulifera is supported by the absence of longitudinal wing veins, marginal wing hairs without sockets, the tube-shaped abdominal segment X, and an eversible, chute-like ovipositor. The monophyly of Terebrantia is disputed as a subordinate placement of Tubulifera within this group cannot be excluded. Potential autapomorphies of Terebrantia are the figure-of-eight shaped sockets of the marginal wing setae, and the abdominal segment VIII completely divided longitudinally on the ventral side (Mound & Morris 2003).

Diagnosis. Small to very small insects with an asymmetric head, a mouthcone, a largely reduced right mandible, an arolium modified as a protrusible vesicle, and narrow strap-like wings with dense fringes of hairs ("cilia") (Fig. 6.20.1).

Morphology. The maximum size known is 14 mm but the length of most species ranges between 1.2 and 2.5 mm. Thysanoptera are usually slender and elongate insects. The coloration is dark in many species and the cuticle is usually well sclerotized.

The head (Figs 6.20.2, 6.20.3) is hypognathous or opisthognathous, with the frons and clypeus on the ventral side and the mouthparts directed posteriorly. It is usually elongate and dorsoventrally flattened particularly in phlaeothripids. The head sutures and ridges are largely reduced, but the dorsal part of the postoccipital suture is always present. The postclypeus is fused with the frons, whereas a separate, more or less membranous anteclypeus is present in terebrantians; the postclypeus is separated from the genal region by a suture on the left side (*Aeolothrips*; Mickoleit 1963) (Fig. 6.20.3). The dorsal head region is separated from the ventral side by the compound eyes and the antennal insertion areas, which are more or less closely adjacent. Three ocelli are primarily present on an elevation between the compound eyes on the dorsal side (Fig. 6.20.1), but they are sometimes reduced or missing, especially in flightless forms; the compound eyes vary in size; the relatively large ommatidia are often separated by distinct chitinous bridges. The short and inconspicuous antennae are composed of scapus, pedicellus and 2–7 unmodified flagellomeres. The labrum is elongate-triangular and sclerotized (Figs 6.20.2, 6.20.3); it articulates with the anterior anteclypeal margin and forms the anterior (ventral) part of the mouthcone, which is posteriorly directed and in close contact with the prosternum in the resting position; the anterolateral labral margins are inflected, thus forming an almost closed tube for the left mandible and the laciniae. The left mandible is stylet-like (potential synapomorphy with Hemiptera); the right mandible is largely reduced (Figs 6.20.2, 6.20.3A). The cardo is primarily present and exposed (*Aeolothrips*) (Fig. 6.20.3B) but usually reduced and fused to the base of the stipes; the stipites form the lateral walls of the mouthcone (Fig. 6.20.3B); the laciniae are modified as stylets and internally grooved; a galea is not recognizable; the palp is 3-segmented in the groundplan (*Aeolothrips*) but up to eight palpomeres can occur. The labium forms the posterior (dorsal) wall of the mouthcone and also the closure of the sucking apparatus formed by the mandibular and maxillary stylets; it is composed of the postmentum and prementum; the palp comprises 1–5 segments

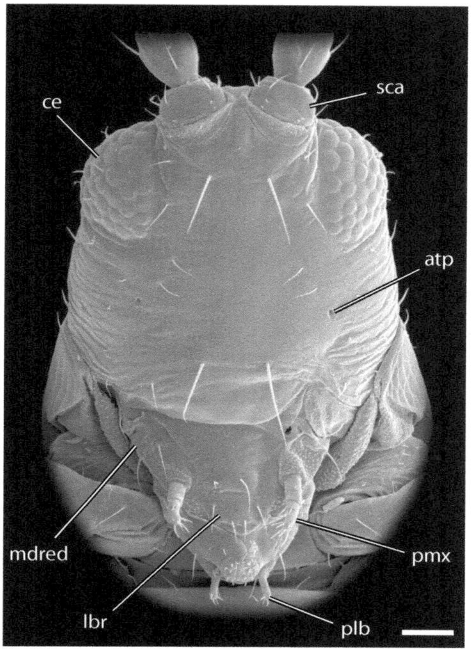

Fig. 6.20.2: *Frankliniella* sp., (Thripidae), head, frontal view, SEM micrograph. Abbr.: atp – anterior tentorial pit, ce – compound eye, lbr – labrum, mdred – reduced right mandible, plb – palpus labialis, pmx – palpus maxillaris, sca – scapus. Scale bar: 25 µm. Courtesy H. Pohl.

and like the paramedial paraglossae it bears sensilla on the distal palpomere; the glossae are fused and form the distal part of the hind wall of the mouthcone; varying degrees of reduction of the labium occur in Tubulifera. The tentorium is relatively complete in the groundplan (*Aeolothrips*), but the dorsal arms are always missing and the posterior parts of the anterior arms are thin and ligament-like; the tentorium is largely reduced in Phlaeothripidae, with all parts missing except for vestiges of the anterior arms (Mickoleit 1963). The mandibular retractor (adductor) is present on both sides, whereas a tentoriomandibular muscles and an unusual M. lacinio-mandibularis are only present on the left side; the musculature of the lacinia is strongly developed; it comprises three stipital protractors and two retractors originating on the genal region; the largest dilator of the cibarium is attached to the epipharynx by an unpaired tendon. A mandibular gland is only present on the left side in terebrantians. The brain is large in relation to body size, especially the protocerebrum. It fills out large parts of the dorsal head lumen. The suboesophageal complex is shifted to the prothorax. The posteriorly directed mouthcone is part of a connecting mecha-

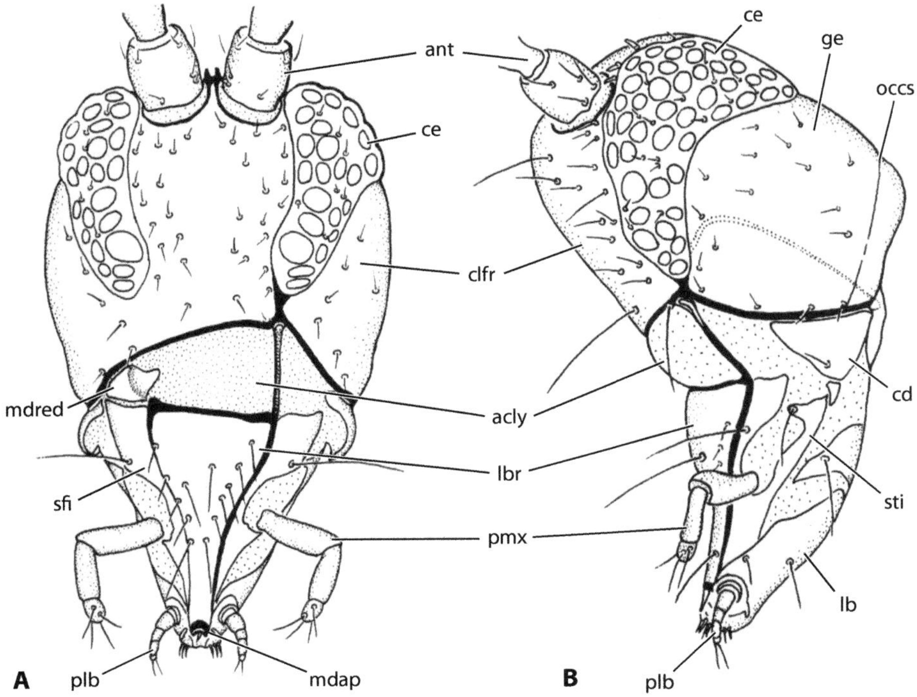

Fig. 6.20.3: *Aeolothrips fasciatus* (Aeolothripidae), head. A, frontal view; B, lateral view. Abbr.: acly – anteclypeus, ant – antenna, cd – cardo, ce – compound eye, clfr – clypeofrons, ge – gena, lb – labium, lbr – labrum, mdap – apex of left mandible apex, mdred – reduced right mandible, occs – occipital sulcus, plb – palpus labialis, pmx – palpus maxillaris, sti – stipes. Redrawn from Kaestner (2005), after E. Mickoleit (1963).

nism between the head and prothorax (see above). The cervical membrane is relatively wide; triangular cervical sclerites are present ventrolaterally.

The thorax of Tubulifera is characterized by a stronger degree of sclerotization with distinctly reduced membranous areas, whereas a greater flexibility is retained in Terebrantia (Mickoleit 1961). The prothorax is large and movable to a certain degree. The pronotum is a well-developed plate; it bears a vestiture of short and few longer setae in *Thrips* (Mickoleit 1961). Large parts of the pleural and sternal regions are membranous. The eusternum and spinasternum are present as distinct sclerites. A protrochantin is present between the coxa, the cervicalia and the lateral pronotal margin. The meso- and metathorax form a closely connected functional unit (pterothorax); the stability is mainly provided by the extensive and undivided sternal plates (eusternum), which are fused with each other in *Phloeothrips*, thus forming a "plastron" (Mickoleit 1961 [note: this term is widely used for the film of air on the outside of the body in aquatic insects]); the spinasternum is missing in both these segments.

The pleural regions vary strongly within Thysanoptera; the large episterna are separated from the epimera by distinct pleural ridges; the very large metepisternum is completely fused with the metasternum in *Thrips* (Mickoleit 1961). The mesonotum is a transverse plate and distinctly smaller than the pronotum. The metanotum is distinctly longer; a separation into a metanotum caudale and craniale is present or absent (Mickoleit 1961). The wings are modified as narrow strap-like structures; the venation is largely or almost completely (Tubulifera) reduced; dense fringes of hairs are present at least at the posterior margin (Fig. 6.20.1); an additional vestiture of microtrichia is present in Terebrantia; different degrees of wing reduction occur, sometimes within species, and sometimes only in males: the wings can be shortened, transformed into small scale-like structures, or absent. The legs are usually short; the forelegs are sometimes thickened and raptorial, and a stridulatory file is present in some Neotropical species; the tarsi are composed of one or two segments; the claws are distinctly reduced in size; the arolium is modified as a protrusible vesicle and associated with an adhesive gland in the tibia (Fig. 6.20.4).

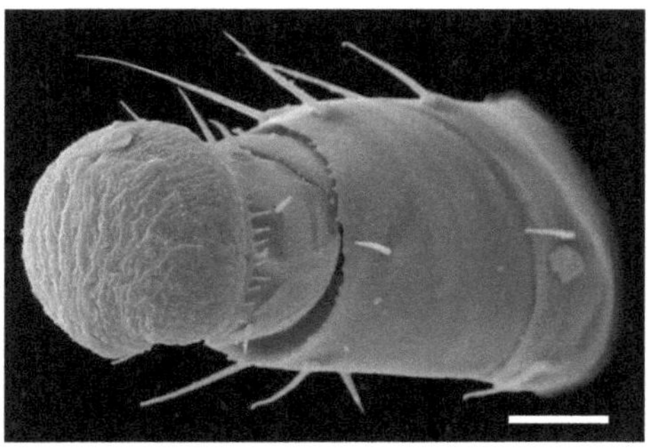

Fig. 6.20.4: *Frankliniella* sp., (Thripidae), distal tarsus and pretarsus, SEM micrograph. Scale bar: 10 µm. Courtesy H. Pohl.

The abdomen is 11-segmented, but segment I and XI are strongly reduced. An ovipositor of two pairs of elongate gonapophyses VIII and IX is preserved in Terebrantia; the paired 1st valves are ventrally serrate and articulate with sternite VIII. Segment X of Terebrantia is longitudinally divided, thus providing space for reception of the ovipositor in its resting position. The ovipositor is more or less strongly reduced in some Terebrantia, and reduced to a small chute-like structure implanted behind sternum VIII and above the subgenital plate in Tubulifera (Heming 1970); it has a paired origin resulting from fusion of valve primordia of segment VIII in the prepupa (Heming 1970: fig. 23F). Segment X is cone-shaped in Terebrantia but tube-shaped in Tubulifera. Segment XI is represented by the unpaired epiproct and paired periprocts. The complex male copulatory organs are located in segment IX and composed of a sclerotized base (periandrum), two pairs of parameres (only in *Aeolothrips* and related genera), the aedeagus with pseudovirga, a membranous epiphallus, and a navicula in some groups; reductions and fusions occur in Tubulifera (z. Strassen & Göllner-Scheiding 2005). Only two pairs of abdominal spiracles are present on segment I and VIII in adults and usually on II and VIII in larvae (Heming 1991).

The heart lies within abdominal segments VII and VIII and is equipped with six muscles arranged in a star-shaped pattern; the aorta cephalica is enclosed by a well-developed muscularis. A valvula cardiaca is present at the foregut-midgut border; the midgut forms a loop in the posterior abdomen. The abdominal ganglia form a single complex as in the other acercarian lineages (autapomorphy of Acercaria). Only four Malpighian tubules are present (autapomorphy of Acercaria).

Biology. Due to their small or very small size Thysanoptera form a characteristic element of aerial plankton and drift passively over distances of hundreds or even thousands of kilometers. Adults and larvae can occur in extreme densities. Both are often found on flowers, but also on leaves, under bark of recently-dead branches, in leaf litter, and associated with grass (Heming 1991). Monophagous and oligophagous species occur, but many thrips are generalists. Feeding habits vary greatly. Most species use their specialized mouthparts to suck contents of cells of flowers or other plant parts or also pollen but not of vascular tissues. Other food sources are algae, fungal hyphae, and lichens. Many species of Phlaeothripidae (Idolothripinae) are specialized to feed on fungal spores by means of very broad laciniae. Some species are predaceous (ca. 6%), especially in the families Aeolothripidae and Phlaeothripidae. Prey includes eggs of insects and mites, smaller thrips, plant lice, mites, very small lepidopteran caterpillars, and minute nematodes. Predators of Thysanoptera are carnivorous beetles, ants, heteropterans (e.g., Miridae), spiders, pseudoscorpions, and others. They are also affected by endoparasitic hymenopterans, nematodes and fungi.

An interesting and not commonly known aspect of the biology of Thysanoptera is the very broad spectrum of behavior from solitary to gregarious, colonial, subsocial to eusocial, as pointed out by Grimaldi & Engel (2005) “a sophistication unexpected for insects commonly assumed to spend their life merely sucking plant juices”. A lichen-feeding species from Panama forms communal “bivouac” sites with up to 200 individuals and adults protecting larvae and eggs. Advanced social behavior with different castes occurs in the phlaeothripid sistergroups *Oncothrips* and *Kladothrips* (Kranz et al. 2002).

Reproduction and development. Thysanoptera are primarily bisexual, but very often have haploid-diploid reproduction (=arrhenotoky) in which males develop from unfertilized, haploid eggs (Heming 1995). An alternative also frequently occurring within the group is thelytoky, with only females hatching from the eggs. Copulation

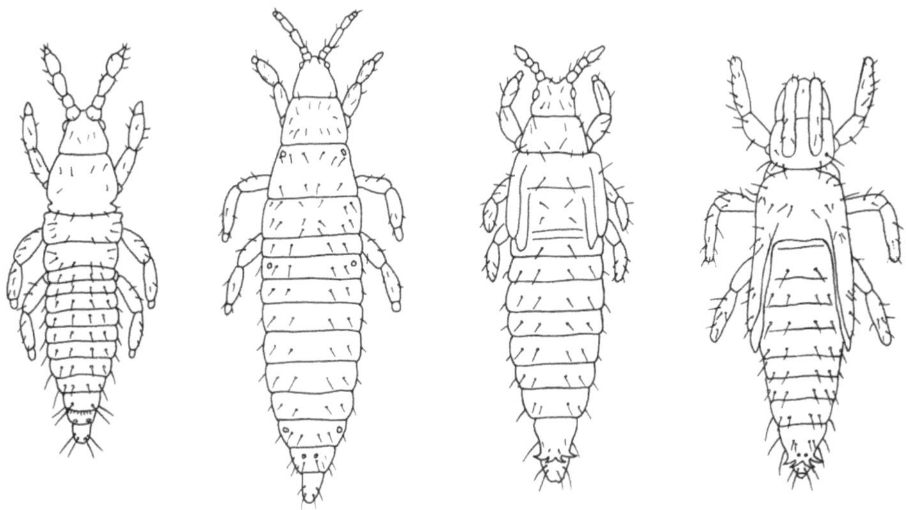

Fig. 6.20.5: Thysanoptera, immature stages. Redrawn from z. Strassen & Göllner-Scheiding (2005).

can last for up to one hour. In some species females reach sexual maturity and copulate before the adult stage (Bournier 1956). Eggs are deposited in plant tissue using the ovipositor by females of Terebrantia. Tubuliferan eggs, which are protected by a well-developed and distinctly sculptured chorion, are deposited on the surface of plants or deposited in narrow crevices of bark or other suitable substrates (z. Strassen & Göllner-Scheiding 2005). Thysanoptera are almost always oviparous, but vivipary is known from species of tropical Phlaeothripidae.

The rate of development in Thysanoptera can vary strongly depending on environmental conditions, especially temperature and nutrient quality of the food sources. The first instars hatch from the exceptionally small eggs (ca. 0.2–0.5 mm) after two to ten days in most species depending on temperature. The postembryonic development of Thysanoptera differs very distinctly from that of most other hemimetabolan lineages but shows affinities with patterns found in subgroups of Sternorrhyncha (Coccina, Aleyrodina) and Holometabola (Heming 1991, 1993, 2003; Moritz 1991; Grimaldi & Engel 2005) (Fig. 6.20.5). The first two instars are apterous, slender and agile larvae. Stage III is a so-called propupa (sometimes referred to as prepupa) with external wing buds or wing buds within the propupal cuticle in the case of Phlaeothripidae (Heming 2003). Like the following instar IV – the "pupa" – it is quiescent, non-feeding and largely immobilized (Fig. 6.20.5). A cocoon is spun at the end of instar II in some species (Heming 1970: anal gland opening in 2nd instar larvae). An additional quiescent instar V (pupa II) occurs in Phlaeothripidae. The mouthparts of the "pupal" stages are non-functional and the antennae are more or less completely reduced (Heming 1975, 1991).

Fossil record. The oldest fossils associated with thrips are Lophioneuridae, a group documented from the Permian to the Cretaceous. However, as pointed out by Grimaldi & Engel (2005), they form the paraphyletic stem group of Thysanoptera. The oldest true thysanopterans, *Triassothrips virginicus* and *Kazachothrips triassicus*, are approximately 220 Ma old (Grimaldi & Engel 2005). In their wing morphology they have preserved some plesiomorphic features compared to extant members of the group. Lower Cretaceous representatives resembling extant thrips are preserved in Lebanese amber. Most fossil thysanopteran species are known from Baltic amber, most of them belonging to extant families.

Economic importance. Thysanoptera play a positive role as pollinators, but many phytophagous thripid species can seriously affect commercial crops including cereal, rice, citrus, coffee, cocoa, banana, onions, cotton, tobacco, and others. Feeding of thrips results in discoloration and deformities and additional damage can be caused by soiling plant surfaces with feces and the formation of galls. Especially the transmission of different plant diseases can have very negative effects. Thrips are vectors of more than 20 plant infecting viruses and some of them belong to the most damaging known plant pathogens. Secondary infections by pathogenic fungi can cause additional damage. Well known viruses transmitted by thysanopteran species are the tomato spotted wilt virus (TSWV) and the Impatiens necrotic spot virus (INSV). The cosmopolitan western flower thrips *Frankliniella occidentalis* is considered the primary vector of plant diseases caused by Tospoviruses.

[G. Mickoleit (1961); E. Mickoleit (1963); Priesner (1968); Heming (1970, 1975, 1991, 1993, 1995, 2003); Lewis (1973, 1997); Moritz (1991, 2006); Childers (1997); Kirk (1997); Kranz et al. (2002); Mound & Morris (2003); Grimaldi & Engel (2005); zur Strassen & Göllner-Scheiding (2005)]

Hemiptera (Auchenorrhyncha, Sternorrhyncha, Coleorrhyncha, Heteroptera)

6.21 **Auchenorrhyncha** (Cicadina, Greek *auchen* = neck, *rhynchos* = snout, English common names: Cicadas, leafhoppers, planthoppers, froghoppers or spittle bugs, treehoppers)

Diversity and Distribution. Auchenorrhyncha are the largest subgroup of Acercaria, with more than 45,000 described species. They occur worldwide and many species reach the margins of the distribution of vascular plants in montane, Arctic and Antarctic regions. However, the diversity is highest in the tropics and subtropics.

Autapomorphies. The monophyly is presently disputed but appears likely considering the morphological evidence. The following features are potential synapomorphies of the two subgroups Fulgoromorpha and Cicadomorpha.
- Complex tymbal acoustic system (tymbal and tympanicum)
- Bristle-shaped antennal flagellum (aristate antenna) (Figs 6.21.1, 6.21.2)
- Median plate at the wing base membranized or reduced

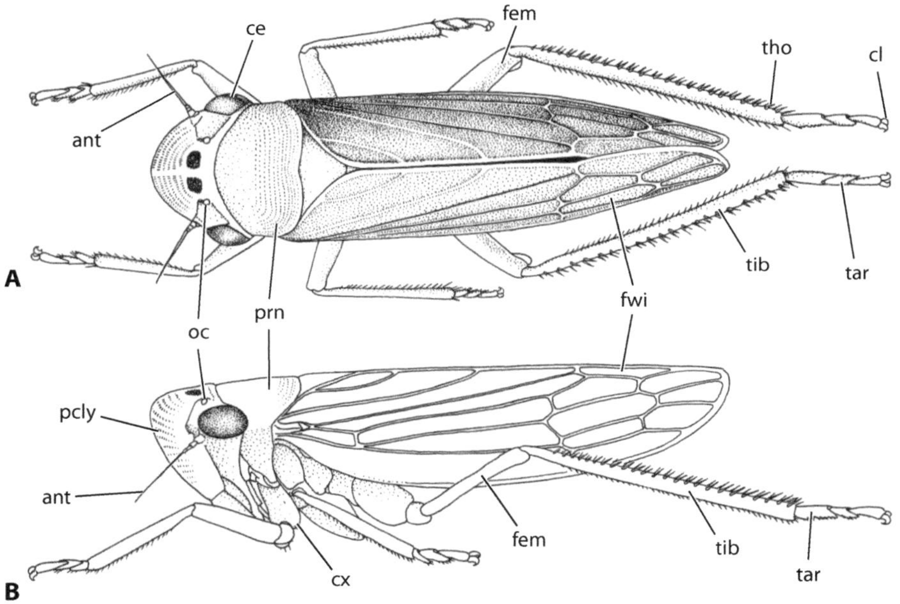

Fig. 6.21.1: *Cicadella viridis* (Cicadellidae), habitus. A, dorsal view; B, lateral view. Abbr.: ant – antenna, ce – compound eye, cl – claw, cx- coxa, fem – femur, fwi – forewing, oc – ocellus, pcly – postclypeus, prn – pronotum, tar – tarsus, tib – tibia, tho – thorns. Redrawn from Strümpel (2005).

Taxonomy. The Auchenorrhyncha are divided into the two suborders Fulgoromorpha (ca. 12,000 spp.) and Cicadomorpha (ca. 33,500 spp.). The former contains a single superfamily Fulgoroidea with ca. 20 families, such as for instance Delphacidae (more than 2,000 spp.), Cixiidae (ca. 2,000 spp.), Flatidae (ca. 1,000 spp.), Fulgoridae (ca. 700 spp.), Derbidae (ca. 1,500 spp.), Issidae (ca. 1,500 spp. [probably paraphyletic]), Gengidae (only two African species), and Hypochthonellidae (only one blind endogean African species). Cicadomorpha comprise the three superfamilies Cercopoidea (e.g., Cercopidae [ca. 1,500 spp.]), Membracoidea (=Cicadelloidea) (e.g., Cicadellidae [=Jassidae] [ca. 25,500 spp.], Membracidae [ca. 3,300 spp.]) and Cicadoidea (Cicadidae [ca. 2,500 spp.], Tettigarctidae [two species]) (Strümpel 2005; Moulds 2005).

Diagnosis. Hypognathous head with large clypeus and rostrum. Antenna bristle-like. Pronotum folded down laterally and posteriorly overlapping more or less extensive parts of the mesonotum. Wings held in a roof-like position (Fig. 6.21.1). Forewings almost always membranous.

Morphology of adults. The size varies between 1 and 95 mm. Most species are small (usually less than 10 mm) but large or very large forms occur in some groups (e.g., Cicadidae, Fulgoridae). The body shape is usually oval or moderately oblong. Conspicuous outgrowths on the head and prothorax are present in some groups. Wing reductions occur frequently and sometimes within one species. Many species are inconspicuous in their appearance but forms with very conspicuous patterns of shape and coloration occur.

The head is hypognathous and more or less immovably connected with the prothorax (Fig. 6.21.1). The posterior part of the head capsule is incompletely sclerotized. A distinctly delimited frons and a coronal suture are present in the groundplan, but the former is often fused with the posterior clypeus, especially in Membracidae. The clypeus is enlarged in correlation with the presence of a cibarial sucking pump with

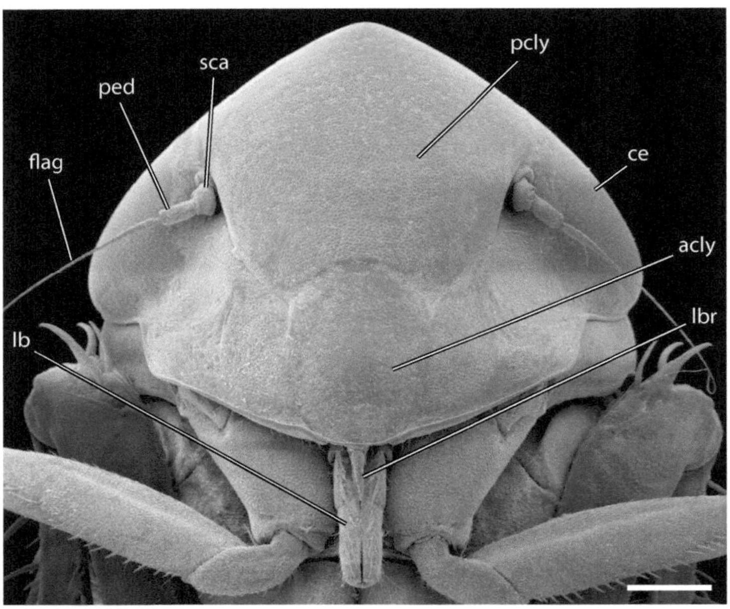

Fig. 6.21.2: *Cicadellidae* (Cicadomorpha), head, anteroventral view, SEM micrograph. Abbr.: acly – anteclypeus, ce – compound eye, flag – flagellum, lb – labium, lbr – labrum, pcly – postclypeus, ped – pedicellus, sca – scapus. Scale bar: 250 µm. Courtesy H. Pohl.

strongly developed muscles; it forms a part of the roof of the head capsule and is subdivided into a postclypeus and anteclypeus (Fig. 6.21.2); the homology of the mandibular and maxillary plates, sclerites or lobe-like structures associated with the clypeus and genae, respectively, is unclear. The compound eyes are usually well-developed but reductions occur in cave-dwelling species; three ocelli are present in most cases; the ocelli and antennal bases lie below the compound eyes in Fulgoromorpha but between or above them in Cicadomorpha (Fig. 6.21.1A). The scapus is usually small; the pedicellus is often enlarged and densely covered with sensilla placodea in Fulgoromorpha (sensilla placodea absent in Cicadomorpha); the antennal flagellum is strongly shortened and bristle-shaped. The labrum covers the base of the rostrum (Fig. 6.21.2). A rim on its ventral side (epipharynx) is continuous with the pharynx posteriorly; the mandibles are transformed into long stylets like the laciniae (autapomorphy of Hemiptera) (Fig. 6.C.2B); two retractor muscles insert on the mandibular base. The laciniae enclose a dorsal food canal and a narrower ventral salivary canal; the maxillary palps are completely reduced (autapomorphy of Hemiptera). The hypopharynx between the proximal parts of the stylets contains the salivary pump. The labium is usually composed of four segments and modified as a rostrum enclosing the four stylets (autapomorphy of Hemiptera); apically it bears mechanoreceptors; labial palps are absent (autapomorphy of Hemiptera); the proximal labial margin is in contact with the cervical membrane. A gula is not present. The tentorium is usually composed of posterior and anterior arms and a bridge; the dorsal arms are present in some cases.

The mesothorax is usually the largest thoracic segment but the prothorax is very strongly extended in Membracidae and Machaerotidae. The pronotum is relatively short in Fulgoromorpha but distinctly enlarged in Cicadomorpha; it covers at least the entire anterior mesonotal margin in many Cicadomorpha and often the entire mesonotum or even the entire abdomen (some Membracidae). The large mesotergum is convex and subdivided into the prescutum, scutum, scutellum and postnotum; the prescutum forms the internal anterior mesophragma and the postnotum the mesopostphragma. The wings are held in a roof-like position at rest; the forewings are usually membranous but an increased degree of sclerotization occurs in some groups (Cercopidae, Membracidae part., Fulgoridae part.); a tegula is present at their base in Fulgoromorpha; the anal field is distinctly separated from the main part of the forewing and referred to as clavus (groundplan autapomorphy of Hemiptera); both wing pairs are usually connected by small hooks during flight (functional dipterism); the indirect flight muscles are usually strongly developed, especially in the mesothorax; the smaller metathorax and the hindwings play a minor role as elements of the flight apparatus. The coxae are elongated and widely separated in Fulgoromorpha; all tarsi are 3-segmented; an arolium is present in some groups (Cercopidae, Cicadidae); the hindlegs are usually modified as jumping legs (not in Cicadidae); whorls of tibial thorns and thorns arranged in a row on the ventral side of the tarsi are present in jumping forms (Fig. 6.21.1).

The abdomen is primarily composed of 11 segments but only ten or nine are visible in adults. The anterior segments are modified in correlation with the presence of the tymbal acoustic organs composed of a tympanum and a tymbal formed by thin plates that vibrate; it is located on either sides of tergite I or II and associated with strongly developed dorsoventral muscles, which are involved in the sound production; the sound is usually inaudible but very loud in Cicadidae, which possess elaborate and large tymbals; typical tympanic organs are always present in Cicadoidea in paired grooves on the ventral side of the abdomen. The terminal abdominal segments are usually fused and form a moveable anal cone, which is used for spraying feces. Segment IX is distinctly modified in males and bears the external genital organs; lateral pygophor lobes and ventral subgenital plates delimit the genital chamber of this segment. The copulatory organ is formed by an unpaired penis and usually by paired parameres (homology rather unclear), which are sometimes replaced by harpagones ("styli") as clasping organs. The ovipositor of the female genital segments VIII and IX is usually well-developed and composed of three connected pairs of valves; it is shortened or vestigial in some groups of Fulgoromorpha.

The central nervous system is strongly concentrated like in all other groups of Acercaria; the abdominal complex is fused with the metathoracic ganglion. The strongly modified digestive tract (Fig. 6.21.3) comprises the unpaired maxillary food channel, the strongly developed cibarium with its pumping apparatus, and the gut; the saliva, which is composed of two components, plays a very important role in the food uptake; many species possess a so-called filter chamber, which forms a "bypass" between the anterior and posterior end of the midgut (closely adjacent epithelia); this results in a rapid and selective transport of the liquid food substrate in the case of phloem feeders (similar mechanisms in phloem feeding sternorrhynchans); the apical parts of the Malpighian tubules are also involved in this mechanism. The paired testes comprise between three and 35 follicles. Different accessory glands occur in females. Females of some groups (e.g., many Cicadoidea) possess a separate copulatory opening on sternum VIII (copulaporus) and an oviporus at the base of the ovipositor. The number of the telotrophic ovarioles varies strongly.

Biology. Cicadina feed almost exclusively on vascular fluids of plants. Only very few species feed on fungi (Grimaldi & Engel 2005). Endosymbioses with bacteria and yeasts are obligatory (with the exception of Typhlocybinae). The nymphs of several groups live in soil substrate and feed on roots (e.g., Cicadoidea, Cixiidae). The nymphs of Cercopidae produce a cover of frothed-up plant sap resembling spit (spittle bugs). This substrate prevents drying up and provides protection against predators. Nymphs of Fulgoroidea produce long strands of wax from the abdomen as protective device (Grimaldi & Engel 2005).

Sphecids (Hymenoptera) use species of Cicadellidae, Cercopidae and Cicadoidea as food for their larvae. Important parasitoids of auchenorrhynchans belong to Dryinidae, Embolemidae (Hymenoptera) and Pipunculidae (Diptera). Females and larvae

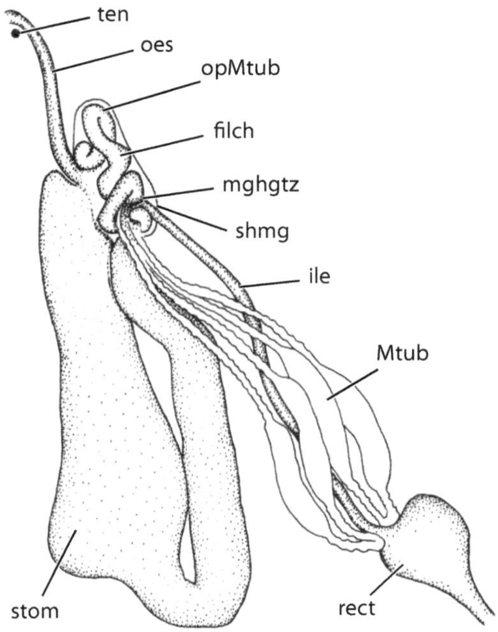

Fig. 6.21.3: Digestive tract of *Gargara* sp. (Membracidae). Abbr.: filch – filter chamber, ile – ileum, mghgtz – midgut-hindgut transition zone, Mtub – Malpighian tubules, oes – oesophagus, opMtub – opening area of Malpighian tubules, rect – rectum, shmg – sheath around part of midgut, stom – stomach, ten – tentorial bridge. Redrawn from Fick (1981).

of the strepsipteran families Halictophagidae und Elenchidae parasitize species of Delphacidae and Cicadellidae. Eggs are parasitized by strongly miniaturized hymenopterans of the families Mymaridae and Trichogrammatidae.

Reproduction and development. Only few species are parthenogenetic. Males and females usually copulate in a V-shaped position. In the typical case the eggs are deposited in the tissue of the host plant, but in some groups in soil or exposed on the surface of stones or rocks. The postembryonic development comprises five stages. The nymphs are usually very similar to the adults. Exceptions are the burrowing legs of the nymphs of Cicadoidea or the air channels of the nymphs of Cercopoidea. Two generations per year are the typical pattern for species of the temperate zones, but up to ten generations occur in some tropical forms. *Magicicada septendecim* (North America) is known for its developmental cycle of 17 years! The adults hatch in extreme masses thus ensuring the survival of enough males and females despite of strong predation pressure.

Fossil record. The oldest fossils from Jurassic deposits belong to Cixiidae. The closely related family †Lalacidae was extremely diverse in the Cretaceous (Grimaldi & Engel 2005). Auchenorrhynchans are also represented in amber from Lebanon, Burma or other countries.

Economic importance. Species of several groups, especially in Fulgoromorpha, are very serious pests of cultivated plants. The greatest damage is caused by the transmission of plant viruses (e.g., Delphacidae: *Javesella pellucida*) and mycoplasmosis.

[Fick (1981); Strümpel (1983); Grimaldi & Engel (2005); Moulds (2005); Strümpel (2005); Cryan & Urban (2012)]

6.22 Sternorrhyncha (Greek *sternos* = chest, *rhyncos* = snout, English common names: plantlice)

Diversity and Distribution. About 16,400 species are described. The distribution is worldwide.

Autapomorphies
- Stylets elongated, forming a loop (Fig. 6.22.1)
- Base of the rostrum shifted posteriorly, between procoxal bases (Fig. 6.C.2)
- Two tarsomeres (groundplan)
- Clavus of forewing reduced (at most one anal vein)
- Frons of embryos with eggburster

The entire group and the four suborders (Figs 6.22.1.1, 6.22.2.1, 6.22.3.1, 6.22.4.1) are well-supported as a clade (e.g., Strümpel 2005; Cryan & Urban 2012). A sistergroup relationship between Sternorrhyncha and the remaining groups of Hemiptera has been recently suggested based on molecular data (Cryan & Urban 2012). Psylloidea strongly resemble smaller species of Auchenorrhyncha. They are either the sistergroup of Aleyrodoidea (Psyllomorpha) (morphological evidence; e.g., Strümpel 2005) or the sistergroup of the entire remaining Sternorrhyncha (molecular data; see e.g., Grimaldi & Engel 2005). Aphidoidea and Coccoidea likely form a clade Aphidomorpha. Potential synapomorphies are the reduction of the ovipositor and the transverse, slit-like oviporus.

A characteristic feature of Sternorrhyncha are strongly elongated stylets forming a loop which are usually kept in a caudally directed pouch-like extension of the preoral space (crumena) at rest (Fig. 6.22.1: crum). The insects are often protected by wax-like substances or secrete hard external shells (females). They have a strong tendency towards a specialized reproductive biology, with parthenogenesis and ovovivipary or vivipary. They are often strongly dimorphic or polymorphic, usually

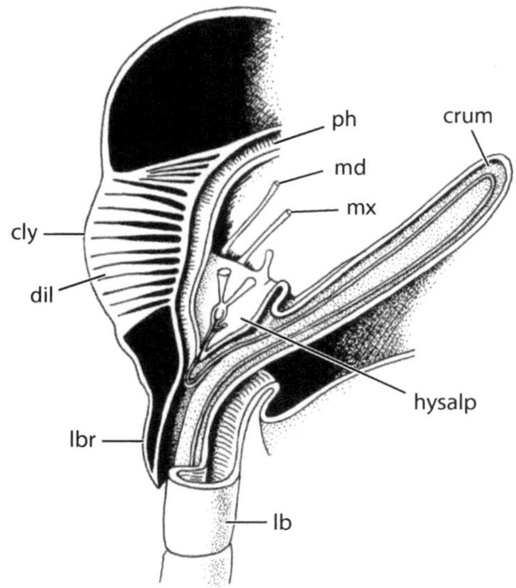

Fig. 6.22.1: Sternorrhyncha, head, sagittal section. Abbr.: cly – clypeus, crum – crumena, dil – dilator, hysalp – hypopharynx with salivary pump, lb – labium, lbr – labrum, md – mandibular stylet, mx – maxillary stylet, ph – pharynx. Redrawn from Stümpel (2005).

with very strongly modified and more or less sedentary females. A typical feature of phloem-feeders (including Auchenorrhyncha) is the secretion of honeydew. This sugar-rich sticky substance is released by the anus in order to dispose of excessive liquid absorbed from the phloem. On plant surfaces it is an ideal medium for the growth of certain fungi (Capnodiales, sooty mold) which can cause great damage on cultivated plants. The amount of honeydew produced by sternorrhynchan phloem feeders can be excessive, exceeding the body weight several times per day (Grimaldi & Engel 2005). Honeydew is intensively collected by ants, which in exchange offer protection against predators and parasitoids of sternorrhynchan colonies, especially to the immobile females (Grimaldi & Engel 2005). This mutual benefit has resulted in quite intensive ecological and evolutionary interactions between both non-related groups of insects.

Despite of the undisputed monophyly, Sternorrhyncha is very heterogeneous morphologically and also in other aspects (Strümpel 1983, 2005). Therefore the four suborders are treated separately in the following.

6.22.1 **Psyllina, Psylloidea** (English common names: psyllids, jumping plantlice, lerp insects)

Diversity and Distribution. About 4,000 described species are arranged in eight families. It is estimated that the group contains at least twice as many species (Burckhardt & Ouvrard 2012; D. Burckhardt, pers. comm.). The species rich families Aphalaridae, Liviidae, Psyllidae (Fig. 6.22.1.1) and Triozidae have a worldwide distribution from the tropics to subarctic and subantarctic regions. The small families Phacopteronidae, Carsidaridae and Homotomidae are predominantly pantropical, and Calophyidae are restricted to the New World as well as tropical and subtropical Asia and Australia (Burckhardt & Basset 2000; D. Burckhardt, pers. comm.).

Fig. 6.22.1.1: Psyllidae (Psyllina), habitus, lateral view, SEM micrograph. Abbr.: ant – antenna, ce – compound eye, froc – frontal cone, fwi – forewing, prn – pronotum. Scale bar: 250 μm. Courtesy H. Pohl.

Autapomorphies
- Very good jumping capacity (?) (arguably a synapomorphy of Psylloidea and Aleyrodoidea)
- Nymphs with cephalothorax (Fig. 6.22.1.4)
- Legs of nymphs modified as short clasping organs
- Sperm pump (?) (arguably a synapomorphy of Psylloidea and Aleyrodoidea)

Taxonomy. The group contains the species rich Aphalaridae, Liviidae, Psyllidae and Triozidae, and four families with a low diversity (Phacopteronidae, Homotomidae, Carsidaridae, Calophyidae). Most species occur in the tropics.

Diagnosis. Superficially resembling small cicadas and less strongly modified than other groups of Sternorrhyncha (Fig. 6.22.1.1). Males and females often differ in size, sometimes also in coloration, and rarely in other characters (forewing venation, male sexual characters on the legs). Wings usually present, but hindwings sometimes shortened or rarely absent.

Morphology of adults. Between 1 and 10 mm, usually 2–4 mm. Often brownish or yellow mixed with green, but there are also very colorful species.

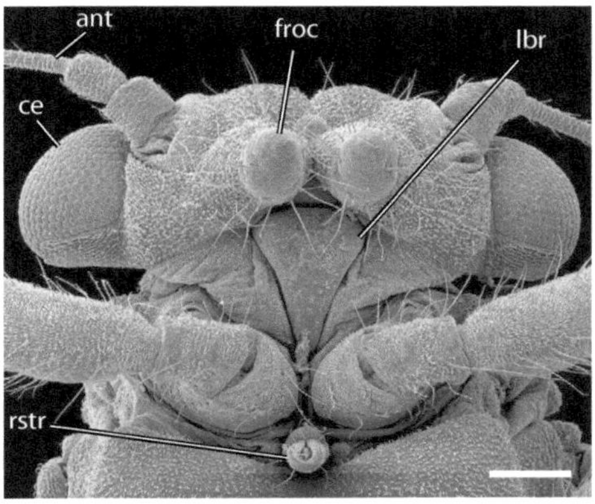

Fig. 6.22.1.2: Psyllidae (Psyllina), head, ventral view, SEM micrograph. Abbr.: ant – antenna, ce – compound eye, froc – frontal cone, lbr – labrum, rstr – rostrum. Scale bar: 100 µm. Courtesy H. Pohl.

The head is very distinctly hypognathous (Fig. 6.22.1.2). Well-developed compound eyes and three ocelli are present; the median ocellus lies on the frons which is often strongly reduced. The ventral side of the head is mainly formed by the genae, which often bear conspicuous cone-shaped protuberances (genal processes) with setae (probably sensorial); the frons is sometimes also included. The antennae are usually 10-segmented; the flagellomeres are equipped with different sensilla and olfactory organs (rhinaria). The stylets are distinctly longer than the labial rostrum and lodged in a ventral fold at rest (crumena). The labial base is seemingly inserted posterior to the procoxae (Fig. 6.22.1.2).

The thorax is dorsally flat or strongly convex (Fig. 6.22.1.1). The mesothorax is larger than the pro- and metathorax and contains well-developed flight muscles. The large forewings are connected with the hindwings (often distinctly smaller) by small hooks during flight (functional dipterism); the venation of the hindwings is strongly reduced. The forewings are always well-developed; partial or complete reduction of the hindwings is an exception. The hindlegs are jumping legs with enlarged coxae which are fused with the segment; the tarsi are 2-segmented; apically a bilobed arolium is present.

The abdominal base is slightly constricted due to a partial reduction of the basal segments. A sperm pump is present in males (Fig. 6.22.1.3) and a complicated receptaculum seminis in females. Mycetomes close to the gut contain endosymbiotic bacteria.

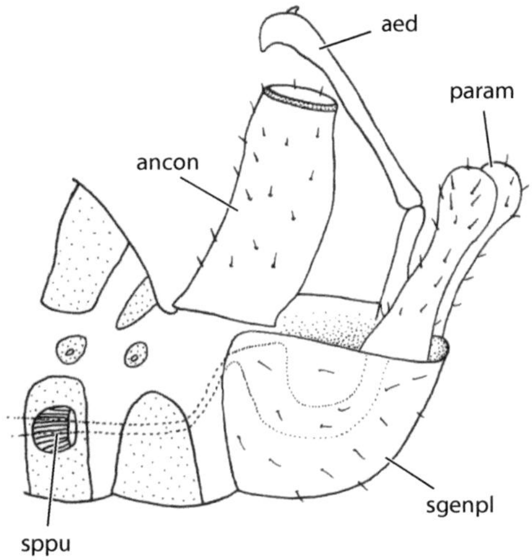

Fig. 6.22.1.3: Psyllina, male postabdomen with external genital organs, lateral view. Abbr.: aed – aedeagus, ancon – anal cone, param – paramere, sgenpl – subgenital plate, sppu – sperm pump. Redrawn from Hodkinson (1971).

Biology. In some respects Psylloidea are less specialized than the other groups of Sternorrhyncha, but they are more host specific (D. Burckhardt, pers. comm.). Tropical and south temperate psyllids are mostly polyvoltine, whereas those of the northern hemisphere are often bi- or monovoltine. The adults of most species are mobile and sometimes leave their host plant after molting from the last immature stage. Some Holarctic species overwinter as adults on conifers and return to their host plants in spring for mating and oviposition. Most psylloid species are monophagous or oligophagous on woody eudicots with only a few species associated with monocots or conifers. Many Holarctic species are associated with herbaceous plants. Often related psyllid species develop on related host plants. Several families or subfamilies of psyllids are entirely or predominantly restricted to a single host taxon. Anthocoridae (Heteroptera), Coccinellidae (Coleoptera), Chrysopidae (Neuroptera) and Syrphidae (Diptera) are important predators, and species of Encyrtidae and Eulophidae (Hymenoptera) are known as parasitoids. Many psyllid species induce galls, particularly in the tropics.

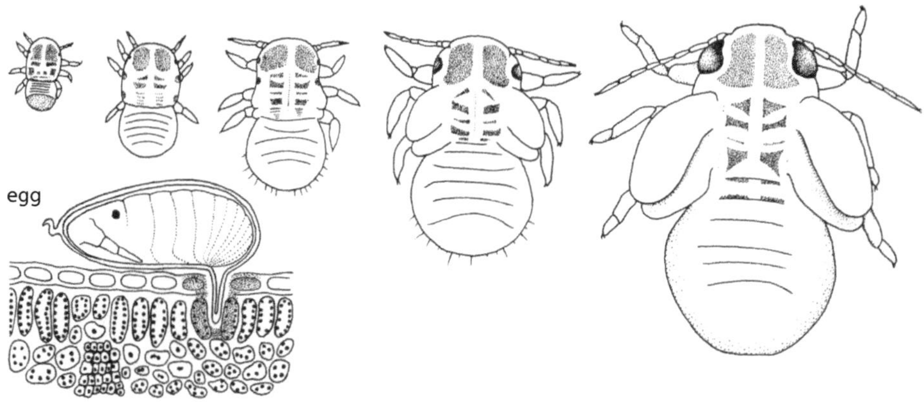

Fig. 6.22.1.4: *Cacopsylla mali* (Psyllina, Psyllidae), 5 nymphal stages, egg, with egg stalk anchored in plant tissue. Redrawn from Strümpel (2005).

Reproduction and development. The reproduction is usually sexual but facultative parthenogenesis has been reported in a few cases (few species of *Cacopsylla* and *Glycaspis*). Eggs are deposited on plant surfaces or in buds and the pedicel (egg stalk) is usually inserted in the plant tissue. Five nymphal stages occur. The nymphs (larvae) differ distinctly from the adults in being strongly dorso-ventrally flattened. The anterior part of the body forms a cephalothorax and the legs are short clasping organs (Fig. 6.22.1.4). They often possess specialized setae, some of them connected with wax secretion. The abundance varies depending on the group. Some species use secretions to produce hardened shells (lerps) providing better microclimatic conditions.

Fossil record. The fossil record of true Psyllina is poor, and the oldest fossils are known from the Eocene. Early fossils described from the Jurassic and Cretaceous (†Liadopsyllidae, †Malmopsyllidae) constitute the stem or sister group of recent Psylloidea (Ouvrard et al. 2010).

Economic importance. A few species are serious pests of cultivated plants as vectors of plant pathogens (bacteria: *Candidatus* Liberibacter and Phytoplasma species). Other species, especially in the nymphal stages, damage their hosts by consuming large quantities of plant sap, by secretion of honeydew soiling fruits or forming the substrate of sooty mold, by toxic substances in their saliva, and by the induction of galls.

6.22.2 Aleyrodina, Aleyrodoidea (English common name: whiteflies)

Diversity and Distribution. Approximately 1,500 species are described. The distribution is world wide with the highest diversity in the tropical regions and only few species in the cooler temperate zones. Some pest species are cosmopolitan. Those occurring in the temperate regions are usually found in green houses.

Autapomorphies

- Only egg larva mobile, legs of the following nymphal stages strongly reduced
- Fourth nymphal stage forms puparium
- Vasiform orifice for expelling honeydew present at the abdominal apex of nymphs

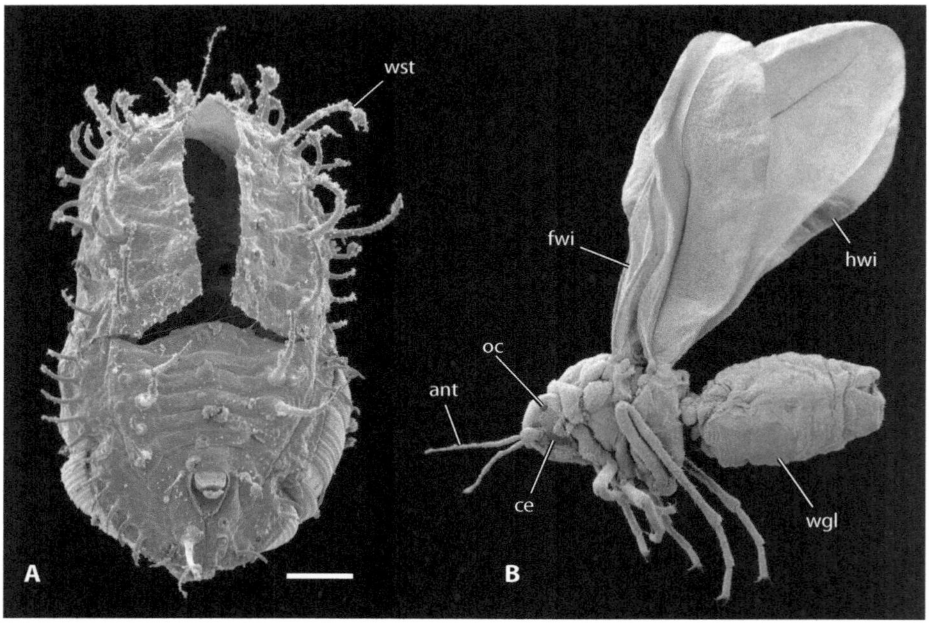

Fig. 6.22.2.1: *Trialeurodes vaporarium* (Aleyrodina, Aleyrodidae), SEM micrographs. A, puparium after hatching, dorsal view. B, adult, habitus, lateral view. Abbr.: ant – antenna, ce – compound eye, fwi – forewing, hwi – hindwing, oc – ocellus, wgl – wax glands, wst – wax sticks. Scale bar: 100 µm. Courtesy H. Pohl.

Taxonomy. The only family Aleyrodidae is subdivided into the two subfamilies Aleurodicinae and Aleyrodinae. It contains approximately 120 genera.

Diagnosis. Weakly sclerotized small or very small insects (Fig. 6.22.2.1B). Body and wings usually covered by white powdery waxy secretions. Wings with strongly reduced venation.

Morphology of adults. Most species are smaller than 2 mm; the males are usually smaller than the females but otherwise similar. The cuticle is thin and weakly sclerotized. The coloration appears usually whitish due to waxy secretions (see above) but the true color of the cuticle is yellow, brownish or red.

The head (Fig. 6.22.2.2) is orthognathous in adults (hypognathous in immature stages) and distinctly separated from the prothorax. The compound eyes are often divided into an upper and a lower portion by a chitinous bridge; two ocelli are present close to the upper margin of the compound eyes; they are not covered with wax particles. The 7-segmented antennae bear a terminal bristle; sensorial setae and rhinaria are present on several antennomeres. The 4-segmented labium can be telescoped during the feeding process; it is extended by the contraction of strongly developed muscles inserting on a tube-shaped cuticular apodeme at the anterior labial margin; this results in the extraction of the long stylets from the plant tissue.

The meso- and metathorax are of similar size. The wing venation is strongly reduced; the fore- and hindwings are not attached to each other during flight. The legs are slender, with long coxae and 2-segmented tarsi; apically, an unpaired, elongate and membranous attachment structure is present, representing a modified arolium

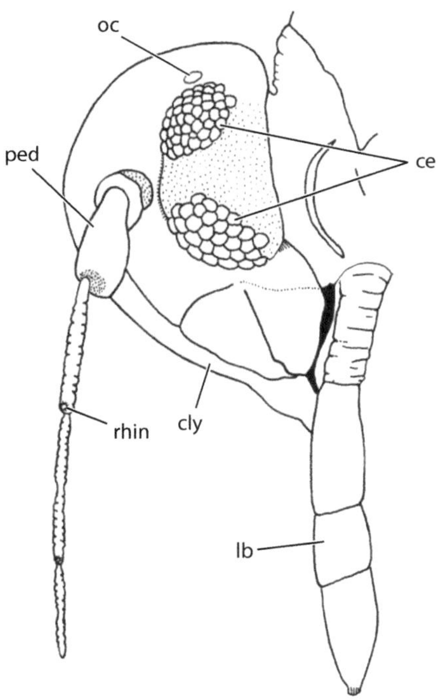

Fig. 6.22.2.2: *Aleyrodes* sp. (Aleyrodina, Aleyrodidae), female, head, lateral view. Abbr.: ce – compound eyes, cly – clypeus, lb – labial rostrum, oc – ocellus, ped – pedicellus, rhin – rhinarium. Redrawn from Weber (1931).

or empodium; basally it is covered with waxy particles and long microtrichia; the hindlegs are modified as jumping legs.

A very distinct constriction at the base of the abdomen results in its high movability. Extensive wax glands (gland plates) are present on sternites III–VI (Fig. 6.22.2.1B). Only two Malpighian tubules insert at the midgut-hindgut border.

Biology. The surface layer of wax reduces water loss via evaporation. The particles are spread on the body surface using combs of hairs on the hindlegs. White flies are restricted to dicotyledons. They prefer woody plants and are usually rather unspecialized. All species feed on phloem sap and large amounts of honeydew can be produced by some species. In contrast to aphids, seasonal dimorphism is uncommon. It occurs only in the puparia of some species of the temperate regions. Whiteflies fly only over short distances but distribution with the host plants is common. The most important natural enemies (and biological control agents) are larvae and adults of green lacewings (Neuroptera, Chrysopidae), ladybirds (Coleoptera, Coccinellidae) and some predaceous heteropterans.

Reproduction and development. The reproduction is sexual or parthenogenetic (arrhenotoky and thelytoky). The postembryonic development comprises four nymphal stages. The first instar hatching from the egg is mobile whereas the following stages have reduced appendages and are sedentary on the host plants. Like in Coccoidea (and Thysanoptera) the penultimate and ultimate instars are quiescent ("prepupa" and "pupa") (Fig. 6.22.2.1A: puparium). The compound eyes, wing buds and genitalia are formed in the second period of the fourth stage.

Fossil record. Whiteflies are preserved in Early Cretaceous (Lebanon) and Tertiary amber (Grimaldi & Engel 2005). Earlier fossils ascribed to Aleyrodidae are questionable. Whiteflies can be abundant in ambers from Lebanon, Burma and New Jersey (see below: Aphidoidea).

Economic importance. In tropical and subtropical countries species of Aleyrodina can be a very serious problem in agriculture. Whiteflies can have disastrous effects on citrus, tomatoes, cotton, and other crops. The damage is mainly caused by honeydew, which is an ideal growth medium for sooty mold, but the transmission of plant diseases can also have a very negative impact. The control is impeded by the fact that whiteflies and the viruses they carry can infect many different host plants.

6.22.3 Aphidina, Aphidoidea (English common name: aphids)

Diversity and distribution. About 5,000 species are described and the group has a worldwide distribution, with the highest diversity in the temperate zones.

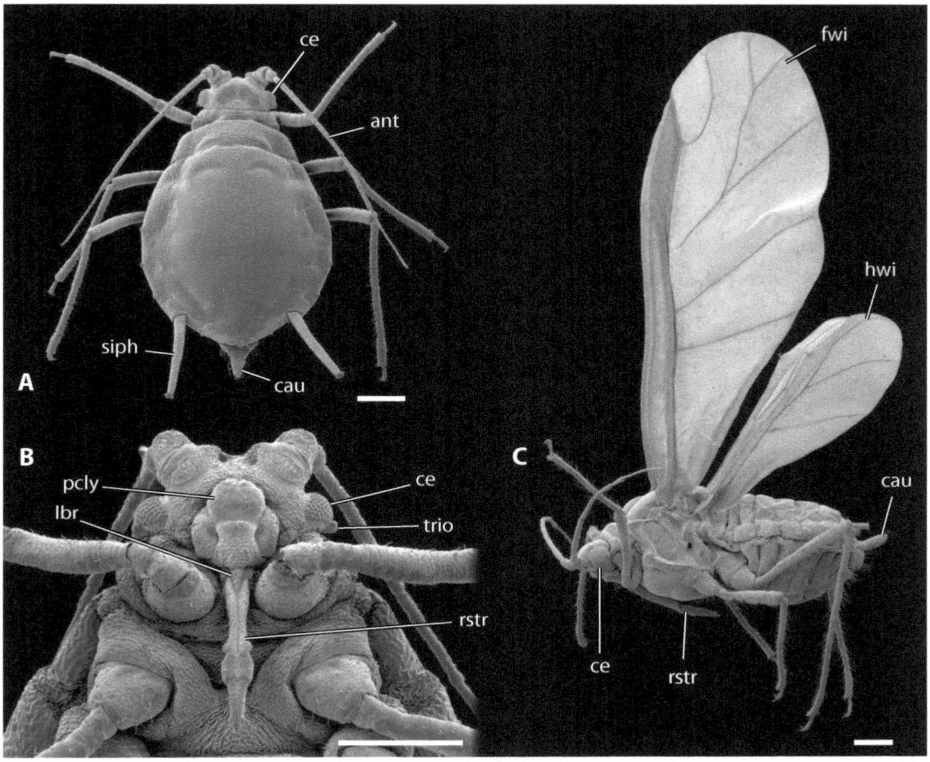

Fig. 6.22.3.1: *Aphis* sp. (Aphidina, Aphidiidae), SEM micrographs. A, wingless morph, dorsal view, B, wingless morph, head, ventral view, C, winged morph, lateral view.
Abbr.: ant – antenna, cau – cauda, ce – compound eye, fwi –forewing, hwi – hindwing, lbr – labrum, pcly – postclypeus, rstr – rostrum, siph – siphunculus, trio – triommatidium.
Scale bar: 250 µm. Courtesy H. Pohl.

Autapomorphies

- Change between sexual and asexual (parthenogenetic) reproduction (groundplan)
- Switches between herbaceous and woody host plants (groundplan)
- Malpighian tubules completely lost
- Only egg larva mobile, legs of the following nymphal stages strongly reduced (also in Aleyrodoidea) (?)
- Fourth nymphal stage forms puparium (see Aleyrodoidea) (?)

Taxonomy. The group is subdivided into two superfamilies (Phylloxeroidea and Aphidoidea) and 12 families in Strümpel (2005). In alternative classifications only three families (Adelgidae, Phylloxeridae, Aphididae) are recognized (Engel and Grimaldi

2005). Aphididae (s.l.) are by far the most successful lineage of Aphidina whereas the diversity of Adelgidae (ca. 50 spp.) and Phylloxeridae (ca. 60 spp.) is low.

Diagnosis. Body less elongate than in Aleyrodina, often distinctly widening posteriorly (Fig. 6.22.3.1, 6.22.3.2). With two pairs of transparent wings. Hindlegs not modified as jumping legs. With winged (alatae) and wingless forms (apterae) within one species. Often with conspicuous tube-like cornicles or siphunculi (Aphididae), paired tube-like structures inserting on or close to abdominal tergite V. The cornicles do not release honeydew but fluids containing alarm pheromones.

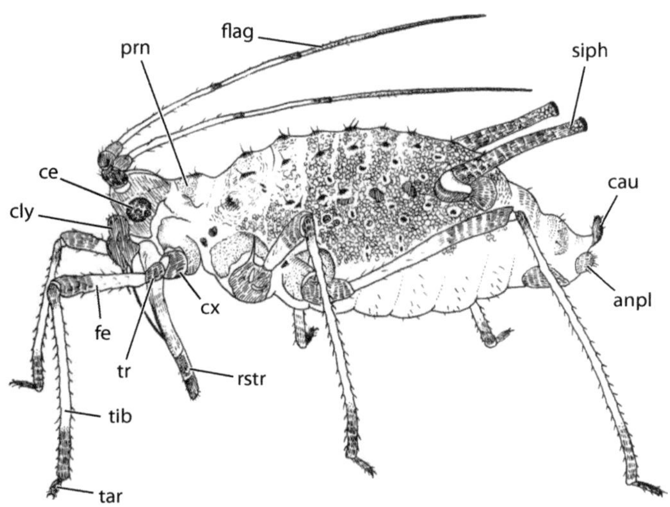

Fig. 6.22.3.2: *Macromyzus woodwardiae* (Aphidina, Aphidiidae), lateral view of viviparous apterous female. Abbr.: anpl – anal plate, cau – cauda, ce – compound, cly – clypeus, cx – coxa, fe – femur, flag – flagellum, prn – pronotum, rstr – rostrum, siph – siphunculus, tar – tarsus, tib – tibia, tr – trochanter. Redrawn from Strümpel (2005), after Miyazaki (1987).

Morphology. The size range is 0.5–8 mm, with most species between 1.5 and 3.5 mm. The body is more or less oval and the abdomen often distinctly widening in its middle or posterior region (Fig. 6.22.3.1). The cuticle is weakly sclerotized. Wax glands on the head, thorax or abdomen occur in many species, usually more strongly developed in the wingless morphs (apterae). The coloration is caused by pigments in the cuticle or by wax particles; most species are yellow, green, red or black.

The hypognathous head (Figs 6.22.3.1. 6.22.3.2) is immovably connected with the thorax. The compound eyes are usually larger in the winged forms; a protuberance bearing three ommatidia is present at the posterior margin of the main part of the compound eyes of adults (triommatidium) (Fig. 6.22.3.1B); in some nymphs and apterae the triommatidium is the only light sense organs; three ocelli are present in the alatae, two of them close to the mesal margin of the compound eyes and one between the antennal bases. The maximum number of antennomeres is six; the scapus and pedicellus are usually short and stout, whereas the flagellomeres are slender (Fig. 6.22.3.2). The labial rostrum is usually 4-segmented; it is completely reduced in some taxa but can be longer than the entire body in others (Lachnidae, Adelgidae); the labium and stylets are often not equally long; very long stylets can be present in species with a short rostrum; in some adelgids they are extremely elongated and coiled.

The thorax of the wingless forms (Figs 6.22.3.1, 6.22.3.2) is less strongly sclerotized and smaller, and the three segments are of similar size. In contrast, the mesothorax of alatae is distinctly larger than the pro- and metathorax, and the pro and metanotum are only narrow transverse sclerites. The forewings are distinctly larger than the hindwings (Fig. 6.22.3.1C); both pairs are connected during flight: a fold at the hind margin of the forewings interacts with some hook-shaped spines at the anterior hindwing margin; the wings are either held in a roof-like position (e.g., Phylloxeridae) or horizontally; the venation is reduced to varying degrees; the mesothoracic flight muscles are strongly developed. The tarsi of the slender legs (Fig. 6.22.3.2) are 2-segmented in adult alatae (1-segmented or reduced in some apterae and nymphs); distally they bear paired claws and membranous thread-like pulvilli as attachment devices.

The abdomen is composed of nine visible normally developed segments. Segment X forms a tail-like cauda. Long paired siphunculi are inserted on tergites V or VI in most species of Aphididae (Fig. 6.22.3.2). Malpighian tubules are always absent. Abdominal wax glands are usually present.

Biology. A very characteristic feature of Aphididae is the dimorphism with distinctly different alatae and apterae. The flight is mostly passive. Alatae can drift over distances of several kilometers with the wind and can reach heights of up to 1,500 m. Dispersion and migration are important mechanisms to avoid unfavorable environmental conditions (Strümpel 2005). Adults and nymphs usually passively feed on sap of phloem vessels and produce large amounts of honey dew. Many species are monophagous but some can feed on numerous plant species. Aphids usually harbor the bacterium *Buchnera aphidicola* (Enterobacteriaceae) in specific cells (bacteriocytes). In some species they are secondarily lost and replaced by other endosymbionts. The endosymbionts are transmitted from generation to generation. The honey dew produced by aphids attracts ants which offer protection against predators, especially larvae and adults of coccinellid beetles (see above). Aphids are generally attacked by numerous predators and parasitoids.

Reproduction and development. Aphids are characterized by an unusually high reproductive rate and are typical r-strategists (Strümpel 2005). A typical feature of the group is the alternation between parthenogenetic and sexual reproduction, often linked with a change of hosts (ca. 10% of all aphid species; Strümpel 2005). The developmental cycle is often complicated. A pattern characteristic for the group was outlined by Strümpel (2005): in spring female fundatrices hatch from fertilized eggs deposited on the primary host (usually woody plants). They are viviparous and parthenogenetic. After several generations, which can comprise alate and wingless morphs, migrantes alatae move to the summer host (usually herbaceous plants). Several parthenogenetic generations (also alatae or apterae) follow. These morphs are referred to as exsulis virgo. The last generation produces alate gynoparae and wingless androparae. The former return to the primary host in the fall. They are viviparous and produce larvae which develop into wingless sexuales females. These are fertilized by alate males produced by the androparae on the secondary hosts.

Fossil record. The earliest stem group representatives of Aphidina appear in the fossil record in the Early to Late Jurassic (170–150 Ma) (Grimaldi & Engel 2005). They display some plesiomorphic features of the wing venation not found in the extant subgroups. The high numbers of inclusions in Cretaceous ambers of Siberia and western Canada shows that aphid species were very abundant in these regions. The abundance of basal groups of Aphididae (and Coccoidea) in coniferous amber suggests that these insects may have primarily fed on conifers (Grimaldi & Engel 2005).

Economic importance. Many species are important plant pests. They cause damage by sucking fluid, saliva with toxic effects, and the growth of sooty mold on plant surfaces covered by honeydew.

6.22.4 Coccina, Coccoidea (English common name: scale insects; mealy bugs)

Diversity and distribution. Approximately 7,800 species are described presently. The distribution is worldwide, with the majority of species in the tropical and subtropical regions.

Autapomorphies

- Advanced sexual dimorphism with wingless and larviform females (Fig. 6.22.4.1)
- Mouthparts of males vestigial, non-functional
- Tarsi of females (if present) and males 1-segmented, with unpaired claw
- Hindwings of males modified as curved, hook-like halteres

Taxonomy. A generally recognized classification does not exist. Strümpel (2005) divided the group into the two superfamilies Orthezioidea (Ortheziidae, Margarodidae) and Coccoidea (ca. 20 families, e.g., Pseudococcidae [ca. 2,000 spp.], Coccidae [ca. 1,130 spp.], Diaspididae [ca. 2,650 spp.]). Margarodidae is possibly the most ancestral group (e.g., females with distinct segmentation, not sedentary).

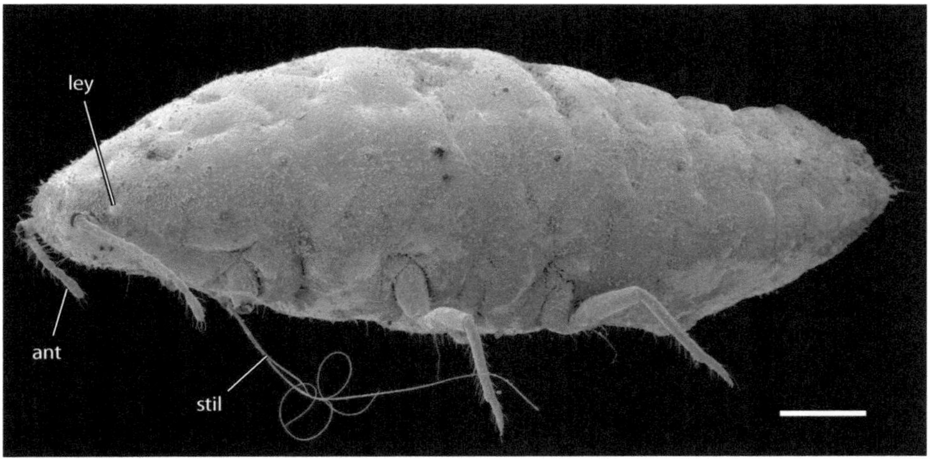

Fig. 6.22.4.1: *Pseudococcus* sp., (Coccina, Pseudococcidae), female, anterolateral view, SEM micrograph. Abbr.: ant – antenna, ley – persistent larval eye, stil – stylet. Scale bar: 250 µm. Courtesy R. Spangenberg and K. Friedemann.

Diagnosis. Mainly characterized by the extreme sexual dimorphism. Males with vestigial mouthparts and hindwings transformed into halteres. Females often sedentary and with strongly reduced antennae and legs. Usually between 1 and 7 mm, maximum size 35 mm (females of *Aspidoproctus maximus*).

Morphology of males. Males are only 1–2 mm long. In contrast to the females they have an insect-like habitus and wings are usually present. The coloration is often yellow, orange or red. Typical compound eyes are only preserved in some Margarodidae; most coccoid species possess unicorneal double eyes, often with additional eyes between them. The mouthparts are strongly reduced and non-functional. The venation of the transparent forewings is strongly reduced (two distinct longitudinal veins). Brachypterous forms occur and few species are completely wingless; the hindwings are usually modified as hook-shaped halteres but completely reduced in some groups (e.g., Aclerdidae). The legs are long and slender; the tarsus is usually 1-segmented and bears a single claw and membranous, thread-like pulvilli. Two pairs of thoracic spiracles are generally present. The abdomen is composed of nine distinctly developed segments. Its apex is often formed by a straight or hook-shaped penis. Two or several terminal appendages formed by wax threads are often present. They are secreted by numerous gland cells.

Morphology of females. The body is only indistinctly divided into head, thorax and abdomen and its shape is highly variable (Fig. 6.22.4.1). The number of antennomeres varies between one and 16 or the antennae are completely reduced. The larval eyes persist in female adults (Fig. 6.22.4.1) or the eyes are completely absent. The stylets are often extremely long (Fig. 6.22.4.1) and retracted in a sack-shaped crumena at rest. The labial rostrum is vestigial. Wings are always absent. The abdomen comprises 8–10 segments. Abdominal spiracles are only present in Margarodidae and Ortheziidae. A characteristic protective shield is often present, usually formed by secretions of wax glands.

Biology. The biology of Coccina is mainly characterized by the extreme sexual dimorphism with non-feeding males and strongly modified larviform females, which are usually sedentary (not in mealy bugs [Pseudococcidae (Fig. 6.22.4.1)], Ortheziidae, and Margarodidae). Females usually coat themselves with a cover of waxy secretions resembling a scale (scale insects). Most species are polyphagous. Hosts are almost always vascular plants. Only few species feed on fungal mats or fungi (e.g., species of *Newsteadia* [Ortheziidae]). Coccoidea are found on exposed organs of plants or they feed on roots. Cyclic migrations from leaves to woody branches occur. Species in South America and Australia produce sexually dimorphic galls.

Reproduction and development. The reproduction is sexual or parthenogenetic. The only known hermaphroditic insects belong to Coccina (e.g., *Icerya purchasi*). The females carry sperm in tissue passed down over generations. The number of chromosomes is highly variable (2n: 8–64). The females are usually oviparous. The first instar is mobile ("crawlers"), with well-developed antennae, mouthparts and legs. After finding a suitable location it becomes sedentary and is permanently attached to the host plant. The usual number of nymphal stages is three in females and four in males.

Fossil record. Permian and Triassic fossils assigned to Coccina are disputable (Grimaldi & Engel 2005). Generally the chance of preservation of the short-lived, very small and fragile males and the sedentary females are minimal. The earliest definite coccoid fossils in rock are from the Early Cretaceous of England (Grimaldi & Engel 2005). Males are often embedded in amber, but females and "crawlers" only rarely. In contrast to aphids, coccoids (and also aleyrodids) are abundant in Cretaceous ambers from Lebanon, Burma and New Jersey (Grimaldi & Engel 2005).

Economic importance. Many species are very important pests of cultivated plants. The main damage is often caused by toxic saliva. Some species are economically valuable. The cochineal (*Dactylopius coccus*) from South America and Mexico produces the red dye carmine. Species of the tropical family Kerriidae (*Kerria*) produce lac, a resinous substance used for dyeing textiles, as a cosmetic or as a medical drug. Secretions of *Kerria lacca* were used to produce shellac. Historically, red dye was obtained from dried exemplars of *Kermes vermilio* (e.g., Roman Empire, Greece, Syria).

[Weber, (1931); Hodkinson (1971); Strümpel (1983, 2005, 2010); Miyazaki (1987); Burckhardt & Basset (2000); Grimaldi & Engel (2005); Moulds (2005); Ouvrard et al. (2010); Cryan & Urban (2012)]

6.23 Coleorrhyncha (Greek *coleos* = sheath, *pteron* = wing, English common name: moss bugs)

Diversity and Distribution. Coleorrhyncha are one of the smallest orders of insects with only 36 extant species. They are restricted to Australia (incl. Tasmania and Lord Howe Island), New Zealand, New Caledonia, southern Argentina, and southern Chile. This suggests a Gondwanan origin, but the fossil record shows that the group is older than the split of Pangaea (see below). Peloridiid species mainly occur in wet moss in temperate and subantarctic rainforests dominated by *Nothofagus* (Nothofagaceae) (Burckhardt 2009, 2010). However, some species also occur in regions where southern beeches are lacking.

Autapomorphies. All known species of Coleorrhyncha display a highly modified and largely uniform habitus (Fig. 6.23.1).
- Body strongly flattened
- Head subtriangular and strongly compressed dorsoventrally
- Compound eyes very widely separated
- Oval areolae on anterior head region (groundplan?)
- Two rows of sensilla at the apex of the labial rostrum
- Three-segmented antennae with spindle-shaped flagellomere
- Pronotum with wide and flat duplicature (paranota)
- Forewings with reticulate pattern
- Jumping capacity (groundplan?)

All extant species except for *Peloridium hammoniorum* lack ocelli and are flightless with completely reduced hindwings.

Taxonomy. The 17 extant genera are placed in the single family Peloridiidae (Larivière et al. 2011). The four genera from New Zealand (*Xenophysella* + (*Oiophysella* [New Caledonia] + (*Xenophyes* + *Oiophysa*))) form the sistergroup of a monophylum comprising the clades from South America (*Kuscheloides* + (*Peloridium* + (*Pantinia* + (*Idophysa* + *Peloridora*)))) and the Australian region (*Hemiodoecellus* + ((*Hemiowoodwardia* + *Hemiodoecus*) + (*Rhacophysa* + (*Howeria* + (*Craspedophysa* + *Peltophysa*) + *Hackeriella*)))) (Burckhardt 2009: fig. 305). Recently erected genera are *Craspedophysa*, *Peltophysa* (both Northern Queensland), *Idophysa* (Southern Chile) and *Rhacophysa* (for *Hackeriella taylori*, Northern Queensland). An excellent overview is given in Burckhardt (2009). The sistergroup relationship between Coleorrhyncha and Heteroptera (Heteropterodea or Prosorrhyncha) is likely. Shared derived features are the presence of a distinct mandibular sulcus, the reduced number of antennomeres, the absence of clasping organs in the labial groove, coiled accessory salivary ducts, and the presence of several newly acquired cephalic muscles (Spangenberg et

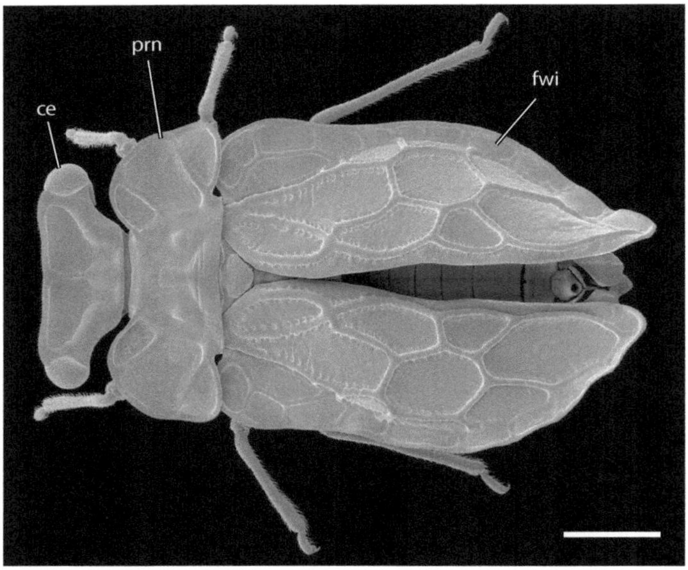

Fig. 6.23.1: *Pantinia darwini* (Peloridiidae), dorsal view, SEM micrograph. Abbr.: ce – compound eye, fwi – forewing, prn – pronotum. Scale bar: 500 µm. Courtesy H. Pohl.

al. 2013). Additional potential synapomorphies are modifications of the forewings, the horizontal position of the forewings at rest, and the fusion of the two segments forming the anal cone.

Diagnosis. Small, strongly flattened and broad (Fig. 6.23.1). Transverse head with widely separated compound eyes and oval transparent zones of weakness (areolae). Pronotum with broad and flat duplicatures. Forewings with reticulate pattern. Ocelli and hindwings usually absent.

Morphology. The body size ranges between 2 and 5 mm. The yellowish or brownish insects often display patterns of lighter and darker coloration. Transparent areas on the anterior head region and on the pronotal duplicatures are usually present (Fig. 6.23.1).

The distinctly hypognathous head is conspicuously flattened and broad; its shape is oval, subtriangular, subtrapezoidal or V-shaped (Fig. 6.23.1). The foramen occipitale is large and triangular; a sclerotized gula is not developed; the postocciput lacks protrusions. The region of the vertex is flattened or moderately convex; oval transparent areolae on the anterior region are present or absent. The moderately large and hemispherical compound eyes are very widely separated and placed anterolaterally; ocelli are usually absent, but two well-developed paired ocelli and a small unpaired ocellus are present in winged individuals of *Peloridium hammoniorum*. The frontal and coronal sutures are absent and the subgenal ridge is also missing. The posteroventral region of the anterior part of the head capsule between the bases of the antennae and the compound eyes is formed by the genal area. The anterior tentorial grooves and the transverse frontoclypeal strengthening ridge are distinct. The posteroventrally located tripartite clypeus is composed of a flattened postclypeus, an elongated and convex median anteclypeus, and two lateral paraclypeal lobes; the dorsal side of the anteclypeus is separated from the remaining clypeus by an incision and bears the opening of the prepharynx. The small tongue-shaped labrum is distinctly separated from the clypeus; it forms the dorsal closure of the base of the feeding tube; the maxillary plates (or maxillary stipital lobes) laterad the anteclypeus are tapering caudally and end at the distal part of the anteclypeus; the mesal area of the maxillary plates, the hypopharynx and the anteclypeus form a guiding device for the mandibular and maxillary stylets; the mandibular plates (lora) are crescent-shaped and fused with the paraclypeal lobes. The 3-segmented antennae are inserted on a prominent socket posteroventrad the compound eye; the cylindrical scapus is slightly bent posterad and twice longer than the pedicellus, which tapers distally to one third of its basal width; distinct cuticular folds characterize its distal half (observed in *Hackeriella*); the single flagellomere is spindle-shaped. The mandibles are a pair of slender, posteroventrally directed stylets; they are articulated within the head capsule; the basal part is plate-like, whereas the posterior part is transformed into a funnel; the external surface is largely glabrous, but distinct cuticular scales are present on the distal region (observed in *Hackeriella*); a distinct mandibular sac, a mandibular lever, and mandibular glands are absent; close to the labral base, each of the mandibular stylets forms a spur, which is directed inwards and mechanically supports the enclosure of the smaller maxillae; the apices of the stylets almost reach the metacoxae posteriorly in their resting position. The maxillae are mainly represented by slender, posteroventrally directed, distally tapering stylets; like the mandibles they articulate within the head capsule; the base is a funnel-shaped sclerite with a thin layer of membrane wrapped around it; a distinct maxillary sac and a maxillary lever are not developed; the palp and galea are absent and the lacinia is fused with the stipes; the stylets reach the metacoxae posteriorly; maxillary glands are absent; the tube-like, 4-segmented labium forms a sheath for the mandibular and maxillary stylets (feeding tube or suctorial beak); it has a posteroventral orientation; the labial groove lies on the ventral side of segments two to four; the basal segment is directly attached to the ventral base of the postocciput and is partly covered by the two prothoracic katepisterna; it is less sclerotized than the distal ones; the apical segment is about as long as the 2nd and 3rd combined; its tip is rounded and bears two almost circular rows of sensilla. The tentorium is well-developed except for the absence of dorsal arms; the tentorial bridge (corpotentorium) is plate-like and rectangular; the posterior arms are short and stout; the minute posterior tentorial pits are shifted anterodorsad of the maxillary plates. Two strongly sclerotized plates (hypopharyngeal wings or suspensorial sclerites) are present laterad the tentorium; their dorsal sides merge with the corpotentorium and the posterior arms while their bases are fused with each other, but provide a gap for the transition to the salivary channel; two clamps of the inner lateral sections of the base of this funnel partly enclose the lateral sides of the pumping chamber; the posterior parts of the suspensorial sclerites merge with the head capsule; the hypopharynx is associated with the ventral base of the fused suspensorial sclerites and located between the mandibular and maxillary stylets; its bifurcated ventral side covers the anterior part of the prepharynx and is also connected with the salivary pump; the entire functional complex comprises the suspensorial sclerites, a wing-like free standing piston, the membranous pumping chamber and two muscles; the salivary pump is dorsoventrally oriented. Two pairs of salivary glands are present on both sides of the gut. Both are enclosed by a thin tunica propria. The brain and the suboesophageal ganglion form a compact unit with a narrow passage for the foregut (possibly an autapomorphy of Hemiptera).

In forms lacking hindwings, the thorax as a whole is distinctly flattened and the three segments are of similar size. The prothorax is mainly characterized by the conspicuous flat lateral paranota (Fig. 6.23.1); in the nymphs they are separated from the main part of the pronotum by sutures and equipped with tracheae; more or less distinct transparent paranotal areolae are present or absent; the shape of the paranota is rounded, angular or triangular. The basisternum and eusternum are combined (Evans 1937: 4th and 5th instar of *Hemiodoecus fidelis*). In adults a pair of prothoracic spiracles is placed in the intersegmental membrane posterior to the pleura and partly covered by a backwardly-projecting flange of the proepimeron; in 3rd instars they are placed on small areas of tergal origin (Evans 1937). The pterothoracic segments are similar with the exception of the wings. On the ventral side only the basisternum is present as a defined sclerotized element (Evans 1937: 4th instar of *Hemiodoecus fidelis*). The shape and size of the mesothoracic scutellar shield varies strongly; it is large, broad and triangular in *P. hammoniorum* (presumptive groundplan condition); the mesoscutum bears two large transverse fields of spines (Burckhardt 2009: fig. 4). The mesothoracic spiracle is placed in the intersegmental membrane and partly covered by the metepisternum (Pendergrast 1962). The metanotum is shorter than the mesonotum and simple (Burckhardt 2009: fig. 4). A rim for reception of the rostrum and stylets is present on the ventral side of the thorax (Fig. 6.23.1). In all three segments distinct transverse antecoxal sclerites (more oblique in the prothorax) and trochantins are present anterior to the coxae. The pro-, meso- and metafurcae arise from separate bases. Short transverse pleural sutures separate the episterna from the epimera. The fore- and hindwings (if present) lie horizontally on the abdomen in their resting position. The forewings, which always lack a corium, are referred to as tegmina (e.g., Burckhardt 2009); in the typical case they are characterized by a reticulate pattern (Fig. 6.23.1); the most complete venation is preserved in fully winged morphs of *P. hammoniorum*; the veins of the forewings of brachypterous species are raised in relief and the enclosed hyaline cells are margined to a varying extent by deep punctures (Evans 1981); different features of the tegmina displayed by some of the species are possibly associated with the brachypterous condition (Evans 1981): this can include a size reduction of the costal cells or their elimination, the expansion of the anterior costal area, the development of a sinuate costal margin, a reduction in cell numbers accompanied by an increase in the size of those retained, a very pronounced reticulation, and an elytriform condition (Evans 1981); the hindwings are reduced in all species except for *P. hammoniorum*, which has macropterous and brachypterous morphs; the well-developed hindwings of the former are shorter and smaller than the unmodified forewings and display five longitudinal veins and a lobate anal field; closed cells are lacking; in flight the meso- and metathoracic wings are coupled (functional dipterism) by rigid bristles on the hind margin of the tegmina. The legs are slender and largely unmodified; the procoxae are more or less cylindrical and mobile, whereas the middle and hind coxae are very broad basally and largely fixed in the coxal cavities; the trochanters are indistinctly separated from the femora which are cylindrical, largely glabrous, and slightly curved in the hindlegs; the long tibiae are cylindrical and irregularly covered with short, stout spines; the metatibia is distinctly longer than the femur; the tarsi are 2-segmented; tarsomere 1 is distinctly shorter than 2 and has an oblique apex; an arolium and an unguitractor plate are present between the simple short claws. The thoracic musculature is extremely reduced in the flightless forms (China 1962).

The abdomen is flat on the dorsal side and moderately convex ventrally; it is broad in its middle region, and narrowing towards the base and its more or less rounded apex (China 1962: fig. 5). It is composed of nine distinctly developed segments and an anal cone. Tergites I and II are distinctly reduced, especially the former (Burckhardt 2009: figs 8–10). The lateral margins of segments III–VII form flat and broadened connexiva. Eight abdominal spiracles are present; spiracles I and II are placed near the anterior margin of the partly reduced terga, spiracles III–VIII ventrolaterally on the sternites; the latter open into a simple atrium and lack a closing muscle (Pendergrast 1962). The genital apparatus of males is formed by a large genital capsule (pygophore) of segment IX, which bears a simple aedeagus and parameres (also referred to as harpagones), associated basally with a triangular or Y-shaped connective or basal plate (Evans 1981). A well-developed ovipositor is present in females.

The dorsal vessel is an inconspicuous colorless tube which ends posteriorly in segment VIII and reaches at least the prothorax anteriorly; the part in the abdomen constitutes the heart (Pendergrast 1962). The three thoracic ganglia are separated from each other and from the suboesophageal complex by short connectives; the metathoracic ganglion is fused with the compact unit formed by the abdominal ganglia. The pharynx and oesophagus are short; the digestive tract lacks an ingluvies, a proventriculus, and a filter chamber; the anteriormost region of the midgut is enclosed by ring muscles, whereas longitudinally arranged fibers are present along its widened main part; posteriorly it gradually narrows, forms a bent and passes forward to meet the hindgut near the anterior end of the abdomen; the midgut cells contain granules

(Pendergrast 1962); the hindgut comprises a dilated rectal sack, a narrow rectum, and a sclerotized anal tube; the rectum has an epithelium of tall cells, a conspicuous cuticular lining, and a strongly developed muscularis (Pendergrast 1962). Two pairs of short and wide Malpighian tubules open into the digestive tract at the end of the midgut, shortly before the rectal sack; they are as long as the sack and usually applied to its surface (Pendergrast 1962). Mycetomes are present as two large masses in nymphs and adults on both sides of the abdomen (Pendergrast 1962 [observed in *Hemiodoecellus fidelis*]); they contain endosymbiotic bacteria. Scent glands are absent. The large fusiform paired testes are formed by a single follicle; the short vas deferens turns forward and dilates to form a long seminal vesicle; anteriorly the vesicle narrows and is surrounded by a muscularis; it enters the common ejaculatory duct near its anterior end and is connected with it by a common laver of muscle fibers; the common duct also receives the secretions of the long accessory glands, which are placed mesad the testes (Pendergrast 1962: fig. 16). The internal female organs comprise paired ovaries with 2–5 ovarioles, paired oviducts, a common oviduct, a vagina, a spermatheca, and a large median vaginal sac with a very thin wall; the oviducts are enclosed by a strongly developed muscularis (Pendergrast 1962: fig. 20).

Biology. The biology of Coleorrhyncha is not well known (Burckhardt 2009). They are characterized by a cryptic lifestyle. Adults and nymphs are found among wet mosses and liverworts in temperate and subantarctic rain forests of the southern hemisphere. They feed on cell contents of moss using their stylet-like paired mouthparts. Probably they are not specialized on particular moss species (D. Burckhardt pers. comm.). Peloridiids are moving slowly with awkward movements and are almost always covered with self-made incrustations of excretions blended with soil particles. The flight capacity is only retained in winged morphs of *Peloridium hammoniorum*. Whether the jumping capacity belongs to the groundplan is unclear. Sound production was described for *Hackeriella veitchii* (Evans 1981; Hoch et al. 2006; Burckhardt 2009, 2010).

Reproduction and development. Females of Coleorrhyncha are oviparous. The eggs are of ovoid shape and lack a distinctly sculptured chorion. The postembryonic development comprises five nymphal stages.

Fossil record. The oldest known fossils unambiguously classified as Coleorrhyncha are placed in the family †Karabasiidae and the peloridiid subfamily †Hoploridiinae. They were found in Laurasian deposits of the Late Jurassic and Early Cretaceous. This implies that the group was originally Pangaean, with only few southern hemisphere representatives persisting until today. The assessment of the distinctly older †Progonocimicidae (Late Permian to Early Cretaceous) is problematic. This group, which was widely distributed in Laurasia, is tentatively assigned to the stem group of Coleorrhyncha (Popov & Shcherbakov 1991; s. also Grimaldi & Engel 2005). Coleorrhyncha are apparently one of the oldest groups of Hemiptera and have changed little since the Jurassic (Pendergrast 1962; Burckhardt 2009).

Economic importance. Coleorrhyncha are economically irrelevant.

[Evans (1937, 1981); China (1962); Pendergrast (1962); Popov & Shcherbakov (1991); Grimaldi & Engel (2005); Hoch et al. (2006); Burckhardt (2009, 2010); Larivière et al. (2011); Spangenberg et al. (2013)]

6.24 Heteroptera (Greek *hetero* = different, *pteron* = wing, refers to the hemelytra, English common name: true bugs [in the narrow sense])

Diversity and Distribution. Heteroptera – commonly known as true bugs or bugs – are the second largest subgroup of Acercaria. They comprise approximately 40,000 described species. The distribution is worldwide. It reaches beyond the Arctic Circle in the north and southward almost to the Antarctic Circle. Heteropteran species occur in a very broad variety of different habitats including for instance tropical rain forests, deserts, alpine regions and the open ocean (species of *Halobates*). They are most numerous in the tropics and subtropics. Some species are cosmopolitan (e.g., *Cimex lectularius* [bed bugs]) (Fig. 6.24.1).

Autapomorphies
- Head prognathous, foramen occipilale narrowed by gula
- 4-segmented antenna with intersegmental sclerites
- Base of labial rostrum shifted anteriorly
- Compound eyes with open rhabdom
- Metathoracic scent glands of adults
- Nymphs with scent glands opening on abdominal tergites

The characteristic hemelytra (Figs 6.24.2, 6.24.3) with forewings divided into an anterior sclerotized corium and a posterior membrane are probably not a groundplan feature.

Taxonomy. Heteroptera are divided into seven monophyletic lineages often referred to as infraorders (Wheeler et al. 1993; Schuh & Slater 1995; Weirauch & Schuh 2011). The Enicocephalomorpha comprise only ca. 450 species in two closely related families. They are likely the sistergroup of the remaining Heteroptera (Panheteroptera). The species are predaceous and mostly found under logs, stones, or bark, or in leaf litter. The Dipsocoromorpha comprise about 300 species in five families. The smallest heteropteran species belong to this group. They are ground-oriented and more or less cryptic like Enicocephalomorpha. They are mostly found in leaf litter or moss.

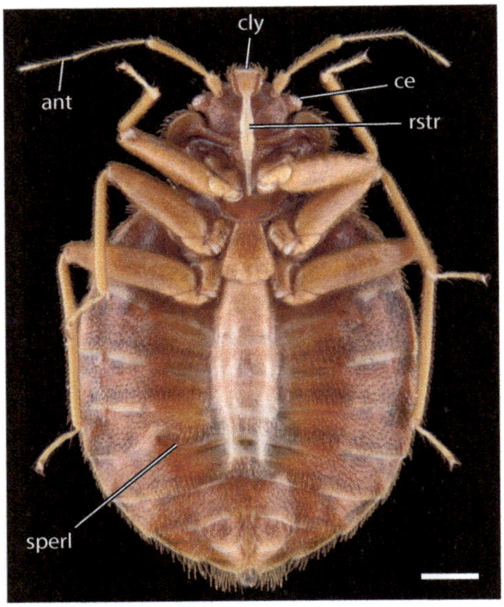

Fig. 6.24.1: *Cimex lectularius* (Cimicidae), female, habitus, ventral view. Abbr.: ant – antenna, ce – compound eye, cly – clypeus, rstr – rostrum, sperl – spermalege (= Ribagasches Organ). Scale bar: 250 µm. Photo courtesy H. Pohl.

The aquatic and predaceous Gerromorpha (water striders) comprise ca. 1,860 species (Fig. 6.24.3B). They are likely the sistergroup of the following lineages. Typically they are specialized on skating on the water surface. The Nepomorpha (Fig. 6.24.3A) are fully aquatic and comprise ca. 2,000 species in eleven families. They are likely the sistergroup of all the remaining Heteroptera (Wheeler et al. 1993). With approximately 300 known species the Leptodomorpha belong to the smallest heteropteran lineages. Some species live in the intertidal zone and others along the margins of rivers or lakes. They are likely the sistertaxon of a very species-rich clade Cimicomorpha + Pentatomomorpha (Fig. 6.24.3C) (Wheeler et al. 1993). Cimicomorpha comprise 17 families with a total of ca. 20,000 described species. The life habits are very heterogeneous, with predaceous or phytophagous species, and also ectoparasites specialized on sucking blood of vertebrates (Cimicidae, Polyctenidae) (e.g., Schuh et al. 2009). Pentatomomorpha are mostly phytophagous and also a very successful group with ca. 16,000 described species placed in 40 families.

Diagnosis. Head prognathous, with gula. Antenna usually 4-segmented. Wings horizontally placed on abdomen in resting position (with few exceptions). Forewings almost always with sclerotized anterior corium and posterior membrane (Figs 6.24.2,

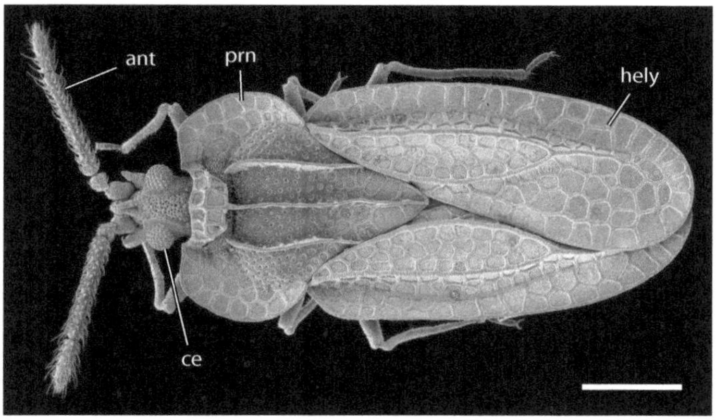

Fig. 6.24.2: *Kalama* sp. (Tingidae), habitus, dorsal view, SEM micrograph. Abbr.: ant – antenna, ce – compound eye, hely – hemelytra, prn – pronotum. Scale bar: 500 µm. Courtesy H. Pohl.

6.24.3). Adults with thoracic scent glands. Nymphs with scent glands associated with abdominal tergites.

Morphology of adults. The smallest species of true bugs measure 0.5 mm (e.g., *Myrmedobia coleoptrata* [Microphysidae]) – 0.8 mm (Schizopteridae); the largest can reach a length of ca. 11 cm (Belostomatidae); most species are medium sized (often 5–8 mm) and distinctly flattened. The wings are almost always held horizontally (Fig. 6.24.3). The thickness, surface sculpture and coloration of the cuticle vary strongly. The color patterns are usually due to pigments in the exocuticle (e.g., melanins, varying from reddish-yellow to black), but metallic interference colors also occur, especially in tropical species. Sexual dimorphism is a common phenomenon in Heteroptera; the males are often smaller and narrower than the females.

The head (Fig. 6.24.1, 6.24.4A, 6.24.5) is usually prognathous. The foramen occipitale is narrowed by a gula. The shape of the head capsule varies strongly; it can appear semicircular, polygonal, narrow, cone-shaped, or shield-like. Vertex, frons, postgenae and a part of the genae are closely connected, forming a structural unit referred to as epicranium. Sutures and ridges are largely reduced in adults, but the frontal and coronal sutures are often present in nymphs. The transverse frontoclypeal strengthening ridge is probably always absent (Denis & Bitsch 1973). The clypeus is entirely placed on the dorsal side of the head. It is subdivided into a postclypeus, which is firmly integrated into the head capsule, and a free anterior anteclypeus, which is continuous with the labrum anteriorly; paired structures of unclear homology are two pairs of paraclypeal lobes, the mandibular plates (jugal plates) and the maxillary plates (loral plates) (see also **6.23. Coleorrhyncha**); they are inserted on the lateral head region anterior to the compound eyes. A concavity on the ventral

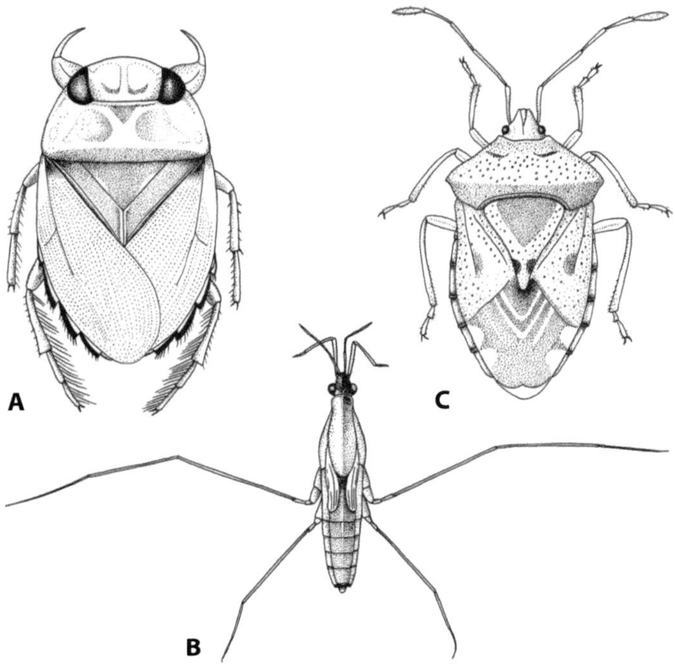

Fig. 6.24.3: Aquatic and terrestrial heteropterans, dorsal view. A, *Ilyocoris cimicoides* (Naucoridae); B, *Gerris lacustris* (Gerridae); C, *Elasmus lateralis* (Acanthosomatidae). Redrawn from Deckert & Göllner-Scheiding (2005), after Jordan (1950, 1952) and Weber (1930).

side of the head, referred to as gena or peristomium, encloses the base of the labial rostrum (Fig. 6.24.5); it is partly covered by the apical part of the clypeus (tylus). The baccula (or buccula) are additional ventral lobes inserted on the gena (or even on a part of the gula); they are present in different groups and placed on both sides of the rostral base. The acone apposition eyes are usually well-developed; the ommatidia are composed of six peripheral and two central rhabdomeres; the open rhabdom is likely an autapomorphic groundplan feature of Heteroptera; partial reductions of the compound eyes occur in Vianaidinae (Tingidae) and some other groups (e.g., Dipsocoromorpha, Cimicidae); they are absent in Polyctenidae, which parasitize bats, and also in the termitophilic Termitaphididae; two ocelli are present in the groundplan; they lie between or slightly posterior to the compound eyes; they are generally absent in nymphs (all groups of Acercaria and also in holometabolan larvae), in most aquatic groups, in Miridae (excl. Isometopinae) and Pyrrhocoridae, and also in some species of other families. The labrum varies considerably within the group; in the groundplan it is probably a simple lobe with a broad base (Schuh 1979) (Fig. 6.24.4), but it can be elongated, narrowed and retracted in different lineages; in some groups,

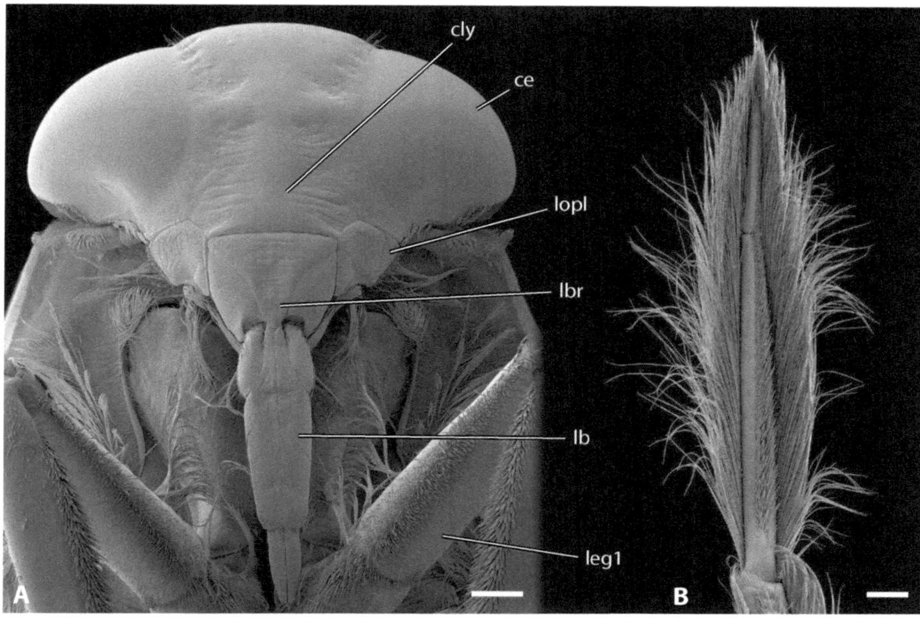

Fig. 6.24.4: *Notonecta glauca* (Notonectidae). A, head, ventral view: B, tarsus of hindleg with swimming hairs. Abbr.: ce – compound eye, cly – clypeus, lb – labial stylet, lbr – labrum, leg1 – fore leg, lopl – loral plate. Scale bars: 400 µm.

an epipharyngeal projection can be extended beyond the labral apex (Gerromorpha, Aradidae). The antennae are composed of four segments in the groundplan (autapomorphy), the scapus, pedicellus, basiflagellum and distiflagellum, plus intersegmental sclerites (Zrzavy 1990; Weirauch & Schuh 2011); an additional secondary antennomere can appear in the postembryonic development (five antennomeres in adult Pentatomidae and Pachynomidae) and reductions also occur in some groups; the antennae are usually longer than the head and filiform or cylinder-shaped; different vestitures of setae and spines occur and also flattened antennomeres; a secondary annulation occurring in one subfamily of Reduviidae (Hammacerinae) results in seemingly multisegmented antennae; the antennae of aquatic groups are shortened (possible autapomorphy of Nepomorpha) and not visible form above due to their position below the large compound eyes. The tubular, stylet-like mandibles articulate with the jugal plates; they are placed laterad the stylet-like laciniae and movably connected with them by a tongue and groove mechanism (possibly not in the heteropteran groundplan); mandibular teeth or barbs occur in some groups. The maxillae articulate with the loral plates; the stylets formed by the laciniae are connected with each other ventrally and dorsally, and they enclose a food canal and a salivary canal. The labium forms a rostrum similar to the one in the other hemipteran lineages

(Figs 6.24.1, 6.24.4A, 6.25.5); it is 4-segmented in most groups (groundplan) and may vary in length depending on the food substrate (with many exceptions); it is usually short in predators but elongated in phytophagous species; the proximal element is usually more or less distinctly reduced and closely attached to the gula; dorsally it bears a sclerite referred to as suspensorial plate (plaque suspenseur; Denis & Bitsch 1973); the second and the following segments form the sheath enclosing the paired stylets. The distal part of the hypopharynx is connected with the upper surface of the labium. A salivarium as a well-defined pouch is absent like in the other hemipteran groups; however, a well-developed salivary pump with a pistil and a pistil-retractor is present. The endoskeleton of the head is strongly reduced; the tentorium is usually absent (not in Nepomorpha; Mahner 1993) and in most groups even the tentorial grooves. The anteriormost part of the alimentary tract is characterized by a strongly developed cibarial pumping apparatus, between the bases of the stylets and the anatomical mouth opening. Like in the other hemipteran groups well-developed and complex salivary glands are present. They extend to the thorax posteriorly. The main part of the bipartite glands produces proteases, amylases and lipases. The function of the second portion is unknown.

The prothorax is the largest thoracic segment and movably connected with the mesothorax (Fig. 6.24.5). The pronotal shield varies strongly in shape; it can be quadrangular, hexagonal, heart-shaped, or semi-cylindrical; different modifications occur, such as for instance keeled or serrate edges, thorn-like projections, or drawn out posterior corners; it can be subdivided by a furrow into an anterior main part and a posterior plate-like projection; anterior calli marking the attachment sites of muscles of the forelegs are often present; a narrow anterior collar-region occurs in several groups; the pronotum covers the mesoscutum and also encloses it laterally; consequently it is more convex and broader in forms with well-developed flight organs. The prosternum is relatively simple; a rim for the rostrum is often present medially and concavities for the forelegs laterally; a prosternal process (xiphus) is often present between the procoxal cavities. The two pterothoracic segments are closely connected and form a compact functional unit. A conspicuous feature of the mesothorax is the mesoscutellar shield; in contrast to the mesoscutum it is almost always exposed (Fig. 6.24.3); in Pentatomidae it is usually trapezoid (pentagonal); in Cydnidae it is greatly enlarged and entirely covers the wings and the dorsal side of the abdomen, which results in a beetle-like habitus; in few groups (Corixidae, Tingidae) it is secondarily concealed; medially it often bears a longitudinal rim; in most groups a lateral rim (frenum) forms an arresting mechanism interacting with an enlarged posterior vein of the forewings (homologous to that of Auchenorrhyncha) (Weirauch & Cassis 2009). The metathorax is slightly smaller than the mesothorax in most groups and its flight musculature is less strongly developed. The meso- and metasternites are usually undivided plates, but separate sternal elements are present in some groups (e.g., Corixidae, Notonectidae, Cimicidae); a secondary subdivision by furrows and ridges is a characteristic of Pentatomidae. An autapomorphy of Heteroptera is the presence of paired metatho-

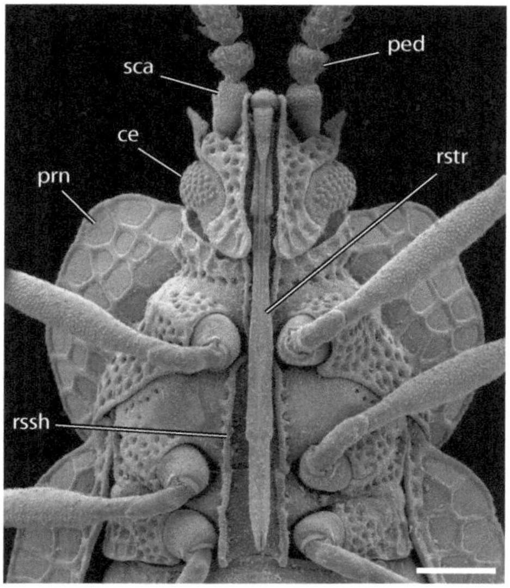

Fig. 6.24.5: *Kalama* sp. (Tingidae), head, ventral view, SEM micrograph. Abbr.: ce – compound eye, ped – pedicellus, prn – pronotum, rssh – rostral sheet, rstr – rostrum, sca – scapus. Scale bar: 200 μm. Courtesy H. Pohl.

racic scent glands (see above), which are more or less vestigial in semiaquatic and aquatic groups, but strongly developed in phytophagous lineages (Schuh & Slater 1995); they usually open on both sides in front of the metacoxae, but the opening is secondarily unpaired in Gerridae and Veliidae, and placed posteromedially on the metasternum (synapomorphy); the paired glands are composed of one simple lobe or several tube-like elements, often with branched channels, and a relatively large paired or unpaired (most groups) reservoir. The wings are almost always held horizontally in the resting position and the membranous hindwings are completely covered by the partly sclerotized forewings (hemelytra) (Fig. 6.24.2); the roof-like position in Notonectidae and Pleidae is a derived condition in Heteroptera; Heteroptera excluding Enicocephalomorpha (and possibly Dipsocoromorpha) are characterized by a distinct wing venation and forewings transformed into hemelytra; the sclerotized, leathery proximal part is called corium and the transparent distal part membrane; both areas are separated by the nodal line; the small sclerotized anal field, the clavus, is separated from the main part of the corium by a distinct fold; in some groups (Miridae, Anthocoridae) a movable triangular cuneus is separated from the remaining anterior corium by a costal furrow (costal fracture); an additional anterior subdivision of the corium occurring in Ceratocombidae and Anthocoridae is the embolium (exocorium); the transverse veins of the forewings are largely or completely reduced; the membrane bears between one and five longitudinal veins in most groups; they are completely absent in Corixidae; a dense network occurs in some members of Nepomorpha and Pentatomomorpha; the hindwings are wider than the forewings and the venation often greatly reduced; they are folded one or two times in order to fit below the hemelytra at rest; during flight they are coupled with the forewings and move more or less passively; consequently, the flight muscles are less strongly developed in the metathorax; the wings are usually well-developed but different degrees of reduction occur in almost all families. The legs are primarily slender and unmodified but many modifications occur, especially related with aquatic and predaceous habits; the coxae are often distinctly separated medially, thus creating space for a sternal rim for reception of the rostrum; two different types occur; the coxae carinatae (e.g., Naucoridae, Notonectidae, Corixidae, Cimicidae, Anthocoridae, Miridae) are elongate and triangular in cross section; they are broadly fused with the segment or at least distinctly restricted in their movability; the smaller and more globular coxae rotatoriae (e.g., Gerridae, Pentatomidae, Coreidae, Reduviidae, Nabidae) can rotate in all directions; they are inserted in deep coxal cavities; usually the type of coxa is taxon- specific and uniform; the co-occurrence of both types in one species is an extremely rare exception occurring in the genus *Gerris*; the femora are often laterally compressed, with distinct dorsal and ventral edges; the latter is set with hairs and thorns in some groups; the femora of raptorial forelegs (chelate and subchelate) of some predaceous forms are equipped with a rim and a row of spines (*Phymata* [Phymatinae]); the tibiae are usually cylindrical; distinct edges are present in some groups and the tibiae are rarely broadened and leaf-shaped; regular rows of setae or spines and terminal spurs are usually present; in some of the aquatic groups the tibiae are distinctly flattened and equipped with fringes of swimming hairs; comb-like protibial antennal cleaning organs occur in Pentatomidae, Pyrrhocoridae and other lineages; the tarsus is usually 3-segmented; the proximal tarsomere is very small in some groups; only two are present in Plataspidae, Piesmidae, Tingidae, Aradidae, Microphysidae, Gerridae, and Hebridae, and only one in Nepidae, Corixidae and some other taxa; heteronomous tarsi occur rarely (e.g., Dipsocoromorpha, Corixidae); the pretarsus usually bears two movable, curved claws of equal size; a tooth like basal process is often present; one claw (e.g., Enicocephalomorpha [partim], *Nepa*) or both (e.g., *Corixa*, *Ranatra*) are reduced on the forelegs of some predaceous aquatic bugs, and reductions also occur on hindlegs modified as swimming legs (e.g., *Corixa*, *Notonecta*) (Fig. 6.24.4A); Heteroptera display an unusual diversity of attachment structures; an arolium is present in Slaterellinae (Lygaeidae), an arolium and smooth euplantulae in Mirinae, smooth, lobe-like pulvilli in members of Coreidae, Pentatomidae and Pyrrhocoridae, and smooth pulvilli and a tibial fossula spongiosa in Peiratinae (Beutel & Gorb 2001); specialized legs occurring in Heteroptera are raptorial forelegs (Phymatidae, Emesinae) (or rarely hindlegs), flattened swimming legs with fringes of hairs (e.g., Corixidae, Notonectidae), and burrowing legs with a broadened tibia (e.g., *Cydnus*). Spiracles are present at the anterior margin of the meso- and metathorax; a closing apparatus with a muscle is only present in terrestrial groups (Weber 1930).

The abdomen is composed of ten (or possibly 11) segments and broadly connected with the metathorax. Tergite I is usually very narrow and sternite I vestigial or absent. Seven tergites and six sternites are usually exposed. The three terminal genital segments are usually exposed but retracted into segment VII in some groups (males of Pentatomomorpha, Reduviidae). A characteristic feature is the division of the tergites into an extensive mediotergite and dorsal and ventral laterotergites (paratergites); the laterotergites bear the abdominal spiracles; together they form the flattened connexivum, which usually extends over segments III–VII. The abdominal sternites are usually undivided plates; in Trichophora, a subgroup of Pentatomomorpha, they bear specifically arranged trichobothria on their lateral region. Segments VIII and IX are the genital segments of females, and segment IX in males. Segment X forms the anal cone around the anus; it bears two small, partly invaginated plates, which possibly represent vestiges of segment XI and form a closing apparatus; segment X is strongly reduced in the aquatic groups; their anal cone bears a large dorsal and a small ventral element. Cerci are absent in all developmental stages. The enlarged sternum IX of males forms a genital capsule (pygophore), which is retracted into segment VII. Segment VIII is ring-shaped and only everted during copulation in some groups. Tergite IX is reduced. The aedeagus articulates with a ring-shaped basal sclerite within the pygophore; the sclerotized proximal part of the copulatory organ, the phallobase (phallotheca), contains the membranous, extrusible endosoma (vesica); paired hook-shaped or sickle-shaped parameres are present in the groundplan; they are laterally inserted on segment IX; in some groups they are asymmetric or reduced. In females an orthopteroid ovipositor is present in the groundplan; it is composed of paired gonocoxites (valvifers) VIII and IX, which bear the elongated or rounded gonapophyses (valvulae) VIII and IX; third valvulae, which are also inserted on the gonocoxites IX, are missing in some groups.

The postcephalic digestive tract is characterized by a midgut which is usually subdivided into two or three differentiated regions; the anterior part is always distinctly dilated; in Pentatomomorpha the middle section is usually equipped with gastral caeca or crypts, which contain endosymbiotic bacteria; the posterior part of the midgut is the more or less globular pylorus, which bears the four free Malpighian tubules (six in *Kokeshia* [Dipsocoromorpha, Schizopteridae]). Like in the other groups of Acercaria only one complex of ganglia is present in the abdomen. The testes are usually composed of 2–8 follicles, but the maximum number is 17. The ovaries are composed of seven ovarioles in most groups. A receptaculum seminis (spermatheca) is usually present on the dorsal side of the vagina. It varies strongly in different lineages and is functionally replaced by a pouch formed by the oviduct in some groups (Deckert & Göllner-Scheiding 2005).

Biology. Heteroptera are a highly successful group with an unusually broad spectrum of habitats, feeding habits, and attachment devices. The two basal groups Enicocephalomorpha and Dipsocoromorpha have a rather cryptic lifestyle in moist habitats among moss or in leaf litter. Phytophagous bugs (e.g., Pentatomidae, Miridae) are usually found on flowers or other parts of plants. Subcortical habitats are characteristic for species of the strongly flattened Aradidae, especially those occurring in the temperate regions. Termitaphididae, highly modified eyeless and wingless inquilines of termites, are possibly a specialized subgroup of Aradidae. They also suck fluids from fungal mycelia with long and hair-like stylets, which are coiled inside the head at rest (Grimaldi & Engel 2005). Associations with ants occur in basal lineages of the cimicomorph Tingidae. These species feed on roots within their host's nests.

A remarkable feature of heteropteran evolution is the very successful colonization of a broad variety of aquatic habitats. Ochteridae and Gelastocoridae (toad bugs) are two closely related semiaquatic families of Nepomorpha. Both are specialized on riparian habitats and the latter group is characterized by its jumping capacity. The more advanced lineages of Nepomorpha display evolutionary parallels to aquatic groups of Coleoptera (Crowson 1995), using their modified forewings in a similar way as beetles their elytra, i.e. enclosing a breathing air storage area. Some of them possess a plastron, a film of air held on the body surface by a dense vestiture of microtrichia. Members of the highly specialized Gerromorpha use the surface water film in a unique way, gliding on it rapidly in search of prey (e.g., Gerridae [water striders, pond skaters], Hydrometridae [water measurers, water treaders], Veliidae [riffle bugs]). Interestingly, at least 14 independent colonizations of the marine environment took place within this group (Andersen 1998). Some species of the gerrid genus *Halobates* live on the open sea, a unique feature in Hexapoda. Their females deposit the eggs on feathers and other swimming objects.

Predaceous habits are very likely a groundplan feature of Heteroptera (e.g., Cobben 1978). This is maintained in the basal heteropteran lineages, some taxa of the highly diverse Cimicomorpha (e.g., Nabidae, Reduviidae, Miridae [partim]) and Pentatomomorpha (e.g., Asopinae), and generally in the aquatic groups (Gerromorpha, Naucoridae, Notonectidae [backswimmers], Pleidae, Belostomatidae, Nepidae etc.). Corixidae (water boatmen) have an unusually broad food spectrum. They are primarily carnivorous but also feed on algae and detritus. Large species of Belostomatidae even attack small fish and amphibians. They possess large, strongly convex compound eyes and raptorial forelegs. Species of the belostomatid genus *Limnogeton* are specialized on aquatic snails. They have secondarily lost the raptorial forelegs, which belong to the groundplan of the family.

Many terrestrial species with primarily predaceous habits are facultative or obligatory consumers of plant sap. Phytophagy arose many times independently in Heteroptera. The majority of species, especially in Pentatomomorpha, primarily feed on plant juices, but many of them consume also animal matter, for instance from dead insects. In some species of Miridae the earlier nymphal stages are predaceous, whereas the older instars and the adults are phytophagous. Vascular fluids of angiosperms are the preferred food source (young shoots, buds, leaves, blossoms, fruit), but few groups are specialized on ferns. Specializations on seeds occur, for instance in Lygaeidae and Pyrrhocoridae, and flower-visiting species of Miridae consume nectar and pollen. Juices of fungal mycelia are used as food substrate by the species-rich Aradidae and the highly specialized Termitaphididae (see above).

Feeding on blood has evolved several times independently from predaceous habits. Members of Polyctenidae, Cimicidae and Triatominae (Reduviidae) suck blood of warm-blooded vertebrates. Triatomine species, which transfer the Chagas disease in South America, use various mammals (mostly rodents) and birds as hosts (Weirauch & Schuh 2011), whereas most species of the other two groups are specialized on bats and specific bird species (e.g., swallows). Ectoparasitic bugs usually leave their host after the blood meal, but this is not the case in Polyctenidae, which occur exclusively on bats (Deckert & Göllner-Scheiding 2005).

Heteropterans are mainly protected by their scent glands. The secretions can also play a role in the intraspecific communication, for instance as alarm or sex pheromones. Camouflage patterns and different forms of mimicry occur in different groups. Body shapes resembling ants have evolved in several families. Some reduviids (*Phonoctonus*) are similar to species of the pyrrhocorid genus *Dysdercus*, which are their preferred prey (Deckert & Göllner-Scheiding 2005).

Migrations occur in many species, either searching for food plants or suitable sites for overwintering. *Eurygaster integriceps* (Scutelleridae [shield bugs]), a cereal pest, covers distances of up to 200 km. Mass migrations occur in the aquatic families Corixidae and Notonectidae (Deckert & Göllner-Scheiding 2005).

Reproduction and development. The reproduction is generally sexual. Parthenogenesis is only know in one species of Miridae (*Campyloneura virgula*), where the males are very rare and have degenerated genital organs. Vivipary occurs in Polyctenidae which parasitize bats, and ovovivipary in some Lygaeidae. The position of copulation varies strongly within the group. Sperm is usually transferred via a spermatophore. This is stored in the spermatheca in most groups, but this is not the case in Cimicomorpha (possible autapomorphy). The reduced spermatheca is arguably a precursor of the traumatic insemination, which is an unusual characteristic of Cimicidae (Grimaldi & Engel 2005). In this group males pierce the female abdomen at the intersegmental membranes. Sperm cells are injected and migrate to the ovaries in the female body cavity.

Single eggs or batches are usually deposited in plant tissue or soil with the ovipositor. In groups with a reduced ovipositor eggs are glued to substrates, more or less randomly deposited on exposed surfaces, or buried in soil. Brood care occurs in some groups (e.g., Cimicidae, Acanthosomatidae, Pentatomidae). Males of the giant water bugs (Belostomatidae) carry the eggs on their back. The egg chorion is usually ruptured at the cephalic region. Preformed zones of weakness occur in Corixidae and Piesmatidae. An operculum is discarded by Nabidae, Cimicidae and Reduviidae. The postembryonic development is of the typical hemimetabolous type (Fig. 2.3.1). The number of nymphal stages is usually five, but few groups have only four (e.g., *Mes-*

ovelia furcata). The nymphs are similar to the adults but lack ocelli and fully developed wings and genitalia. They generally live in the same habitats. The wing buds are indistinct in the first two stages (Deckert & Göllner-Scheiding 2005).

Fossil record. The oldest heteropteran fossil is †*Paraknightia magnifica* from the Upper Permian of Australia. It is characterized by large paranotal lobes, a long ovipositor, a distinct clavus, and a strongly reduced venation of the forewing. The basal Enicocephalomorpha and Dipsocoromorpha are known from Cretaceous amber of Lebanon and Myanmar, and also from other amber deposits. Very few specimens of Gerromorpha are known from Jurassic (Karatau) and Cretaceous deposits, whereas Nepomorpha are well recorded as a successful group as early as in the mid-Jurassic (ca. 180 Ma). Pentatomomorpha and Cimicomorpha are represented by Mesozoic compression fossils of species of Cydnidae (Late Jurassic), Anthocoridae, Coreidae, Reduviidae, and Tingidae (Early Cretaceous) (Grimaldi & Engel 2005).

Economic importance. Some of the predaceous aquatic bugs can cause damage to fish populations in ponds, especially in pisciculture. The ectoparasitic bed bug *Cimex lectularius* (Cimicidae) is an increasing problem in the hotel business. They can be an extreme nuisance and also cause skin rashes, allergic reactions and psychological symptoms. Another species of the family, *Haematosiphon inodorus*, is known as poultry bug or "Mexican chicken" (Essig 1926). It is typically associated with birds, especially with chicken, but occasionally invades human habitations. It is reported that in some cases the inhabitants preferred to burn down and rebuild the entire village (Essig 1926). Serious medical problems are caused by species of the genus *Rhodnius* and some other triatomines (e.g., *Triatoma*), which are important vectors of Chagas disease (American trypanosomiasis) in Mexico, Central America and South America. Other reduviids have a positive impact as predators of pest insects. Some predaceous species (e.g., Lygaeidae, Miridae) are used as biological pest control agents. Numerous phytophagous species can cause great damage, especially in tropical and subtropical regions. Cultivated plants affected by true bugs are sweat potatoes, tomatoes, coconut (Coreidae), banana (Cydnidae), corn, cotton (Lygaeidae, Pyrrhocoridae), rice (Pentatomidae), cocoa, tea (Miridae) and others. Heteropteran species play a limited role as pollinators of plants (e.g., True sago palm) (Deckert & Göllner-Scheiding 2005).

[Essig (1926); Weber (1930); Jordan (1950, 1952, 1972); Cobben (1968, 1978); Dennis & Bitsch (1973); Schuh (1979); Zrzavy (1990); Mahner (1993); Wheeler et al. (1993); Crowson (1995); Schuh & Slater (1995); Andersen (1998); Beutel & Gorb (2001); Deckert & Göllner-Scheiding (2005); Grimaldi & Engel (2005); Schuh et al. (2009); Weirauch & Cassis (2009); Weirauch & Schuh (2011)]

Holometabola (Endopterygota)

The monophyly of the megadiverse Holometabola (ca. 770,000 spp.) is unambiguously confirmed by morphological and molecular data (Trautwein et al. 2012) (Fig. 6.D: 1). The holometabolous development as such is very likely an autapomorphy (with partial reversal in Strepsiptera; Niehuis et al. 2012) and one of the factors contributing to the unparalleled diversification of the group. Larvae and adults differ fundamentally in their morphology and usually also in their life habits. They can use different microhabitats and exploit different food resources, which means reduced intraspecific competition. The pupa is very likely not a groundplan apomorphy of the group as very different forms occur. Very plesiomorphic mobile pupae with movable mandibles (pupa dectica) occur in Raphidioptera.

The phylogenetic relationships of the holometabolan orders can be considered as largely clarified as results of recent studies based on extensive morphological and molecular studies are almost completely congruent (Wiegmann et al. 2009; Beutel et al. 2011; Niehuis et al. 2012; see also Trautwein et al. 2012). The recently established phylogenetic concept is largely consistent with the branching pattern suggested by Hinton (1958) and Hennig (1969), but Hymenoptera are not placed as sistergroup of Mecopterida ("panorpid orders" of Hinton 1958), but as the sistergroup of all the remaining lineages as already suggested by Kukalová-Peck & Lawrence (2003) and Rasnitsyn & Quicke (2002) (Fig. 6.D: 2). Hymenoptera are the only group of Holometabola with fully developed labial endite lobes and a complete and functional orthopteroid ovipositor (groundplan).

A large clade comprises the Neuropterida (Neuroptera, Raphidioptera, Megaloptera), the highly specialized and enigmatic Strepsiptera, and the megadiverse Coleoptera (Fig. 6.D: 3). The interrelationships within the neuropterid orders are not fully settled yet but a sistergroup relationship between Megaloptera and Neuroptera appears more likely presently (e.g., Aspöck & Aspöck 2008) (Fig. 6.D: 4, 5). A primarily aquatic lifestyle of larvae is a potential synapomorphy of both groups (larvae secondarily terrestrial in most groups of Neuroptera). After a long controversy (e.g., Carpenter et al. 1997; Wheeler et al. 2001; s. also Beutel & Pohl 2006) the sistergroup relationship between monophyletic Coleoptera and Strepsiptera (Coleopterida) is well established by morphological data (Beutel et al. 2011), single gene analyses (Wiegmann et al. 2009) and an analysis of entire genomes (Niehuis et al. 2012) (Fig. 6.D: 6). A hypothesized group "Halteria" comprising Strepsiptera and Diptera (e.g., Wheeler et al. 2001) can be considered as obsolete and also a subordinate placement of Strepsiptera within polyphagan beetles as suggested by Crowson (1981).

Mecopterida comprising the two clades Amphiesmenoptera (Trichoptera + Lepidoptera) and Antliophora (Mecoptera, Siphonaptera, Diptera) is not unambiguously supported by morphological data (Fig. 6.D: 7), but it was clearly confirmed in recent molecular studies (Wiegmann et al. 2009; Niehuis et al. 2012). Amphiesmenoptera and Antliophora are well-supported as clades (Fig. 6.D: 8, 9). Within Antliophora a sistergroup relationship between Mecoptera (Fig. 6.D: 10) and the ectoparasitic Siphonaptera is suggested by the results of most recent studies. In contrast to a suggested sistergroup relationship between the mecopteran family Boreidae and Siphonaptera (Fig. 6.D: 11; Whiting 2002) the monophyletic origin of Mecoptera was confirmed in recent morphology-based and molecular studies (Wiegmann et al. 2009; Beutel et al. 2011).

Apomorphies of supraordinal taxa (Fig. 6.D; mainly based on cladogram of Wiegmann et al. [2009] and morphological data matrix of Beutel et al. [2011])

Holometabola
- Holometabolous development
- Larvae with reduced glossae and paraglossae
- Larvae with simplified antennae
- Larval tarsus always undivided
- Larval cerci absent (reversal in Strepsiptera?)
- Pterothoracic sternites of adults invaginated
- Pterothoracic coxae medially adjacent
- Appearance of fully developed compound eyes including external apparatus in the pupal stage (reversal in Strepsiptera)
- External wing buds absent in larvae (reversal in Strepsiptera)

Holometabola excl. Hymenoptera
- Paraglossae and their muscles reduced

The glossae and their muscles are also missing in almost all groups but are preserved in Micropterigidae (Hannemann 1956).
- Orthopteroid ovipositor distinctly modified or reduced
- Larvae prognathous (groundplan)
- Well-developed larval stemmata (groundplan)

Neuropteroidea (Coleopterida + Neuropterida). This group is supported by molecular data (Wiegmann et al. 2009) and tentatively by characters of the ovipositor (Mickoleit 1976) (Fig. 6.D.3). However, this structure is completely reduced in Strepsiptera.
- Adult head prognathous (groundplan, several reversals) (?)
- Modifications of axillary sclerites (Hörnschemeyer 1998) (?)
- Cerci of adults largely or completely reduced (?)
- Modifications of the female postabdomen and their appendages (Mickoleit 1973) (ovipositor completely reduced in Strepsiptera) (?)

Coleopterida (Coleoptera + Strepsiptera)
- Reduced number of antennomeres (maximum 13 in Permian beetles)
- Posteromotorism (flight with hindwings)
- Mesothorax distinctly reduced in size (correlated with posteromotorism)

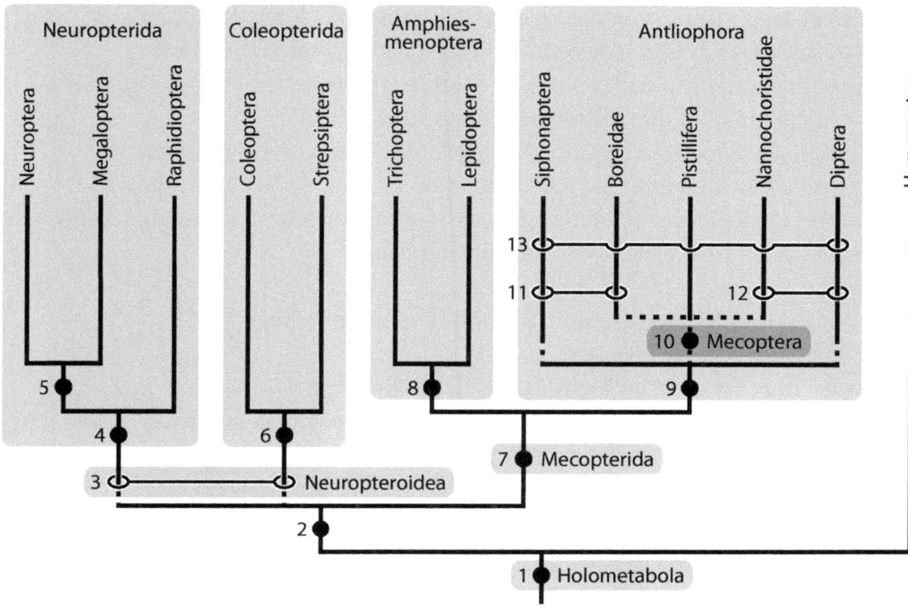

Fig. 6.D: Holometabolan relationships, with selected potential apomorphies. **1**, holometabolous development, larval glossae and paraglossae reduced, simplified larval antennae, larval tarsus undivided, larval cerci absent (Strepsiptera?), adult pterothoracic sternites invaginated, compound eyes appear in pupal stage (reversal in Strepsiptera), external wing buds absent in larvae (partial reversal in Strepsiptera); **2**, paraglossae reduced, orthopteroid ovipositor partly reduced, larvae prognathous (groundplan), larval stemmata; **3**, adult head prognathous (groundplan?), modifications of axillary sclerites, cerci reduced (?), modifications of female postabdomen; **4**, increased number of retinula cells in stemmata, metapostnotum medially divided, more than 5 costal crossveins; **5**, primarily aquatic larvae, mesothoracic prealare present, accumulated trichobothria on anal segment forming rosette; **6**, reduced number of antennomeres, posteromotorism and related features, pronotum and propleura connected, profurcal arm and propleuron separate, membranous area between mesoscutellum and mesopostnotum, ventral meso- and metasternal processes absent, cerci of adults absent, larval M. frontolabralis absent; **7**, part of pleurotergal muscle shifted to axillary 1, orthopteroid ovipositor absent, telescoping female postabdomen (groundplan), ventral metasternal process elongated, larval dorsal tentorial arm reduced, less than 3 larval antennomeres, larval stipes divided into basistipes and dististipes, larval galea and lacinia fused, loss of cephalic larval muscles, larval M. craniodististipitalis present; **8**, movable epiphysis of fore tibia, mesofurca and mesepimeron fused, metafurca and metepimeron fused, extrusion of female terminal segments by muscle force, field of sensilla trichodea in membrane between subalare and dorsal epimeral margin, larval maxillolabial complex, spermathecal gland, male sex chromosomes homogametic; **9**, M. craniocardinalis absent, 2 labial palpomeres (reversal in Siphonaptera), transverse epipharyngeal muscles absent, brain and suboesophageal complex form compact mass around pharynx, loss of cephalic muscles in larvae; **10**, male genital segments form genital capsule.

- Loss of several mesothoracic muscles (correlated with posteromotorism)
- Metathorax distinctly enlarged (correlated with posteromotorism)
- Pronotum and propleuron partly or completely connected (also in Diptera)
- Profurcal arm and propleuron not connected
- Median mesonotal suture absent (parallel evolution in Antliophora)
- Membranous area between mesoscutellum and mesopostnotum present
- Ventral meso- and metasternal processes (forming sterno-coxal joint below corresponding furca) absent (secondarily reduced)
- Cerci of adults absent
- Larval M. frontolabralis absent (parallel evolution in Neuroptera)

Neuropterida (Neuroptera, Megaloptera, Raphidioptera)
- Distinctly increased number of retinula cells in the larval stemmata
- Metapostnotum medially divided
- More than 5 costal crossveins (polarity?)

Megaloptera + Neuroptera
- Primarily aquatic larvae
- Mesothoracic prealare present (also in Amphiesmenoptera)
- Muscular connection between metafurcal arm and epimeral apophysis
- Accumulated trichobothria on anal segment forming a rosette (groundplan of Neuropterida ambiguous)

This sistergroup relationship (e.g., Aspöck & Aspöck 2007) is also supported by transcriptome data (Peters et al., unpubl. data).

Megaloptera + Raphidioptera (?)
- Adults with gula (also in Coleoptera)
- Metathoracic postepimeron present and connected with metafurca by an apodeme and a muscle (Willmann 2005a)
- Tarsomeres with vestiture of adhesive microtrichia
- Telotrophic ovarioles with secondary nutritive chamber
- Specific cleaning behavior

Mecopterida (Amphiesmenoptera + Antliophora)
- Part of pleurotergal muscle shifted to axillary 1 (Mickoleit 1966, 1969)
- Orthopteroid ovipositor completely reduced
- Telescoping female postabdomen (groundplan; Hünefeld & Beutel 2012)
- Ventral metasternal process elongated (Friedrich & Beutel 2010b)
- Larval dorsal tentorial arm strongly reduced or absent
- Less than 3 larval antennomeres (reversal to 3 in some groups) (?)

- Larval stipes divided in basistipes and dististipes (e.g., Kristensen 1999a; Beutel et al. 2009)
- Larval galea and lacinia extensively or completely fused (also missing as separate structures in Neuroptera and Strepsiptera)
- Loss of maxillary and labial muscles in larvae
- Larval M. craniodististipitalis present

Amphiesmenoptera (Trichoptera + Lepidoptera)
(see also Kristensen [1975], Kristensen [1999b: fig. 2.1] and Kristensen et al. [2007])
- Movable epiphysis of fore tibia present
- Mesofurca and mesepimeron fused
- Metafurca and metepimeron fused
- Extrusion of terminal segments of females by muscle force
- Ventral sclerites of female segment IX absent
- Field of sensilla trichodea in membrane between subalare and dorsal epimeral margin present
- Larval maxillolabial complex present
- Spermathecal gland present
- Male sex chromosomes homogametic

Antliophora (Mecoptera, Siphonaptera, Diptera). The name Antliophora refers to the presence of a sperm pump (Greek *antlia* = pump). Whether such a specialized apparatus belongs to the groundplan of this lineage was already questioned by Hennig (1969) and it was shown by Hünefeld & Beutel (2005) that this is not the case.
- M. craniocardinalis absent
- Two or less labial palpomeres (reversal in Siphonaptera)
- M. clypeopalatalis enlarged (?)
- Transverse epipharyngeal muscles absent
- Brain and suboesophageal complex forming compact mass around pharynx
- Median mesonotal suture absent (parallel evolution in Coleopterida)
- Loss of cephalic muscles in larvae

Mecoptera + Siphonaptera
- Acanthae of proventriculus close-set, prominently elongated
- Intraprofurcal muscle present, connecting profurcal arms
- 9 + 2 microtubili pattern in the sperm axoneme

[Hinton (1958); Mickoleit (1966, 1969, 1976); Hennig (1969); Kristensen (1975, 1981, 1999a, b); Crowson (1981); Hannemann (1986); Kukalová-Peck & Lawrence (1993); Carpenter et al. (1997); Hörnschemeyer (1998); Wheeler et al. (2001); Whiting (2002); Hünefeld & Beutel (2005, 2012); Rasnitsyn & Quicke (2002); Willmann (2005a); Beutel & Pohl (2006); Aspöck & Aspöck (2007, 2008); Kristensen et al. (2007); Beutel et al. (2009, 2011); Wiegmann et al. (2009); Friedrich & Beutel (2010b); Niehuis et al. (2012); Trautwein et al. (2012)]

6.25 Hymenoptera (Greek *hymen* = membrane, *pteron* = wing, English common names for subgroups: sawflies, wood wasps, bees, wasps, ants)

Diversity and distribution. Hymenoptera are one of the extremely species rich groups of Holometabola and presently comprise ca. 132,000 recognized species. Conservative estimates of the real species number vary between 600,000 and 1,200,000 (Grimaldi & Engel 2005). The group has a worldwide distribution with the exception of Antarctica. Hymenopteran species reach 83 degrees latitude north (Greenland) and ants were collected at an elevation of 4,800 m in the Himalayas. Ants in the Sahara are able to cope with extremely high temperatures of up to 46°C.

Autapomorphies
- Clypeus inflected (Fig. 6.25.1)
- Maxillolabial complex (Fig. 6.25.1)
- Deep infrabuccal pouch
- Strongly developed longitudinal muscle of the epipharynx
- Protibial spur modified as cleaning device
- Distinctly simplified wing venation (Fig. 6.25.2)
- Smaller hindwings connected with forewings by small hooks (hamuli)
- Abdominal tergite I at least partly fused with metanotum
- Males haploid (haplodiploidy)

Taxonomy. Hymenoptera is traditionally subdivided into the two suborders "Symphyta" (e.g., Xyelidae, Tenthredinidae [ca. 5,000 spp.], Cephidae, Siricidae) and Apocrita. The former are a paraphyletic group comprising the families without or with a vaguely indicated (Orussidae) wasp-waist, and a largely unmodified orthopteroid ovipositor suitable for egg deposition in plant tissue. The small family Xyelidae (ca. 70 spp.) is the sistergroup of all remaining hymenopteran families, and the species rich Tenthredinoidea the second branch (Vilhelmsen 2001). The parasitoid Orussidae and Apocrita form a clade Euhymenoptera. Presumptive synapomorphies are reduced larval eyes and legs. The adaptations to a parasitoid lifestyle include a far-reaching reduction of the larval mouthparts. Apocrita comprise more than 90% of all known hymenopteran species. They are characterized as a clade by a distinct wasp-waist (Fig. 6.25.3B). Within Apocrita, the "Hymenoptera parasitica" or "Terebrantia" (ca. 65,000 spp., e.g., Chalcidoidea, Cynipoidea, Ichneumonoidea) are a traditional, non-monophyletic lineage, which is paraphyletic with respect to Aculeata. Aculeata (ca. 60,000 spp.; e.g., Chrysidoidea, Vespoidea, Apoidea) are a large monophyletic unit within Apocrita. It is characterized by the transformation of the ovipositor into a defensive stinging device ("stinging wasps").

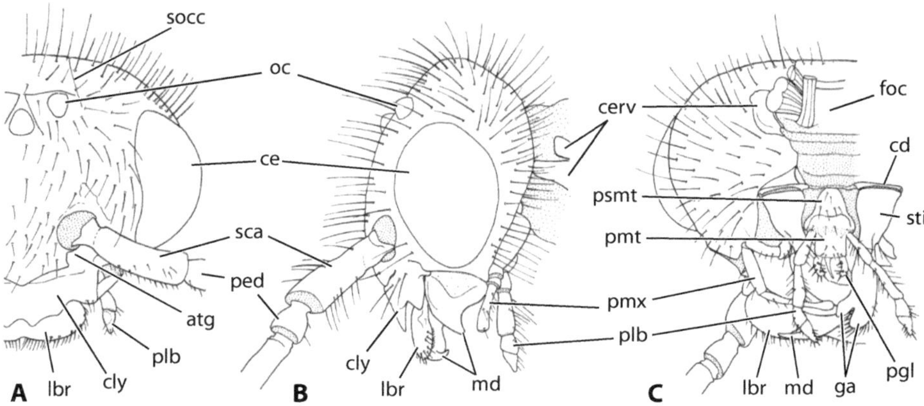

Fig. 6.25.1: *Macroxyela* (Xyelidae), adult head. A, frontal view; B, lateral view; C, posterior view. Abbr.: atg – anterior tentorial grooves; cd – cardo, ce – compound eye, cerv – cervical sclerite, cly – clypeus, foc – foramen occipitale, ga – galea, lbr – labrum, md – mandible, oc – ocellus, ped – pedicellus, pgl – paraglossa, plb – palpus labialis, pmx – palpus maxillaris, pmt – prementum, psmt – postmentum, sca – scapus, socc – sulcus occipitalis, sti – stipes. From Beutel & Vilhelmsen (2007), modified.

Diagnosis. Extremely small to large, with transparent wings with reduced venation (Fig. 6.25.2). Hindwings distinctly smaller than forewings. Head orthognathous, fully exposed and highly movable. Mouthparts comprising biting mandibles and a maxillolabial complex (Fig. 6.25.1), often modified for the uptake of liquid food. Abdominal base more or less fused with metathorax. Larvae resemble caterpillars ("Symphyta" minus Orussidae) or are structurally strongly simplified (Euhymenoptera [=Orussidae + Apocrita]).

Morphology, adults. The size of the adults ranges between 0.25 mm (Trichogrammatidae) and 70 mm (e.g., some Scoliidae, Pompilidae). The smallest winged species are parasitoid Hymenoptera. The cuticle is usually well sclerotized but can be very thin in miniaturized forms (e.g., Mymaridae); a strengthened cuticle occurs in some parasitoids. The basic colors are brown, yellow and black, but conspicuous patterns occur, often yellow combined with black, and species of different groups display more or less conspicuous metallic colorations (e.g., Chalcidoidea, Chrysididae, Sphecidae). The cuticular surface is often densely covered with setae (e.g., Apidae) but it is largely glabrous and shiny in some groups (e.g., Vespidae); a vestiture resembling velvet is present in Mutillidae (velvet ants).

The head (Fig. 6.25.1) is almost always fully exposed (closely associated with prothorax in Siricidae) and orthognathous; it is usually compressed between its frontal and posterior surface and the posterior side is distinctly concave. The foramen occipitale is narrow. A hypostomal bridge is usually present but missing in the groundplan (Xyelidae, Tenthredinoidea); a postgenal bridge has independently evolved in Siric-

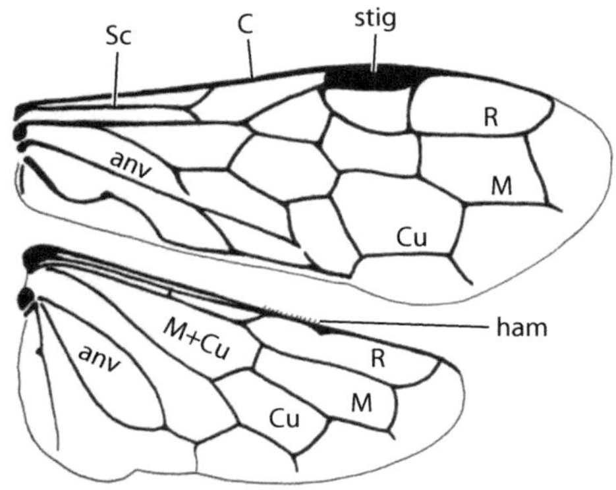

Fig. 6.25.2: Fore- and hindwing, *Pamphilius* sp. (Megalodontesidae). Abbr.: anv – anal vein, C – costa, Cu – cubitus, ham – hamuli, M – media, R – radius, Sc – subcosta, stig – stigma. Redrawn from Dathe (2005).

idae and the clade comprising Orussidae and Apocrita. The narrowed cervical region results in an increased movability of the head. Anteriorly diverging occipital sulci are present in the groundplan. The Y-shaped epicranial suture (frontal and coronal sutures) is always reduced. The compound eyes are almost always well-developed and often larger in males (e.g., 7,500 ommatidia in males of the honeybee versus 6,300 in females); they are dorsally adjacent in some cases (e.g., male honeybee) and kidney-shaped in most Vespidae; in some cases the size of the ommatidia is different on the ventral and dorsal side; ocelli are primarily present but often partly or fully reduced in flightless forms or enlarged in nocturnal species. The clypeus is always inflected (autapomorphy). The labrum is free and exposed in Xyelidae and Tenthredinoidea, but inflected, displaced posteriorly to the tips of the mandibles, and often reduced in size in the other groups of Hymenoptera. The antennal insertion areas (toruli) are more or less closely adjacent on the anterior side of the head capsule. The antennae are primarily multisegmented and filiform, with up to 90 segments (some Braconidae), and usually densely covered with sensilla; the scapus is often elongated; an elongated antennomere 3 (e.g., Xyelidae, Tenthredinidae, Cimbicidae) is likely a groundplan autapomorphy of Hymenoptera; serrate, clubbed (Cimbicidae) or geniculate (Formicidae) antennae occur; flabellate antennae are present in males of few groups of Tenthredinoidea (e.g., Diprionidae) and Apocrita (e.g., Eulophidae). The mouthparts are orthopteroid in the groundplan. The mandibles are generally well-developed but modified in some groups depending on the function (e.g., Apidae); they are distinctly asymmetric in Xyelidae, but only slightly asymmetric or symmetric in the other groups; a membranous basolateral area occurs in basal lineages; the mola is well-developed in Xyelidae, but distinctly reduced or absent in the other groups. The maxillae and labium are closely connected, thus forming a maxillolabial complex. The maxillae are primarily composed of a well-developed cardo and stipes, normally developed galea and lacinia, and a 5-segmented palp, but far-reaching modifications occur; in *Apis* and its relatives the cardo is strongly narrowed and connected by a transverse rod (lorum) with the small postmentum; the palp is minute, the lacinia reduced, and the galeae are enlarged and form a sheath together with the labial palps; a tendency to reduce the lacinia, the maxillary palp, and the paraglossae is characteristic for Hymenoptera. The labium is composed of an undivided postmentum and a prementum, which bears well-developed glossae and paraglossae and 3-segmented palps in the basal groups (e.g., Xyelidae, Tenthredinidae); the glossae of bees are strongly elongated, fused with each other, and densely pubescent, thus forming a device for taking up liquid food. The anterior epipharynx is covered by microtrichia; a prominent buccal lobe is formed by the intermediate section; it is laterally fused with the posterior hypopharynx, thus forming a prepharyngeal tube; sclerotized posterolateral epipharyngeal lobes with spines reach into the pharynx (groundplan). The hypopharynx forms a structural and functional complex with the prelabium. The salivarium is only represented by the salivary duct. The floor of the cibarium and prepharynx is formed by the sclerotized sitophore plate; the distal part of the hypopharynx covers the opening of the salivary duct; the large anterodorsal part forms the wall of a deep infrabuccal pouch. A very extensive and strongly sclerotized vertical collar-like structure is formed by the posterior tentorial arms and the tentorial bridge (autapomorphy); the anterior and dorsal arms are usually well-developed. The head musculature is well-developed including the muscles of the glossae and paraglossae, which are usually reduced in holometabolous insects; a very strongly developed longitudinal muscle of the epipharynx is likely an autapomorphy of Hymenoptera. Lateral cervical sclerites are present in the groundplan of Hymenoptera and articulate with propleural elements.

The pro- and metathorax are reduced in size whereas the mesothorax with its strongly developed flight muscles is distinctly enlarged (Fig. 6.26.3). The pronotum varies greatly in shape and is reduced to a very short, curved sclerite in different groups (e.g., Tenthredinidae); it is often firmly connected with the mesothorax, whereas the prosternal elements are free. The scale-like tegulae are posterolaterally adjacent with the pronotum. The basisternum is the predominant sclerite on the ventral side of the thoracic segments, especially in the mesothorax. The pleura of the pterothoracic segments are divided into an anterior anepisternum and a posterior epimeron (groundplan), but the metapleural suture and ridge are reduced in Apocrita. The mesonotum is subdivided into an anteromedian prescutum (often triangular and with a median furrow), an extensive scutal region, and a posterior mesoscutellum varying greatly in shape and size; a transverse articulatory furrow (transscutal articulation) separates the mesoscutum from the mesoscutellar-axillar complex in Xiphydriidae, Orussidae and Apocrita; in these groups, the transscutal articulation separates paired, triangular regions from the posterior mesoscutum, which lay anterad the scutoscutellar sulcus

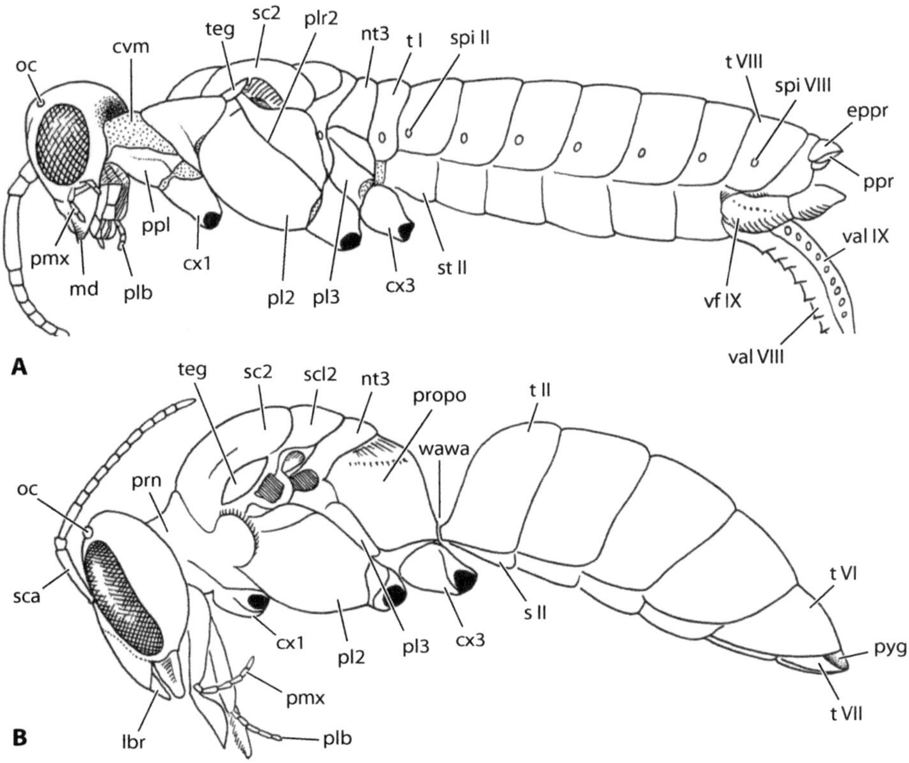

Fig. 6.25.3: "Symphytan" and apocritan body organization, lateral view. A, Tenthredinidae, female, generalized; B, *Melitta* sp. (Apidae). Abbr.: cvm – cervical membrane, cx1/3 – pro-, metacoxa, eppr – epiproct, lbr – labrum, md – mandible, nt3 – metanotum, oc – ocellus, plb – palpus labialis, pl2/3 – meso-, metapleuron, plr2 – mesothoracic pleural ridge, pmx – palpus maxillaris, ppl – propleura, ppr – paraproct, prn – pronotum, propo – propodeum (=abdominal tergum I), pyg – pygidium, sca – scapus, sc2 – mesoscutum, scl2 – mesoscutellum, spiII/VIII – abdominal spiracles II/VIII, sII – sternum II, teg – tegula, tI/II/VI-VIII – tergum I/II/VI-VIII, valVIII/IX – valvulae VIII/IX, vfIX – valviver IX; wawa – wasp waist. Redrawn from Dathe (2005), after Weber (1966) and Mitchell (1960).

and are termed axillae, and the mesonotum is subdivided into the mesoscutum and the mesoscutellar-axillar complex; at its posterior margin the mesoscutellum bears the mesopostnotum. The metanotum is much smaller than the mesonotum; in almost all symphytan groups it bears knob-shaped protuberances, the cenchri, which hold the folded wings in their resting position. The abdominal tergite I is at least partly fused with the metapostnotum (autapomorphy of Hymenoptera) (Fig. 6.25.3A); it is referred to as propodeum in Apocrita and forms a rigid functional and structural unit with the metapostnotum and metapleura and the remaining parts of the thorax; this entire unit is called mesosoma (Fig. 6.25.3B). The membranous wings are characterized by a distinctly simplified venation (autapomorphy) (Fig. 6.25.2); they are usually well-developed, but reductions occur in few groups (e.g., *Gelis* spp. [Ichneumonidae], males of Mutillidae, *Olixon* spp. [Rhopalosomatidae]); the smaller hindwings are connected with the inflected hind margin of the forewings by small hook-shaped structures (hamuli); both pairs are moved simultaneously (functional dipterism); the wing base of the forewings is covered and protected by tegulae, small sclerites anterolaterally attached to the mesonotum; costa, subcosta, radius, media cubitus and three anal veins are usually present (groundplan), and an additional short jugal vein occurs in some symphytan and sphecid species (the homologisation of the veins is difficult due to the distinctly reduced condition); the maximum number of closed cells of the forewing is 17; a pterostigma at the anterior forewing margin is present in the groundplan; it is separated from the costa by a more or less distinct articulation; the anal veins (if present) end before they reach the posterior wing margin and form at least one closed cell; the plesiomorphic number of three is only preserved in symphytan lineages. The legs are composed of the usual elements; the coxae are undivided, i.e. they lack a distinctly separated posterior meron; a small proximal piece is often separated from the rest of the femur (trochantellus); one of the protibial spurs is modified as a cleaning device (groundplan autapomorphy); the second protibial spur is absent in Apocrita and the remaining spur forms a specialized cleaning apparatus for the antenna, together with a notch of the proximal protarsomere; a similar structural modification occurs on the hindlegs of some aculeate lineages (Sphecidae, Vespidae, Pompilidae); the tarsi are usually 5-segmented but legs with only four or three tarsomeres occur in some groups (e.g., Chalcidoidea partim, Trichogrammatidae, Eulophidae); adhesive tarsal thorns are present in the symphytan lineages; a membranous and smooth arolium is generally present as a pretarsal attachment device; forelegs with tibiae and tarsi equipped with strongly developed spines are used for digging nests by females of some groups referred to as "fossores" (e.g., Mutillidae, Sphecoidea, Pompilidae); other modifications of legs occur in different lineages of Hymenoptera, such as for instance densely pubescent hindlegs of Apidae used for collecting pollen, jumping hindlegs of some Chalcididae, or highly specialized clasping forelegs of females of Dryinidae.

The anterior abdomen is characterized by the secondary tagmosis with a more or less distinct borderline or incision between the abdominal segments I and II. The unit composed of the second and the following abdominal segments is referred to as metasoma. The distinct wasp-waist of Apocrita (less distinct in Orussidae) with a reduced sternite I greatly increases the movability of the abdomen and the ovipositor (Fig. 6.25.3D); a petiolus, a narrow, stalk-like element connecting the metasoma with the mesosoma occurs in different apocritan lineages (e.g., Formicidae) and is sometimes strongly elongated and tube-shaped (e.g., Sphecidae). Nine segments are exposed in males and only eight in females (groundplan) (Fig. 6.25.4A). The intersegmental membranes are concealed as the tergites and sternites overlap with those of the following segments and the lateral tergal regions partly cover the sternites. A pair of unsegmented socii is present at the abdominal apex (groundplan); it is unclear

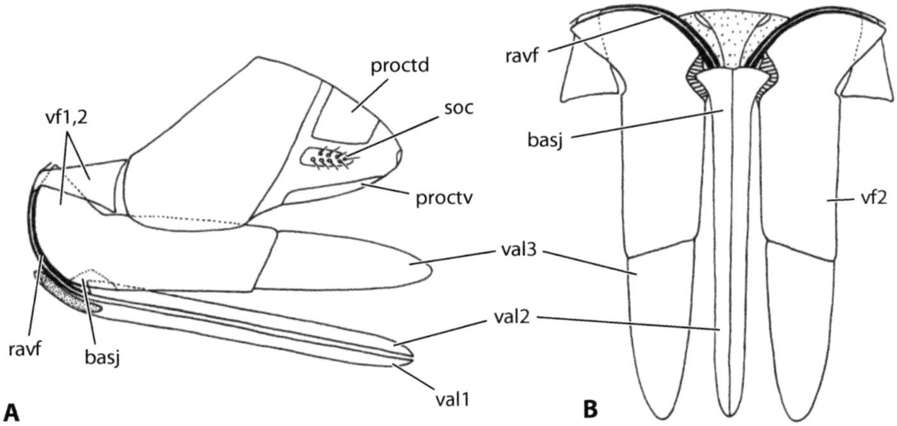

Fig. 6.25.4: Female postabdomen, ovipositor (groundplan). A, lateral view; B, dorsal view, without tergite IX. Abbr.: basj – basal joint, proctd – proctiger (dorsal), proctv – proctiger (ventral), ravf – rami of valvifers, soc – socius, val 1/2/3 – valvulae 1/2/3, vf 1/2 – valvifer 1/2. Redrawn from Oeser (1961).

whether they are appendages of segment X (pygopods) or segment XI (cerci). Segment VII and the following are internalized in Aculeata (excl. Bethylidae), leaving only six segments of the metasoma exposed; a cavity formed during the invagination process contains the stinging apparatus. An extreme degree of simplification is found in Cheloninae (Braconidae), where all metasomal tergites form a single plate. Unpaired stridulatory organs occur between the 2nd and 3rd metasomal segments in Mutillidae and many Formicidae. The ovipositor is largely exposed or invaginated in its resting position; in the groundplan of Hymenoptera (symphytan groups) it has retained its original function of depositing eggs (Fig. 6.25.4); it is composed of the gonocoxae VIII and XI (valvifer 1 and 2), the gonapophyses VIII and IX (valvulae 1 and 2), and the third valvula inserted on valvifer 2; the valvulae 1 are slender and acuminate; the valvifers 2 (oblongum plate) articulate with tergite IX and bear the valvulae 2 which are usually medially fused; the paired 3rd valvulae form a sheath; the ovipositor has often a double function in parasitoid wasps, deposition of eggs and the injection of poison; in Aculeata the eggs do not pass through the ovipositor but are released at its base (with few exceptions: e.g., Chrysididae). The male copulatory apparatus is composed of the penis, the parameres, and the volsellae inserted between them; the penis and parameres are connected by the gonobase or basal ring (more or less reduced in different group); the volsellae, pincer-shaped sclerites equipped with muscles, are a characteristic feature of males of Hymenoptera; the groundplan condition is a protandric copulatory apparatus, with the volsellae on the ventral side; a torsion resulting in a dorsal position of the volsellae and intercrossing seminal ducts occurs in different symphytan lineages.

Fig. 6.25.5: Mymaridae, female, habitus, lateral view, SEM micrograph. Abbr.: ant – antenna, ce – compound eye, fwi – forewing, hwi – hindwing, ovip – ovipositor. Scale bar: 100 µm. Courtesy H. Pohl.

The digestive tract of Hymenoptera varies strongly; an ingluvies is often present; it lies within the abdomen and can be extremely developed in forms which feed other members of colonies (e.g., honeypot ants); a specialized valve-like proventriculus is present between the ingluvies and the mid-gut. A plesiomorphic feature of Hymenoptera is the high number of Malpighian tubules (usually between 20 and 40) (Dathe 2005). The ovarioles are polytrophic.

Morphology, larvae. The polypodous eruciform larval type found in the symphytan lineages (Fig. 6.25.6) represents the groundplan condition of Hymenoptera. The head capsule is well-developed, thoracic legs are present, and the cylindrical abdomen bears a series of short prolegs. A progressive reduction from free living larvae through wood and stem boring symphytan immature stages to the strongly simplified larvae of Apocrita is characteristic for the order. Apocritan larvae usually feed in concealed microhabitats and are completely apodous and eyeless. The head capsule is frequently reduced and the body unsclerotized and unpigmented.

The larval head is well sclerotized, orthognathous and globular in the groundplan of Hymenoptera (Fig. 6.25.6), but weakly sclerotized and lightly pigmented if at all in Apocrita. A longitudinal furrow on the posterior head capsule is present in most symphytan lineages. Frontal sutures and a coronal suture are usually present, but often incomplete and not or scarcely recognizable in smaller larvae. A frontoclypeal transverse strengthening ridge is present in Xyelidae, but absent in other groups. The clypeus is divided into an anterior anteclypeus without muscle attachment and a posterior postclypeus (symphytan larvae). Relatively small simplified

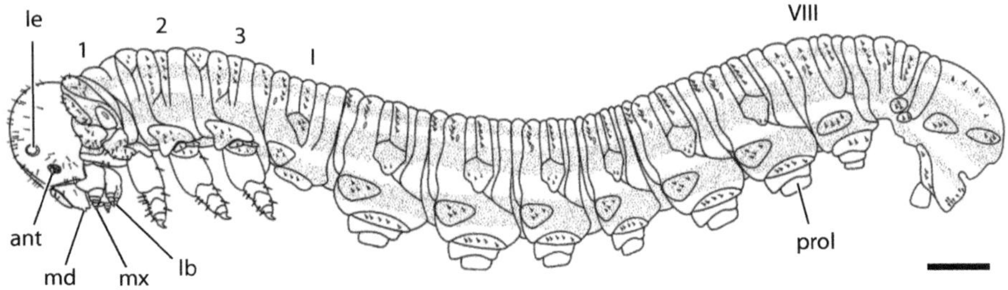

Fig. 6.25.6: *Neodiprion sertifer* (Diprionidae), larva, habitus. Abbr.: I/VIII – abdominal segments I/VIII, 1–3 – pro-/meso-/metathorax, ant – antenna, lb – labium, le – larval eye, md – mandible, mx – maxilla, prol – abdominal proleg. Scalebar: 1 mm. From Grabarek (2008), modified.

compound eyes with several rhabdoms but without individual corneal lenses and crystalline cones (unicorneal composed eyes) are present in the hymenopteran groundplan (Fig. 6.25.6); they are absent in symphytan groups with endophytic (e.g., *Xyela*), wood-boring (Siricidae) or parasitoid habits (Orussidae); the lateral eyes are also reduced in Apocrita, even though pigmented streaks (parietal bands) occur in some groups; ocelli are always absent (synapomorphy of Acercaria and Holometabola [=Eumetabola]). The labrum is free and well equipped with intrinsic and extrinsic muscles. The antennae are short and composed of six (Xyelidae) or seven (Pamphiliidae) segments in the groundplan; less segments are present in most symphytan lineages, and the antennae are 1-segmented or vestigial in Orussidae and Apocrita; the antennal muscles are usually weakly developed. The mandibles are always well-defined and functional; they are strongly developed in the symphytan groups and sometimes palmate or scoop-shaped; a mola with grinding surface is lacking. The ventral mouthparts are strongly retracted and form a maxillolabial complex without an articulatory membrane between them. This structure is mainly moved in vertical direction. The cardo bears a lateral lobe in symphytan larvae (possible groundplan autapomorphy of Hymenoptera); the stipes bears the segmented palp and usually a distinctly developed galea and lacinia (groundplan). The labium is composed of the undivided postmentum and the prementum, which bears the 3-segmented palps (groundplan, 1- or 2-segmented in some symphytan groups) and the opening of the silk-glands (prelabio-hypopharyngeal lobe); glossae and paraglossae are always missing. The maxillae and labium of apocritan larvae (and larvae of Orussidae) are distinctly simplified; they are usually present as more or less prominent fleshy lobes; the palps are represented by small discs or protuberances with sensilla; the endite lobes of the maxilla are reduced. A strongly developed X-shaped tentorium is present in symphytan larvae, with dorsal arms (groundplan) and a massive tentorial bridge (corpotentorium). An additional caudal tentorial arm occurs in different symphytan groups (e.g., Xyelidae).

The thoracic segments of symphytan larvae are divided into two or more annulets and bear a varying number of lateral lobes. The legs are usually short (very short in *Xyela*) but distinctly developed, segmented and equipped with a single claw (Fig. 6.25.6). Different degrees of reduction occur in internally living symphytan larvae (e.g., Cephidae). The legs are represented by small protuberances in the parasitoid Orussidae and always reduced in Apocrita. Spiracles are present between the pro- and mesothorax (usually the largest) and between the meso- and metathorax.

The abdomen is 10-segmented. Segments I–IX of symphytan larvae are usually similar to those of the thorax. They are subdivided into 2–7 annulets (usually 3–7) and bear series of abdominal prolegs (Fig. 6.25.6), on segments II–VI, II–VII, II–VIII, on IV–VIII + X, or on all segments (Xyelidae). The prolegs always lack apical crochets and they are reduced or completely absent in some internally feeding forms. Tergite X is rounded and usually displays ornamentation similar to that of the preceding segments; sometimes it bears fleshy protuberances or a postcornus; sternite X lacks appendages or bears fleshy lobes or simple or segmented appendages. The ornamentation of the postcephalic body may include patterns of setae, tubercles, warts, dark plates, or stout or slender or branched spines; it is usually lacking in internally feeding larvae and also in larvae of Apocrita (Smith & Middlekauff 1987).

Morphology, pupae. The pupa adectica libera is usually unpigmented and unsclerotized. A pupa obtecta occurs in some groups of Chalcidoidea. A cocoon is formed in some groups using secretions of the silk glands (see below). It can be parchment-like in texture, a thin, silken lining within the larval cell (Aculeata), or made using soil particles. Some ants do not form any cocoon.

Biology. Adult hymenopterans occur in almost all terrestrial habitats. They are often encountered on plants, especially on flowers, but they also occur in leave litter or the upper layers of soil substrate. Most species have an excellent flying capacity and the dispersal ability is generally good. Whereas larger species disperse by active flight, smaller forms may rely on air currents or phoresis. Aquatic habits are an extremely rare exception, but some parasitoid species swim in fresh water habitats to find suitable hosts. Some species specialized on seashores build nests immediately above the high tide line. One species is associated with corals.

The majority of species is diurnal. The adults usually feed on nectar, pollen or honeydew, but predaceous species occur in some groups (e.g., Vespidae). Adults of some species of Tenthredinidae and some ants actively prey on other insects, and females of some parasitoid species feed on body parts or the haemolymph of their hosts (host feeding). Similarly, some nest-building aculeates (e.g., Vespidae) feed on body parts or haemolymph of the insects they also use to feed their offspring. Sound production (stridulation) occurs frequently and has evolved independently in several lineages. Possible functions are defense or intraspecific communication.

The abundance of Hymenoptera is mainly regulated by climatic factors and the availability of resources, suitable hosts and nest sites. Hymenopteran larvae use specific secretions or haemolymph for their defense or live in nearly inaccessible or cryptic habitats. Many species are repugnant or inedible to vertebrates and the stinging apparatus of aculeates is a highly efficient defensive device. Aposematic color patterns occur frequently (usually yellow and black). Wasps and ants are very often models for mimicry complexes.

Hymenoptera are especially successful as ecto- or endoparasitoids. In this context they do not only play an immensely important role in natural ecosystem, but also in integrated pest management. Many species are hyperparasitoids parasitizing other parasitoids (e.g., Tachinidae) on or within a primary host (Grimaldi & Engel 2005).

Within Aculeata, highly developed social systems have evolved several times independently, within Vespidae, in Formicidae (ants) and in Apidae (bees) (see below). In the eusocial taxa nests are either founded by one or several fertile females of the same species, and several or numerous individuals feed and protect the offspring. There is always a division of labor between fertile individuals in charge of the reproduction and at least temporarily sterile workers, which take care of the brood, procure food and defend the colony. The co-existence and co-operation is regulated by chemical communication but also mechanical (ants) or visual signals (bees). Individuals in charge of defense can be morphologically modified (see also Isoptera in **6.17 Blattodea**), whereas an improved sense of direction and the capacity to remember specific locations can occur in those specialized on providing food. They use chemical signals or landmarks and polarized light for their orientation. Nests of eusocial species can be simple and abandoned after a short time or elaborate and used for a prolonged time span. A specialization occurs in hymenopteran social parasites: the queen enters nests and usurps the position of the original queen. The workers of the host species accept the new queen and take care of its offspring.

Reproduction and development. The main characteristics of the reproductive system of Hymenoptera are the absence of sex chromosomes, females usually developing from fertilized diploid eggs, and haploid males developing from unfertilized haploid eggs (arrhenotoky). Whether fertilization takes place or not is under the control of the egg-laying female. The sex is primarily determined by the chromosome set (i.e. haploid versus diploid), but in many hymenopterans the actual sex determination is controlled by a single locus with many alleles. Diploid males occur only very rarely and play no role in the reproduction. Thelytoky, with females developing from unfertilized diploid eggs, is relatively common. In groups of parasitoid wasps males can be very rare (e.g., Ichneumonidae, Cynipidae). The genetic variability is reduced by the specialized hymenopteran reproductive system, which also includes ameiotic spermatogenesis.

The reproductive behavior is often influenced by either parasitism or social systems, which evolved several times independently. Mating places are often determined by nest sites, hosts, or abiotic structures. Pheromones play often a role in finding partners. Complex courtship behavior involving antennal contact and vibrations of legs and wings occurs in different groups, especially within Apocrita. In some cases copulation takes place within galls (Cynipidae) or nests (*Bombus* spp.). Mating flights occur in bees and ants. Males often die after copulating.

The eggs of Hymenoptera are usually elongate and slightly curved. Eggs with a length of 16.5 mm produced by females of *Xylocopa* species (Apidae) are probably the largest occurring in the entire Hexapoda. Stalked eggs are common in the parasitoid Cynipidae, Chalcidoidea and Ichneumonoidea. Symphytan females lay 50–150 eggs, whereas queens of the honey bee can produce up to 750,000 during their lifetime.

Females of species with non-parthenogenetic sexual reproduction store sperm in their spermatheca. This allows controlling the fertilization of eggs depending on external conditions, i.e. controlling the timing of the production of offspring and the sex ratio. Under unfavorable conditions females can hold back the eggs in the oviduct and even reabsorb them to secure their own survival. Five different categories of larvae can be distinguished in Hymenoptera: 1) free-living, phytophagous eruciform larvae (Fig. 6.25.6) of most symphytan groups (e.g., Tenthredinidae); 2) symphytan larvae boring in different plant tissues including wood (e.g., Siricidae); 3) parasitoid larvae or larvae developing in plant galls; this can include hyperparasitism (most parasitized parasitoids also belong to Hymenoptera or to Diptera [Tachinidae]); 4) non-parasitoid and non-social apocritan larvae which are supplied with parts of insects, pollen or nectar; 5) eusocial apocritan larvae which are fed continuously with nectar containing animal matter, pollen or nectar (mixed with gland secretions in Apidae), or prey mixed with nectar, seeds or fungi (ants). Resulting from their habits of feeding seeds to their offspring ants play an important role in the dispersal of spermatophytes.

The length of the larval development varies greatly. In parasitoid species it can take several years. The larvae are always the main feeding stage. They are primarily phytophagous and more or less exposed ("Symphyta") but frequently feed within the body of a host, in a nest, in wood or stems, or in galls or leave mines. Hypermetamorphosis occurs frequently in the parasitoid lineages and also in some groups of Aculeata. Pupal cocoons are often produced using secretions of the labial glands. In other cases, protection is provided by the hosts integument or cocoons or webs produced by the host.

Fossil record. The distinctly reduced wing venation is a feature by which hymenopteran fossils can be easily recognized (Grimaldi & Engel 2005). The oldest definite members of the group are from the Triassic of Australia, Central Asia and Africa. Stem lineage hymenopterans have not been identified so far. The likely sistergroup relationship between Hymenoptera and the remaining holometabolan orders implies an origin in the early Permian or late Carboniferous. Xyelidae, the sistergroup of the remaining Hymenoptera, belong to the oldest extant insect families (Triassic, ca. 230 Ma). The first apocritans are documented from the Upper Jurassic and an earlier diversification of Hymenoptera was likely linked with the rise of Spermatophyta. Aculeates appear in the Cretaceous (ca. 100–125 Ma) and their immense diversification was apparently closely connected with the "angiosperm revolution". Definite fossils of Apidae are ca. 65 Ma old. Specimens preserved in Baltic amber can be assigned to extant families, often even to genera (Dathe 2005).

Economic importance. The economic importance of Hymenoptera can hardly be overestimated. They play an extremely important role as pollinators of angiosperms. Honey production using *Apis mellifera* is established since thousands of years and is a sizeable economic factor in some countries. Hymenoptera parasitica play a very important role in decimating populations of harmful insects, especially forest pests.

[Mitchel (1960), Oeser (1961), Weber (1966), Schedl (1991); Smith & Middlekauff (1987); Vilhelmsen (2001); Beutel & Vilhelmsen (2007); Beutel et al. (2008); Dathe (2005); Grimaldi & Engel (2005); Graberek (2008)]

Neuropterida

6.26 Neuroptera (=Planipennia, Greek *neuro* = nerve, *pteron* = wing, refers to the wing venation with numerous crossveins, English common name: net-winged insects)

Diversity and distribution. About 6,000 species of Neuroptera are described. The group has a worldwide distribution and occurs in the temperate, subtropical and tropical regions. The family Dilaridae is missing in Australia, and Osmylidae and Nemopteridae are absent in North America. Rhachiberothidae is restricted to Sub-Saharan Africa, Nymphidae is only found in the Australian region, and Polystoechotidae only in North and South America. The presumably basal family Nevrorthidae has an interesting disjunctive distribution. It occurs in the Mediterranean region, in Japan, Taiwan, and in southeastern Australia (Aspöck & Aspöck 2005; Liu et al. 2012a).

Autapomorphies. Neuroptera are almost exclusively characterized by larval autapomorphies.
- Larval mandibles and maxillae together form sucking jaws (Figs 6.26.1, 6.26.2)
- Larval poison glands associated with the sucking jaws
- Maxillary palp of larvae completely reduced
- Larval labrum fused with head capsule
- Larval cervix (specialized neck region) (Fig. 6.25.1)
- Larval midgut terminally closed
- Mandibles of adults with molar process
- Left mandible of adults with shovel-like extension

Fig. 6.26.1: *Austroneurortus* sp. (Nevrorthidae), larva, habitus, lateral view, SEM micrograph. Abbr.: ant – antenna, cv – cervix, mms – mandibulo-maxillary stylets, plb – palpus labialis, prn – pronotum. Scale bar: 1 mm. From Beutel et al. (2010), modified.

Taxonomy. The group is subdivided into three suborders and 18 families. Nevrorthidae (ca. 30 spp.) is the only family of Nevrorthiformia and probably the sistergroup of the remaining Neuroptera. Hemerobiiformia comprise 12 families, among them Ithonidae, Chrysopidae (green lacewings), Osmylidae, Sisyridae (spongillaflies), Coniopterygidae (dustywings), and Mantispidae (mantidflies). The well known Myrmeleontidae (antlions) and Ascalaphidae (owlflies) are two of the five families of the strongly supported clade Myrmeleontiformia.

Diagnosis. Small to very large insects with an orthognathous head and largely unmodified mouthparts. Wings large, usually nearly equally sized, with numerous transverse veins and a series of crossveins meeting the anterior wing margin. Larvae terrestrial or aquatic, with sucking jaws formed by the mandibles and maxillae, and

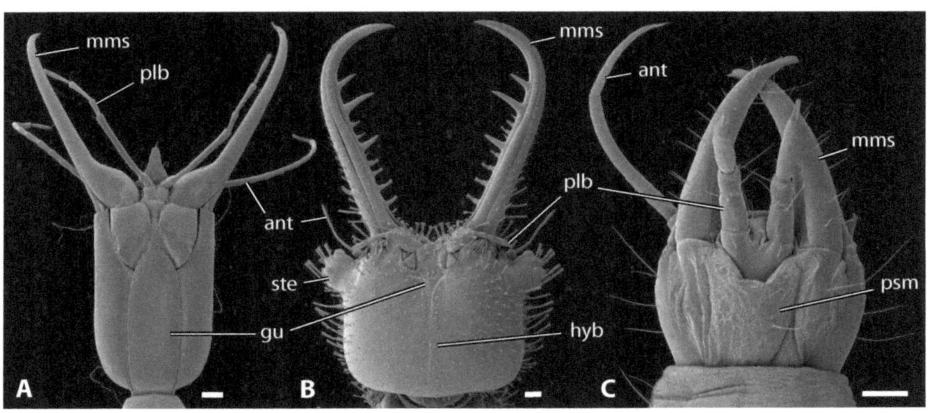

Fig. 6.26.2: Neuroptera, larval heads, ventral view. SEM micrographs. A, *Austroneurortus* sp. (Nevrorthidae); B, Hemerobiidae; C, *Libelloides* sp. (Ascalaphidae). Abbr.: ant – antenna, gu – gula, hyb – hypostomal bridge, mms – mandibulo-maxillary stylets, plb – palpus labialis, psm – postmentum, ste – stemmata. Scale bar: 100 µm. From Beutel et al. (2010), modified.

lacking maxillary palps. The habitus can vary strongly. Mantispidae superficially resemble mantids, Ascalaphidae butterflies, and Myrmeleontidae dragonflies.

Morphology, adults. Neuropterans are small (length of forewing of Coniopterygidae 1.8–5 mm) to very large insects (maximum length of forewings of Myrmeleontidae 80 mm). Most species are medium sized, with a total length of 5–20 mm (Aspöck & Aspöck 2005). The shape and size of the pterothoracic segments is usually similar. The cuticle is rather weakly sclerotized. The thin cuticle bears a more or less dense vestiture of setae.

The head (Fig. 6.26.3) is moderately sized to large, usually more or less round or triangular in frontal view, only rarely elongated (Nemopteridae partim). It is always orthognathous, and a gula is always absent. A hypostomal bridge occurs in Coniopterygidae and *Plega* (Mantispidae). The postoccipital region is exposed or slightly retracted into the prothorax. The vertex is sometimes strongly domed. A separate dorsomedian neck sclerite delimited by a distinct transverse furrow is present in some groups (e.g., Osmylidae, Chrysopidae); the lateral occipital areas form very conspicuous, hemispherical lobes in *Osmylus* and some other genera; a distinct posterolateral occipital furrow is present in the neuropteran groundplan (e.g., *Osmylus*, *Myrmeleon*). Extensive areas of the cuticle are glabrous but a rather sparse vestiture of longer or smaller setae is present, and dense tufts of hairs also occur (Ascalaphidae). The coronal and frontal sutures are primarily present but usually poorly defined or

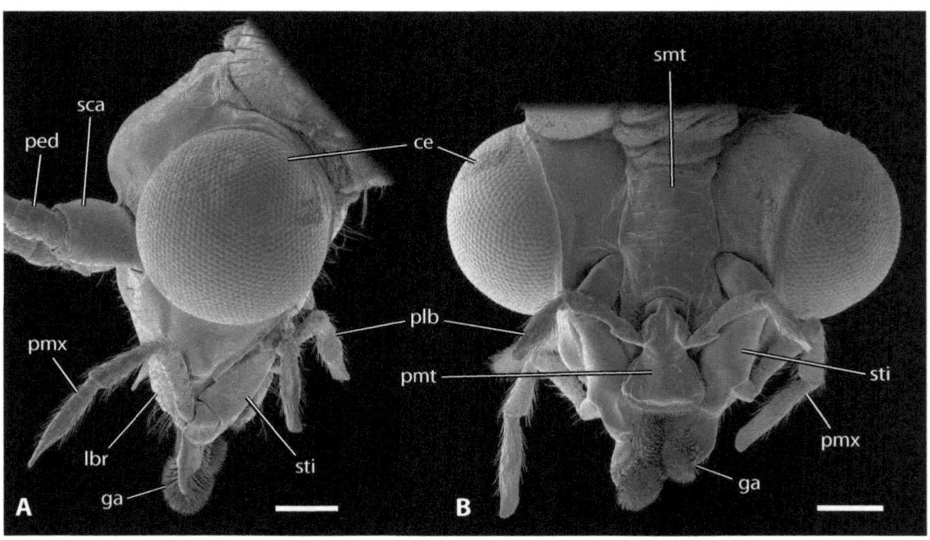

Fig. 6.26.3: *Chrysoperla carnea* (Chrysopidae), head, SEM micrographs. A, lateral view; B, posteroventral view. Abbr.: ce – compound eye, ga – galea, lbr – labrum, ped – pedicellus, plb – palpus labialis, pmt – prementum, pmx – palpus maxillaris, sca – scapus, smt – submentum, sti – stipes. Scale bar: 200 µm.

even absent. More or less long dorsolateral furrows occur in Osmylidae and several other groups and probably belong to the groundplan of the order; a more or less distinct suture extending from the anterior tentorial pit to the ventrolateral edge of the antennal foramen is also present in many groups including *Osmylus*. The eucone compound eyes are always well-developed and can be strongly convex (e.g., Nemopteridae) and very large; well-developed ocelli are only present in Osmylidae, but ocellar or cranial pulvinae (setose convexities) occur in several groups. The frontoclypeal strengthening ridge is present (e.g., *Chrysopa*) or absent (e.g., Osmylidae); the clypeal area is almost always undivided; the entire frontoclypeal region is strongly elongated in some Nemopteridae (see above); the posterior margin of the clypeus or clypeal region is marked by more or less distinct anterior tentorial pits. The labrum is free (Fig. 6.26.3A); its anterior margin is rounded, almost truncate or more or less distinctly convex medially. The antennae are filiform or moniliform in most groups, but sometimes clubbed (Ascalaphidae) or pectinate; the antennomeres are usually cylindrical, but in some species globular or disc-shaped. The mandibles are well-developed and slightly to distinctly asymmetric; the mesal walls are shortened; this modification and a characteristic molar process and a shovel-like extension of the left mandible are probably autapomorphies of Neuroptera, and possibly also the presence of a narrow field of stiff hairs on the posterior mandibular surface (e.g., *Osmylus*, *Myrmeleon*). The maxillae are largely unmodified (Fig. 6.26.3B); the lacinia is often elongated and strongly setose; the galea is composed of two segments; an additional proximal subgalea is present (e.g., *Osmylus*,) or absent; the apical galeomere (distigalea) can be moderately sized, small and peg-like (e.g., *Myrmeleon*), or strongly elongated (Nemopteridae partim); in some groups it bears a dense vestiture of microtrichia; the palp is almost always 5-segmented. The labium is composed of a plate-like submentum, a very narrow mentum, and a prementum in the groundplan (e.g., *Osmylus*); glossae and paraglossae are absent; the ligula is sometimes membranous and elongated; the palps are usually 3-segmented; the apical palpomere bears a field of sensilla (palpimacula) in most groups. The hypopharynx forms a structural and functional unit with the anterior labium. The tentorium is complete, with well-developed anterior, dorsal and posterior arms, and a tentorial bridge (corpotentorium). One or two (*Nemopterella*) cervical sclerites, likely of preepisternal origin, are present, and a dorsal cervical sclerite also occurs (e.g., *Osmylus*); the lateral sclerite bears a balloon-shaped, membranous gland structure at its posterior end in Osmylidae. The muscle system of the head is nearly complete, including a well-developed tentoriomandibular muscle. Large sac-like reservoirs of the salivary glands are present in the posterior head region (*Osmylus*); tube-like glandular lobes are present in the mandibles and the lateral head region, and more compact gland tissue between the extrinsic tentorial muscles of the maxilla and the premental retractors; the former are partly embedded in loose tissue resembling fat body elements (*Osmylus*). The salivary glands are very long. An unpaired nervus connectivus connecting the protocerebrum and frontal ganglion is present.

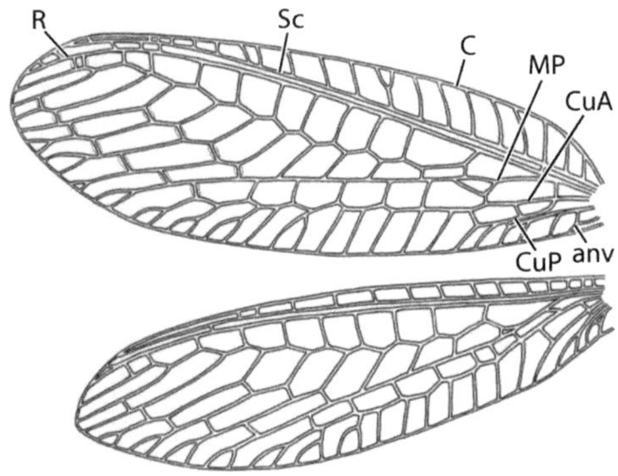

Fig. 6.26.4: *Borniochrysa* sp. (Chrysopidae), wings.
Abbr.: anv – anal vein, C – costa, CuA – cubitus anterior, CuP – cubitus posterior, MP – media posterior, R – radius, Sc – subcosta. Redrawn from Aspöck & Aspöck (2005), after Tjeder (1966).

The prothorax is more or less freely movable, in contrast to the pterothoracic segments, which form a compact unit. It is shortened in Coniopterygidae and Ascalaphidae, and elongated in Mantispidae and some Nemopteridae. A sclerotized pronotum is present but large parts of the prothorax are membranous; the tubular pronotum of some Mantispidae appears subdivided, and may be in fact a composite structure. The propleura is small and subdivided into an episternum and epimeron. The tip of the profurcal arm is firmly fused with the propleural apophysis. A precoxal bridge and basisternal area anterior to it are absent. A prothoracic spina is present. The protrochantin is weakly developed. The pterothorax is broader than the prothorax and the meso- and metathorax are of about equal size. The major dorsal regions scutum, scutellum, and postnotum are usually well-defined, but the metanotum of Nemopteridae is transverse and simplified. A distinct mesothoracic prealare is present. A median mesonotal suture is distinctly developed and a well-defined membranous field between the mesoscutellum and mesopostnotum is usually present (absent in *Osmylus*); the middle part of the mesoscutellum is triangular; a mesothoracic postalar bridge is usually absent but present in Nemopteridae. The lateral scutal areas of the metanotum of some families have well-developed "velvet zones" (New 1989); they are likely used to hold the wings in the resting position. The pleurae of both pterothoracic segments are distinctly divided into an anterior anepisternum and a posterior epimeron by a well-developed pleural ridge; the katepisternum anterior to the coxae is well-developed. There are no exposed true sternal elements except for the ventral mesosternal process (below the mesofurca) which forms the sterno-coxal joint. The meso- and metatrochantins are present. The membranous wings (Fig. 6.26.4) are well-developed and usually held in a roof like position when at rest; they are usually more or less transparent ('lacewings') but conspicuous color patterns also occur (e.g., Ascalaphidae), and they are covered with waxy scale-like structures in Coniopterygidae; the pattern of longitudinal and transverse veins is well-developed; the fore- and hindwings are usually similar in size, shape and degree of sclerotization; exceptions are the elongated hindwings of Nemopteridae, the leathery forewings of some Hemerobiidae, size reductions of the hindwings in some Coniopterygidae, winglessness in one coniopterygid species, and simplifications of the venation in the same family; the costal cell is subdivided by a series of crossveins; a pterostigma is often present; a pronounced anal field is missing (Fig. 6.26.4). Spiracles are present between the pro- and mesothorax, and between the meso- and metathorax. The legs are composed of the usual elements; they are usually largely unmodified walking legs, but the forelegs of Mantispidae and Rhachiberothidae are raptorial; the coxae are closely adjacent medially; a distinct meron is present; the tarsi are 5-segmented; an arolium is present and a simple or bilobed empodium; ventral hairy soles of the tarsomeres are absent in contrast to Raphidioptera and Megaloptera; the claws sometimes bear a subapical tooth. The thoracic muscle system is complex and always well-developed; it is similar in both pterothoracic segments.

The more or less elongated and cylindrical abdomen is composed of ten distinctly developed segments. It is usually well sclerotized but largely membranous in Coniopterygidae and Sisyridae. Like in the other neuropterid orders, tergite I and sternite II are subdivided by a transverse suture (Achtelig 1975). Sternum I can be strongly reduced (Myrmeleontidae, Ascalaphidae). Spiracles are almost always present in the pleural membranes of segments I–VIII. Different types of fusion occur in the posterior abdominal segments (e.g., Chrysopidae, males of Osmylidae). The male genital segments vary strongly. In the groundplan, tergite IX (sometimes medially divided) and sternite IX (sometimes elongated or shortened) are simple, arcuate plates; the gonocoxites IX are connected with the gonarcus or reduced; the gonapophyses are usually obliterated. Tergites X and XI are fused to form the epiproct; in the middle region of this structure trichobothria are arranged in a rosette-like manner; the shape of the epiproct, which can be paired or unpaired, is of great taxonomic importance. The parameres are paired, unpaired or obliterated; the paired or unpaired mediuncus is distinct or amalgamated with the parameres, forming a simple or complex structure; the gonarcus can be paired or unpaired and posteriorly visible or internalized; it varies greatly in shape; the arcessus is a free sclerite or medially connected with the gonarcus; the hypandrium internum is a triangular, keeled, small sclerite at the terminal opening of the ejaculatory duct (Aspöck & Aspöck 2005). The female genital segments (Fig. 6.26.5) are less heterogenous; sternite VII may be elongated and has a simple or modified hind margin; sternite VIII is a well-developed plate, reduced to a narrow band, represented by several separate sclerites, or obliterated; tergite IX reaches far ventrad and articulates with the club-shaped or ovoid gonocoxites; a well sclerotized bursa copulatrix, a paired receptaculum seminis, and a ductus seminalis originating ventrally on the bursa are probably ancestral, but many modifications occur within the group (Aspöck & Aspöck 2005).

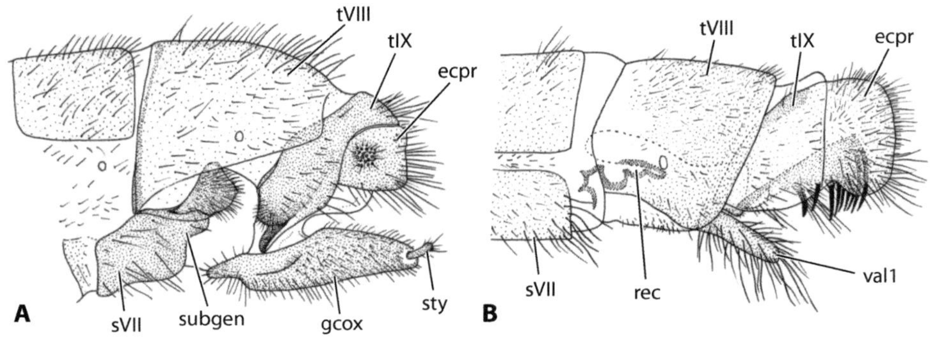

Fig. 6.26.5: Female postabdomen, lateral view. A, *Osmylus fulvicephalus* (Osmylidae); B, *Cueta beieri* (Myrmeleontidae). Abbr.: ecpr – ectoproct, gcox – gonocoxite, rec – receptaculum seminis, sty-stylus, subgen – subgenitale, sVII – sternum VII, tVIII/IX – tergum VIII, IX, val1 – valvula 1. Redrawn from Aspöck & Aspöck (2005), after Aspöck et al. (1980).

Different exocrine glands occur in different parts of the thorax and abdomen in representatives of different families (e.g., Nevrorthidae [eversible pleuritocavae], Osmylidae, Nemopteridae). The foregut is characterized by a large, dorsal diverticulum (ingluvies) and a funnel-shaped proventriculus with a strong musculature and with or without acanthae. Eight free Malpighian tubules are present in most groups but only six in Coniopterygidae. The testes are usually ovoid and simple and lie dorsolaterally within segments VI or VII. Approximately ten meroistic-polytrophic ovarioles are present in females.

Morphology, larvae. The larval body is more or less slender (Fig. 6.26.1) in the groundplan (e.g., Nevrorthidae, Sisyridae, Osmylidae), but scarabaeiform in Ithonidae and Polystoechotidae, and shortened, broad and stout in several myrmeleontiform lineages (e.g., Myrmeleontidae, Ascalaphidae). A characteristic feature of Mantispidae and Berothidae is hypermetamorphosis, i.e. distinctly differing larval stages. The first instar is slender and campodeiform, whereas at least the 2nd instar (Berothidae) is scarabaeiform. Tubercles and a broad variety of setae occur in different groups. Specialized thick, stiff, apically frayed dolichasterine setae are a characteristic feature of larvae of Myrmeleontiformia.

The head is prognathous and parallel-sided (Nevrorthidae) (Fig. 6.26.1) or laterally rounded (Fig. 6.26.2B, C); it can be strongly flattened (e.g., Nevrorthidae, Dilaridae, Berothidae) or strongly convex ventrally (Ithonidae and Polystoechotidae); it is well sclerotized, at least dorsally and laterally. A well-developed gula as it is present in Nevrorthidae is probably ancestral; it is completely absent in larvae of Hemerobiiformia were the ventral side of the head is largely formed by maxillary and labial elements ("maxillary head"; Aspöck et al. 2001) (Fig. 6.26.2C); a sclerotized hypostomal bridge and a small, anteriorly shifted triangular gula are present in myrmeleontiform larvae. The frontal and coronal sutures are well-developed in the groundplan (e.g., Nevrorthidae). A distinctly reduced frontoclypeal sulcus is present in Nevrorthidae and in most other groups. The number of stemmata varies between seven (several groups of Myrmeleontiformia) and a single eyespot without a distinct cuticular lens (Nevrorthidae); a strongly protruding ocular region is present in several myrmeleontiform families (Fig. 6.26.2B). The labrum is always fused with the clypeus. The antennal insertion area is completely enclosed by the sclerotized dorsal wall of the head capsule in Nevrorthidae, Sisyridae, Mantispidae, and Psychopsidae. The antennae are probably 3-segmented in the groundplan, but appear multisegmented in some groups (e.g., Nevrorthidae, Psychopsidae); they are characterized by a fixed curvature in Ithonidae and Polystoechotidae, and are very short in several groups of Myrmeleontiformia (Nymphidae, Myrmeleontidae, Ascalaphidae); a long, specialized terminal antennal seta is present in most groups of Hemerobiiformia (absent in Ithonidae and Polystoechotidae). The mandibles are relatively short and compact in Ithonidae and Polystoechotidae, but more or less strongly elongated in the other groups (Figs 6.26.1, 6.26.2); they can be extremely slender (e.g., Sisyridae) or set with mesally directed sclerotized teeth (Myrmeleontidae, Ascalaphidae) (Fig. 6.26.2B); they are usually curved inwards (groundplan), but sometimes straight (e.g., Dilaridae) or even bent outwards (Osmylidae). The maxillae are likely composed of the cardo, stipes and lacinia; the galea and the palp are absent. The most unusual autapomorphy of Neuroptera is the formation of specialized sucking jaws by the mandibles and the distal part of the maxillae (probably the laciniae); the mandibular and maxillary elements enclose a sucking channel and an additional mesal poison channel; three paired glands (poison and salivary glands) are associated with this highly specialized food uptake apparatus. The labium is composed of submentum, mentum and prementum in Nevrorthidae (groundplan) but only by a postmentum and prementum in the other groups; glossae and paraglossae are absent; labial palps are usually present and composed of two 2–7 segments (two in Coniopterygidae and Crocinae); they are missing in Sisyridae. The presence of a distinct collar-like cervix connecting the head and prothorax is an autapomorphy of Neuroptera (strongly elongated in Nevrorthidae and Nemopteridae); a highly specialized articulation of the neck is present in Nevrorthidae; a finger-like mid-dorsal apodeme extends from the posterior margin of the postoccipital process far into the prothorax in Ascalaphidae, Myrmeleontidae and Nemopteridae.

The prothorax is usually anteriorly converging (e.g., Coniopterygidae, Sisyridae, Berothidae); it is distinctly elongated in Nevrorthidae (Fig. 6.26.1) and extremely long in some Nemopteridae. A distinctly developed, medially divided pronotum is usually present. The meso- and metanota are less distinct. Three well-developed legs are almost always present. They are distinctly reduced after the 1st larval stage in the parasitic larvae of Mantispidae and Berothidae, and also in larvae of Ithonidae; the coxae are large and widely separated; the trochanter is usually small and the femur and tibia often long and slender (e.g., some Nemopteridae); the tarsi are fused with the tibia in Ithonidae (all legs) and also in Ascalaphidae and Myrmeleontidae (mid- and hindlegs); a tube-shaped empodium functioning as attachment device occurs in several groups and possibly belongs to the groundplan of the order.

The abdomen is composed of ten segments but the posterior ones are retracted in Myrmeleontidae (segment IX and X) and Ascalaphidae (X). Segment I is usually narrow. Distinctly developed tergites are usually missing on all abdominal segments and the ventral side is usually also unsclerotized. Ventral gills occur in Sisyridae. The abdominal apex is sclerotized or lobe-like and functioning as an adhesive organ (e.g., Osmylidae [paired and equipped with numerous small spines]). The larval midgut is not linked with the hindgut. The Malpighian tubules (usually eight, seven in Osmylidae, six in Coniopterygidae) produce cocoon silk; at least some of them are usually cryptonephric (only one in Sisyridae), but this is not the case in the aquatic larvae of Nevrorthidae.

Morphology, pupae. The neuropteran pupa is decticous and exarate and always formed within a cocoon (New 1989).

Biology. The adults are diurnal or nocturnal and usually relatively short-lived, with a life span ranging between few days and several weeks. Only few species overwinter in the adult stage. The flying abilities are usually modest but some neuropterans are strong fliers (Ascalaphidae). Swarming occurs but is a rare exception (e.g., Coniopterygidae). Most species are carnivorous and specialized predators of aphids occur in some groups (e.g., Hemerobiidae, Chrysopidae). Some species are almost exclusively found on the host plant of the preferred prey species. Other food sources of the adults are pollen, nectar and honey dew. The larvae are exclusively or almost exclusively carnivorous, with highly specialized sucking jaws associated with poison glands and salivary glands. The digestion is extraoral. Larvae of Nevrorthidae (Fig. 6.26.1) and Sisyridae are aquatic. The latter develop in freshwater sponges or Bryozoa after the first larval stage. Larvae of Osmylidae are semiaquatic and those of the other families terrestrial. The immature stages of Ithonidae and Polystoechotidae live in soil. It is assumed that they feed on roots, but this is unconfirmed. Other terrestrial larvae are found in the vegetation, under bark or in soil. Larvae of *Myrmeleon* and some other myrmeleontid genera use funnel-shaped pitfall traps in sandy substrate to catch small insects, mainly ants. Other larvae in this family hunt prey actively. Larvae of Ascalaphidae conceal themselves under stones or on tree trunks, and some of them camouflage themselves with debris, as it also occurs in the closely related family Myrmeleontidae. Larvae of Berothidae are termitophilous and some species of Mantispidae parasitize in egg sacs of spiders.

Reproduction and development. Neuropterans are generally bisexual. Pheromones or species specific substrate vibrations play a role in finding a mating partner. Copulation can last between few minutes and several hours and the postures vary strongly within the order. Spermatophores were observed in few cases (Chrysopidae, Osmylidae). The globular or oblong eggs have a smooth or sculptured surface and are sometimes covered with sand particles. They are deposited individually or in smaller or larger clusters (up to several thousands) on or in the substrate or on suitable parts of plants. Eggs of Sisyridae are enclosed in a silk web and females of Mantispidae and Nymphalidae produce egg stalks using secretions of their Malpighian tubules. The embryonic development is completed after only few days in most groups. Depending on the temperature it varies between 2.5 and 13 days in Chrysopidae. An 'ovirup-tor' is involved in the hatching process in some groups and a circular zone of weakness may be present. Three larval stages occur in most groups, but 5–9 in Ithonidae and even ten in Dilaridae (under laboratory conditions). Neuropterans hibernate as larva, prepupa or pupa, or rarely as adult. The pupal stage lasts between few days and several months. The pupa is enclosed in a cocoon formed by several layers of silk produced by the Malpighian tubules (Aspöck & Aspöck 2005). The pupa cuts its way out of the cocoon using its mandibles. Like in Raphidioptera it may move around actively before molting to the adult stage.

Fossil record. The oldest known definite neuropteran fossils from Eurasian Early Permian deposits were assigned to the family †Permithonidae (e.g., Novokschonov & Novokschonova 1996). The oldest known members of an extant family are Triassic fossils of Psychopsidae. The highest biodiversity was reached in the Jurassic, with some very large forms in the extinct family †Kalligrammatidae. The oldest clearly identifiable members of the presumably basal Nevrorthidae were found in Baltic amber (Aspöck & Aspöck 2005).

Economic importance. Neuropteran predators of aphids and mites, notably Chrysopidae, Hemerobiidae and Coniopterygidae, play an increasing role in integrated pest management.

[Tjedev (1966), Achtelig (1975); Aspöck et al. (1980, 2001); New (1989); Novokschonov & Novokschonova (1996); Aspöck & Aspöck (2005, 2007); Grimaldi & Engel (2005); Beutel et al. (2010); Liu et al. (2012a)]

6.27 **Megaloptera** (Greek *megalo* = large, *pteron* = wing, English common names: alderflies, dobsonflies, fishflies)

Diversity and distribution. Megaloptera are a small group with approximately 325 described species. They occur in the Holarctic and Neotropical regions, the Orientalis, in eastern Australia and the southern margin of the African continent. Sialidae are largely restricted to the Holarctis, whereas Corydalidae mainly occur in temperate and tropical regions of the northern hemisphere and in South America.

Autapomorphies

- Sensorial appendage of larval antenna located on antepenultimate segment
- Lateral abdominal gills of larvae (Fig. 6.27.1)
- Eversible sacks in the terminal area of the male genital segment

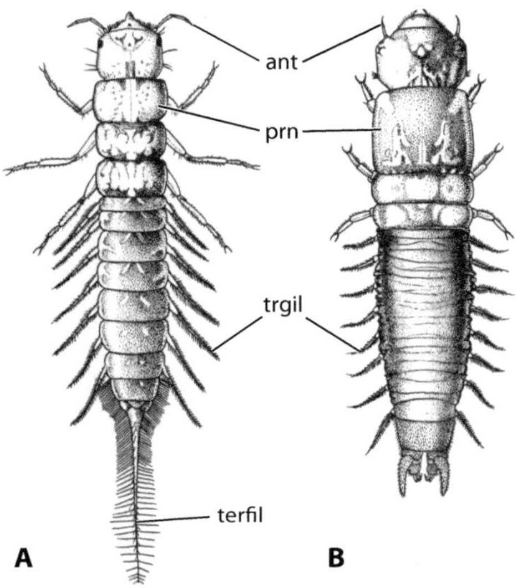

Fig. 6.27.1: Megaloptera, larvae, dorsal view.
A, *Sialis chilensis* (Sialidae); B, *Archichauliodes* (?) (Corydalidae). Abbr.: ant – antenna, prn – pronotum, terfil – terminal filament, trgil – tracheal gills.
Redrawn from Aspöck & Aspöck (2005), after Flint (1973).

Taxonomy. The group is subdivided into the two families Sialidae and Corydalidae, and the latter in the subfamilies Chauliodinae and Corydalinae.

Diagnosis. Medium sized to very large insects. Approximately equally sized wings held in a roof-like position above the abdomen. Aquatic larvae with lateral abdominal gills.

Morphology, adults. Megaloptera are medium sized to very large insects. The length of the forewings ranges between 8 and 85 mm.

The head (Figs 6.27.2, 6.27.3) is prognathous and posteriorly closed by a sclerotized gula. It is distinctly flattened, especially in Corydalinae, and laterally rounded (Sialidae), gradually narrowing towards the foramen occipitale (Chauliodinae), or characterized by a conspicuous postocular ridge and abruptly narrowed posteriorly (Corydalinae). The postoccipital region is very slightly retracted into the prothorax.

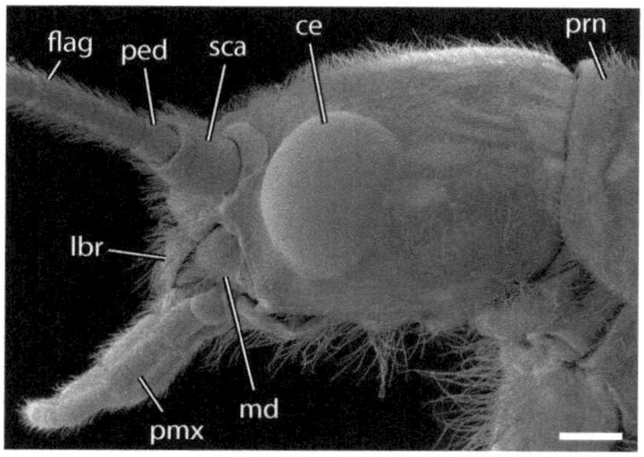

Fig. 6.27.2: *Sialis lutaria* (Sialidae), adult head (male), lateral view, SEM micrograph. Abbr.: ce – compound eye, flag – flagellum, lbr – labrum, md – mandible, ped – pedicellus, pmx – palpus maxillaris, prn – pronotum, sca – scapus. Scale bar: 300 μm.

The compound eyes are always well-developed and strongly convex in Chauliodinae; three ocelli are present in Corydalidae (Fig. 6.27.3A) but completely reduced in Sialidae. The frontal and coronal sutures are distinct in *Sialis* but more or less strongly reduced in Corydalidae. A deep median excavation of the frons is present in *Sialis flavilatera* (Röber 1942). The frontoclypeal strengthening ridge is generally present but medially interrupted in *Corydalus*. The anterior clypeal margin is medially incised to different degrees in some Corydalidae. The labrum is free; it is rounded anteriorly or more or less tapered and triangular, and covered with setae. The filiform, serrate or pectinate antennae articulate close to the anterior margin of the head capsule, laterally in Sialidae but on the dorsal side of the head in Corydalidae (Figs 6.27.2, 6.27.3); they are long and composed of 30–85 flagellomeres. The mandibles always lack a mola and movable appendages; two mesal teeth are present in Sialidae and Chauliodinae and three occur in Corydalinae, but teeth are missing in the genus *Corydalus*; the mandibles are extremely elongated in males of some species of Corydalinae. The maxillae are largely unmodified; the palp comprises only four segments in *Sialis* and *Corydalus*; the galea is densely setose; it is short in Sialidae and long and tapered in Corydalidae. The submentum is fused with the gula posteriorly; the mentum lies between the stipites; the prementum bears the palp which is composed of three or four segments, and a bilobed membranous ligula. The tentorium is primarily complete but lacks a bridge in *Corydalus*. Cervical sclerites are present but differ in the two families; a large sclerotized ventral plate and a pair of dorsal plates are present in *Corydalus* and *Chauliodes*, both derived from the prothoracic presternum; the lateral cervical sclerites of Sialidae are apparently of preepisternal origin and anteriorly connected with the postocciput. The muscle system of the head is nearly complete. A single pair of tubular salivary glands is present and mandibular glands are described for *Sialis*.

The pronotum is transverse in Sialidae but elongate in Corydalidae, where it is distinctly arched and covers parts of the pleural region. The main element of the sternal region is a large undivided basisternum, which extends posteriorly between the procoxae; the small furcasternum bears relatively large lateral furcal pits; a small spina is present. The pterothoracic segments are similar in size and shape. The prescutum is not distinctly separated from the scutum, whereas the scutellum is always large and distinct. Three axillary sclerites are associated with the forewings but four with the hindwings. The ventral sclerites show the typical holometabolous pattern. The membranous wings are well-developed and held in a roof like position at rest; the forewings are narrower than the hindwings, especially in Corydalidae, which possess an expanded anal field of the hindwing; the venation is largely complete; the costal cell is subdivided by crossveins, less than 12 in Sialidae, but up to 60 in Corydalidae. The legs are composed of the usual elements; the tarsi are 5-segmented, with a heart-shaped tarsomere 4 in Sialidae; ventral hairy soles of the tarsomeres serve as attachment devices; paired claws and a small subapical tooth occur in Chauliodinae. Spiracles are present on the meso- and metathorax. The muscle system is strongly developed and similar in both pterothoracic segments.

The elongated abdomen is composed of ten distinctly developed segments and relatively weakly sclerotized; its anterior part is cylindrical and segment I is shorter than the following pregenital segments. Tergite I and sternite II are subdivided by a transverse suture or sternal seam, respectively (Achtelig 1975). A membranous pouch is present posterad sternite VI. Segment VIII is sometimes shorter than the other pregenital segments. Segment IX and X are strongly modified (Figs 6.27.4, 6.27.5); tergites

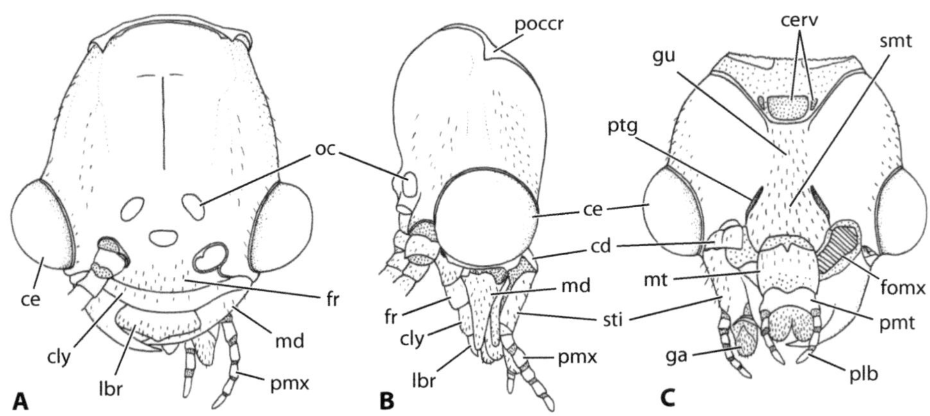

Fig. 6.27.3: Corydalidae, head, adults. A, dorsal view; B, lateral view; C. ventral view. Abbr.: cd – cardo, ce – compound eye, cerv – cervical sclerites, cly – clypeus, fomx – fossa maxillaris, fr – frons, ga – galea, gu – gula, lbr – labrum, md – mandible, mt – mentum, oc – ocellus, plb – palpus labialis, pmt – prementum, pmx – palpus maxillaris, poccr – postoccipital ridge, ptg – posterior tentorial grooves, smt – submentum, sti – stipes. Redrawn from Aspöck & Aspöck (2005), after Maki (1936).

and sternites IX of males are distinctly developed and separated. The gonocoxites IX of Sialidae are developed as paired plates; in Corydalidae they are partly fused with tergite IX, obliterated, or form a processus which is approximated to the ectoproct together with the gonarcus; gonostyli occur only rarely. The ectoproct, likely a product of fusion of tergites X and XI, is variously modified within the group; in Corydalidae it is a large lateral sclerite or process, including or excluding the group of trichobothria arranged as a rosette. The parameres are large and connected by a transverse structure in Corydalidae; they are apparently reduced in Sialidae; the gonarcus is present as a large and variously modified caudal plate (Sialidae) or a narrow transverse bar or obliterated (Corydalidae); the endophallus is inconspicuous. In females (Fig. 6.27.5) sternite VIII is plate-like or reduced to a narrow sclerite (Sialidae). A short ovipositor formed by the gonocoxites IX bears small, paired styli; it is equipped with intrinsic muscles.

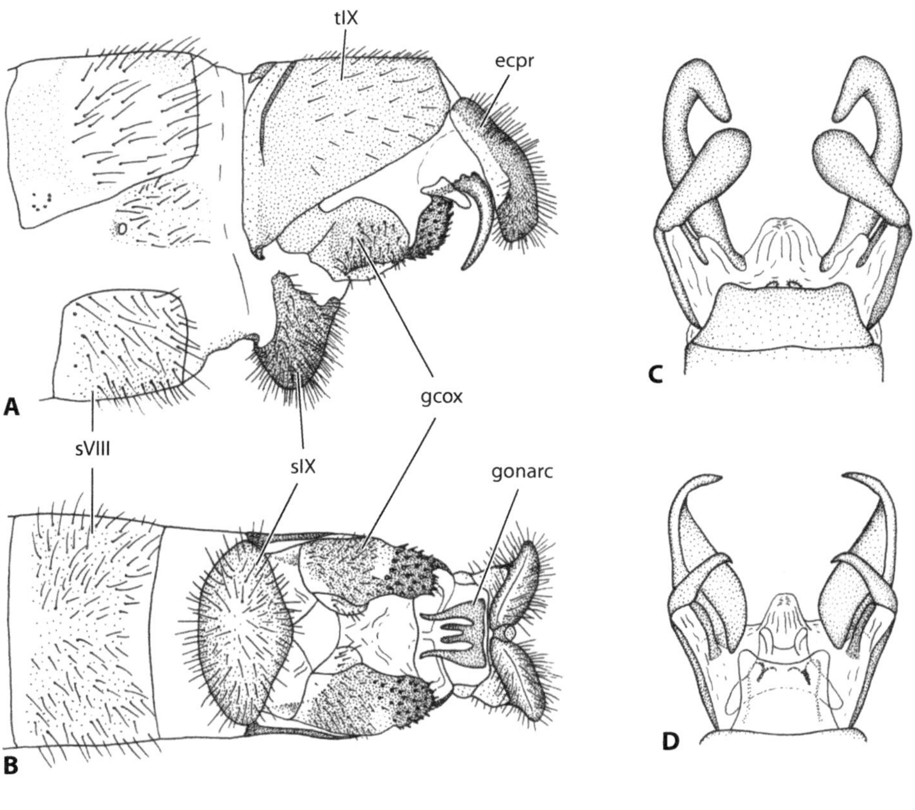

Fig. 6.27.4: Male genital segments. A, B, *Sialis nigripes* (Sialidae), lateral and ventral view; C, D, Corydalidae, ventral view. C, *Corydalus cornutus*; D, *Nevromus intimus*. Abbr.: ecpr – ectoproct, gonarc – gonarcus, gcox – gonocoxae, sVIII/IX – sternum VIII/IX, tIX – tergum IX. Redrawn from Aspöck & Aspöck (2005), after Aspöck et al. (1980) and Glorioso (1981).

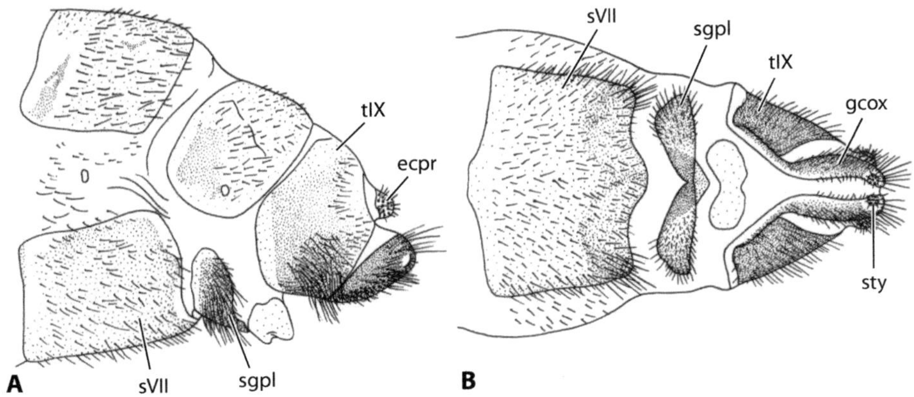

Fig. 6.27.5: *Sialis nigripes* (Sialidae), female genital segments. A, lateral view; B, ventral view. Abbr.: ecpr – ectoproct, gcox – gonocoxite, sgpl – subgenital plate, sty – stylus, sVII – sternum VII, tIX – tergum IX. Redrawn from Aspöck & Aspöck (2005), after Aspöck et al. (1980).

The posterior foregut is characterized by a food reservoir (diverticulum); a proventriculus with six small longitudinal folds with acanthae is described for *Corydalus*; the foregut of adults may be filled with air and non-functional as an alimentary organ (*Sialis flavilatera*), or the entire digestive tract may be obliterated; the midgut has a deeply folded lining. Six free Malpighian tubules are present in Sialidae and eight in Corydalidae.

Morphology, larvae. The aquatic larvae of Megaloptera are campodeiform and characterized by lateral tracheal abdominal gills (Fig. 6.27.1). The body is moderately flattened and the thoracic legs are well-developed.

The more or less parallel-sided head (Fig. 6.27.6) is always strongly sclerotized. It is subprognathous and moderately flattened (Sialidae) or prognathous and strongly flattened (Corydalidae). The gular area (Fig. 6.27.6B, C) is trapezoid and semimembranous or membranous in Sialidae; a sclerotized, nearly parallel-sided gula and a postgular sclerite are present in Corydalidae; the gula is distinctly delimited laterally and anteriorly fused with the submentum. The coronal and frontal sutures are distinct, the latter enclosing a relatively short triangular frons. A very distinct circular (occipital) ridge delimits a neck region, which is distinctly narrowed in Corydalidae, also resulting in a narrowed foramen occipitale. The frontoclypeal transverse ridge is present in Sialidae but absent in Corydalidae, where the trapezoid clypeal area is subdivided into a transparent anterior anteclypeus and a posterior postclypeus; the clypeus is undivided and transverse in Sialidae. Six well-developed stemmata are present, with a greatly increased number of retinula cells. The labrum is free and fully equipped with extrinsic and intrinsic muscles; it is triangular in *Sialis* but transverse

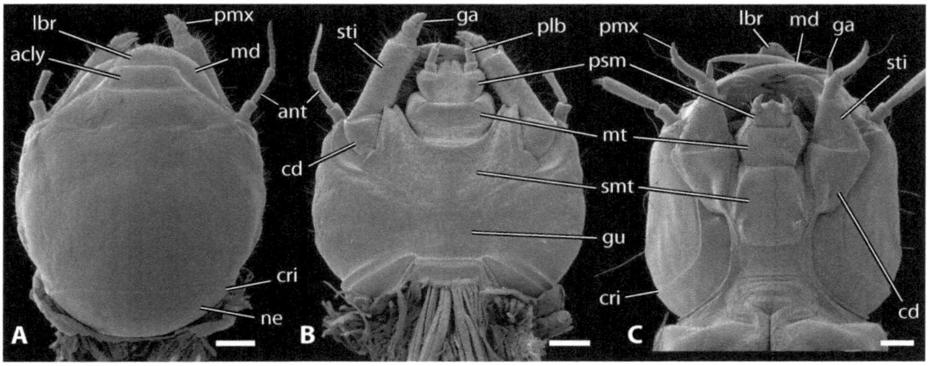

Fig. 6.27.6: Larval heads, SEM micrographs. A, *Neohermes* sp. (Corydalidae), dorsal view, B, *Neohermes* sp. (Corydalidae), ventral view, C, *Sialis* sp. (Sialidae), ventral view. Abbr.: acly – anteclypeus, ant – antenna, cd – cardo, cri – circular ridge (delimiting neck), ga – galea, gu – gula, lbr – labrum, md – mandible, mt – mentum, ne – neck, plb – palpus labialis, pmx – palpus maxillaris, psm – postmentum, smt – submentum, sti – stipes. Scale bars: A, B: 500 μm, C: 200 μm.

or semicircular or elongated Corydalidae. The antennae are inserted anterolaterally and 4- or 5-segmented (*Corydalus*); a sensorial appendage inserts on the antepenultimate antennomere. The mandibles are largely symmetrical and lack a mola and prostheca; several symmetric or asymmetric teeth occur on the mesal margin. The ventral mouthparts (Fig. 6.27.6) are distinctly separated from each other and largely unmodified. The maxillary grooves are deep in Sialidae but distinctly shortened in Corydalidae. The cardo is larger than the stipes in *Sialis*, whereas the stipes is strongly elongated in Corydalidae, where it bears a broad distal membranous collar; the galea and lacinia are distinctly separated; the lacinia is a large hook-shaped structure in Sialidae, but inconspicuous in Corydalidae; the palp is 4-segmented and strongly shortened in Corydalidae; the labium is composed of submentum, mentum and prementum; glossae and paraglossae are reduced; the ligula is bilobed in Corydalidae; the palp is 3-segmented. The tentorium is fully developed in Corydalidae but the dorsal arm is very thin in Sialidae; the posterior tentorial grooves are strongly shifted anteriorly in Corydalidae. Two thin dorsal sclerites (jugulare) and a large ventral cervical sclerite are present in Corydalidae.

The prothorax is the largest thoracic segment. The large sclerotized pronotum is medially divided by an ecdysial suture. On the ventral side a narrow, curved basisternum is present anterior to the coxae in Sialidae, and a ventral shield formed by the basisternum and episternum in Corydalidae. The meso- and metanota are smaller and less strongly sclerotized than their prothoracic equivalent (Fig. 6.27.1). Ventral sclerotized elements are absent from these segments. The 5-segmented legs are well-developed and equipped with a more or less dense vestiture of setae. Spiracles are present anterior to the mesocoxae and a second more or less vestigial pair occurs anterior to the metacoxae; they are annular in Sialidae and cribriform in Corydalidae.

The elongate 10-segmented abdomen is dorsoventrally flattened; it is distinctly tapering posteriorly in Sialidae and less so in Corydalidae. Segments I–VII (Sialidae) or I–VIII (Corydalidae) laterally bear long, setose tracheal gills (filaments) (Fig. 6.27.1); they are distinctly divided into five subsections in Sialidae; tufts of tracheal gills are present at the base of the filaments in Corydalinae. Segment X of Sialidae bears a long unpaired, setose terminal filament (Fig. 6.27.1A), whereas a pair of prolegs is present in Corydalidae; the prolegs are retractile and bear an outer lateral filament (stylus) and a pair of hooked claws. Spiracles VIII located on long respiratory tubes occur in Chauliodinae; they probably function as snorkels.

Morphology, pupae. Megaloptera are characterized by a very plesiomorphic pupa dectica exarata, with rows of strong setae across the abdominal tergites. The antennae are comparatively short and the compound eyes large. The mandibles are not enlarged (New & Theischinger 1993).

Biology. The short-lived adults are nocturnal (Corydalidae), diurnal or crepuscular (Sialidae). They are usually only found close to the aquatic habitats of the larvae. The flying abilities are modest. Food is consumed in liquid form if at all. The aquatic larvae are predaceous with a very broad prey spectrum. Cannibalism occurs occasionally. The larvae live in lakes, ponds, rivers, creeks, and marshes. Some Corydalidae develop in temporary water bodies and *Corydalus cornutus* can occur in cold mountain streams but also in warm springs (32°C). Larvae of *Sialis* dig in mud and can form J- or U-shaped burrows. Pupation takes place outside of the water in mud, moss, under stones or logs, or among low growing vegetation.

Reproduction and development. Megaloptera are bisexual. Mating and the related behavior were described in detail for *Sialis*. Males and females produce species specific vibrations of the abdomen, partly also involving the wings. The signals are transmitted by the legs to the substrate and received by the legs of the mating partner. Antennal vibrations initiate the copulation, which takes only a few minutes (*Sialis*) or even less. The male of *Sialis* pushes underneath the female's abdomen and pushes upwards, folds the wings to one side, and twists the abdomen upwards to copulate. A large, white spermatophore is transferred. Its main body is taken gradually into the female genital tract. Cases of females chewing on the spermatophore after mating have been observed. Eggs are deposited very close to water bodies, in the vegetation or on rocks, branches or other suitable objects. The egg batches (usually 1,000–3,000 eggs) can be one-layered or composed of up to five layers. They are covered by protective, hardening secretions. The primary larvae enter the aquatic habitat after hatching, often simultaneously at night. The number of instars varies between 10 and 12.

The entire developmental cycle takes one or two years in Sialidae and 2–5 in Corydalidae.

Fossil record. The oldest putative megalopteran fossils are from Late Permian deposits (†Parasialidae) (Grimaldi & Engel 2005). A Permian larva described as †*Permosialis* possibly belongs to the beetle family Gyrinidae. Definite fossils of Sialidae are restricted to the Tertiary but wing fragments from the Early Jurassic might belong to a sialid (†*Dobbertinia recticulata*; Ansorge 2001). Corydalid adults are known from the Middle Jurassic of China (Liu et al. 2012b) and from Baltic amber (Grimaldi & Engel 2005).

Economic importance. A limited economical effect may be caused by predaceous megalopteran larvae occurring in large numbers.

[Maki (1936); Röber (1942); Kelsey (1954, 1957); Flint (1973); Achtelig (1975); Aspöck et al. (1980); Glorioso (1981); New & Theischinger (1993); Aspöck & Aspöck (2005); Grimaldi & Engel (2005); Beutel & Friedrich (2008); Yang & Liu (2010); Ansorge (2001); Liu et al. (2012b)]

6.28 Raphidioptera (Greek *raphio* = needle, *pteron* = wing, English common names: snakeflies, camelneck flies)

Diversity and Distribution. Raphidioptera are the smallest order of Holometabola and comprise only slightly more than 235 described extant species. The group is largely restricted to certain arboreal parts of the Holarctic. Snakeflies occur in the Palearctic including North Africa, Syria, Iraq, Iran, Pakistan, and some areas of the Orientalis including northern India, Thailand, and also Taiwan, and the southwestern Nearctic down to the Mexican/Guatemalan border. Forests are the typical habitat but few species occur in open grassland with shrubs. The maximum altitude is about 3000 m (Aspöck & Aspöck 2005).

Taxonomy. The group is divided into the two extant families Inocelliidae and Raphidiidae. The former comprises seven genera (35 spp.) and the latter 26 (ca. 200 spp.). Inocelliidae are characterized by lacking ocelli.

Autapomorphies
- Head elongated, flat and slender
- Prothorax distinctly elongated and movably connected with the mesothorax
- Bilobed third tarsomere

Diagnosis. Medium sized insects with elongated head and prothorax and almost equally sized pairs of transparent wings. Ovipositor almost as long as the rest of the body or considerably longer.

Morphology, adults. Raphidioptera are medium-sized, slender insects. The length of the forewings ranges between 5 mm and 21 mm.

The well sclerotized head is elongated, prognathous, and more or less strongly flattened (Fig. 6.28.1A); it is distinctly narrowed posteriorly in Raphidiidae, but relatively broad at its hind margin in Inocelliidae. The foramen occipitale is narrow. A well-developed postoccipital ridge is present. The coronal and frontal sutures are absent. The clypeus is separated from the frons by a narrow unpigmented area; it is subdivided into the postclypeus and an anterior unpigemented anteclypeus. The compound eyes are well-developed, distinctly convex, and round in lateral view; three small ocelli are present in Raphidiidae, whereas they are completely lacking in Inocelliidae. The labrum is free (Fig. 6.28.1A) and rounded at its anterior margin or slightly emarginated anteromedially. The antennal insertion areas lie on the dorsal side of the head and are relatively closely adjacent in Inocelliidae; an antennifer is present anterolaterally. The antennae are filiform or moniliform and moderately long in Raphidiidae but strongly elongated in Inocelliidae; the scapus is by far the largest segment; the number of antennomeres varies between 25 and 80 and each of them bears two whorls of setae and additional sensorial hairs. The mandibles are biting and equipped with 3–4 well-developed, acuminate teeth; a mola and prostheca are lacking. The maxillae are inserted at the anterior head margin with a transverse cardo; the moderately sized stipes bears the 5-segmented palp and the two endite lobes; a palpifer is absent; the lacinia bears a regular row of spines along its mesal margin; the galea is composed of a membranous basal stalk, an intermediate sclerotized element, and a large membranous distal part. The posterior main element of the labium (submentum or postmentum) is adjacent with the gula posteriorly and strongly widened anteriorly; a medially divided intermediate sclerite either represents a separate posterior part of the prementum (e.g., Matsuda 1956) or the mentum (see discussion in Achtelig [1967]); the (anterior) premental sclerotization bears 3-segmented palps and a large membranous ligula, which is overtopped by the anteriormost part of the hypopharynx; the three main sclerotized labial elements are separated from each other by extensive membranes; glossae and paraglossae are not present as clearly defined structures, but more or less integrated into a large ligula. The salivarium (unpaired salivary duct) and salivary glands are well-developed. A fairly narrow gula is present in the central region of the ventral head capsule and the posteroventrally confluent genae form a postgenal bridge as an additional ventral closure. The tentorium is weakly developed; the anterior pits are not recognizable externally; the anterior and dorsal arms are slender; the latter are attached to the dorsal head capsule; the posterior tentorial grooves are fissure-shaped; the posterior arms are closely adjacent at their base, relatively broad but strongly flattened and weakly sclerotized; they are connected by a curved, thin tentorial bridge. The muscle system of the head is largely unmodified. The muscle of the glossa is absent.

The prothorax is slender and strongly elongated (Fig. 6.28.1A); it is capable of rapid downward strokes for catching prey. The pronotum is well sclerotized; it is shield-like

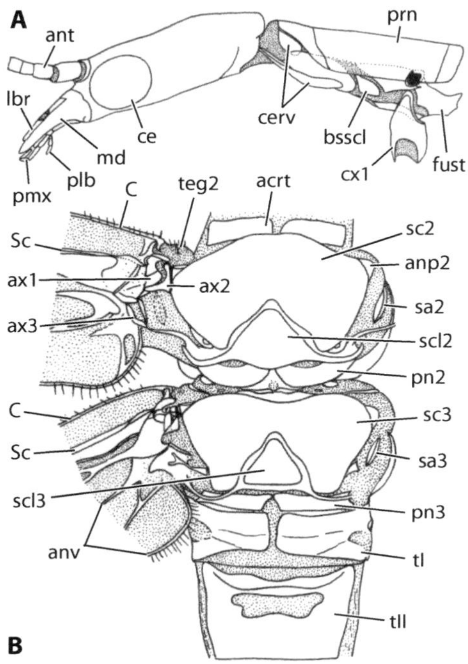

Fig. 6.28.1: *Parainocellia ressli* (Inocellidae), adult. A, head and prothorax, lateral view; B, pterothorax, dorsal view. Abbr.: acrt – acrotergite, anp2 – mesothoracic anterior notal wing process, ant – antenna, anv – anal vein, ax1/2/3 – axillary 1/2/3, bsscl – basal sclerite, C – costa, ce – compound eye, cerv – cervical sclerites, cx1 – procoxa, fust – furcasternite, lbr – labrum, md – mandible, plb – palpus labialis, pmx – palpus maxillaris, pn2/3 – meso-, metapostnotum, prn – pronotum, Sc – subcosta, sc2/3 – meso-, metascutum, scl2/3 – meso-, metascutellum, sa2/3 – meso-/metathoracic subalare, tI/II – abdominal tergites I/II, teg2 – mesothoracic tegula. Redrawn from Aspöck & Aspöck (2005), after Aspöck et al. (1991).

in Inocelliidae but almost tube-shaped in Raphidiidae, and covering the pleura which are not divided into an episternum and epimeron. The coxae are inserted at the hind margin of the prothorax. Two large unpaired plates are present anterad the procoxae in Raphidiidae; the anterior one forms a specific type of unpaired cervical sclerite and articulates with the head capsule with its anterolateral apices; the posterior one is interpreted as a basisternum connected with subcoxal elements; the interpretation of the four ventral elements occurring in Inocelliidae is uncertain (Aspöck & Aspöck 1971). The pterothoracic segments are almost equally sized and very similar in their general construction (Fig. 6.28.1B). The metathorax is only slightly smaller than the mesothorax. The main elements of the pterothorac nota are the large scutum and the triangular scutellum, which bears the postnotum posteriorly; the metapostnotum is completely divided medially. The pleura are divided by a diagonal pleural suture and ridge; the epipmera are distinctly larger than the anepisterna; below the pleural wing articulation, a deeply invaginated prealar apophysis is present, which functionally replaces the basalare. The true sternal parts are invaginated as in the other holometabolan groups; the ventral side is covered by the anterior preepisternum and the posterior katepisternum; both elements are shorter in the metathorax. The trochantins are laterally adjacent with the katepisterna, which form the apically pointed ventral coxal joints posteromesally. Spiracles are present between the pro- and mesothorax, and between the meso- and metathorax. The wings are slender and transparent (Fig. 6.28.2); the venation is well-developed, but the number of transverse veins is reduced; the hindwings are slightly smaller; a pterostigma is present with a brown or yellow coloration or a combination of both; a series of parallel costal veins connect the costa and subcosta. The legs are normally developed walking legs and similarly developed on all three segments; the meso- and metacoxae are distinctly divided into an anterior part and a posterior meron; the tarsi are 5-segmented; tarsomere 3 is heart-shaped and tarsomere 4 strongly reduced in size; dense brushes of adhesive hairs are present on the ventral side of the tarsi; the claws are paired and slightly curved; an arolium and pulvilli are lacking.

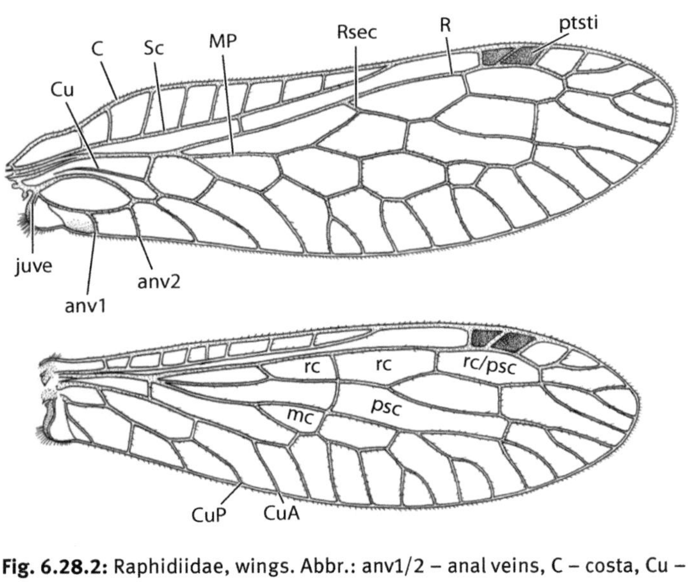

Fig. 6.28.2: Raphidiidae, wings. Abbr.: anv1/2 – anal veins, C – costa, Cu – cubitus, CuA – cubitus anterior, CuP – cubitus posterior, juve – jugal veins, ptsti – pterostigma, mc – medial cell, MP – media posterior, psc – poststigma cell, R – radius, rc – radial cell, Rsec – radius sector, Sc – subcosta.
Redrawn from Aspöck & Aspöck (2005), after Aspöck et al. (1991).

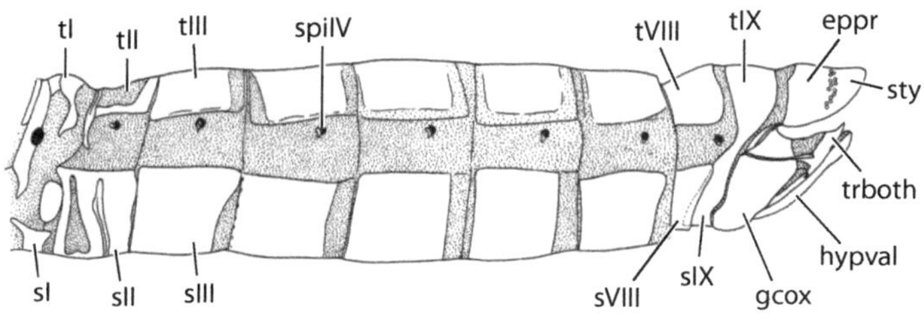

Fig. 6.28.3: *Harraphidia laufferi* (Raphidiidae), male abdomen, lateral view. Abbr.: ecpr – ectoproct, hypval – hypovalva, gcox – gonocoxae, sI–III/VIII/IX – sternum I–III/VIII/IX, spiIV – spiracle IV, sty – styli, tI–III/VIII/IX – tergum I–III/VIII/IX, trboth – trichobothria. Redrawn from Aspöck & Aspöck (2005), after Aspöck et al. (1991).

The abdomen is composed of 11 segments, but segment XI is only represented by the tergite, which is largely fused with tergite X. The tergites and sternites III–VII (females) or III–VIII (males) are simple and plate-like. The sclerites of segments I and II are fragmented and tergite I is completely interrupted medially (autapomorphy of Neuropterida). Small spiracles are present in the lateral pleural membranes of segments I–VIII. The male genital segments vary extremely (Fig. 6.28.3). Segment IX is usually ring-shaped. The gonocoxites IX can be simple and shell-shaped or highly modified and complex, with an articulated stylus, often with a prominent apex (Raphidiidae), and ventrally with paired or unpaired gonapophyes IX (Aspöck & Aspöck 2008). The ectoproct representing the tergites X and XI bears trichobothria arranged as an irregular band. The parameres (elements of gonocoxites X) are paired or unpaired, simple and rod-shaped or complex, and often reduced to varying degrees; the keel-shaped or trifurcated hypandrium internum is adjacent to the opening of the ejaculatory duct; the endophalllus varies in length; it is membranous and often equipped with small chitinous teeth or spines or small accessory sclerites; a shield-shaped gonarcus (fused gonocoxites XI) usually extends caudally between the gonocoxites IX, but it is membranous or reduced in some cases. The female genital segments are characterized by the very long ovipositor (Fig. 6.28.4); it is formed by paired elements of segment IX (gonocoxites) which are connected with each other; apically they bear a stylus equipped with sensorial setae; the "functional" ventral closure is formed by an elongate unpaired sclerite of segment VIII (fused gonapophyses). A bursa copulatrix is present and composed of an atrium bursae and a sacculus bursae.

The digestive tract (*Phaeostigma notata*) is associated with a salivary pump and salivary glands which reach the metathorax posteriorly; it is characterized by a large dorsal diverticulum of the ingluvies. Six Malpighian tubules are present, two of them anteriorly oriented and four posteriorly towards the rectum. The paired testes contain 12 follicles. The ovaries are composed of ca. 40 telotrophic meroistic ovarioles (Aspöck & Aspöck 1971, 2005, 2008; Aspöck et al. 1991).

Morphology, larvae. The terrestrial campodeiform larvae are slender, subparallel, and characterized by the elongate prothorax and the unsclerotized abdomen. The cuticle is smooth and bears a sparse vestiture of long setae. The 1st instar is unpigmented whereas the following stages are characterized by a brown and cream-white color pattern of the postcephalic body. Mature larvae can reach a length of 25 mm.

The head is well sclerotized, pronouncedly prognathous, parallel-sided, and very distinctly flattened, with nearly parallel dorsal and ventral walls. A very distinct neck region is present; it is slightly retracted into the prothorax. A moderately distinct parietal ridge is present. The frontoclypeal transverse strengthening ridge is absent. The V-shaped frons is moderately long and fairly narrow. The Y-shaped ecdysial suture is distinct; the frontal sutures are lyriform and distinctly diverging anteriorly; the coronal suture is about half as long as the head capsule. Seven or six stemmata are present in Raphidiidae and four in Inocelliidae. The small, rounded labrum is connected with the anterior clypeal margin by a membrane. The antenna is 4-segmented, fairly short, and inserted below the secondary mandibular joint and anterad the ocular region; antennomeres 1, 3, and 4 are elongate and about equally long, whereas segment 2 is very short; the distal antennomere is cylindrical and nearly truncate apically; the apical region of each segment is unsclerotized. The mandibles are very slightly asymmetric; they are falciform and elongate, with an apical tooth and additional subapical teeth; a mola and a prostheca are lacking. The maxilla is strongly protracted, i.e. inserted at the anteroventral margin of the head capsule; the base is partly covered and retractile; the cardo is transverse and short; the stipes is undivided and narrowing distally; it bears the common base of the galea and lacinia; the lacinia is an unsclerotized, rounded mesal lobe with two short, curved spines; it is enclosed by the ear-shaped unsclerotized galea; the mesal edge of the galea is set

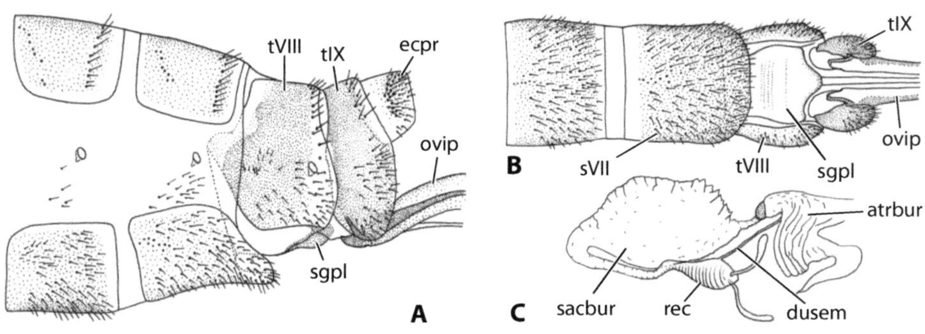

Fig. 6.28.4: *Phaeostigma notata* (Raphidiidae), female abdomen, A, lateral view; B, ventral view, ovipositor cut off close to its base; C, genitalia, lateral view. Abbr.: atrbur – atrium bursae, dusem – ductus seminalis, ecpr – ectoproct, ovip – ovipositor, rec – receptaculum seminis, sacbur – sacculus bursae, sgpl – subgenital plate, sVII – sternum VII, tVIII/IX – tergum VIII/IX. Redrawn from Aspöck & Aspöck (2005), after Aspöck et al. (1991).

with a curved row of very short microtrichia; the 5-segmented palp is inserted laterally on the basal part of the stipes. The submentum is laterally bordered by a faint suture but integrated into the ventral wall of the head capsule; posteriorly it is fused with the gula; the mentum is represented by an unpigmented and weakly sclerotized area at the anteromedian edge of the head capsule; the small, short prementum is inserted between the maxillary bases; anterodorsally it bears an undivided, roughly quadrangular sclerotized ligula; the 3-segmented labial palps are inserted below it; a palpiger, glossa and paraglossae are absent. A well-developed salivary tube and its opening are present between the prelabium and the anterior hypopharynx. An undivided moderately broad gula is present; it is almost completely fused with the ventral wall of the head capsule. The posterior tentorial grooves are strongly shifted anteriorly; the posterior tentorial arms are sclerotized, flat and moderately broad; they are continuous with the weakly sclerotized dorsal arms, which are attached to the frons; they are also connected with the dorsal wall of the head capsule by a flat ligamentous cord posterior to the brain; this flexible structure is also in close contact with gland-like structures below the pharynx; the tentorial bridge connects the posterior arms; its main part is well-developed, sclerotized and almost straight; the lateral part connecting it with the posterior arms is unsclerotized and slender; the anterior arms are very thin; they arise from a very strong transverse internal sulcus, which is intersected by the anteriormost part of the frontal suture; the anterior tentorial grooves are long and fissure-shaped (Beutel & Ge 2008)

The flattened prothorax is almost as long as the meso- and metathorax combined. The parallel-sided pronotum is strongly sclerotized with a shiny cuticle and a dark brown coloration. The meso- and metathorax are largely unsclerotized. The legs are well-developed; the coxae are widely separated; femur and tibia are about equally long; the slender, unsegmented tarsus is slightly tapering distally and bears well-developed paired claws at its apex; an enlarged spiracle is present between the meso- and metathorax.

The unsclerotized abdomen is composed of ten segments. A cream-white and brown color pattern of segments I–VIII is present in both families (Aspöck et al. 1991). Segment III–IV are wider than the other segments. The abdomen is distinctly tapering towards its apex. Segment IX bears the pygopod formed by the narrowed segment X; this movable structure has the shape of an elongate cone and enables the larvae to change their direction very swiftly when moving on the ground. Spiracle I is enlarged and shifted towards the anterior margin of the segment, whereas spiracles II–VIII are placed in the middle region of the membranous pleura of the respective segments.

Morphology, pupae. The freely moving pupa exarata dectica of Raphidioptera represents the most ancestral pupal type known in Holometabola.

Biology. The adults are mainly active in the day time, particularly in bright sunshine. They are poor flyers, usually sitting or running on the vegetation, preferably on trees and shrubs but (some species) also on low vegetation, mainly in open forested areas. The preferred prey of Raphidiidae is adults and immatures of soft bodied arthropods such as aphids, springtails, barklice, mites, and spiders, but they consume also insect eggs, pollen, and nectar. Adult Inocelliidae do not take up food, except sometimes pollen. A characteristic feature is the intensive cleaning behavior after feeding, mating, or other contact with conspecifics. The larvae of all Inocelliidae but only some Raphidiidae develop under bark. Larvae of most raphidiid species live in superficial layers of soil or in detritus around the roots of shrubs. The larvae of both families are carnivorous, feeding on soft-bodied arthropods. In contrast to other groups of Holometabola the pupae are capable of moving around actively (see above).

Reproduction and development. Males and females find each other using olfactory signals. Before the copulation the abdomen starts vibrating. During the copula the males are connected with the female genital apparatus upside-down. Two positions of copulation occur. In the "wrecking position" the males hang head first from the females and are carried by them (Raphidiidae). In the "tandem position" males crawl under the females and attach their head to the female sternum V (Inocelliidae) (Aspöck 2002). Approximately 800 eggs are laid by the females singly or in batches. They are deposited with the elongate ovipositor in cracks in bark or in rotting wood. The eggs are elongate-oval and of a yellowish-whitish coloration. A distinct, knob-like micropyle is present. The number of larval instars varies between ten and 12 (or even 15). A cocoon is not formed after the larval development. The metamorphosis takes place in a simple cavity either in soil or under bark. The entire developmental cycle is usually completed in two years, but occasionally also three years (and more) or only one. The larvae or pupae overwinter. Low temperatures are an essential trigger for the prepupal stage or for eclosion.

Fossil record. Even though a Permian (or at least Triassic) origin of the group is likely, the oldest definite snakeflies are described from Early Jurassic deposits (Grimaldi & Engel 2005). The described species already show the features typical of extant members of Raphidioptera. The former occurrence in moist tropical environments is documented by specimens of †*Nanoraphidia* (†Mesoraphidiidae) embedded in Burmese amber (Cenomanian forests) (Grimaldi & Engel 2005).

Economic importance. Snakeflies can play a positive role by decimating plant pests, especially aphids.

[Matsuda (1956); Achtelig (1967); Aspöck & Aspöck (1971, 2005, 2008); Aspöck et al. (1991); Aspöck (2002); Grimaldi & Engel (2005); Beutel & Ge (2008)]

Coleopterida (Coleoptera + Strepsiptera)

6.29 Coleoptera (Greek *coleos* = sheath, *pteron* = wing, English common name: beetles)

Diversity and distribution. With approximately 355,000 described species Coleoptera are by far the most species rich insect order. Beetles occur on all continents with the exception of Antarctica.

Taxonomy. Coleoptera are classified into four extant suborders (Archostemata, Adephaga, Myxophaga, Polyphaga) and ca. 168 families. The interrelationships of the four main lineages are not fully clarified. A considerable number of plesiomorphies preserved in Archostemata (Fig. 6.29.1) suggest a basal placement of this group (e.g., Beutel et al. 2008; Friedrich et al. 2009). An alternative concept is a basal placement of Polyphaga (Kukalová-Peck & Lawrence 1993, 2004).

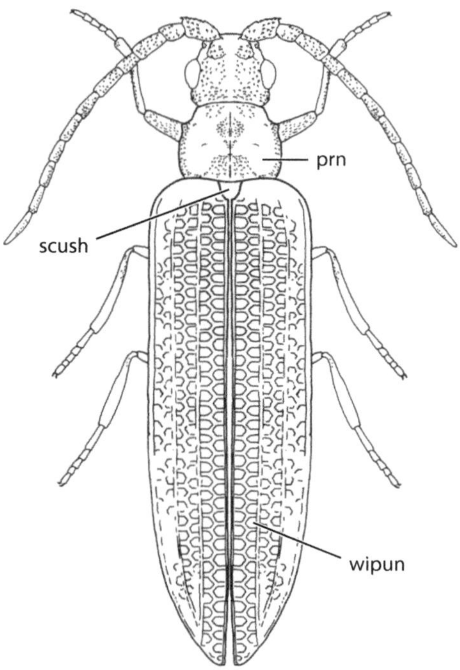

Fig. 6.29.1: *Distocupes varians* (Cupedidae, Archostemata), dorsal view. Abbr.: prn – pronotum, scush – scutellar shield, wipun – window punctures of elytra. Redrawn from Moore (1982), modified.

Archostemata is the smallest suborder, with only slightly more than 30 species and four or five extant families. All Permian fossil groups (e.g., †Tshekardocoleidae, †Permocupedidae) and also †Triadocupedinae do not belong to Archostemata but are members of the stem group of the entire Coleoptera (Fig. 5.1.1). Myxophaga (Fig. 6.29.5) presently comprise four families and ca. 100 species. Adephaga are classified into 10 or 11 families. Two or three are terrestrial (the family status of the wood-associated Rhysodidae is disputed) (Fig. 6.29.4), six fully aquatic, and two recently introduced very small families (Aspidytidae, Meruidae) live in hygropetric habitats. With ca. 315,000 described species, 16 superfamilies and 144 families the suborder Polyphaga is an extremely successful insect lineage.

Autapomorphies. The following derived features belong to the groundplan of the extant lineages of Coleoptera (crown group; see also Fig. 5.1.1). Some plesiomorphies such as for instance antennae with 13 segments are preserved in some of the extinct Permian groups.

- Strongly sclerotized, no exposed membranes (Figs 6.29.1, 6.29.4, 6.29.5)
- Cuticle with tubercles and scales (groundplan feature, only maintained in two archostematan families) (Fig. 6.29.2)
- Head prognathous and wedge shaped (Figs 6.29.1–6.29.4)
- Antenna max. 11-segmented (number of antennomeres also reduced in Strepsiptera) (?)
- Forewings transformed into sclerotized elytra (also in stem group fossils) (Fig. 6.29.1)
- Mesothorax distinctly reduced in size (also in Strepsiptera) (?) (Figs 6.29.4, 6.29.5)
- Elytra locking mechanisms involving mesothoracic scutellar shield and metathoracic alacristae
- Longitudinal and transverse hindwing folding
- Terminal segments invaginated

Diagnose. Strongly sclerotized. Head almost always prognathous (Figs 6.29.2–6.29.4). Antenna with 11 segments or less. Pronotum shield-like. Mesothorax shortened, forming a compact unit with the metathorax. Forewings transformed into sclerotized elytra.

Morphology, adults. Beetles range in size between 0.3 mm (Ptiliidae) and 170 mm (*Titanus giganteus* [Cerambycidae, longhorn beetles]). They are almost always strongly sclerotized, without externally exposed membranes; the dorsal side of the pterothorax, the hindwings and the abdominal tergites are usually covered by the forewings, which are transformed into sclerotized elytra (Fig. 6.29.1). The thick cuticle is primarily covered with tubercles and scales (Fig. 6.29.2) but this pattern is only preserved in the archostematan families Cupedidae and Ommatidae. Numerous variations of the surface sculpture and coloration occur.

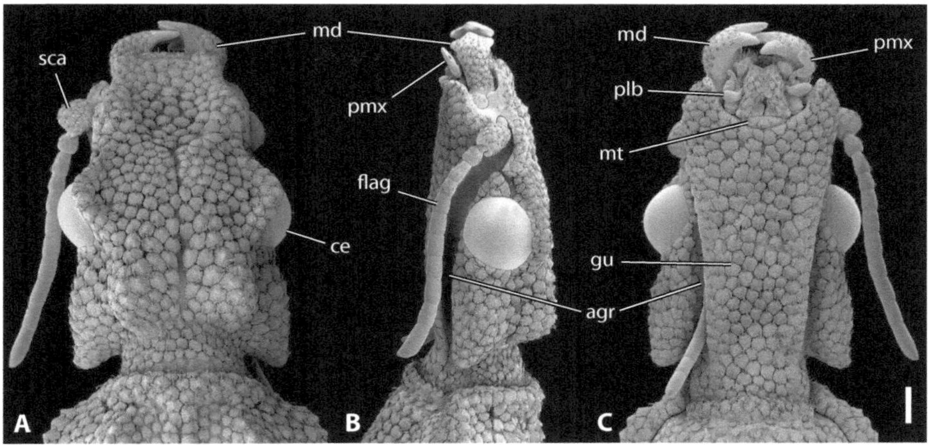

Fig. 6.29.2: *Tetraphalerus bruchi* (Archostemata, Ommatidae), head, SEM micrographs. A, dorsal view, B, lateral view, C, ventral view. Abbr.: agr – antennal groove, ce – compound eye, flag – flagellum, gu – gula, md – mandible, mt – mentum, plb – palpus labialis, pmx – palpus maxillaris, sca – scapus. Scale bar: 250 µm. From Beutel et al. (2008), modified.

The head (Figs 6.29.2, 6.29.3) is almost always prognathous and wedge-shaped in lateral view, apparently an adaptation to penetrating narrow crevices, especially subcortical spaces. The frontal and coronal sutures are absent (Figs 6.29.1, 6.29.2A); a dorsomedian furrow occurs occasionally (Hydrophiloidea). The frontoclypeal transverse ridge is present or absent; the clypeus is not subdivided. The compound eyes vary in shape and size; they can be subdivided (e.g., Gyrinidae) or partly or completely reduced; ocelli are usually missing, but three occur in *Sikhotealinia* (Jurodidae) (not yet confirmed by anatomical data), two in Hydraenidae, few Staphylin-

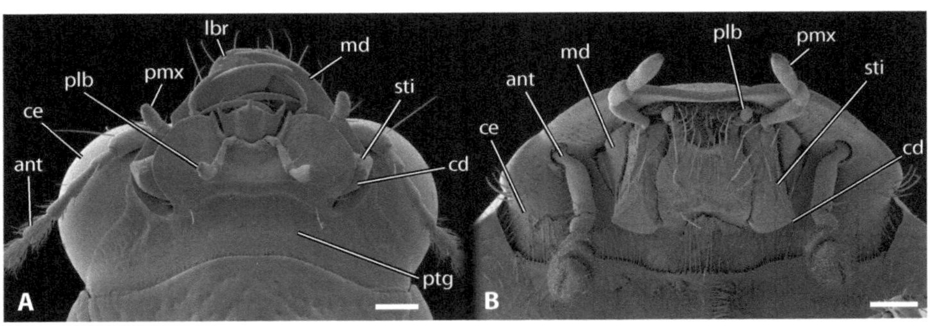

Fig. 6.29.3: Adult heads, ventral view, SEM micrographs. A, *Notiophilus* sp. (Adephaga, Carabidae), B, *Aphodius* sp. (Polyphaga, Scarabaeidae). Abbr.: ant – antenna, ce – compound eye, cd – cardo, lbr – labrum, md – mandible, plb – palpus labialis, pmx – palpus maxillaris, ptg – posterior tentorial groove, sti – stipes. Scale bars: 200 µm.

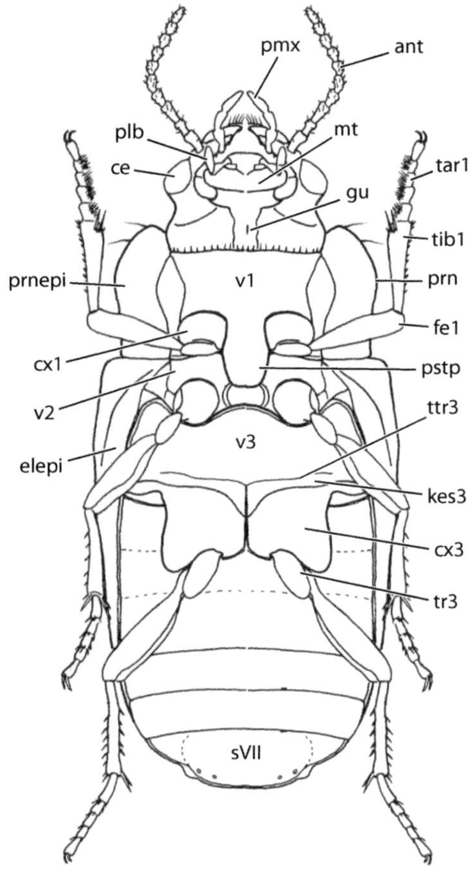

Fig. 6.29.4: *Trachypachus holmbergi* (Adephaga, Trachypachidae), ventral view. Abbr.: ant – antenna, ce – compound eye, cx1/3, pro-, metacoxa, elepi – elytral epipleuron, fe1 – metathoracic femur, gu – gula, kes3 – metathoracic katepisternum, mt – mentum, plb – palpus labialis, pmx – palpus maxillaris, prn – pronotum, prnepi – pronotal epipleuron, pstp – prosternal process, tar1 – prothoracic tarsus, tib1 – prothoracic tibia, tr3 – metathoracic trochanter, ttr3 – metathoracic transverse ridge, v1/2/3 – pro-, meso-, metaventrite, sVII – terminal abdominal sternite VII.

idae and Derodontidae, and one in Dermestidae (partim). The antennae are inserted laterally anterad the compound eyes or on the dorsal side of the head (e.g., Cicindelinae); they are primarily 11-segmented; reductions occur in several groups (e.g., Myxophaga, Hydrophiloidea, Cucujoidea); the shape is highly variable, filiform, moniliform, clubbed, serrate, geniculate or flabellate. The labrum is transverse in most groups and usually movably connected with the anterior clypeal margin; the median labral retractor is always missing. The mouthparts are almost always orthopteroid

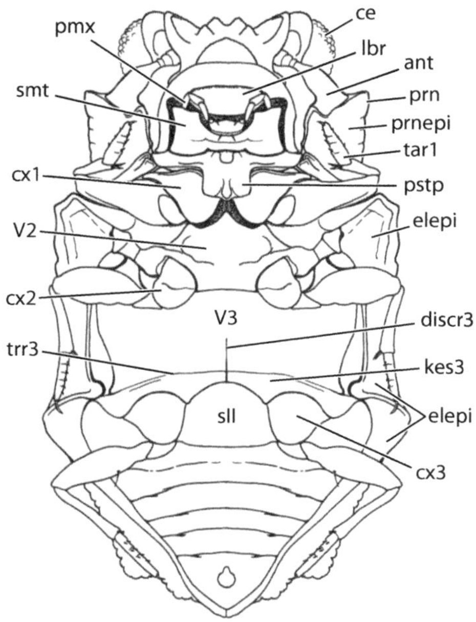

Fig. 6.29.5: *Lepicerus bufo* (Myxophaga, Lepiceridae), ventral view. Abbr.: ant – antenna, ce – compound eye, cx1–3 – pro-, meso-, metacoxa, discr3 – discimen of metaventrite (longitudinal ridge), elepi – elytral epipleuron, kes3 – metathoracic katepisternum, lbr – labrum, pmx – palpus maxillaris, prn – pronotum, prnepi – pronotal epipleuron, pstp – prosternal process, smt – submentum, tar1 – protarsus, ttr3 – metathoracic transverse ridge, v2/3 – meso-, metaventrite, sII – abdominal sternite II.

(Figs 6.29.2, 6.29.3). The biting mandibles lack a mola in Archostemata and Adephaga but it is present in Myxophaga and basal polyphagan lineages (e.g., Hydrophiloidea); a prostheca occurs in beetles feeding on small particles (e.g., Hydraenidae); in few cases the mandibles of adults are stylet-like, reduced (e.g., Rhipiphoridae partim), or strongly enlarged (e.g., males of Lucanidae). The maxillary stipes is divided into a basistipes and mediostipes; the lacinia is fused with the mediostipes, set with spines or strong setae along its mesal edge, and bent inwards distally in most groups; the galea is usually 2-segmented and pubescent, but palp-like in Adephaga; the palp is 4-segmented and often inserted on a more or less distinct palpifer. The labium is primarily composed of submentum, mentum and prementum; paraglossae and glossae are largely or completely reduced; the palp is 3-segmented; a palpiger is present or absent (Fig. 6.29.3). The hypopharynx is fused with the dorsal side of the anterior labium (prelabium); the salivarium, salivary ducts and salivary glands are reduced.

Ventrally the head capsule is closed by a sclerotized gular plate, which is usually enclosed by strongly developed internal gular ridges. The tentorium is usually complete, but reductions occur, especially in Archostemata; the bridge connects the gular ridges, which form a structural unit with the well-developed postoccipital ridge posteriorly and the posterior tentorial arms anteriorly; an accessory anterior tentorial bridge (medially connected laminatentoria) is present (e.g., Meruidae) or absent. Cervical sclerites are present (Polyphaga [major part], vestigial and not exposed in Ommatidae) or absent. The musculature of the head is similar to that of other holometabolan groups. The muscles of the glossae, paraglossae and salivarium are absent.

The prothorax forms a functional unit with the head. The pronotum is large and strongly sclerotized, and laterally inflected, usually with a distinct lateral edge (reduced e.g., in Micromalthidae) (Figs 6.29.1, 6.29.4, 6.29.5). The pleura is not divided into an episternal and epimeral part; at least its upper part is internalized (cryptopleura); the entire pleura is invaginated in Polyphaga; it is fused with the protrochantin in Polyphaga and Myxophaga (trochantinopleura). The ventral sclerite (prosternum) is likely composed of true sternal and anapleural elements; it lies anterad and between the procoxae. The median prosternal process is connected with a mesally directed posterior propleural process in some groups (externally closed procoxal cavity); an internal closure of the procoxal cavities does also occur, sometimes combined with the external postcoxal bridge. The profurca is well-developed, usually with both arms separated basally. The meso- and metathorax form a rigid structural and functional unit (Figs 6.29.4, 6.29.5). The mesothorax is distinctly shortened. The lateral wall is formed by the anepisternum and epimeron. The ventrite is mainly formed by the preepisternum and probably not clearly identifiable sternal and katepisternal elements; the transverse ridge, the discrimen (median longitudinal suture, invagination site of the mesosternum) and the ventral coxal joint are absent in almost all beetles (present in the archostematan families Cupedidae and Ommatidae). The scutellar shield, an elevated part of the mesoscutellum, is usually exposed; it is part of a complex elytral locking device. The arms of the mesofurca are usually separated basally and often fused or connected with the well-developed pleural ridge. The forewings are transformed into strongly sclerotized elytra with lateral inflected parts, the epipleurae; regularly arranged window punctures (Fig. 6.29.1, 6.29.6), small zones of thin transparent cuticle, are only present in Cupedidae and Ommatidae (groundplan of Coleoptera, also found in stem group fossils); the inner surface of the elytra is often equipped with binding patches. The flight function is largely or completely restricted to the large metathorax. Movements of the elytra during flight occur only in archostematan beetles but do not significantly contribute to the forward motion of the body. The metanotum is much more extensive than the mesonotum and distinctly subdivided into different attachment areas of large flight muscles; a small membranous field is present posterad the prescutum; the scutum is the largest element; the paramedian alacristae, longitudinal ridges covered with flattened microtricha, fix the elytra in their resting position, together with the meso-

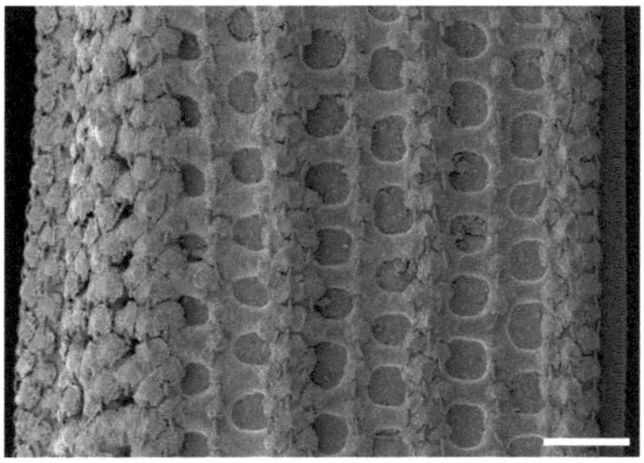

Fig. 6.29.6: *Tetraphalerus bruchi* (Archostemata, Ommatidae), detail of left elytron showing window puncturing, SEM micrograph. Scale bar: 200 µm.

notal scutellar shield; the metascutellum is triangular; the postnotum usually forms a large internal phragma, the attachment area of the large dorsal longitudinal flight muscles. The composition of the metaventrite corresponds with that of the mesoventrite, but a discrimen and a transverse ridge are often present; the transverse ridge separates the short katepisternum from the main part of the ventrite in many groups; an exposed trochantin is only present in Cupedidae and Ommatidae (groundplan of Coleoptera, also present in Permian fossils). The metafurca has a common stalk and well-developed anterior and lateral arms in most groups. The hindwings (Fig. 6.29.7) are usually well-developed but can be partly (brachypterous) or completely reduced; in their resting position they are usually folded longitudinally and transversely, but the apical part is rolled in Archostemata and Myxophaga (and some small aquatic adephagan species); the venation is simplified, especially in the polyphagan Cucujiformia; an oblongum cell is usually preserved in the non-polyphagan groups; binding patches of microtrichia occur in some groups (e.g., Dytiscoidea); the wing margin bears a fringe of microtrichia in very small beetles; the wings are reduced to stalk-like structures with long microtrichia in the extremely small Ptiliidae. The pro- and mesocoxae are usually globular or cone-shaped, whereas the metacoxae are almost always transverse (Figs 6.29.4, 6.29.5); the degrees of freedom at the coxal bases are distinctly reduced and the metacoxae can be plate-like and immobilized (e.g., Dytiscidae, Gyrinidae); the legs can be modified in many different ways; they can be slender walking legs (e.g., Carabidae), swimming legs (e.g., Dytiscidae), modified for burrowing (e.g., Geotrupidae), or jumping legs (e.g., Alticini or Galerucinae [Chrysomelidae]); different cleaning devices occur (e.g., protibial antenna cleaner in Geadephaga); the tarsi are primarily 5-segmented, but the number of tarsomeres varies strongly; the tarsal formula (e.g., 5-5-5, 5-5-4, 4-4-4 or 3-3-3) is an important diagnostic feature; the primary attachment devices are hairy pads on the tarsomeres; arolium and pulvilli are absent; the claws can be equal or unequal, pectinate or modified in other ways. A small spiracle is usually present between the pro- and mesothorax and a larger one between the pterothoracic segments. The number of thoracic muscles is distinctly reduced compared to other holometabolan groups, even though all functions of the locomotor apparatus are preserved; this is apparently related to reduced degrees of freedom resulting from the strongly sclerotized body; the highest number of thoracic muscles is found in Archostemata, the lowest in Polyphaga.

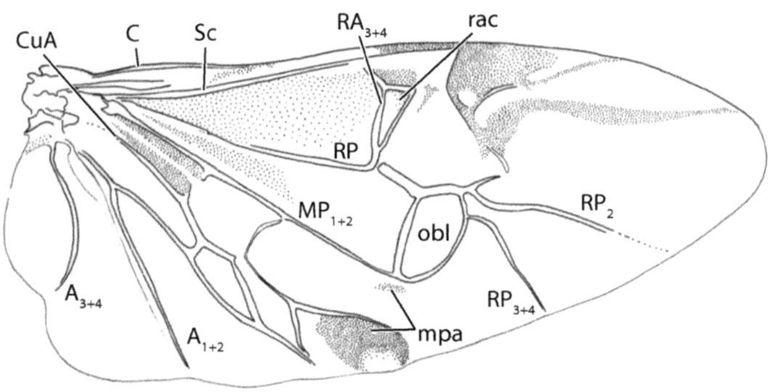

Fig. 6.29.7: *Aspidytes niobe* (Adephaga, Aspidytidae), hindwing. Abbr.: $A_{1+2/3+4}$ – anal veins, C – costa, CuA – cubitus anterior, mpa – medial patch, MP_{1+2} – branches of media posterior, obl – oblongum cell, rac – radial cell, RA_{3+4} – branches of radius anterior, $RP_{2/3+4}$ – branches of radius posterior, Sc – subcosta.

The movability between the abdomen and the metathorax is minimal (Figs 6.29.4, 6.29.5). Ten abdominal segments are usually present in males and nine in females. The tergites and sternites are usually connected by pleural membranes, which contain the spiracles on segments I–VIII in most groups. Tergites I–VI are distinctly less sclerotized than the sternites. Tergite VIII is terminal on the dorsal side and sternite VII ventrally. Segments IX and X are invaginated (Figs 6.29.4, 6.29.5). Sternite I is not present as a recognizable sclerotized element and is covered by the metacoxae. Only the lateral part of sternite II is present in most groups and it is often connected with sternite III; a separate intercoxal piece II occurs in some groups. Sternite VIII is retracted in almost all groups (exposed for instance in Hydraenidae), together with the genital segment IX. The male copulatory apparatus is primarily composed of a basal piece (phallobase), paired symmetrical parameres, a median lobe (penis), and a membranous endophallus (internal sac); numerous modifications occur including asymmetric parameres or loss of one or both. The external female genital apparatus is composed of strongly reduced genital appendages VIII and appendages of segment IX resembling papillae, palps or forceps (products of fusion of the gonocoxite and gonostylus); genital appendages IX flank the simple or double genital opening; the gonapophyses IX are completely reduced (see also Neuropterida).

A proventriculus occurs in Adephaga and Curculionidae; large midgut caeca are often present in Polyphaga (e.g., Buprestidae, Elateridae, Curculionidae); small regenerative crypts (papillae) occur frequently in this suborder; mycetome organs with endosymbionts can also be associated with the midgut; the rectal papillae are normally well-developed. Six or four Malpighian tubules are present; they are closely attached to the hindgut and enclosed by layers of membranes in Bostrichoidea and Cucujiformia (cryptonephric condition). The abdominal ganglionic chain is strongly condensed in some groups; only one complex is present in Histeridae, Scarabaeidae and Curculionidae. The testes are tube-like (e.g., Archostemata) or composed of several follicles (Polyphaga); accessory glands are usually present and spermatophore glands occur. The ovarioles are polytrophic in Adephaga and telotrophic in Archostemata and Polyphaga; usually they are associated with accessory glands.

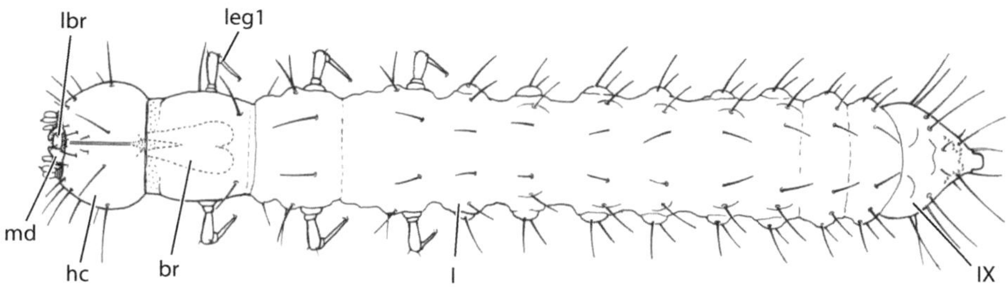

Fig. 6.29.8: *Micromalthus debilis* (Archostemata, Micromalthidae), 1st instar larva. Abbr.: br – brain, hc – head capsule, lbr – labrum, leg1 – fore leg, md – mandible, I/IX – abdominal segments I/IX. From Beutel & Hörnschemeyer (2002), modified.

Morphology, larvae. The head capsule of the larvae is sclerotized and complete with very few exceptions (partly reduced in Lycidae). The body is slender, with well-developed walking legs (campodeiform, e.g., Carabidae, Staphylinidae) (Figs 6.29.8, 6.29.9), distinctly flattened, with distinct lateral tergal projections (onisciform, Silphidae), grub-like, with strongly shortened legs (e.g., Cupedidae, Cerambycidae), C-shaped, with largely unsclerotized postcephalic segments (Scarabaeoidea), or maggot-like, with completely reduced legs (Curculionidae).

The head is primarily subprognathous and posteroventrally closed by a short membranous gular area (e.g., Myxophaga, Hydraenidae); a prognathous head with a sclerotized gula occurs in different groups, usually but not always correlated with predaceous habits (e.g., Adephaga, Hydrophiloidea, Cantharoidea, Cleroidea) (Fig. 6.29.9). Six or five larval eyes (stemmata) are present in most groups; ocelli are always absent. A V-shaped or sinuate frontal suture and a coronal suture are usually present; dorsal strengthening ridges (endocarina) and ventral hypostomal and epicranial ridges occur in different groups (e.g., Archostemata). The labrum is primarily free, but can be fused with the clypeus, especially in groups with a prognathous head and predaceous habits (e.g., Adephaga, Staphylininae, Hydrophiloidea) (Fig. 6.29.9). The antennae are usually composed of four (Archostemata, Adephaga) or three antennomeres (Myxophaga, Polyphaga); a sensorial appendage is usually present on the penultimate antennomere. The mouthparts are primarily biting. The presence of a mandibular grinding mola is a groundplan feature of Coleoptera but it is generally reduced in groups with predaceous larvae (e.g., Adephaga, Hydrophiloidea); a prostheca occurs in some groups with microphagous larvae (e.g., Myxophaga, Cucujoidea). The maxilla is primarily inserted in distinct maxillary grooves but articulates with the anterior margin of the head capsule ('maxillae protracted') in some predaceous groups (e.g., Adephaga, Histeridae); the cardo and stipes are primarily connected by a hinge but the movability between these elements can be strongly reduced in groups with protracted maxillae; the cardo is rarely reduced (Histeridae); the lacinia is absent in some predaceous groups (e.g., Dytiscoidea) and both endite lobes are fused in Cucujiformia, thus forming a single mala; the palp is usually 4-segmented; a palpifer is present or absent. The labium is primarily composed of a submentum, mentum, prementum and a 2-segmented palp; a ligula is present or absent; it is enlarged and strongly sclerotized in larvae of Archostemata (autapomorphy of the suborder); the submentum is fused with the ventral wall of the head capsule in some groups with prognathous larvae (Adephaga, Hydrophiloidea) and the mentum is membranous in Adephaga.

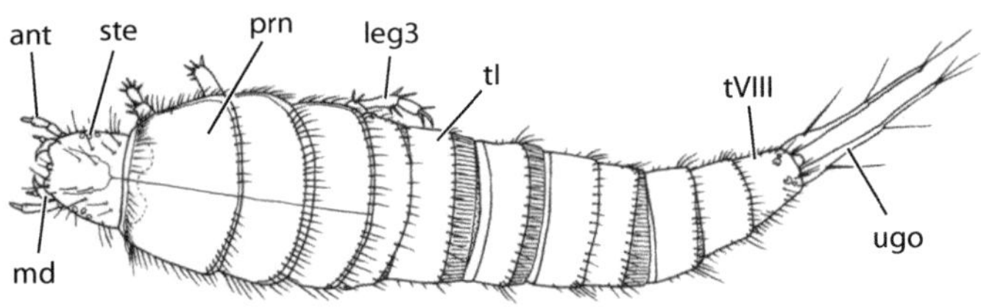

Fig. 6.29.9: *Aspidytes niobe* (Adephaga, Aspidytidae), 3rd instar larva. Abbr.: ant – antenna, leg3 – hindleg, md – mandible, prn – pronotum, ste – stemmata, tI/VIII – abdominal tergites I/VIII, ugo – urogomphi.

The thoracic tergites are primarily well-developed and often divided by a narrow median zone of weakness. The meso- and metatergites are unsclerotized in some groups (e.g., Hydrophiloidea) and the entire postcephalic body wall can be more or less membranous and unpigmented in grub-like larvae (e.g., Archostemata, Scarabaeoidea). The legs are composed of six (Archostemata, Adephaga) (Figs 6.29.8, 6.29.9) or five segments (Myxophaga, Polyphaga), with the tibia and tarsus fused in the latter case; complete reduction occurs in groups with mining larvae (Micromalthidae [later instars], Curculionoidea); the claws are primarily paired but always unpaired in Myxophaga and Polyphaga. Two pairs of thoracic spiracles are present in most groups.

The abdomen is usually composed of ten segments. The terminal segments IX and X are reduced in some groups (Noteridae, Dytiscidae, Paelobiidae, Amphizoidae) (Fig. 6.29.9). Tergite IX bears articulated (Adephaga part., basal polyphagan groups) or fixed urogomphi (e.g., *Trachypachus*, Elateroidea, Cleroidea, Cucujoidea, Tenebrionoidea) (Fig. 6.29.9: ugo). Segment X is usually present as a ventrally directed pygopodium, often with eversible terminal vesicles, and with terminal hooks in few groups (e.g., Gyrinidae, Hydraenidae). Spiracles are usually present on segments I–VIII but different modifications and reductions occur. Abdominal tracheal gills occur in Gyrinidae and as an exception in Dytiscidae (*Coptotomus*) and Hydrophilidae (*Berosus*); spiracular gills are present in Myxophaga (excl. Lepiceridae) and microtracheal gills in Haliplidae (excl. *Peltodytes*).

Morphology, pupae. The mandibles are immobilized. The legs and wing sheaths are usually free (pupa adectica exarata). A pupa obtecta occurs in Staphylininae. Nine abdominal tergites and eight sternites are usually present. The number of spiracles is lower than in larvae. The spiracles VIII are always closed. The head and body are equipped with a specific vestiture of setae which prevent direct contact with the walls of the pupal chamber.

Biology. The biology of the megadiverse Coleoptera is difficult to cover in a short chapter. A comprehensive treatment was presented by the outstanding coleopterist of the 20th century, the late Prof. Dr. Roy A. Crowson (Crowson 1981), and extensive treatments are also found in the Coleoptera volumes of the Handbook of Zoology (Beutel & Leschen 2005; Leschen et al. 2010).

Beetles are found in nearly all types of terrestrial and limnic habitats including deserts, edges of glaciers, tree canopies, caves, soil, and fast flowing rivers. Some aquatic species live in brackish water and some species occur in tide pools (e.g., *Ochthebius* [Hydraenidae]). Aquatic habits were acquired at least ten times independently (e.g., Gyrinidae, Dytiscoidea, Hydraenidae, Hydrophiloidea, Elmidae, Lampyridae [partim], Donaciinae). The adults can be aquatic (Dryopidae, Helophorinae) or the larvae (Scirtidae, Eulichadidae) or both (most groups). Hygropetric habits, i.e. life on rocks covered with algae or moss and a thin film of water, also evolved several times independently (Myxophaga [major part], Meruidae, Aspidytidae, Hydraenidae [partim]). The primary habitat of beetles is probably the subcortical space and the wedge-shaped head and the elytra are apparently adaptations to this specific microhabitat. The larvae are usually more cryptic than adults. Larvae mining in wood are possibly the ancestral condition in Coleoptera (Archostemata), but this life style was also secondarily acquired by very successful polyphagan lineages (e.g., Buprestidae, Cerambycidae, Scolytinae).

Like the habitats, the food sources of beetles are extremely diverse. In some groups the feeding habits of larvae and adults differ distinctly (e.g., Hydrophiloidea). A diet of wood infested with microorganisms or fungal spores or hyphae is probably ancestral for the order. The wood associated geadephagan rhysodids are specialized on slime molds (Myxomycetes). Saprophagous and microphagous feeding habits are common (e.g., Staphylinoidea [partim]). Predaceous habits have evolved in different lineages (e.g., Adephaga, Staphylininae, Hydrophiloidea [larvae], Cantharoidea). Larvae of some carnivorous groups possess mandibular sucking channels (Gyrinidae, Dytiscidae [major part], Lampyridae) and larvae of others have evolved specialized ambush strategies (Cicindelinae, Metriinae). Carrion feeders occur in several polyphagan lineages (e.g., Silphidae, Trogidae, Dermestidae) and dung-feeding has also evolved in different groups (e.g., Sphaeridiinae [Hydrophilidae], Geotrupidae, Aphodiinae [Scarabaeoidea]). Haliplidae and non-adephagan hygropetric beetles and their larvae feed on algae, and many aquatic beetles use fresh or decaying parts of water plants as food source (e.g., Hydrophilidae, *Bagous* [Curculionidae]). Very successful beetle lineages are more or less specialized plant feeders, notably the megadiverse Chrysomelidae and Curculionidae. Feeding on gymnosperms is a rare exception. Ca. 99% of the phytophagous beetles feed on angiosperms. Feeding on the surface of exposed plant parts is common. Larvae of some groups feed on roots (e.g., Melolonthinae). The legless larvae of Curculionidae mine in plant tissue and some of them produce galls.

Interestingly bioluminescence, which is best known from fire flies, has evolved independently in several groups of Elateroidea such as Pyrophorini, *Balgus* (both Elateridae), Lampyridae, Phengodidae, and Rhagophthalmidae.

Inquilines occur in several groups of beetles. The adephagan paussines (flanged bombardier beetles) are usually obligate myrmecophiles and prey upon workers and larvae of the ants. As a defense mechanism their pygidial glands produce explosive secretions like in brachinine Carabidae (bombardier beetles). Inquilinism has also evolved in representatives of other lineages such as Staphylinidae (Aleocharinae, Tachyporinae, Pselaphinae), Leiodidae, Ptiliidae, Histeridae, Scarabaeoidea (Aphodiinae, Cetoniinae), Tenebrionidae, Anobiidae, Salpingidae, and Chrysomelidae.

Symbiosis with fungi and microorganisms also occurs in different groups. Bark associated beetles of different groups are known to carry fungal spores in cuticular concavities (e.g., Sphindidae, *Dendrophagus*, [Silvanidae]). Mycetome organs occur in Nosodendridae and Chrysomelidae and other families. Nitrogen-fixing bacteria and yeasts are known from ambrosia beetles (Scolytinae, Platypodinae).

Reproduction, development and immature stages. Beetles are usually bisexual but parthenogenesis occurs in phytophagous lineages (Curculionidae, Chrysomelidae) and also in some other groups (Ptiliidae, Ciidae, Bothrideridae). In plant feeding species this allows a strongly accelerated growth of populations under favorable circumstances.

Pheromones can play a role in finding a mating partner (e.g., Scolytinae). This is often correlated with enlarged and modified antennae in males (e.g., Melolonthinae). Complex courtship behavior is only known from members of few groups (Melyridae,

Meloidae, Cerambycidae). Strongly enlarged mandibles (Lucanidae) or horn-like structures (e.g., Dynastinae [Scarabaeidae]) of males are often used by males for fighting over females. Modified tarsi can ensure a firm grip on the mating partner. Highly specialized sucking devices are present on the fore tarsi of males of Dytiscinae.

Eggs are normally oval and whitish. The presence of a distinctly developed net-like chorion is possibly ancestral for Coleoptera (Archostemata, Gyrinidae), even though it is absent in most groups. The number of deposited eggs varies greatly. Very small beetles (e.g., Sphaeriusidae, Ptiliidae) deposit only a single egg. Cephalic (e.g., Carabidae, Dytiscoidea, Hydrophiloidea) or thoraco-abdominal egg bursters (e.g., Histeridae, Scarabaeidae), which facilitate the hatching of the first instar larvae, have evolved independently in several groups. The number of larval instars varies between one (e.g., *Systolosoma* [Trachypachidae]) and 14, but three stages are likely ancestral for beetles (e.g., Adephaga, Silphidae, Histeridae, Staphylinidae, Scarabaeidae). In many groups pupal chambers are built in soil and pupation under bark or in wood is typical for xylophagous taxa. Pupation within the last larval exuvia occurs in surface feeding phytophagous beetles and the hygropetric Myxophaga, but it takes place within the tissue of the host plants in groups with mining larvae. A pupal cocoon occurs only in few groups.

In some groups, especially those with parasitic larvae, a complex developmental cycle has evolved (e.g., Passandridae, Bothrideridae, Meloidae, Rhipiphoridae), and *Micromalthus debilis* (Micromalthidae [Archostemata]) has arguably the most complicated life cycle of all insects, with parthenogenesis, thelytoky, hypermetamorphosis, and viviparous larvae which are consumed by their offspring. Cocoons containing several eggs are characteristic for Hydrophiloidea and also occur in some groups of Chrysomelidae. Vivipary is not only known from Micromalthidae but also occurs in symbiotic aleocharines (Staphylinidae). Brood care is a rare exception in beetles. It occurs in Scarabaeoidea and in a more elaborate form in Nicrophorinae (Silphidae).

The duration of the development from the egg to the adult varies widely depending on the taxon, food resources, temperature and other factors. It can take only few weeks in predaceous forms (e.g., *Calosoma*, Carabidae; *Cybister*, Dytiscidae) or several years in the case of phytophagous species (e.g., *Melolontha*, Scarabaeidae; *Lucanus*, Lucanidae). A one-year life-cycle is most common.

Fossil record. The fossil record of the strongly sclerotized beetles is unusually rich, but in many cases only elytra are preserved. A Carboniferous fossil recently assigned to Coleoptera (Bethoux 2009) is apparently not a beetle but belongs to an extinct basal holometabolan lineage (Kukalová-Peck & Beutel 2012). Comparatively few specimens are described from Permian deposits. They were all assigned to Archostemata in traditional classifications (e.g., Ponomarenko 1969) but as recently demonstrated belong in fact to the stem lineage of the entire Coleoptera (e.g., †Tshekardocoleidae, †Permocupedidae) (e.g., Beutel 1997; Beutel et al. 2008). In the Late Triassic (240-220 Ma) beetles had already undergone a distinct radiation (20 families). Several aquatic and terrestrial families of Adephaga are recorded and the first groups of Polyphaga appear (e.g., Hydrophilidae, Staphylinidae). In contrast to the Permian beetles and the extant archostematan families Cupedidae and Ommatidae, extant and extinct representatives of Adephaga and Polyphaga lack the plesiomorphic elytral window punctures and the typical cuticular surface pattern with tubercles and scales. Another radiation is documented by the rich fossil record of Jurassic beetles (especially Central and East Asian deposits, e.g., Madygen series, Kara Tau, Xixian formation). The diversity is distinctly increased with 35 families and ca. 600 species (Grimaldi & Engel 2005). Cretaceous fossils are less numerous than those from Jurassic deposits, but the preservation of amber fossils (e.g., Lebanon, Taimyr Peninsula [Siberia]) give much deeper insights in the morphology. A dramatic increase of the beetle diversity took place in the Late Cretaceous. Whether this was mainly due to a successful co-evolution with angiosperms is still a matter of dispute. Numerous tertiary beetle fossils are known. Brown coal deposits in Germany are famous and numerous species are beautifully preserved in Baltic amber.

Economic importance. Numerous beetle species are predaceous and many of them have an indirect positive economic impact. Some large carabids (e.g., *Cychrus*, *Carabus* species) are specialized on snails as food and species of *Calosoma* are successful predators of harmful caterpillars of lepidopteran species. Coccinellid species are very efficient in decimation populations of aphids and scale insects. Phytophagous beetles feeding on weeds by can be considered beneficial. The aquatic curculionid *Cyrtobagous salviniae* was successfully introduced in Australia to suppress the rampant growth of the water plant *Salvinia molesta*. Beetles belonging to several families play a positive role by processing rotting plant materials, carrion, and dung (see above).

Numerous beetle species have a negative impact on forestry (especially Scolytinae, Cerambycidae and Buprestidae), agriculture (e.g., Chrysomelidae, Curculionidae), and storage of products (e.g., *Stegobium paniceum*, Ptinidae). Specialized species can also damage other materials used by humans such as textiles or leather (e.g., Ptinidae, Dermestidae). Devastations can be caused in neglected natural history collections by species of *Anthrenus* ("museum beetle") (Dermestidae) and by *Ptinus fur* (Ptinidae). Beetles play no role as vectors of diseases or parasites of humans or domestic animals. Species of *Paederus* (Staphylininae) can cause serious lesions (whip-lash dermatitis). Oedemerid and meloid beetles contain cantharidin, a toxic substance, which causes blisters ("blister beetles") and poisoning.

[Ponomarenko (1969, 1995); Crowson (1981); Lawrence & Newton (1982); Moor (1982); Kukalová-Peck & Lawrence (1993, 2004); Pakaluk & Ślipiński (1995); Beutel (1997); Beutel & Hörnschemeyer (2002); Beutel & Leschen (2005); Grimaldi & Engel (2005); Beutel et al. (2008); Friedrich et al. (2009); Tan & Ren (2009); Bethoux (2009); Leschen et al. (2010); Lawrence et al. (2011); Kukalová-Peck & Beutel (2012)]

6.30 **Strepsiptera** (Greek *strephein* = to twist, *pteron* = wing, English common name: twisted wing parasites) (Hans Pohl & Rolf G. Beutel)

Diversity and distribution. About 600 species are described. Strepsipterans occur on all continents except for Antarctica. The diversity increases towards the tropical region and is likely highest in Africa. The distribution of strepsipterans and their hosts is strongly correlated. However, some species are absent in parts of the area where the host occurs, whereas others may choose an alternative host outside the main host's distribution range.

Autapomorphies (Figs 6.30.1, 6.30.2)
- Endoparasitic secondary instar larvae
- Hypermetamorphosis with very different larval stages
- Vivipary
- First instar larvae strongly miniaturized (average size ca. 230 µm)
- Pretarsal adhesive pads of 1st instar larvae
- Antennae of 1st instar larvae vestigial or absent
- Maxillae of 1st instar larvae medially fused (maxillary bridge)
- Ventral side of 1st instar larvae with posteriorly directed spinulae and setae
- Jumping capacity of 1st instar larvae (based on caudal bristles [cerci?])
- Secondary larvae with distinctly simplified head appendages and legs
- Cuticle of adult males densely covered with microtrichia
- Antennae of males antler-shaped, maximum number of antennomeres eight
- Antennae with sensorial groove on 4th antennomere (Hofeneder's organ) and numerous dome-shaped chemoreceptors
- 'Raspberry compound eyes', large ommatidia separated by chitinous bridges with dense vestiture of microtrichia
- Maxilla strongly simplified, without lacinia and with 1-segmented palp
- Labium not present as a defined structure
- Mouthfield sclerite (probably formed by epipharynx and prementum)
- Forewings modified as halteres
- Hindwings fan-shaped
- Metathorax greatly enlarged, with very large dorsal indirect flight muscles
- Metapostnotum enlarged, shield-like
- Copulatory organ simplified, simple blade-like or hook-shaped penis
- Malpighian tubules and fat body greatly reduced or absent
- Female wingless and larviform
- Females with birth organs
- Ovaries reduced, oocytes floating in haemolymph

Fig. 6.30.1: Life stages of Mengenillidae, SEM micrographs (except C). A, first instar larva of *Eoxenos laboulbenei*, lateral view, length: 0.3 mm; B, secondary instar larva of *E. laboulbenei*, lateral view, length: 0.8 mm; C, male tertiary larva of *E. laboulbenei*, lateral view, length: 4.6 mm; D, male puparium of *Mengenilla chobauti*, lateral view, length: 5.4 mm; E, female puparium of *E. laboulbenei*, lateral view, length: 5.8 mm; F, female of *E. laboulbenei*, ventral view, length: 4.2 mm; G, male of *M. moldrzyki*, semifrontal view, length: 4.2 mm. Courtesy H. Pohl.

The numerous autapomorphies (those listed above are only a selection) greatly impeded the systematic placement of Strepsiptera, which turned out as one of the most persistent problems in systematic entomology ("The Strepsiptera problem"). Recently this question was solved. Extensive morphological character sets and phylogenomic data unambiguously support a placement as sistergroup of Coleoptera. Both groups together form the supraordinal unit Coleopterida.

Taxonomy. Three genera are placed in the stem group of Strepsiptera, †*Protoxenos* (†Protoxenidae) as the first branch, †*Cretostylops* (†Cretostylopidae) as the second, and †*Mengea* (†Mengeidae) as the third. The recently described monospecific Bahiaxenidae from Brazil are the sistergroup of all the remaining extant Strepsiptera. Mengenillidae, which still have free-living females, form the next branch (Fig. 6.30.1). Their monophyly is questionable. The genus *Eoxenos* has preserved some groundplan features missing in the other mengenillid genera (presence of a sensorial groove on the maxillary palp, absence of the balloon-gut). Crown group Strepsiptera excl. Bahiaxenidae and Mengenillidae form a strongly supported clade Stylopidia (ca. 97% of all species), mainly characterized by very strongly modified endoparasitic females with a legless sclerotized cephalothorax and a large and sack-shaped membranous posterior body (Fig. 6.30.2). Within Stylopidia Corioxenidae form the sister group of a clade Stylopiformia, which comprises the 90% of all species in seven families, Elenchidae, Halictophagidae, Bohartillidae, Myrmecolacidae, Callipharixenidae, Xenidae and Stylopidae. Halictophagidae (ca. 120 spp.), Xenidae (ca. 110 spp.) and Stylopidae (ca. 160 spp.) are the largest families. The latter parasitize in aculeate hymenopterans and have a worldwide distribution (Pohl & Beutel 2008).

Diagnosis. Males with raspberry compound eyes, antler-shaped antennae, mesothoracic halteres, a strongly enlarged metathorax, and fan-shaped hindwings. Free-living females wingless and larviform. Endoparasitic females strongly simplified, legless, with a sack-shaped posterior body. Primary larvae campodeiform, extremely small. Endoparasitic larval stages strongly simplified morphologically.

Morphology, adult males. The males are small (extant species 0.85–6.5 mm) and fragile insects. The color is usually brown or yellowish. Most parts of the body surface are densely covered with short microtrichia.

The head is orthognathous. Its posterior part is not retracted into the prothorax. The dorsal side is often partly membranous. The foramen occipitale is wide. The coronal suture is usually absent (well-developed in †Protoxenidae); the frontal sutures are always missing. The well-developed compound eyes are composed of 15–150 large round ommatidia with separate retinulae; the ommatidia are separated by distinct chitinous bridges ("raspberry compound eyes"); the compound eyes are similar to the schizochroal eyes of the trilobite family †Phacopidae; ocelli are absent. A well-developed labrum is present in the groundplan (e.g., †Protoxenidae, Bahiaxenidae) but usually absent (vestigial but present in an Australian species of *Mengenilla*). The antler-shaped antennae are 8-segmented in the groundplan (Fig 6.30.1G, 6.30.2E); the flagellomeres are densely set with dome-shaped chemoreceptors; a lateral process (flabellum) is at least present on flagellomere 1, usually also on other flagellomeres; flagellomere 4 bears a sensorial groove (Hofeneder's organ). The mouthparts are strongly reduced. The mandibles are relatively robust in the groundplan but usually blade-like and simple (distinctly reduced in Corioxenidae); the secondary mandibular joint is usually reduced (well-developed in members of the stem group and in Bahiaxenidae). The maxillae are strongly simplified; they are basally fused with the head capsule in *Mengenilla*; a galea is only present in †Protoxenidae; the lacinia is always missing; the palp is 1-segmented; it bears a sensorial groove in the ground-

Fig. 6.30.2: Life stages of Stylopidia (*Xenos vesparum*), SEM micrographs. A, first instar larva, ventral view, length: 0.36 mm; B, secondary instar larva, lateral view, length: 3.0 mm; C, female, lateral view, length: 7.2 mm; D, male pupa, ventral view, length: 5.0 mm; E, male, lateral view, length: 3.9 mm. Courtesy H. Pohl.

plan (extinct groups and *Eoxenos*). The labium is not present as a separate element of the head; endite lobes and palps are absent. A very characteristic cone-shaped mouthfield sclerite is probably formed by the epipharynx and prementum. A gula is not developed. The tentorium is completely reduced. The muscle system is simplified but the pharyngeal dilators are strongly developed. The labial muscles are vestigial (*Stylops*?) or absent. Cervical sclerites are lacking.

The thoracic sclerites are separated by quite extensive membranous areas laterally and on the ventral side. The vestiture of microtrichia is only missing along sutures and in the vicinity of articulations. The pro- and mesothorax are short, whereas the metathorax is strongly enlarged. The prothoracic notum, pleura and ventrite are indistinguishably fused, forming a simple ring-like element which is more heavily sclerotized on the dorsal side. The mesothorax bears the forewings which are modified as halteres, which are highly specialized gyroscopic sense organs; the halteres are distally extended and their venation is vestigial. The small mesonotum is simple and not subdivided into different regions; a small scale-like tegula is present at the base of the halteres (*Mengenilla*). The pleural suture is preserved but a mesepimeron is not recognizable as a sclerotized element. The metanotum is subdivided into an elongated, roughly triangular posterior scutellum and a large scutum, which is fused with the prescutum anteromedially; the metapostnotum is elongated and shield-like (not in †Protoxenidae); it is separated from the metascutellum by a membranous postlumbium. The metanepisternum and metepimeron are strongly extended posteriorly; the pleural suture is almost horizontal. Laterally, a large subalare is embedded in the extensive membranous area below the scutum. The extensive metaventrite is medially divided by a distinct discrimen which reaches the anterior third of the sclerite. The metafurca is very large; its base is firmly connected with the median ridge corresponding with the discrimen; lateral furcal arms are lacking. The fan-shaped foldable hindwings are strongly extended anteroposteriorly; the venation is strongly reduced; only longitudinal veins are present in extant strepsipterans (one transverse vein in †*Mengea*). Three axillary sclerites are present at the wing base; axillary 1 is the largest and posteriorly connected with axillary 3; the elongate axillary 2 articulates with the pleural wing process and basalare anteriorly; posteriorly it is connected with the subalare and the postalar bridge; the small, membranous tegula at the anterior wing base is densely covered with microtrichia but lacks sensilla (observed in *Mengenilla*); the axillary cord at the posterior wing base is connected with the postlumbial joint between the lateral edges of the scutellum and postnotum. The pro- and mesocoxae are relatively large and free; the lateral part of the metacoxa is fused with the ventral margin of the katepimeron; a distinct metacoxal meron is present; the pro- and mesotrochanter are fused with the respective femora; the tibiae are constricted in their proximal fourth in some Halictophagidae, thus forming a "pseudopatella"; the tarsi are slender, 5-segmented and lack specialized adhesive hairs in the groundplan (stem group, Bahiaxenidae, Mengenillidae); only 2–4 tarsomeres are present in most groups; brushes of adhesive hairs are present on one tarsomere in Corioxenidae (groundplan of Stylopidia) and on all of them in Stylopiformia (autapomorphy); the claws are present in the groundplan but usually reduced; pretarsal attachment structures are always missing. The metathoracic musculature is very strongly developed, especially the dorsal indirect flight muscles.

The abdomen is 10-segmented and weakly sclerotised. The tergites are usually less strongly sclerotised than the sternites (not in †*Protoxenos*). Segment I is partly integrated into the metathorax. Tergites I and II are covered by the enlarged metapostnotum. Segment IX bears the strongly sclerotized, blade-like or hook-like penis, which is held below segment X at rest.

The anterior part of the foregut usually forms an air uptake apparatus. The oesophagus and midgut are almost always very voluminous and filled with air ("balloon gut"; not in *Eoxenos*); the "balloon gut" stabilizes the abdomen; the connection between the midgut and the hindgut is interrupted. The Malpighian tubules are almost always reduced (present as small papillae) but distinct in *Xenos*. The nervous system is strongly concentrated. The testes are large. The ductus ejaculatorius with its strongly developed muscularis forms a sperm pump.

Morphology, females. The females are either free-living (Mengenillidae [females of Bahiaxenidae unknown]), wingless and larviform (Fig. 6.30.1), or endoparasitic, legless and generally greatly reduced morphologically (Stylopidia) (Fig. 6.30.2). The large, sack-shaped and completely membranous posterior part of the body of stylopidian females lies within the host's body cavity; the females of this large subgroup of Strepsiptera remain in the exuvia of the last larval stage and the pupa (puparium) and form a functional unit with this cuticular envelope.

The head is relatively normally developed in Mengenillidae, but the appendages are more simplified than in males. The compound eyes are smaller. The antenna lacks flabella and a sensorial groove. The simple mandibles are ventrally directed (Fig. 6.30.1F). The head of females of Stylopidia is extremely simplified and fused with the thorax and parts of abdominal segment I, thus forming a strongly sclerotized cephalothorax. Only this part is protruding from the host's body. The head appendages are distinctly simplified or completely reduced (Fig. 6.30.2C). The mandibles move outwards. A transverse mouth opening serves as birth opening in Corioxenidae, whereas a fissure-shaped transverse secondary birth opening is present in Stylopiformia.

The thorax of Mengenillidae is simplified and completely wingless. Legs are present; the tarsi are 3- or 4-segmented. The muscle system is weakly developed. The thorax of Stylopidia forms a part of the compact and sclerotized cephalothorax (see above). No segmental borders and separate tergal, pleural or sternal sclerites are recognizable. Appendages are completely absent. Strongly developed Nassonoff's glands are present ventrally below the thoracic complex of ganglia.

The abdomen of Mengenillidae is 10-segmented and weakly sclerotised. The birth opening lies at the hind margin of sternite VII. The large, sack-shaped abdomen of stylopidian females remains in the body cavity of the host. The ventral ganglionic chain is strongly condensed. The somatic tissue of the ovaries is almost completely reduced; they are not recognizable as discrete structures; the oocytes float freely in the haemolymph. A single birth organ is present in Mengenillidae and several in Stylopidia, with openings on segments II–VII; a brood canal between the external body wall of the female and the last larval and pupal exuvia is present in Stylopidia.

Morphology, first instar larva. The agile campodeiform first instar larvae belong to the smallest known metazoans (80–850 µm, average size ca. 230 µm). The well sclerotized cuticle is smooth and wears a specialized vestiture of posteriorly directed setae and spinulae. The color varies between light brown and black (Fig. 6.30.1A).

The head is strongly sclerotized and wedge shaped in lateral view. The anterior edge is sharp and the labrum completely fused with the head capsule. Head sutures are absent. Groups of 2–6 well-developed stemmata are present on both sides; ocelli are absent. The antennae are present as vestigial antennal fields or absent. The preoral cavity is enclosed by the upper part of the head capsule and the medially fused maxillae (maxillary bridge); it has two openings, one apically and one on the ventral side. The horizontal mandibles lie within the preoral cavity and can protrude through its apical opening. The hypopharynx and labium are fused; they form the posterior delimitation of the ventral opening of the preoral cavity. Maxillary and labial lobes and palps are usually present. The tentorium is present but incomplete. The head musculature is distinctly simplified.

The thorax is formed by three distinctly developed segments, which are distinctly convex dorsally and bear a conspicuous, posteriorly directed sternal plate on the ventral side. The prothorax is slightly larger than the two other segments but otherwise they are similar. The legs are well-developed and also similar on all three segments; they are composed of coxa, trochanterofemur, tibia and a 1-segmented tarsus; the tarsus is usually leaf-shaped on the fore- and middle legs and bristle-like on the hindleg; apically the legs bear a specialized pretarsal attachment pad. The thoracic muscles system is well-developed.

The abdomen is 11-segmented and subparallel. It is slightly narrowing towards its apex. The well-developed segment XI bears the strongly developed long caudal bristles, which are equipped with extrinsic muscles and possibly homologous to cerci; an additional shorter pair of caudal bristles is normally inserted laterally on the hind margin of the segment.

Morphology, secondary larvae. The secondary larvae are distinctly simplified morphologically (Figs 6.30.1B, 6.30.2B). Simplified legs are retained during the following postembryonic stages in Mengenillidae, but are absent from the 2nd secondary larva in Stylopidia. Anlagen of the legs re-appear in males of Stylopidia in the last stage before the pupation. The compound eyes appear in the first secondary larval stage and external wing buds are recognizable as lateral convexities. The exuvia of the last stage forms the puparium.

Morphology, pupae. The pupa adectica exarata is enclosed in the last larval exuvia (puparium) (Fig. 6.30.1D).

Biology. Strepsiptera are mainly characterized by their endoparasitic habits and numerous derived features related to this specialized life style. Hosts are Zygentoma in the case of Mengenillidae, and neopteran insects in the case of Stylopidia. Corioxenidae parasitize true bugs, Elenchidae Auchenorrhyncha, Halictophagidae roaches, orthopterans, members of all three major hemipteran subgroups, and flies, Myrmecolacidae ants (males), mantodeans and orthopterans (females), and Stylopidae and Xenidae bees and wasps, respectively. The hosts of Bohartillidae are unknown. The damage to the host and host population is very limited. The life cycles are highly synchronized. The development of the host may be accelerated (e.g., *Hylaeus* [Colletidae]) or delayed (e.g., *Andrena* [Andrenidae]).

The agile adult males fly very well. The halteres rotate during flight and function as gyroscopic sense organs. During hatching from the puparium the males fill their digestive tract with air ("balloon gut"). They do not consume food (like the females) and are active only for few hours. The minute primary larvae identify the host with their well-developed stemmata and probably with olfactory organs. They jump on its body using the abdominal jumping mechanism (caudal bristles) and attach on it with their pretarsal pads. The host's body wall is penetrated with the sharp anterior edge of the head and the mandibles, usually at intersegmental or articulatory membranes. After the first molt the secondary larval stages consume haemolymph. The activities of the sluggish free-living females of Mengenillidae are very restricted and a part of the females stay within the puparium (unfertilized). Females of Stylopidia stay within the host with the main part of their body. The strongly sclerotized cephalothorax protrudes from the host's body at an intersegmental membrane of the abdomen.

Reproduction and development. The short-lived males must find the females within few hours, mainly using olfactory signals (Nassonoff's glands). In all groups except for Mengenillidae the endoparasitic females are fertilized through their mouth opening (Corioxenidae) or the secondary birth opening (Stylopiformia). The primary larvae leave the birth organ and are released through the same openings after passing through the brood canal. Their number varies between 1,000 and 750,000. The first molt takes place after the tiny first instar larvae have entered the host's body. The secondary larvae usually molt five times. They are the feeding stage and move around in the host's body cavity and consume haemolymph. After molting the secondary larvae stay within the cuticle of the preceding stage. In a later stage they assume a position in the abdomen which allows the penetration the host's body wall. After that the last larval exuvia forms the puparium. The pupal stage is short. Unlike in the typical holometabolan development different organs develop already in the secondary larval stages (e.g., compound eyes, genital organs). During the hatching process the males cut off the anterior part of the puparium using the elongate mandibles. Skeleton and wings are completely hardened by that time. The male is able to fly immediately after hatching.

Fossil record. The fossil record of strepsipterans is very limited. Adults are only preserved as amber fossils. †*Cretostylops engeli* in Cretaceous amber is the oldest fossil representative of the order. †*Protoxenos* and †*Mengea* are embedded in Baltic amber (ca. 40 Ma). A brown coal impression fossil of a first instar larva (Eocene of Germany) was recently re-described and assigned to Stylopidia (Pohl 2009).

Economic importance. The economic importance of strepsipterans is very limited. In combination with other parasites or external factors they can decimate the populations of certain pest species (e.g., in Delphacidae).

[Kinzelbach (1971a, b); Pohl (2000, 2009); Pohl et al. (2005, 2010); Hünefeld & Beutel (2005); Pohl & Beutel (2005, 2008); Koeth et al. (2012); Niehuis et al. (2012)]

Amphiesmenoptera (Trichoptera + Lepidoptera)

6.31 **Trichoptera** (Greek *trichos* = hair, *pteron* = wing, English common name: caddisflies)

Diversity and Distribution. With approximately 14,500 described species (Morse 1997) Trichoptera belong to the moderately successful insect orders. The group has a worldwide distribution. Caddisflies occur from the arctic region (e.g., Svaalbard) to Tierra del Fuego at the southern tip of South America. They reach their highest diversity in Southeast Asia and South America. Endemic families occur on all continents except for Europe, most of them on the southern hemisphere (Mey 2005).

Autapomorphies. See Kristensen (1999b: p. 9) for a full list of apomorphies.
- Mandibles of adults vestigial and without articulations with the head capsule (Fig. 6.31.2)
- Ventral mouthparts of adults form an evertable haustellum (Fig. 6.31.2)
- Antennae of larva 1-segmented, strongly reduced
- Stemmata (larval eyes) closely aggregated
- Abdominal segment X of larvae with anal claws (Fig. 6.31.7A)
- Maxillolabium of larvae with unpaired spinning tube (Fig. 6.31.8)

Taxonomy. The group presently comprises 49 extant families. The two large suborders Annulipalpia and Integripalpia very likely form clades, whereas the third smaller subgroup, the Spicipalpia, is very likely not monophyletic. The four families

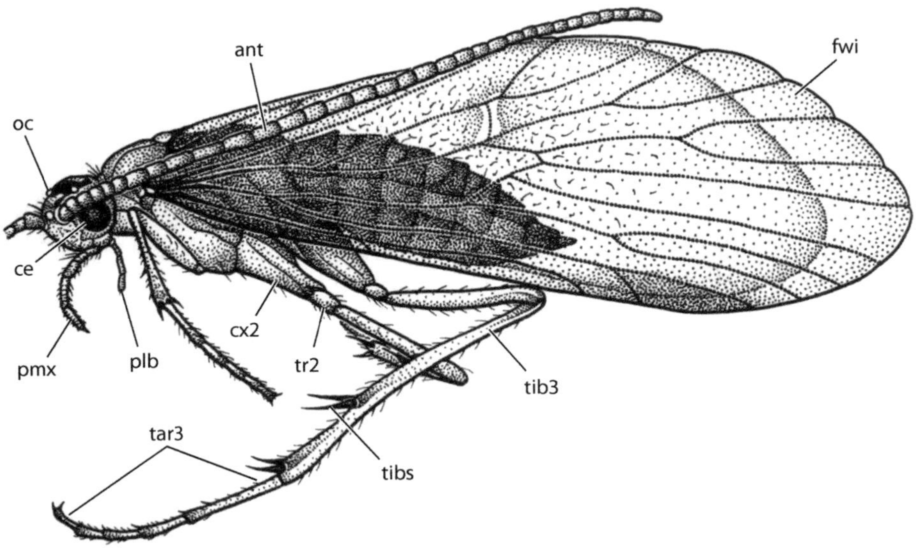

Fig. 6.31.1: *Fansipangana vernalis* (Rhyacophilidae), adult, lateral view. Abbr.: ant – antenna, ce – compound eye, cx2 – mesocoxa, fwi – forewing, oc – ocellus, plb – palpus labialis, pmx – palpus maxillaris, tar3 – metatarsus, tib3 – metatibia, tibs – tibial spur, tr2 – mesotrochanter. Courtesy F. Hünefeld.

traditionally assigned to Spicipalpia (Rhyacophilidae, Hydrobiosidae, Hydroptilidae, Glossosomatidae) are mainly characterized by plesiomorphic features.

Diagnosis. Medium sized insects with two large and densely setose pairs of membranous wings (Fig. 6.31.1). Head of adults with vestigial mandibles, long filiform antennae, and a conspicuous haustellum. Aquatic larvae with well-developed terminal hooks. Protective cases are formed in most groups.

Morphology, adults. Most species are of medium size. The length of the forewings varies between 1 and 43 mm, but 5–20 mm is most common. The densely setose wings are held in a roof-like position above the abdomen.

The main part of the orthognathous head is comparatively small and more or less globular (Fig. 6.31.2). The posterior region, which is in close contact with the anterior prothorax, is glabrous whereas other regions are densely setose. The relatively small, key-hole shaped foramen occipitale is subdivided by the tentorial bridge into a ventral neuroforamen and a dorsal alaforamen. The well-developed, compound eyes are strongly convex; three large ocelli are usually present. The ventrolateral corner of the head below the large compound eye is drawn out as a subgenal process. The membranous antennal articulatory areas lie between the median ocellus and the upper margin of the compound eyes; they are separated from each other by less than

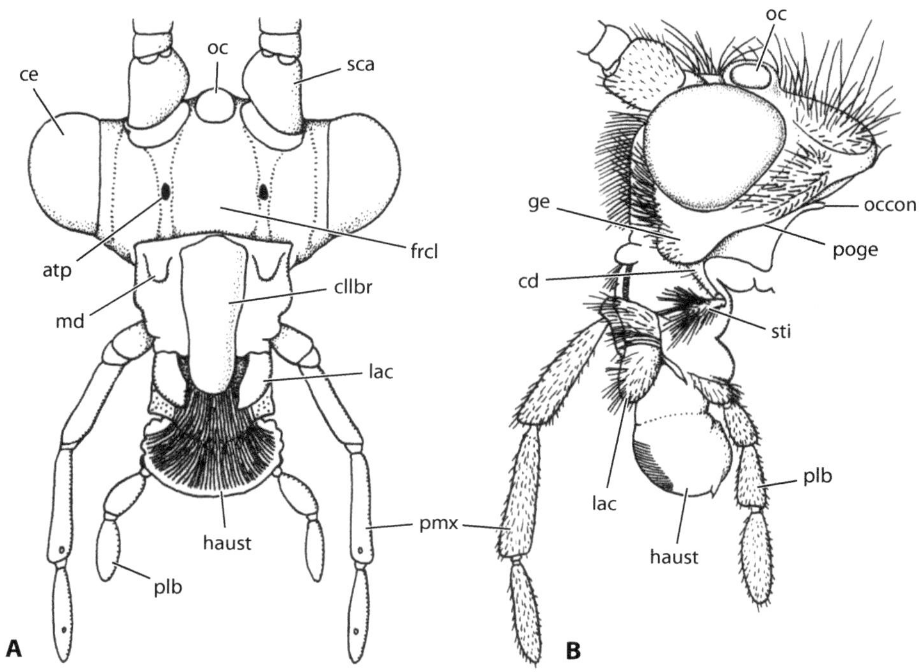

Fig. 6.31.2: *Phryganea bipunctata* (Phryganeidae), head. A, frontal view; B, lateral view. Abbr.: atp – anterior tentorial pit, cd – cardo, ce – compound eye, cllbr – clypeolabrum, frcl – frontoclypeus, ge – gena, haust – haustellum, lac – lacinia, md – mandible, oc –ocellus, occon – occipital condyle, plb – palpus labialis, pmx – palpus maxillaris, poge – postgena, sca – scapus, sti – stipes. Redrawn from Mey (2005), after Crichton (1957).

their diameter. The frons and the dorsal clypeal region (postclypeus) are fused thus forming the frontoclypeus; the boundary between the two areas is marked by the anterior tentorial pits; laterally, the frontoclypeus can be delimited by a sulcus frontogenalis; a very distinct transverse furrow and a corresponding internal ridge (sulcus epistomalis) separates the dorsal clypeal region from the clypeolabrum, which is a product of fusion of the labrum and the lower part of the clypeus (anteclypeus); a border between both elements is still recognizable in some cases (e.g., *Phryganea*). The antennae are always filiform, long and multisegmented, and anteriorly directed in their resting position. The mandibles are either small, weakly sclerotized and lack defined articulations with the head capsule, or greatly reduced lobe-like vestiges. The cardo and stipes form a complex with the labium in few groups (e.g., Hydropsychoidea); the galea is reduced; different degrees of reduction of the maxillary palp occur; the number of segments ranges between six and one, but five are present in most groups; an annulate apical maxillary and labial palpomere is characteristic for Annulipalpia. The postmentum is missing; glossae and paraglossae are absent; the

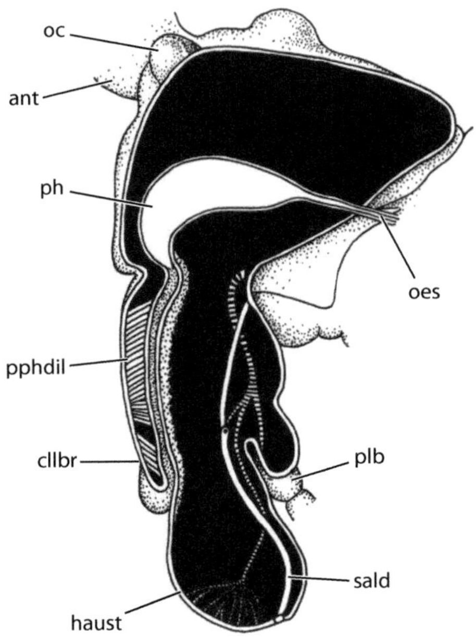

Fig. 6.31.3: *Phryganea bipunctata* (Phryganeidae), head, sagittal section. Abbr.: ant – antenna, cllbr – clypeolabrum, haust – haustellum, oc – ocellus, oes – oesophagus, ph – pharynx, plb – palpus labialis, pphdil – prepharyngeal dilator, sald – salivary duct. Redrawn from Mey (2005), after Crichton (1957).

labial palp is usually 3-segmented. The characteristic haustellum is formed by the hypopharynx dorsally and by the prementum on the ventral side (Figs 6.31.2, 6.31.3); a more or less complex system of parallel, oblique and transverse channels formed by microtrichia is often present on the dorsal side of the haustellum (ligula). The floor of the cibarium is formed by the sitophore plate; at its proximal margin it is thickened as a pistil-like structure which corresponds with an epipharyngeal concavity at the anatomical mouth. A gula is absent. The tentorium is composed of short, stout posterior arms which are connected by the transverse tentorial bridge and the anterior arms; the dorsal arms are vestigial, membranous (e.g., Annulipalpia) or well-developed and strongly sclerotized (Integripalpia). One pair of lateral cervical sclerites is present. They articulate with the condyli occipitales.

The prothorax is usually small and narrower than the head. The pronotum bears tubercles with numerous bristles. The mesothorax is the largest segment. The metathorax is slightly smaller. The wings are more or less densely covered with setae (scales in some Hydroptilidae) (Fig. 6.31.4), and held in a roof-like position at rest (Fig. 6.31.1), like in Megaloptera and some other groups; reductions of one or both pairs are rare exceptions; characteristic color patterns can result from the distribution

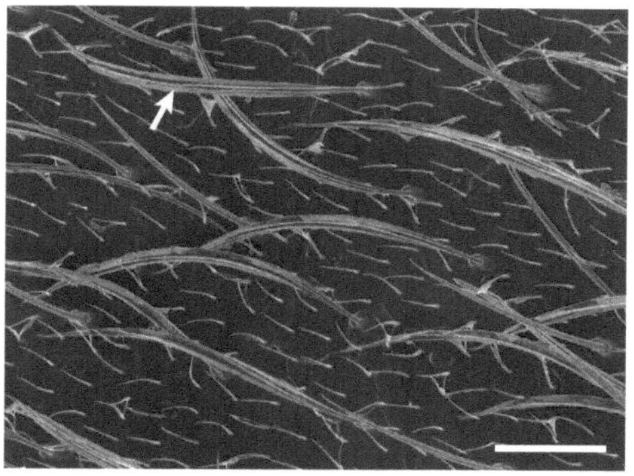

Fig. 6.31.4: *Agrypnia varia* (Phryganeidae), forewing surface, SEM micrograph. Note the typical vestiture of long setae (arrow). Scale bar: 50 µm.

of the hairs; the pattern of longitudinal veins is almost complete, whereas only few transverse veins are present (Fig. 6.31.5); the hindwings are covered by the forewings and sometimes folded or wrapped around the abdomen (e.g., *Molanna*); an additional membranous basal lobe (fibula) is present or absent; the forewings and hindwings are often connected by attachment devices (usually small hooks). The flight musculature is well-developed. The legs are elongate (Fig. 6.31.1) and slender; an extensive meron and an elongate mesal articulatory process (sternal articulation) is present on all coxae; the meso- and metacoxae are larger than the procoxae and posteroventrally directed; the trochanters are small, the femora and tibiae are elongate and slender; the pattern of articulated apical and subapical spurs on the tibiae is an important taxonomic character complex; the maximum number is three on the fore tibia and

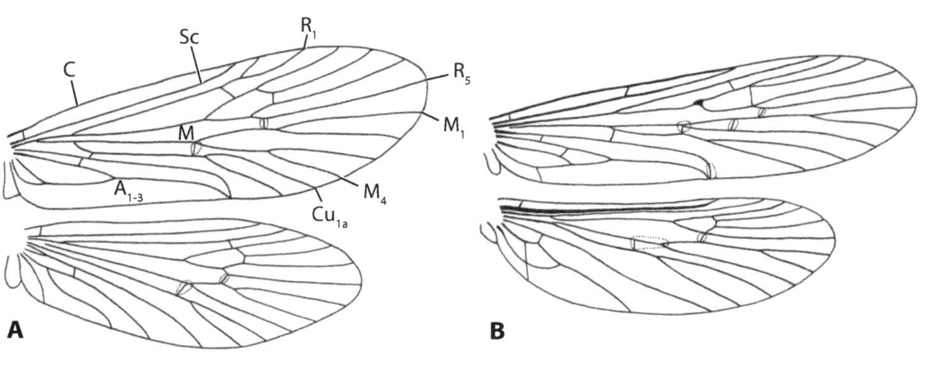

Fig. 6.31.5: Wings. A, *Glossosoma lividum* (Glossosomatidae); B, *Chimarra obscura* (Philopotamidae). Abbr.: A_{1-3} – anal veins, C – costa, Cu_{1a} – cubital branch, $M_{1,4}$ – medial branches, $R_{1,5}$ – radial branches, Sc – subcosta. Redrawn from Mey (2005), after Schmidt (1982, 1983).

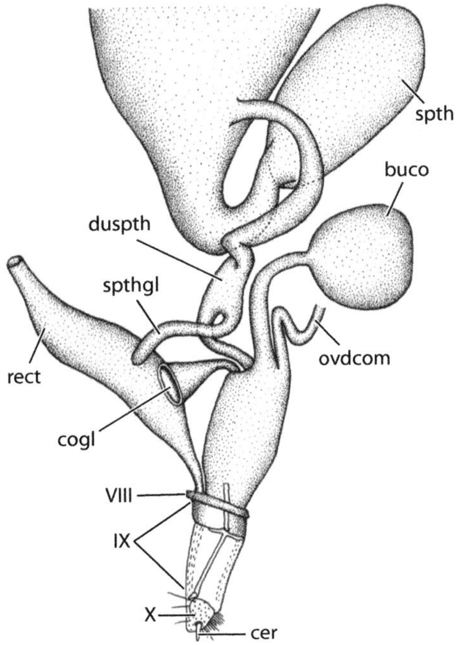

Fig. 6.31.6: *Rhyacophila fasciata* (Rhyacophilidae), female genital organs. Abbr.: buco – bursa copulatrix, cer – cerci, cogl – collateral glands, duspth – ductus spermathecae, ovdcom – oviductus communis, rect – rectum, spth – spermatheca, spthgl – spermathecal gland, VIII/IX/X – segments VIII/IX/X. Redrawn from Nielsen (1980).

four on the middle and hindlegs; the slender tarsi are 5-segmented; a small arolium and pulvilli are present apically.

The abdomen is usually composed of ten distinctly developed segments. Sternite I is largely reduced. The intermediate segments II–VIII are more or less ring-like, with the tergites and sternites connected by lateral pleural membranes. Segment X is distinctly reduced; it bears minute, 1-segmented cerci in females. Segment XI is usually obsolete but distinctly recognizable in Rhyacophilidae. The terminal segments IX–VI can be variously fused in males and bear paired or unpaired appendages; the male copulatory apparatus varies strongly and provides important species specific diagnostic feature; it is either composed of a tube-shaped, sclerotized phallotheca with an evertable endophallus or vesica, or of a phallotheca, endotheca, a pair of parameres, and an aedeagus. The female genital structures (Fig. 6.31.6) also vary greatly; the genital opening lies on the ventral side between segments IX and X or apparently posterior to segment VIII if the sternal element IX is strongly reduced; one to three paired appendages occur in different families; a telescope-like secondary oviposition apparatus is present in the spicipalpian families and in Annulipalpia.

Six Malpighian tubules are present. A paired gland is present in abdominal segment V in many families; the secretions function as sexual pheromones or for defense. The female internal genital apparatus comprises a bursa copulatrix, receptaculum seminis (spermatheca) and accessory glands (Fig. 6.31.6); the ovaries are composed of three ovarioles; the anus is completely separated from the genital opening in Integripalpia, whereas a cloaca is present in the other groups.

Morphology, larvae. The larvae are free-living in Rhyacophilidae and Hydrobiosidae, and construct precocious shelters in Glossosomatidae and Hydroptilidae (Frania & Wiggins 1997) (Fig. 6.31.7). They construct portable tubes in Integripalpia and stationary retreats in Annulipalpia. The ancestral condition in Trichoptera is unkown. A well-developed head capsule and legs are always present. The abdomen is largely unsclerotised.

The head is prognathous (e.g., Rhyacophilidae) or orthognathous (e.g., Limnephilidae). The foramen occipitale is narrowed by a hypostomal bridge, but the formation of the ventral wall of the head capsule varies distinctly among different groups; the ventral ecdysial lines may be paired or unpaired and forked anteriorly or posteriorly or both (anterior and posterior apotome). The frontal and coronal sutures are distinct. The frontoclypeal transverse ridge is present (e.g., *Rhyacophila*) or absent. The clypeus is subdivided into a transparent anterior anteclypeus and a posterior postclypeus. Six stemmata are present in most trichopteran larvae and form a closely aggregated complex, but reductions occur. The labrum is free and fully equipped with extrinsic and intrinsic muscles. It forms an eversible, brush-like net-cleaning device in Philopotamidae. The antennal insertion area lies primarily close to the anterior head margin, but is shifted posteriorly in different groups. The antennae are 1-segmented and strongly simplified or reduced; the single antennomere is cylindrical or flattened and disc-shaped (spicipalpian families); it is variously equipped with sensilla basiconica and sensilla trichodea. The mandibles are often asymmetric; they lack a mola and prostheca; a penicillus is present or absent; depending on the feeding habits the mandibles are either equipped with a scraper-like edge or a cutting edge with one or several teeth. The ventral mouthparts form a maxillolabial complex. The maxillary grooves are small or absent; the maxillae are strongly protracted in some groups (e.g., Hydropsychidae, *Limnephilus*); the size of the cardo and stipes varies greatly; a distinctly defined galea and lacinia are not present; the more or less undivided endite lobe is covered with sensilla and can be largely membranous; the palp is 4-segmented. The labium is primarily composed of submentum, mentum and prementum (e.g., *Limnephilus*); the anterior part of the labium is closely associated with the hypopharynx (prelabio-hypopharynx); it is largely membranous except for the premental part; the palps are usually 1-segmented but can be reduced; the single orifice of the spinning tube lies on a distinct prelabio-hypopharyngeal lobe (spinneret) (Fig. 6.31.8: opst);

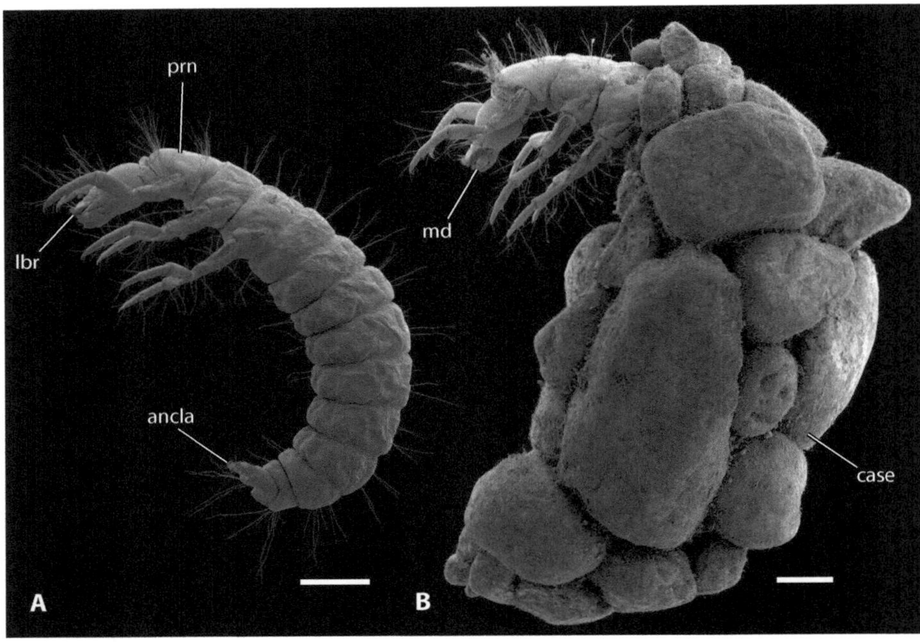

Fig. 6.31.7: *Agapetus ochripes* (Glossosomatidae), larva, lateral view, A, without case, B, with case, SEM micrographs. Abbr.: ancla – anal claw, lbr – labrum, md – mandible, prn – pronotum. Scale bars: 500 μm.

a silk press is probably generally present but differs distinctly in different groups. The tentorium is formed by simple paired rods which include the posterior and anterior arm in *Limnephilus*; a vestigial dorsal arm is present in *Drusus*; the tentorial bridge is well-developed (e.g., Glossosomatidae), very thin (e.g., Rhyacophilidae) or absent. Large mandibular glands open at the dorsal mandibular articulatory membrane, and the maxillary glands between the mandibles and maxillae.

The pronotum is usually narrower than the meso- and metanotum and always sclerotized and divided by a narrow median ecdysial suture. The sclerotisation of the dorsal side of the meso- and metathorax varies; a well-defined mesonotum divided by a median line can be present or absent, or alternatively several smaller sclerites; several small metanotal sclerites are usually present if the mesonotum is well-developed; the metanotum is rarely represented by a single, medially divided sclerite combined with a well-developed mesonotum (e.g., Hydropsychidae); it is often largely or entirely membranous. One or two well-developed pleurites are present on each thoracic segment. Filamentous gills are present in some groups (Rhyacophilidae) but they are usually absent. The legs are well-developed and usually 6-segmented including the single claw (tibia and tarsus fused in Xiphocentronidae); the coxae are prominent, about as long as the trochanters and femora combined; the long trochanter is

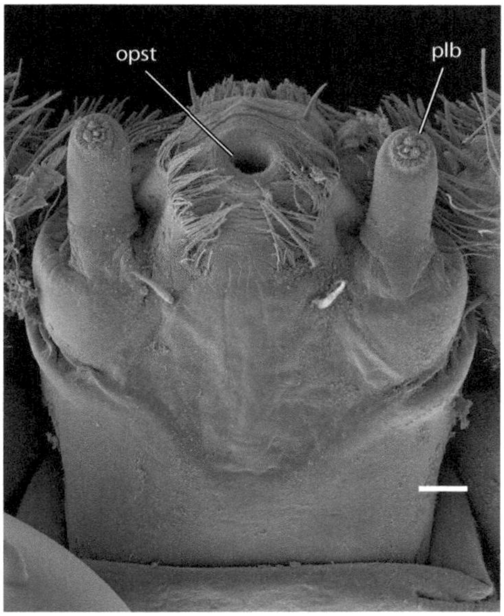

Fig. 6.31.8: *Rhyacophila fasciata* (Rhyacophilidae), spinneret of larva, ventral view, SEM micrograph. Abbr.: plb – palpus labialis, opst – opening of spinning tube. Scale bar: 20 µm.

subdivided by an annular line; the legs are either about equally long or progressively longer from front to rear (Hydroptilidae partim, Integripalpia).

The abdomen is composed of 10 segments (Fig. 6.31.7A). Segments I–VIII are usually largely or entirely membranous; large, single tergites I–VIII occur in Hydroptilidae. Segment IX is rarely entirely membranous but usually bears a large tergal plate or smaller sclerites. The abdominal segments frequently bear filamentous branched or unbranched gills. A line of fine filaments usually extends along the sides of most abdominal segments in Limnephiloidea (lateral fringe). Chloride epithelia, specialized ovoid areas for ion absorption and osmoregulation, occur in Limnephilidae and some other groups. Segment IX bears the anus and a pair of well-developed paired prolegs, which are derivatives of segment X; the prolegs are movable and primarily 3-segmented including the well-developed anal claw (Fig. 6.31.7A); distinct variations occur in different groups.

Morphology, pupae. Pupation usually takes place under water within a pupal case which is fixed at the substrate; it is either newly formed (groups with free-living larvae) or a modified larval case. The pupa dectica libera is equipped with specialized movable mandibles, swimming legs, and a longitudinal fringe of hairs at the abdomen; they swim to the surface before the adults emerge.

Biology. It is likely that trichopteran larvae are primarily adapted to oxygen-rich flowing water. They are generally more common in swift running rivers and streams, but also occur in water with reduced current, lakes, ponds, marshes, small temporary water bodies, saline coastal lakes, or tidal ponds. An important feature of the larvae is the ability to produce silk and related to that the construction of protective cases, shelters or nets. Five different functional groups are distinguished (e.g., Frania & Wiggins 1997): (1) free living larvae constructing only pupal shelters (Rhyacophilidae, Hydrobiosidae); (2) small larvae spinning portable cases only in the 5th instar (Hydroptilidae); (3) saddle-case makers (Glossosomatidae) constructing saddle-shaped cases of small stones (Fig. 6.31.7B); (4) net spinners or retreat makers (Hydropsychoidea) construct fixed shelters of silk and debris and capture food particles from the current or the substrate; (5) tube-case makers (Limnephiloidea, major part) use silk to build portable tubes of organic or mineral particles. The feeding habits of trichopteran larvae are remarkably diverse. They consume bacteria, different types of algae, fungi, diatoms, live or decaying parts of vascular plants, and also aquatic insect larvae or other small invertebrates. Shredders-chewers (e.g., leaves), collectors (fine organic particles), scrapers (e.g., diatoms), piercers (fluids of living plants) and predators can be distinguished.

Pupation takes place in a newly built shelter (e.g., Rhyacophilidae) or in a modified larval case. Silk cocoons within the shelter are woven by mature larvae of Hydrobiosidae and Glossosomatidae. Using strong and movable mandibles (pupa dectica) and hook-bearing abdominal plates the pharate adults leave the case and swim to the surface. The final molt takes place within short time on the water surface, emerged parts of plants or on other objects.

Adults are usually short-lived (usually 1–2 weeks) and feed on liquid substrates or very small particles if at all. They are usually found in the riparian vegetation and are crepuscular or night active. The flight performance depends on the species and temperature.

Reproduction and development. Mating takes place on riparian vegetation, on the ground or in flight. Large aggregations are formed during the mating flight in some species (e.g., *Mystacides*, Leptoceridae). The oviparous females deposit either single eggs or batches in or near a suitable water body, often enclosed in a gelatinous mass. First instar larvae usually hatch from the eggs after about three weeks. The usual number of instars is five.

Fossil record. The oldest definite trichopteran fossil is from the Early Jurassic (Lias). Annulipalpia and Integripalpia are documented from the Middle Jurassic (Grimaldi & Engel 2005). More than 600 species are preserved in Cretaceous and Tertiary amber.

Economic importance. Trichopteran larvae belong to the most abundant organisms in a great variety of fresh water habitats and are a very important element of the food chain. They process decaying plant matters and other organic materials and constitute a major food source for trouts and other freshwater fish. They are also very useful as indicators of water quality. Minor damage can be caused when adults emerge en masse synchronously, and fouling of water can occasionally result from mass occurrence of net-spinning hydropsychid larvae.

[Crichton (1957); Malicky (1973); Nielsen (1980); Schmidt (1982, 1983); Wiggins (1984, 1987); Dean & Cartwright (1987); Frania & Wiggins (1997); Morse (1997); Kristensen (1999b); Kjer et al. (2002); Grimaldi & Engel (2005); Mey (2005)]

6.32 Lepidoptera (Greek *lepís* = scale, *pteron* = wing, refers to the vestiture of scales on the wing, English common names: moths and butterflies)

Diversity and Distribution. Lepidoptera is one of the megadiverse holometabolan orders, with approximately 175,000 described species, and arguably the most popular group of insects. They occur on all continents with the exception of Antarctica. The vast majority of species occurs in the tropical regions. Approximately 48,000 species from the Indo-Australian region are known, versus ca. 22,000 in the Palearctic. The

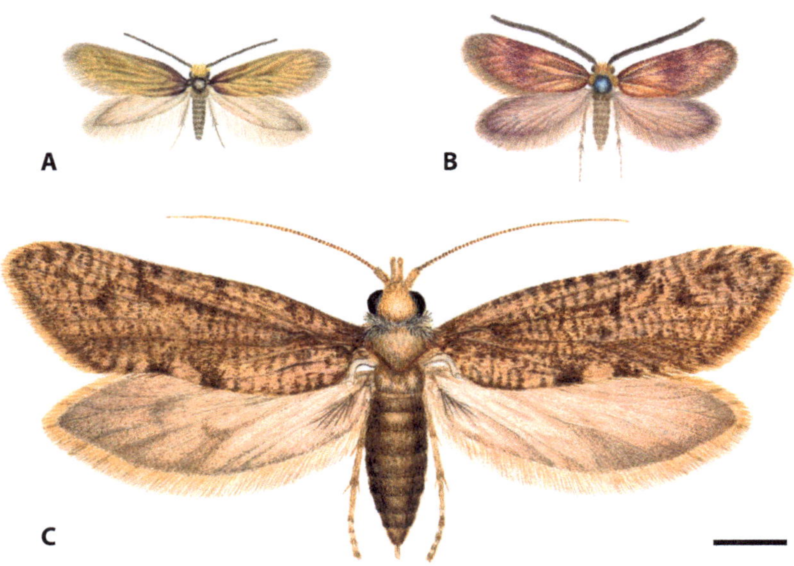

Fig. 6.32.1: Micropterigidae and Agathiphagidae, habitus. A, *Micropterix calthella* (Micropterigidae); B, *Palaeomicroides obscurella* (Micropterigidae); C, *Agathiphaga vitiensis* (Agathiphagidae). Scale bar: 250 µm. Drawings courtesy N.-P. Kristensen.

Fig. 6.32.2: *Sabatinca calliacha* (Micropterigidae). A, larva; B, adult. Photos courtesy G. Gibbs.

Arctic Apollo (*Parnassius arcticus*, Papilionidae) occurs in the Arctic Circle in northeastern Yakutia and other Apollo species (*Parnassius* spp.) were found in altitudes of up to 6,000 m above sea level in the Himalayas.

Autapomorphies. A full list of autapomorphies is given in Kristensen (1999b: p. 7) and Kristensen et al. (2007).

- Median ocellus absent
- Intercalary sclerite in membrane connecting scapus and pedicellus
- Cardo fused with basistipes
- Maxillary palp with points of flexion between segments 1/2 and 3/4
- Postlabium arched and with long piliform scales
- Wings covered with scales (Figs 6.32.2, 6.32.3)
- Laterocervical sclerite with proprioreceptive hair-plate
- Dorsal profurcal arm

Fig. 6.32.3: *Panolis flammea* (Noctuidae), forewing surface covered with scales, SEM micrograph. Scale bar: 50 µm.

- Mesothoracic tergopleural apodeme present
- Metathoracic prescutal arm present
- Protibia with movable 'epiphysis' (antenna cleaning organ)
- Cerci lacking in both sexes

Taxonomy. Lepidoptera are the sistergroup of Trichoptera (Amphiesmenoptera). They comprise 46 superfamilies and 126 families. All groups except for the "ancestral" Micropterigidae (ca. 120 spp.) (Figs 6.32.1A, B, 6.32.2), Agathiphagidae (2 species, Australia, New Caledonia etc.) (Fig. 6.32.1C) and Heterobathmiidae (9 spp., South America) form a well-founded clade Glossata, characterized by the presence of a sucking tube formed by the galeae (proboscis). Within this unit Eriocraniidae (24 Holarctic spp.) are the sistergroup of all remaining families (Coelolepida). Myoglossata comprising Coelolepida excl. Acanthopteroctetidae (4 species in North America) and Lophocoronidae (6 Australian species) are supported as a clade by intrinsic musculature in the proboscis. Neolepidoptera, Myoglossata excluding the family Neopseustidae (10 spp. in East Asia [e.g., Sichuan] and South America), are characterized by the pupa obtecta (chrysalis) and caterpillars with abdominal prolegs on segments III–VI and X. Ditrysia are a well-founded clade comprising the vast majority of lepidopteran superfamilies, among them so well known and successful groups like Zygaenidae, Tortricidae, Pyralidae, Papilionidae, Pieridae, Nymphalidae, Hesperiidae, Noctuidae and Geometridae (ca. 21,000 spp.).

Diagnosis. Adults are characterized by wings densely covered with scales, often displaying conspicuous color patterns (Fig. 6.32.2B), and usually by a long proboscis formed by the galeae. The orthognathous and phytophagous larvae of the vast majority of Lepidoptera are called caterpillars. They are characterized by a largely unsclerotized, cylindrical postcephalic body and usually possess a series of abdominal prolegs.

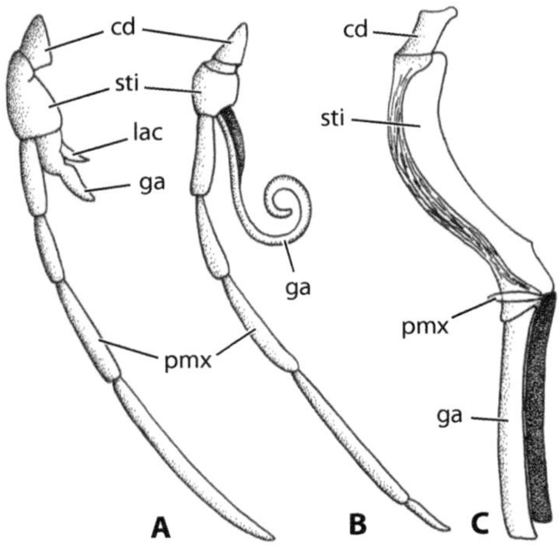

Fig. 6.32.4: Maxillae, schematized, A, Micropterigidae ("Aglossata"); B, Eriocraniidae (Glossata); C, Sphingidae (Ditrysia). Abbr.: cd – cardo, ga – galea, lac – lacinia, pmx – palpus maxillaris, sti – stipes. Redrawn from Gaedike & Häuser (2005).

Morphology of adults. Lepidopterans are small to very large insects. The wing span ranges between 4 mm (Nepticulidae) and 320 mm (*Ornithoptera alexandrae*, Papilionidae). Many species are characterized by conspicuous color patterns of the wings. Polymorphism occurs in different forms, between individuals in one population, and also as sexual dimorphism.

The head is orthognathous, with a vertical, flattened posterior surface and foramen occipitale, but a wide cervical membrane permitting a semiprognathous posture. The primary vestiture was likely dense and composed of hair-scales often concentrated on elevations on the dorsal side (resembling the "trichopteran warts"; s. Kristensen 2003). A unique feature of Lepidoptera are the chaetosema, raised spots or clusters of sensilla near the eyes; they are usually present but secondarily reduced in some groups. The compound eyes are well-developed; internally they are enclosed by a circumocular ridge (ocular apodeme); the paired ocelli lie above the antennal insertions, in the groundplan far mediad the compound eyes, in Ditrysia close to the eye margin; the median ocellus is absent (autapomorphy). In basal lineages, a subgenal process is present below the compound eyes, like in Mecoptera and Trichoptera. A hypostomal bridge is absent in the non-glossatan groups (Micropterigidae, Agathiphagidae, Heterobathmiidae); it may belong to the groundplan of Glossata (present in Eriocraniidae and Acanthopteroctetidae), but is also missing in most groups belonging to this lineage. The coronal suture is present on the anterodorsal head surface; a corresponding internal crest is very high in Agathiphagidae. In non-glossatan moths and Eriocraniidae the ocelli are connected by one or two transverse sulci; the posterior one (postinterocellar) also occurs in different groups of Glossata and likely belongs to the lepidopteran groundplan, like more or less complete occipital, postoccipital and temporal ridges (sulci) (Kristensen 2003). A typical transverse frontoclypeal (epistomal) ridge is preserved in non-glossatan moths, but absent in Glossata (frontoclypeus). Different frontoclypeal processes have evolved independently in ditrysian lineages; a laterofacial sulcus is present in the glossatan groundplan; the anterior tentorial pits are usually located at its lower end, but displaced dorsad in some groups. The broad labrum is movably attached to the anterior clypeal margin in non-glossatan moths, but within Glossata this condition is only preserved in Eriocraniidae; an elaborate, asymmetrical armature of the epipharynx is present in basal lepidopteran groups; in Glossata the more or less strongly modified labrum forms the closure of the food canal at the base of the proboscis; unusual modifications occurring at the lateral labral base of Heteroneura are the lobe-like setiferous pilifers. The antennal articulation areas are usually situated between the upper parts of the mesal eye margin (much lower in Agathiphagidae). The antennae are usually moderately long (ca. 0.5–1 times forewing length), but different degrees of elongation occur, often sexually dimorphic (Adelidae: male antennae ca. 3 times forewing length); the scapus and pedicellus are small and an intercalary sclerite is present in the membrane connecting them; the flagellum is multisegmented and highly variable, filiform, moniliform, dentate, serrate, lamellate, bi-, tri or quadripectinate, or apically clubbed. In the lepidopteran groundplan the mandibles are well-developed and articulated in the typical dicondylic manner; apically they bear distinct teeth which are different on both mandibles (asymmetry); a rugose mola is present mediobasally (Micropterigidae, Heterobathmiidae); in the most basal groups of Glossata they are still functional in the pharate stage but lack apical teeth and a mola, and the articulations are reduced; complete non-function is a typical feature of Neolepidoptera, but complete loss occurs only rarely. All maxillary elements are present in the lepidopteran groundplan (Micropterigidae), but the cardo is fused with the basistipes (basal piece) (Fig. 6.32.4A); in Glossata the lacinia is missing (Fig. 6.32.4B, C); the palp is primarily 5-segmented but reductions of segments occurs in many groups; the presence of points of flexion between segments 1/2 and 3/4 is a groundplan autapomorphy of Lepidoptera; the proboscis, a conspicuous autapomorphy of Glossata, is formed by the modified, mesally grooved dististipites and the connected galeae which enclose the main part of the food canal; intrinsic muscles are present in Myoglossata; in the typical case the proboscis is spirally coiled in repose (Fig. 6.32.4B). The labium is composed of a postlabium, proximal prelabium (only a part is homologous with the mentum), and prementum (distal prelabium); the postlabium is arched and setose in all non-glossatan groups, but this condition is only retained in few groups of Glossata (e.g., Eriocraniidae); the absence of the short proximal prementum is a glossatan autapomorphy; the prementum is consistently bilobed; it bears the labial palp and a median ligula, and a distinct paraglossa in Micropterigidae (groundplan); the palp is primarily 3-segmented and medium sized, but extremely short in Micropterigidae (often 1-segmented); it usually bears a vestiture of scales which is very variable and phylogenetically informative (Kristensen 2003); the ligula (fused glossae) forms the apex of a composite labio-hypopharyngeal lobe, which closes the preoral cavity, and bears the salivary opening in a dorso-subapical position. The functional mouth leads into a composite cibario-pharyngeal sucking pump, the floor of which is formed by the sitophore plate. The tentorium is a pi-shaped configuration in the groundplan, formed by the posterior and anterior arms, which are connected by a bridge with a posteromedian process; the dorsal arms are short processes or ridges in the non-glossatan moths, but often very prominent in Glossata. The wide cervical membrane is equipped with a single pair of approximately triangular cervical sclerites (laterocervicalia).

The prothorax is usually small. The pronotum is characterized by paired anterior elevated areas covered with long hair-scales; they are possibly homologous with the pronotal warts of Trichoptera and become conspicuously enlarged within the "higher" Glossata (patagia); a sublateral pair of proprioreceptors is present at the pronotal hind margin in the basal lineages. The propleuron is an arched plate in the groundplan and subdivided by a sulcus, which is possibly homologous with pterothoracic pleural ridges; ventrally it forms the pleurocoxal articulation; a small katepisternum is separated from the large anepisternum; the katepisternal margin articulates with a narrow, largely unsclerotized protrochantin, which again articulates with the coxa. The small eusternum is partly invaginated, thus forming an internal median ridge (discrimen). The hollow profurcal arms arise from the eusternum and are laterally fused with the propleuron; a strongly developed, dorsally directed arm arises from them (groundplan autapomorphy of Lepidoptera); the prospinasternum is present as an elongate sclerite articulating with the posterior eusternal margin; in "higher" Lepidoptera the propleurosternum becomes profoundly modified, notably by the formation of a firm precoxal bridge; the katepisternum and trochantin become obliterated in some lineages. The mesothorax is the largest thoracic segment but only slightly larger than the metathorax in the groundplan. The small prescutum is laterally demarcated from the large scutum by paired clefts; the subtegula are elongate sclerites between laterally directed prescutal lobes (prealar arms) and the tegula, a specialized structure of the base of the forewings (see below). The scutellum is distinctly smaller than the scutum. The pleural ridge separates the anterior episternum from the epimeron; a rod-like apodeme arising shortly below the pleural wing process just anterad the pleural ridge (tergopleural apodeme, tegular arm) is a lepidopteran autapomorphy; the anapleural cleft separates a smaller dorsal anepisternum from a larger ventral katepisternum; a complete reduction of the anepisternum occurs in several groups. A broad precoxal bridge is probably partly formed by the basisternum (anteromedially), but a clear identification of pleural and sternal elements is not possible; the mesoventrite is medially divided by a high discrimen, which merges with the base of the mesofurca. The furcal arms and processes vary strongly within Lepidoptera; the fusion of the meso- and metafurcal prong with the epimeron is a synapomorphy of Trichoptera and Lepidoptera. A small mesospina is present in the groundplan but reduced in many groups (Heteroneura). The metathorax is similar to the mesothorax in its general organization; it is strongly reduced in size in Heteroneura; the metathoracic subtegula (prescutal arm) is longer than its mesothoracic counterpart and not hidden by a tegula; its loss is an autapomorphy of Heteroneura. A setiferous sclerotized plate above the metepimeron is a proprioreceptor interacting with the hindwing (Kristensen 2003). The wings are usually well-developed, membranous, and held in a roof-like position above the abdomen; a characteristic feature is the formation of conspicuous color patterns; the surfaces are densely covered with two layers of overlapping scales (macrotrichia), each with a short pedicel inserted in a socket; narrowing of wings, longitudinally cleft wings, and reductions (usually in females, e.g., Geometridae, Psychidae, Cossidae) occur in some groups; both pairs are similar in size in the basal lineages, whereas the hindwings are distinctly smaller in "higher" Lepidoptera; fore- and hindwings are held together by different locking-devices, such as for instance a patch of strong microtrichia ("spinarea") of the anal area of the forewings interacting with microtrichia of the metascutum (functional dipterism); the venation of the basal lineages resembles that of basal Trichoptera, whereas a trend towards loss and fusion of veins is characteristic for "higher" Lepidoptera; a specialized structure associated with the forewing is the tegula, a hollow integumental outgrowth, which arises with a narrow membranous stalk above the anterior forewing base and curves around it (Kristensen 2003). The legs are usually well-developed and clothed with scales, but otherwise largely unmodified; reductions of fore- or hindlegs occur in few groups and all legs are degenerate in females of some Psychidae; the protibial epiphysis, a specialized spur functioning as antenna cleaning organ, is an autapomorphy of the order; the pterothoracic coxae are partly immobilized; the trochanters are small; the femora are often covered with long hairs; the mesotibiae bear one pair of apical spurs in most groups, whereas an additional mesal pair is usually present on the metatibiae; an expandable tuft of scent hairs is sometimes present on the metatibia (rarely on the mesotibia) of males; the tarsi are 5-segmented and usually equipped with short spines; the paired claws are simple, toothed, bifid or reduced; the pretarsus comprises an arolium, pseudempodium and hairy pulvilli. The thoracic musculature of Lepidoptera is generally well-developed.

The abdomen is 11-segmented in the groundplan (males) but only ten segments are present in females and segments IX and X are often indistinguishably fused ('terminal unit'; Hünefeld & Kristensen 2010). The abdominal base differs distinctly from the following segments; in the groundplan tergite I is largely restricted to a marginal frame and sternite I is small but distinctly developed. Sternite II bears paired membranous windows behind its concave anterior margin. Segments III–VII (and VIII in males) are usually simple and composed of more or less rectangular sternites and tergites and a rather extensive pleural membrane bearing the spiracles. Different scent organs occur; the gland of sternite V (autapomorphy of Amphiesmenoptera) is present in the groundplan, but is probably generally lacking in Myoglossata. The genital and postgenital segments are highly diverse in both sexes. Sternite VIII can be distinctly or largely reduced (e.g., Micropterigidae) or well-developed; specializations (e.g., prominent internal apodemes) occur in different groups. The sclerotized elements of segment IX are often represented by a dorsal tegumen and a ventral vinculum, probably derived from a closed sclerotized ring in the groundplan (Kristensen 2003). Segment IX is exposed in the basal lineages (e.g., Micropterigidae, Heterobathmiidae, Eriocraniidae) but largely concealed in most groups. The gonopods (valves) of segment IX are flanking the intromittent organ of males and primarily used to clasp the abdomen of the females; they are probably simple hairy lobes in the groundplan; 2-segmented gonopods have secondarily evolved in Ditrysia (Kristensen 2003). The intromittent organ varies strongly within Lepidoptera; according to Kristensen (2003) an aedeagus in the usual morphological sense (sclerotized part of the phallic tube distad the phallobase) occurs only in few groups and a 'phallotheca'-type similar to that of Trichoptera is probably present in the groundplan; the dorsal sclerotisation of segment X is referred to as uncus; it is primarily movably connected with the tegument IX, but often fused to it. The terminal postgenital region posterad or ventrad segment X is the anal cone, which bears the anus on its apex. In the groundplan segment VIII of females is composed of a tergum and a corresponding ventral plate. Segment IX is clearly separated from segment X in the groundplan but this condition is only preserved in females of Micropterigidae (Kristensen 2003). Segment XI is not recognizable as a defined entity and cerci are absent like in males (autapomorphy). The primary orthopteroid ovipositor is functionally replaced by an oviscapt like in the other orders of Mecopterida; a characteristic feature of Lepidoptera and Trichoptera is the presence of internal apodemes of the anterior margins of segments VIII and IX, respectively (anterior and posterior apophyses); in contrast to the groundplan of Mecopterida (e.g., Nannochoristidae) the telescoping terminal segments (oviscapt) are everted by muscle force. A single genital opening is present in the groundplan of Lepidoptera (monotrysian condition), whereas separate openings for copulating (ostium bursae) and egg laying (oviporus) are present in Exoporia and in Ditrysia.

A crop as a strongly extended region of the posterior foregut is usually present; midgut caeca are missing; the midgut is usually short and straight, whereas the hindgut is elongated and coiled. Six free Malpighian tubules are present in the groundplan; they open separately in the digestive tract and do not form a cryptonephric complex.

Morphology, larvae (caterpillars). Typical lepidopteran caterpillars have an orthognathous head and a largely unsclerotized, cylindrical postcephalic body with irregular patterns of secondary setae and abdominal prolegs. This habitus, which is usually found in neolepidopteran larvae which feed exposed on plants (caterpillars), is not a groundplan feature of the order. The unsclerotized integument of the postcephalic body is often characterized by soft, evaginable protuberances, which can emit substances produced in dermal glands. Characteristic patterns of chaetotaxy play a certain role in lepidopteran taxonomy.

The head is primarily prognathous (groundplan) like in Micropterigidae and other basal lineages, but secondarily subprognathous, orthognathous or hypognathous in many groups. The head capsule is strongly retracted in Micropterigidae but usually fully exposed. Seven stemmata are probably present in the groundplan; six stemma arranged in a semicircle are present in most groups but different degrees of reduction occur. The foramen occipitale is generally large. A hypostomal bridge occurs in the basal lineages. A strengthened postoccipital ridge is usually present and the lower margin of the head capsule which bears the mouthparts is also reinforced. A transverse cuticular fold (transclypeal sulcus) separates an anteclypeus from the rest of the head capsule. A frontoclypeal transverse ridge is missing. A dorsal epicranial sulcus is present in the groundplan but absent in Glossata. A characteristic feature of Heterobathmiidae and Glossata is the presence of V-shaped adfrontal strengthening ridges, in addition to the frontal sutures or cleavage lines; the coronal suture is present or absent. The labrum is generally well-developed and movably connected with the anterior clypeal margin. The antennae are usually short and 3-segmented; they are secondarily elongated in Micropterigidae and reductions occur in several groups. The mandibles are well-developed and sclerotized; they usually bear several teeth; a prominent molar region is usually present but varies greatly within the group. As in the related mecopterid orders the ventral mouthparts form a maxillolabial complex. The maxilla is composed of the usual elements; the stipes is divided into a basistipes and dististipes, with the latter probably also containing a basal palp component; the galea and lacinia likely separated in the groundplan but a single endite lobe (lobarium) is present in all groups except for some Micropterigidae; the palp is 3-segmented in the groundplan. The labium is distinctly simplified; it is composed of a postmentum and a prementum bearing small 2-segmented palp; the anterior part (prelabium) forms a functional unit with the hypopharynx; a prominent tubular spinneret on the premento-hypopharyngeal lobe is an autapomorphy of Glossata; the salivary orifice lies on the apex of this lobe. A well-developed tentorium with a broad tentorial bridge and well-developed posterior and anterior arms is only retained in Micropterigidae; the arms and bridge are distinctly reduced in the other groups.

The thorax is largely unsclerotized like the entire postcephalic body. The membranous cuticle often displays elaborate sculptures such as spinules, granules or specialized setae. Transverse and longitudinal integumental folding of the postcephalic segments is also common. Melanized areas demarcating muscle attachment are called tonofibrillary platelets. A sclerotized pronotum and a divided or undivided prosternal sclerotisation are preserved in the groundplan. A sclerotized propleura is always absent. Meso- and metatergal sclerotisations and sternal plates occur in some groups, but are missing in the basal lineages. In the groundplan all segments of the thoracic legs are well-developed, but the distal part of the trochanter is contiguous with the femur; the claws are unpaired; different degrees of leg reductions and specializations occur frequently; the legs of Micropterigidae are small and play a minor role in locomotion and larvae of Agathiphagidae are apodous. The mesothoracic spiracle is usually anteriorly displaced and that of the metathorax is non-functional. The body musculature of the entire postcephalic body is highly complex, like in other groups with larvae with a largely unsclerotized body wall.

The abdomen is 10-segmented. Prolegs are usually present but are probably not part of the groundplan, with the possible exception of the appendages of "anal feet" on segment X (Kristensen 2003); the presence of nearly cylindrical prolegs of segments III–VIII, with an apical planta with a circle of cuticular hooks (crochets), is probably a groundplan autapomorphy of Neolepidoptera; partial reduction or loss of prolegs has occurred many times, especially in lineages with endophytic larvae; a looping (or semilooping) type with reduced anterior prolegs has evolved in several groups (e.g., Geometridae). The dorsal side of the anal segment X is characterized by a sclerotized anal plate in the groundplan of Neolepidoptera; the triangular anal opening is also located on this segment; it is flanked by small lobes, the unpaired epiproct and the paired paraprocts, which possible represent vestiges of segment XI. In Micropterigidae the anal opening is directed ventrally and the perianal membrane functions as an adhesive device; whether a well-developed anal proleg is part of the lepidopteran groundplan is uncertain. Abdominal spiracles are present on segments I–VIII; they are almost always located in the middle of the segments but are shifted anteriorly in Micropterigidae.

The larval midgut is simple and smooth in the groundplan but strongly folded in many groups of Ditrysia (Kristensen 2003: fig. 6.3); goblet cells are characteristic of the ditrysian midgut; they are deeply invaginated, forming a large intracellular cavity (Kristensen 2003). Six Malpighian tubules are present in the groundplan; in Neolepidoptera they form two groups, each of them opening with a common duct with a vesicle in the digestive tract; the tubules are free in the basal lineages, but form a cryptonephridial complex with the rectal wall in higher Lepidoptera.

Morphology, pupae. In the groundplan (non-glossatan Lepidoptera) the pupae (Fig. 6.32.5A) have a soft integument, free appendages, and distinctly developed, movable mandibles (pupa dectica). Pupae with a sclerotized integument, non-functional mandibles (pupa adectica), and appendages glued to the body wall are characteristic for Neolepidoptera; however the degree of sclerotization and immobilization of the antennae and legs (obtectness) differs considerably within the group (Fig. 6.32.5B–D). The anterior and posterior abdominal segments are generally more or less fused, whereas the movability is retained in the middle region. An armature of spines and hooks often facilitates eclosion involving wriggling with the abdomen. A specialized feature of typical obtect pupae (appendages completely glued to body wall) is a terminal abdominal prolongation with hooks and thorns (cremaster). A great diversity of pupal cases, cocoons or shelters occur in higher Lepidoptera (Heteroneura); pupae of Exoporia groups stay in loose silk galleries or webbings; pupae without cocoons or other types of shelters rest in soil or are freely exposed.

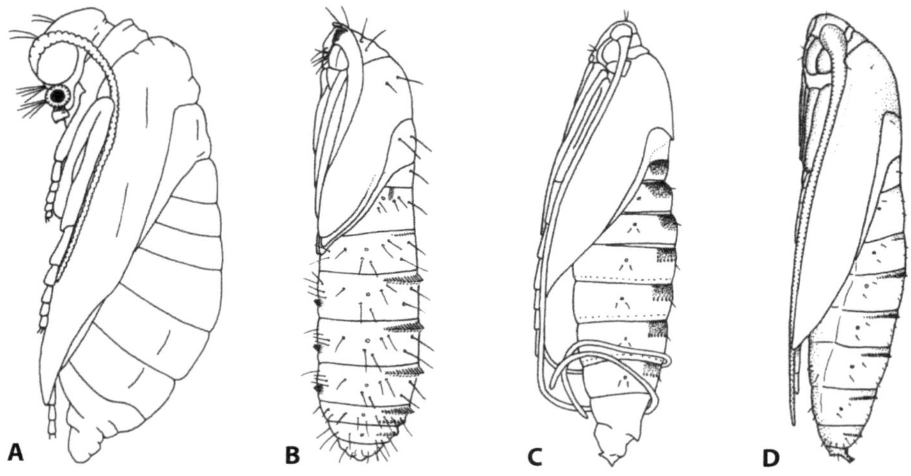

Fig. 6.32.5: Pupae, A, *Agathiphaga* sp. (Agathiphagidae); B, *Fraus* sp. (Hepialidae); C, *Nemophora* sp. (Adelidae); D, *Monopis* sp. (Tineidae). Redrawn from Kristensen (1999b).

Biology. Most species occur in the tropics and only few in the arctic areas. Many butterflies are active during daytime but the majority of lepidopterans is nocturnal (moths). The lifespan of adults is usually short, ranging between one and six weeks, unless they overwinter. An exceptionally long adult life span of ca. 11 months is known from the eurasian Brimstone (*Gonepteryx rhamni*, Pieridae), including an aestivation and a hibernation. Moths and butterflies usually consume liquid food if they feed at all. An exception are the non-glossatan Micropterigidae and Agathiphagidae, which possess distinctly developed mandibles and feed on pollen or "kauri" (*Agathis*, Araucariaceae), respectively. Most species visit flowers to suck nectar and at the same time act as pollinators. Other substances consumed by the adults are tree sap, honey dew, sweat, urine, excrements, decaying carcasses of vertebrates, juices forming in rotting fruit, and water. Sugar contained in the food is often essential for producing fertile eggs and many species can be attracted by artificial baits containing large amounts of sugar. Few species have a rigid proboscis and penetrate banana or *Citrus* (tropical "fruit piercing moths", family Noctuidae) or even the skin of vertebrates (e.g., the noctuid genus *Calyptra*). Sucking on vertebrate tear drops, including human, is known from some tropical species of Pyralidae. The proboscis can be largely or completely reduced in non-feeding adults (e.g., Hepialidae).

Almost all lepidopteran species are capable of flight, which is used for evading predators, searching for food and finding females. The typical lepidopteran flight is characterized by irregular wavering movements but some species, especially of Sphingidae, can reach high speed up to 48 km/h. The ability to hover above food plants is quite common and gliding as a component of flight also occurs in some species. A very high frequency of the wing beat is characteristic for some groups such as for instance Hesperiidae, Sphingidae and Lasiocampidae (40–90s^{-1}). High temperature is usually required (ca. 25 °C) for flight activity. Moths can create increased temperatures of the thorax by vibrations of the flight muscles. However, some winter moths (e.g., family Geometridae) and high-altitude species are active at temperatures just little above zero. The fore- and hindwings are connected and the main propulsive force is created by the mesothorax (anteromotorism). During rest, the wings are usually held in a roof-like position like in Trichoptera, but horizontally in some groups, especially in Geometridae. In other groups the wings are held in an oblique position, twisted, rolled in different ways, or reduced in few cases (females).

Some lepidopteran species undertake long seasonal flights. A well known example is the Monarch butterfly *Danaus plexippus* (Nymphalidae). Its migration between Mexico and Canada spans two to three generations. Lepidopterans can overwinter in all life stages and this can vary even between closely related species. Univoltism (one seasonal generation), bivoltism and multivoltism occur. Some species mate continuously throughout the year. Most species in temperate climates are univoltine, while in tropical climates most have two or more seasonal broods. Diapause usually occurs in the egg or as a reproductive delay in adults. It is obligatory in the last larval stage of Agathiphagidae, where it can last up to 12 years.

Defense of adults comprises camouflage mechanisms, defensive colorations (usually red or yellow combined with black) and mimicry. As an active defense mechanism some species display conspicuous color patterns (e.g., eye spots), which are not visible in the normal resting position. Strongly smelling secretions are produced by Arctiidae and Zygaenidae. The effect can be increased by emitted haemolymph. Sound production with different and often unclear function is quite common, especially in tropical and subtropical species. The most common mechanism is stridulation. Tympanal organs, particularly sensible to ultrasonic sounds, evolved several times independently in Pyralidae, Geometridae and throughout Noctuoidea. An intensive squeaking sound is produced by the Death's-head Hawkmoths (*Acherontia* spp., Sphingidae), probably as a defense mechanism used against bees when entering hives for foraging. It is produced by vibrations of a duplicature of the anterior digestive tract caused by air very rapidly sucked in though the short proboscis and expelled.

The caterpillars are almost exclusively phytophagous. Micropterigidae are associated with moss or foliose liverwort, but the vast majority relies on angiosperms as food plants. Caterpillars of few species feed on dead plant materials, hairs or feathers, animal excretions, carrion, or stored products. Some carnivorous caterpillars feed on larvae of conspecifics or larvae of other lepidopteran species, and very few species develop as parasites of hemipterans. Strictly monophagous larvae are the exception, even though many caterpillars have a very narrow food spectrum. Often plants belonging to one or several closely related plant genera or even families are suitable as food resource. Especially species known as potential plant pests are often polyphagous.

Some species causing serious damage in forestry are able to feed on different species of coniferous and deciduous trees. However, they are usually specialized on specific plant parts, such as for instance roots, leaves, or bark. The amount of consumed food is often enormous. *Bombyx mori* increases its weight by the factor 9,000 during its larval period. The resources accumulated in the fat body by the actively feeding larval stages also provide most of the energy required during the short adult life.

While the caterpillars of higher lepidopterans are almost always feeding exposed on their host plants, leaf mining is widespread amongst the lower families of Glossata. An endophytic life style as borers in more solid parts of the host plants is known from the larvae of Hepialidae (root-boring), Cossidae and Sesiidae (wood-boring), Castniidae (in palms) and several groups within Noctuoidea (borers in a large variety of plants). Larvae of few species are endophytic in aquatic plants below the water surface (e.g., Pyralidae), and caterpillars of some Lycaenidae and Tineidae live in ant's nests as food parasites and/or predators. Nocturnal larvae are often hiding in soil or closely attached to bark during daytime. Gregarious behavior occurs in many families. Nests of the Oak Processionary (*Thaumetopoea processionea*, Thaumetopoeidae), which are covered by silk webs, can reach a length of up to one meter. Lepidopteran larvae are potentially attractive for numerous predators such as for instance small mammals, birds, Hymenoptera, beetles and others. Like in the adults camouflage can play a role and also warning colorations (especially yellow, red, white, black) and mimicry. Defensive postures occur in several groups (e.g., Notodontidae), gyrating movements with the anterior part of the body, or inflation of parts of the thorax. Conspicuous armatures of strong setae can make caterpillars unattractive for predators. Defensive substances are often produced in subepidermal glands. Sequestering from the host plants and complete de novo synthesis of defensive secretions can occur alternatively in closely related species (e.g., Zygaenidae). A defense mechanism occurring in Nymphalidae is the oral emission of smelly gut content. The larvae of some notodontids (e.g., *Cerura* spp.) possess the ability to release formic acid from a groove between the first and second thoracic segment. Lepidopteran larvae are also often attacked and decimated by parasites, especially by Ichneumonidae (Hymenoptera) and Tachinidae (Diptera). The Gypsy Moth (*Lymantria dispar*, Lymantriidae) can be attacked by 13 different parasitic species during its life cycle. Parasitism often down-regulates lepidopteran populations by 90% or more.

Reproduction and development. The reproduction is usually bisexual but parthenogenesis occurs in several groups (e.g., Psychidae). Eclosion usually takes place earlier in males. Sexual dimorphism occurs frequently, with different colorations in the sexes, and a larger body volume (and sometimes larger wings) in the females. Antennae and labial palps are often more strongly developed in males. Partial or total reduction of the wings rarely occurs in females (e.g., Psychidae, some Geometridae and Lymantriidae) and in extreme cases the legs can be vestigial or absent (Psychidae). In diurnal species visual stimuli generally play a crucial role for finding a partner, whereas nocturnal species (including almost all moths) use pheromones, which often function over long distances. In the typical case secretions produced by the females attract the males, which often are equipped with specialized pinnate or comb-shaped olfactory antennae. Different glands occur in both sexes, and different components produced in different glands can play a role in certain stages of the reproduction and stimulate specific patterns of behavior. Epidermal glands and specialized olfactory scales on the wings are typical for the males. Similar specialized scales usually occur in the intersegmental membranes of the abdominal segments VII–VIII and VIII–IX. Acoustic courtship and vibrations also play a role in some groups (e.g., Ctenuchinae). Especially in low-density butterfly species males and females frequent specific meeting places, as conspicuous single trees, to find a suitable partner (e.g., hill-topping and tree-topping in Papilionidae and Nymphalidae).

In the precopulatory stage the male usually sits beside the female. Then it bends its abdomen and attaches to the female genital segments using its copulatory apparatus. The typical copulatory position is with the heads pointing in opposite directions, although exceptions occur (in the Death's heads Hawkmoths the male is riding on the back of the female). Elaborate courtship behavior, including the defense of certain territories (light spots in woodlands, exposed twigs), occurs in several groups (e.g., Nymphalidae, Satyridae).

Oviposition usually takes place on the host plants of the larvae, which are mainly detected using chemical cues. Females of some generalists drop the eggs more or less randomly during flight (e.g., Hepialidae) and they are deposited in crevices in soil by females of Micropterigidae. The total egg number per female normally varies between 20 and 2000, but in few cases it can exceed 10,000. In the lower-grade glossatan families egg insertion into living plant tissue takes place by use of a specialized oviscapt with an 'oviscapt saw' at its tip ('*Eriocrania*-type oviscapt'; Hünefeld & Kristensen 2010). The larvae hatch after a few days for example in the Monarch butterfly *Danaus plexippus* (Danaidae), or after several months if the egg is the overwintering stage as in the underwing moths of temperate climates (*Catocala* spp., Noctuidae).

Butterflies have usually 4–5 larval instars but the number is distinctly higher in moth species, for instance up to 17 in the Clothing Moth (*Tineola biseliella*, Tineidae). The normal larval development can take less than two weeks in many polyvoltine species, or up to several years in wood-boring species (e.g., Cossidae). In the last instar larvae a changing balance of juvenile hormone and ecdysone triggers a change in the behavior and often also in the coloration. The larvae stop feeding and seek a suitable place for pupation. Pupation takes place at or in the soil, in a cocoon at the bottom or between plant parts, or even exposed and attached to plants or other surfaces (free-hanging pupae of most butterflies). The larvae of borers (e.g., Cossidae, Sesiidae, Castniidae) usually pupate in their frass galleries, but often the pupa moves her anterior body half out of the gallery by gyrating movements before eclosion. The time span between pupation and eclosion takes one week (in many bi- or polyvoltine species) up to several years (many species of temperate climate with overwintering pupae).

Fossil record. The fossil record of Lepidoptera is exceptionally sparse. The earliest known fossil is *Archaeolepis mane* from the Jurassic of Dorset in England (ca. 190 Ma). Moths are relatively frequently found in Baltic amber (Grimaldi & Engel 2005).

Economic importance. Lepidopteran caterpillars can cause great damage in agriculture by feeding on cultivated plants and stored products. Human health can be affected by food pests (e.g., Pyralidae) or if caterpillars possess and release venomous hairs (e.g., Thaumetopoeidae, Lymantriidae) causing allergic responses (including anaphylaxis). Some species are traditionally used as human food, as the Bogong moth (*Agrotis infusa*, Noctuidae) of Australia. Silk production (*Bombyx mori*, Bombycidae) plays a very important role since several thousand years in China and other Asian countries. Many lepidopteran species function as pollinators.

A considerable number of lepidopteran species is critically endangered by habitat destruction or fragmentation, and in many countries increasing financial effort is spent in order to save or restore their biotopes.

[Kristensen (1999b, 2003); Gaedike & Häuser (2005); Grimaldi & Engel (2005); Kristensen et al. (2007); Hünefeld & Kristensen (2010); Nieukerken et al. (2011)]

Antliophora (Mecoptera, Siphonaptera, Diptera)

6.33 **Mecoptera** (including Nannomecoptera and Neomecoptera, Greek *meco* = long, *pteron* = wing, English common names for subgroups: scorpionflies, hangingflies etc.)

Diversity and Distribution. Mecoptera are one of the small groups of Holometabola. Approximately 550 species are described. The distribution is worldwide. Extant species of the presumably basal Nannochoristidae (or Nannomecoptera) are restricted to the Southern Hemisphere (seven species in Chile, Australia, and New Zealand) (Fig. 6.33.1), whereas the flightless Boreidae (or Neomecoptera) occur only on the Northern Hemisphere (Fig. 6.33.2, 6.33.6). Choristidae are endemic in Australia. Apteropanorpidae are restricted to Tasmania and southern Australia.

Autapomorphies. The monophyly was strongly disputed (e.g., Whiting 2002; Beutel & Baum 2008) but is clearly supported by recent studies based on an extensive morphological data set (Beutel et al. 2011) and single copy nuclear genes (Wiegmann et al. 2009). However, there is only one convincing morphological autapomorphy of the order (Friedrich et al. 2013).

- Male genital segments forming a genital capsule (secondarily modified in Boreidae) (Fig. 6.33.4)

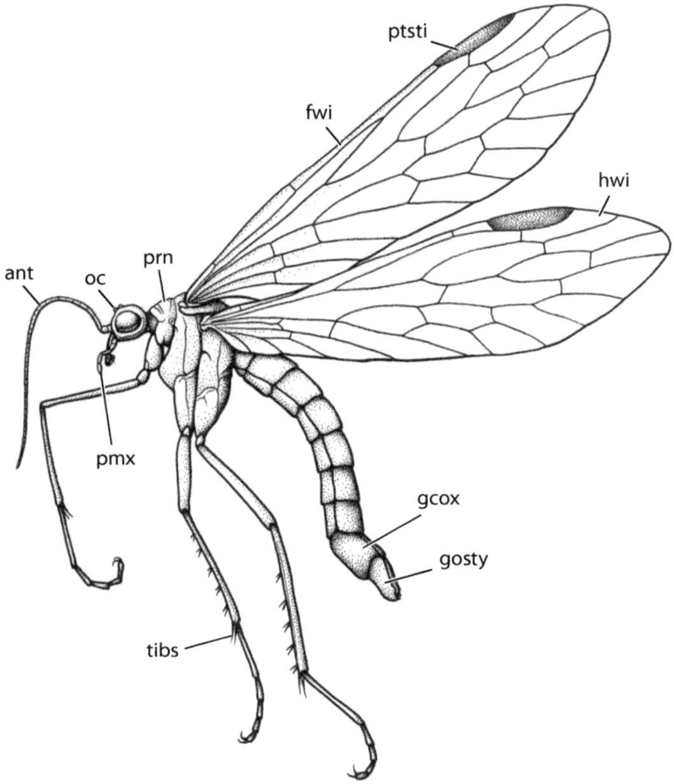

Fig. 6.33.1: *Nannochorista* sp. (Nannochoristidae), female, lateral view. Abbr.: ant – antenna, fwi – forewing, gcox – conocoxa, gosty – gonostylus, hwi – hindwing, oc – ocellus, pmx – palpus maxillaris, prn – pronotum, ptsti – pterostigma, tibs – tibial spur. Courtesy F. Hünefeld.

Other features suggested earlier as mecopteran apomorphies (e.g., clypeolabrum, loss of all hypopharyngeal muscles) have turned out as untenable.

Taxonomy. The order is presently subdivided into the nine extant families Nannochoristidae (Fig. 6.33.1), Boreidae (Figs 6.33.2, 6.33.6), Meropeidae, Bittacidae, Eomeropidae, Choristidae, Apteropanorpidae, Panorpidae (Fig. 6.33.5) and Panorpodidae. Nannochoristidae (=Nannomecoptera) are probably the sistergroup of the remaining Mecoptera which are characterized by a highly unusual intrinsic muscle of the apical salivary duct ("Sekretformer"; Grell 1939). Boreidae (=Neomecoptera) are likely the sister taxon of a clade Pistillifera, which is strongly supported as a clade by a very specific time of sperm pump. A suggested alternative placement of Boreidae as sistergroup of Siphonaptera has been refuted (see above). Pistillifera comprise the

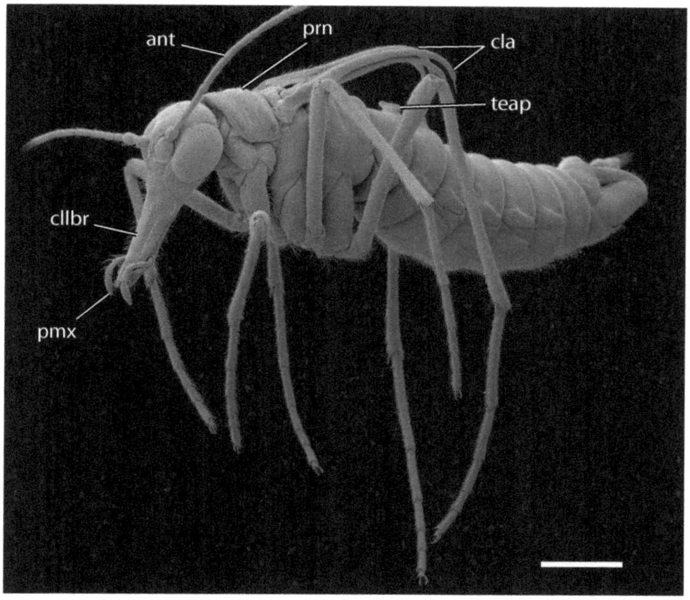

Fig. 6.33.2: *Boreus hyemalis* (Boreidae), male, habitus, anterolateral view, SEM micrograph. Abbr.: ant – antenna, cla – clasps, cllbr – clypeolabrum, pmx – palpus maxillaris, prn – pronotum, teap – tergal apophysis. Scale bar: 100 μm. From Beutel et al. (2008), modified.

remaining seven mecopteran families. Their interrelationships are not fully clarified (e.g., placement of Bittacidae and Meropeidae) (Friedrich & Beutel 2010a; Friedrich et al. 2013). Panorpidae and Panorpodidae are sistergroups.

Diagnosis. Head of adults of most genera with strongly elongated rostrum (Figs 6.33.2). Usually with two pairs of more or less equally sized membranous wings and a conspicuous male genital capsule (Fig. 6.33.1).

Morphology, adults. Mecopterans are small to medium sized insects (Figs 6.33.1, 6.33.2); the body length varies between 1.4 mm (*Caurinus dectes*) and 20 mm; the maximum wing span is ca. 70 mm. Two pairs of wings of approximately equal size are present in most genera (reduced in Boreidae and Apteropanorpidae). The cuticle is usually smooth and shiny and brown and yellow colors are common; it is unusually strongly sclerotized and dark in the boreid genus *Caurinus* ("armored boreid").

The orthognathous head is usually fully exposed, but very slightly retracted into the prothorax in Boreidae (Figs 6.33.2). The compound eyes are well-developed; they are very large and convex in Meropeidae and Nannochoristidae (Fig. 6.33.3), and extremely enlarged in some species of Bittacidae; the ommatidia appear unusually large in the miniaturized *Caurinus*; three large ocelli are usually present but reduc-

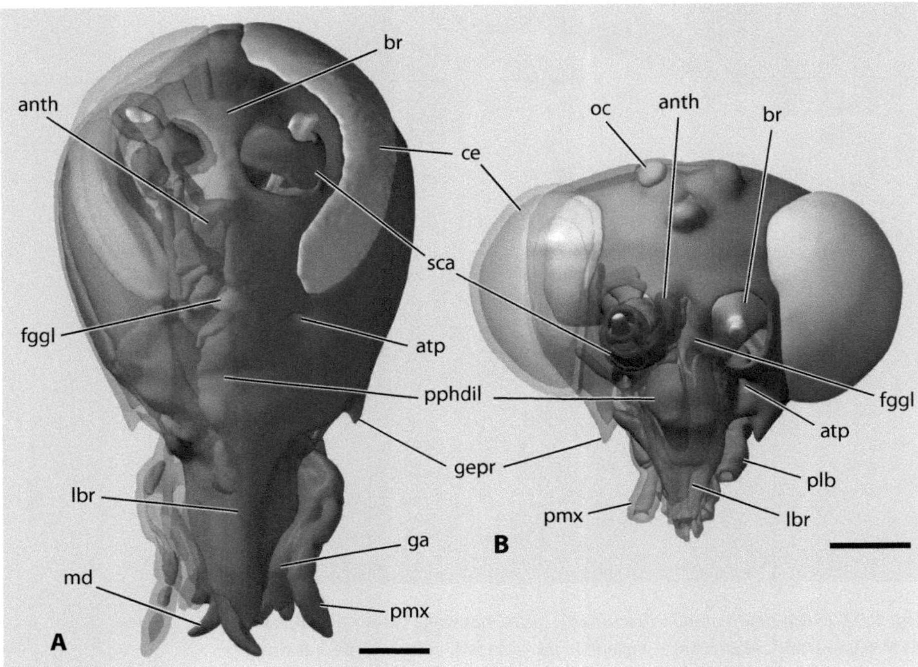

Fig. 6.33.3: Adult heads, anterior view, 3-dimensional reconstructions. A, *Merope tuber* (Pistillifera, Meropeidae) (from Friedrich et al. 2013, modified), B, *Nannochorista neotropica* (Nannochoristidae). Abbr.: anth – antennal heart, atp – anterior tentorial pit, br – brain, ce – compound eye, fggl – frontal ganglion, ga – galea, gepr – genal process, lbr – labrum, md – mandible, oc – ocellus, plb – palpus labialis, pmx – palpus maxillaris, pphdil – prepharyngeal dilator, sca – scapus. Scale bars: 200 μm.

tions occur in Meropeidae (Fig. 6.33.3A) and flightless species (e.g., *Caurinus*). A median internal frontal apodeme is present in *Nannochorista* but absent in all other groups. A temporal suture is usually present on the posterior head region but absent in *Nannochorista*, *Boreus*, *Caurinus* and *Bittacus*. The clypeus is strongly elongated in most genera and forms a conspicuous rostrum together with the genae (Figs 6.33.2); the rostrum is rather short in some genera (e.g., *Merope*) (Figs 6.33.3A) and absent in *Nannochorista* (Figs 6.33.3B) and *Caurinus*. The labrum is usually fused with the clypeus but this is not the case in *Nannochorista* and *Caurinus*; in *Nannochorista* it is subdivided into a posterior region and an elongated anterior part, which bears a food canal on its ventral side. The antennae are multisegmented (minimum 14 antennomeres in Bittacidae) and filiform and inserted on large membranous insertion areas, which are usually closely adjacent on the lower frontal region; large sensorial grooves are present on antennomeres of some species (e.g., *Merope*). The mouthparts are usually biting but strongly modified and only suitable for the uptake of liquids in Nannochoristidae. The primary mandibular joint is absent in *Nannochorista* but present in other groups; the mandibles are greatly elongated and flattened in *Nannochorista*, devoid of any teeth or a molar area; the mandibular base is stout in *Caurinus*, with a large, prominent molar area without grinding surface; the distal part forms an elongate tooth, with narrow channels on the surface; fairly small and simple biting mandibles are present in most other taxa, with small teeth but without molar region or movable appendages (prostheca); the mandibles are very elongate and blade-like in the predaceous Bittacidae. The ventral mouthparts form a very compact maxillolabial complex in Boreidae, and to a lesser degree in the other groups. The stipites and labium are strongly elongated in the group with a rostrum (e.g., *Boreus*, *Panorpa*, *Bittacus*); the galea is usually spoon-shaped and wrapped around the labral region laterally, but is absent in *Nannochorista*; a deep sensorial groove is present on the 3rd maxillary palpomere of *Nannochorista*. The labial palps are 2-segmented; glossae and paraglossae are absent; the prelabium forms a trough-like structure in *Nannochorista* and the palp muscle is strongly enlarged. The tentorium is usually composed of well-developed posterior and anterior arms (hollow in *Nannochorista*) and a transverse bridge; the dorsal arms are usually very thin and weakly sclerotized but well-developed in *Caurinus*. A hypostomal bridge is usually present but missing in *Nannochorista*. A cranial extrinsic antennal muscle occurs in most groups but is absent in *Caurinus*, *Bittacus* and *Merope* (possibly secondarily shifted to the dorsal tentorial arm). The tentoriomandibular muscle is strongly enlarged in *Caurinus* but extremely thin in the other genera; the mandibular flexor and extensor muscles are completely reduced in *Nannochorista*. An elongate prepharyngeal tube with a sclerotized floor is usually present, but it is short in some taxa (e.g., Nannochoristidae, *Caurinus*). The anterior pharynx is usually characterized by a more or less extensive precerebral pumping chamber enclosed by conspicuous sclerotized claspers (indistinct in *Nannochorista*, small in *Boreus*). The typical hypopharyngo-salivary muscle is present in *Nannochorista* but transformed into a compact intrinsic muscle of the sclerotized distal part of the salivary duct in the other groups ("Sekretformer", see above). A strongly developed postcerebral pharyngeal pumping chamber is present in *Nannochorista*. Like in the other antliophoran groups the brain and suboesophageal complex form a compact mass around the pharynx. The cervical membrane is elongated in some groups (Panorpidae, Bittacidae, Choristidae); lateral cervical sclerites are present and additional dorsal cervical sclerites occur in few taxa (e.g., *Boreus*).

The small prothorax is characterized by a saddle-shaped pronotum, which is wider than long and bears transverse ridges in some groups; in some genera it encloses the mesothoracic spiracles. The pterothoracic segments are simplified in the flightless taxa (Boreidae, Apteropanorpidae) (Fig. 6.33.2), but otherwise well-developed and about equally sized, with distinctly separated dorsal elements, i.e. scutum, scutellum and postnotum, and a very distinct pleural suture and internal ridge. The sterna are invaginated and the coxae usually medially adjacent (not in *Apteropanorpa*). The membranous wings are usually well-developed and of about equal

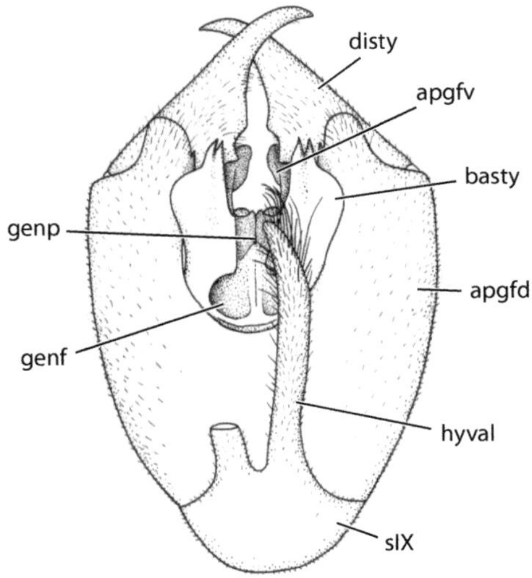

Fig. 6.33.4: *Panorpa communis* (Pistillifera, Panorpidae), male genital apparatus, ventral view. Abbr.: apgfd – dorsal appendage of genital field, apgfv – ventral appendage of genital field, basty – basistylus, disty – dististylus, genf – genital field, genp – genital pore, hyval – hypovalva (process of sIX), sIX – sternum IX. Redrawn from Willmann (2005b).

size (Fig. 6.33.1); the wing base is narrowed in Bittacidae and an increased number of longitudinal and transverse veins results in a reticulate pattern in Eomeropidae and Meropeidae; reductions of the venation occur in Panorpidae and Bittacidae, and Boreidae are always flightless; the forewings of male boreids are modified as claspers and used to hold the females during copulation; the forewings are largely reduced in females and the hindwings are small vestiges in both sexes (Fig. 6.33.2). The thoracic flight musculature is usually well-developed but strongly reduced in the flightless Boreidae. The legs are slender, with elongated conical coxae; long sternal processes form the mesal coxal articulation; the tarsi are 5-segmented; they are markedly elongated in Bittacidae, which use them as catching devices; their tarsomeres 4 and 5 form a subchela as a specialized catching device; small hairy pulvilli and an arolium are usually present but the latter is lacking in Boreidae; the middle and hindlegs are modified as jumping legs in boreids.

The abdomen is composed of 11 segments. It is usually slightly constricted at the connection with the thorax, widest near mid-length, and tapering posteriorly; it is distinctly elongated and subcylindrical in Bittacidae and more compact in Boreidae, especially in *Caurinus*. Tergum I is fused to the metathorax; sternite I is free and often divided into two sclerites. The cerci are almost always 1-segmented in males but missing in Boreidae and *Apteropanorpa*; the highest number of three cercomeres is found in females of *Nannochorista*; only 1-segmented is present in Bittacidae and also in Boreidae; the basal cercomeres of females are connected by a sclerite or fused in females of most groups, but not in Nannochoristidae, Boreidae, and Bittacidae. Segment XI of males usually bears a conspicuous genital capsule formed by the fused basistyli, which bear 1-segmented dististyli as appendages (Fig. 6.33.4); the capsule contains a complex sperm pump with a pistill in males of Pistillifera; the sperm pump is not fully developed in Nannochoristidae and is absent in Boreidae, which do not transmit liquid sperm but a spermatophore. The opening of the genital chamber of females lies on the ventral side between segments VIII and IX in *Nannochorista*, but is shifted posteriorly in the other groups. A secondary ovipositor formed by the 1-segmented cerci is present in *Hesperoboreus* (short) and *Boreus* (absent in *Caurinus*) .

A proventriculus equipped with a strong musculature and numerous thin acanthae is usually located in the prothorax; midgut caeca are absent; the length of the midgut varies greatly; six free Malpighian tubules are present and six rectal papillae. The testes are composed of three or four fusiform follicles in Panorpidae and Bittacidae. The number of serially arranged polytrophic ovarioles in females varies even within genera (6–10 in *Boreus*). The accessory glands are elongate and often connected basally.

Morphology, larvae. The larvae of Nannochoristidae are aquatic and extremely elongated and slender (campodeiform). They are terrestrial and more or less resembling caterpillars in the other groups (scarabaeiform in Panorpodidae and Boreidae, eruciform in the other groups) (Figs 6.33.5, 6.33.6).

The head is distinctly prognathous in Nannochoristidae but orthognathous in all other groups (Fig. 6.33.5). The simplified compound eyes are usually well-developed; they are distinctly reduced in Nannochoristidae (maximum of 16 ommatidia), without externally recognizable cuticular lenses, and distinctly separated stemmata are present in Boreidae (Fig. 6.33.6: stem); a median frontal ocellus is described for larvae of *Bittacus* and Choristidae but the homology is not confirmed by anatomical data. The frontal and coronal sutures are present and also an oblique dorsal sulcus ("occipital suture"); a V-shaped adfrontal strengthening ridge is described for *Panorpa*. The clypeus is divided into a posterior postclypeus and an anterior transparent anteclypeus; it is transverse in *Nannochorista* but trapezoid and longer than wide in the other groups (synapomorphy of Boreidae and Pistillifera). The labrum is free; in *Nannochorista* it bears a conspicuous tripartite field of sensorial papillae and groups of microtrichia; the epipharynx is exposed in *Nannochorista* but not in the other groups. The antenna is 3-segmented and moderately long in *Nannochorista* and Pistillifera, but 2-segmented and very short in Boreidae; the cylindrical or club-shaped pedicellus contains a Johnston's organ. The mandibles are equipped with an articulated prostheca and acanthae (prosthecal brush) in *Nannochorista*; a mola is present in Pistillifera. The maxillae are inserted in distinct maxillary grooves in Boreidae (Fig. 6.33.6) and *Nannochorista*, but inserted at the anterior margin of the head

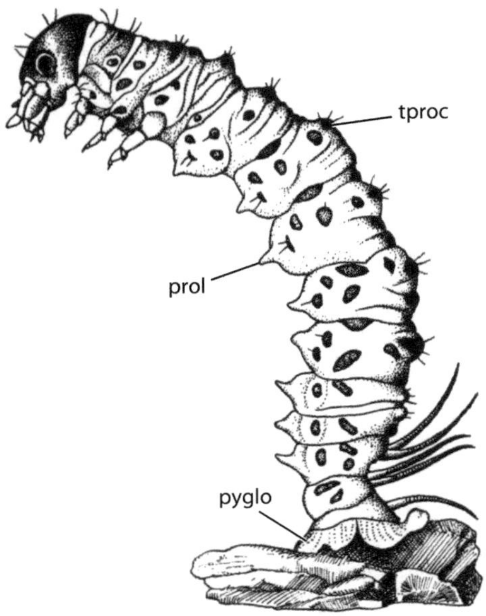

Fig. 6.33.5: *Panorpa communis* (Pistillifera, Panorpidae), larva. Abbr.: prol – abdominal proleg, pyglo – pygidial lobe (adhesive device), tproc – tergal process. Redrawn from Willmann 2005b).

capsule in Pistillifera; a discrete and large cardo is present in Boreidae (observed in 1st instar) but not in the other groups, where it is fused with the basistipes; this combined sclerite is transverse in Pistillifera; a dististipes is probably generally present but the homology is problematic; galea and lacinia are either separated by a furrow, or a division into discrete endite lobes is not recognizable. The labium is composed by of a prelabium and postlabium in *Nannochorista* and Boreidae; these elements are not clearly separated from each other in the former group and the postlabium is not present as a separate recognizable structure in Pistillifera; it is likely represented by the membranous area connecting the sclerotized anterior margin of the head capsule with the prementum; the 2-segmented labial palps are inserted on palpomere-like prelabial processes adjacent to the salivary orifice in Pistillifera but not in the basal lineages; glossae and paraglossae are always absent. A medially divided hypostomal bridge is present in *Nannochorista* and Pistillifera but absent in Boreidae; it is very distinctly delimited laterally in *Nannochorista* but not in the other groups.

The thorax of *Nannochorista* is cylindrical and slender. The prothorax is the smallest segment and bears a distinctly sclerotized and pigmented pronotum. The meso- and mesothorax of *Nannochorista* are similar to the abdominal segments, except for the presence of slender 5-segmented legs, which are short but otherwise well-developed. The surface of the segments is smooth in *Nannochorista*, whereas

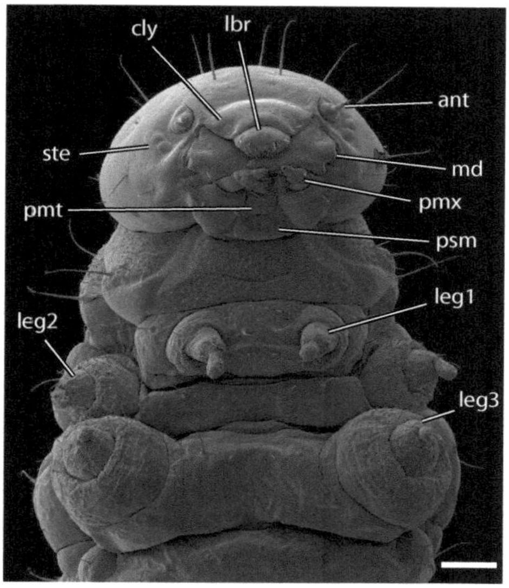

Fig. 6.33.6: *Hesperoboreus brevicaudus* (Boreidae), larva, ventral view, SEM micrograph. Abbr.: ant – antenna, cly – clypeus, lbr – labrum, md – mandible, pmt – prementum, pmx – palpus maxillaris, psm – postmentum, ste – stemmata. Scale bar: 100 μm.

variously shaped tubercles, conspicuous folds, and fleshy processes (mesonotum of Bittacidae) occur in the other groups (Fig. 6.33.5); larvae of Boreidae and Pistillifera are also characterized by a more compact thorax and a weak degree of sclerotization of the postcephalic segments with the exception of the pronotum and legs (Figs 6.33.5, 6.33.6). The legs are 3-segmented in boreid (Fig. 6.33.6) and pistilliferan larvae; the forelegs are strongly modified in larvae of *Boreus* and *Hesperoboreus* (synapomorphy) (Fig. 6.33.6) but not in *Caurinus*.

The abdomen is 11-segmented in Boreidae but apparently 10-segmented in the other groups (pygidial lobes see below). It is extremely slender and tapering posteriorly in *Nannochorista* and prolegs are completely missing; apically it bears a fleshy lobe with a pair of sclerotized hooks and dorsal and ventral pairs of apically pointed anal appendages; prolegs are also absent in Boreidae and Bittacidae (scarabaeiform larva), but present as ventral fleshy lobes on segments I–VIII in the other groups (Fig. 6.33.5). Different tergal processes (peg-like, filament-like or star-shaped) occur in pistilliferan larvae; together with dirt particles they form a camouflage mechanism in Bittacidae. Four finger-like fleshy pygidial lobes, possibly derivatives of segment XI, are present in larvae of Pistillifera (autapomorphy); they serve as attachment devices (Fig. 6.33.5).

Morphology, pupae. The pupa dectica exarata resembles the adults in its habitus. The rostrum, if present, is shorter than in the adult stage.

Biology. Most mecopterans occur in warmer regions and in habitats with dense vegetation. Few species occur in semideserts and *Boreus* prefers very cold habitats, especially moss in the vicinity of snow or ice. The boreid *Caurinus dectes* is associated with liverworts (Russell 1979). Adults of Nannochoristidae are only found close to running water. Their larvae are fully aquatic and they pupate close to the river edge. The diet of the larvae is unknown and the highly modified mouthparts of the adults strongly suggest liquid feeding. Bittacidae prefer shrubs and suspend themselves with their elongate hindlegs. The specialized legs (see above) form a basket for catching insect prey. Panorpidae, which are often found on flowers, are probably mainly feeding on carrion, and are apparently also facultative cleptoparasites in spider nets. Arguably the elongated rostrum has evolved as an adaptation to feeding on insect carcasses. The flying abilities of mecopterans are moderate or the flight organs are reduced (Boreidae, Apteropanorpidae). Larvae of Boreidae live among moss or liverworts (*Caurinus*). Larvae of *Panorpa* dig tunnels in soft and moist soil.

Reproduction and development. Pheromones play a role in finding partners in members of Pistillifera. Before mating, males and females of *Panorpa* vibrate with their wings and carry out vertical movements with the abdomen. Mecoptera are a classical example for female sexual selection. Males offer either a droplet of sugary secretions to the females (*Panorpa*) or insect prey (Bittacidae). Females estimate the fitness of males based on the size of the nuptial present. Spermatophores are produced by Boreidae, but liquid sperm is transferred by a sperm pump in Pistillifera and apparently also in *Nannochorista* ("Spermienauspressvorrichtung"; Mickoleit 2008). The females of Boreidae ride on the back of the males during copulation, and are held in this position by the claspers formed by the highly modified forewings of the males. Females deposit single eggs or small groups. Typical oviposition sites are roots of moss plants. The embryonic development takes place within 6–16 days in Panorpidae, whereas the first instar larvae of Bittacidae hatch in the year following the deposition of the egg (216–300 days). Nannochoristidae and Pistillifera have probably generally four larval stages. Larvae of *Boreus* overwinter and pupate towards the end of the summer in the following year (Willmann 2005b). A prepupal stage with a quiescent 4th instar larva precedes the pupation.

Fossil record. The number of fossil species approximately equals the recent diversity of Mecoptera. In contrast to that, the number of extinct higher ranking taxa assigned to the order (families and genera) is nearly three times as high as the number of corresponding extant supraspecific groups. The placement of fossils in the mecopteran crown group is often problematic. Due to the plesiomorphic condition of the wings it cannot be excluded in many cases that the fossils in question belong to the stem group of the entire Antliophora (Willmann 1989).

The oldest mecopteran fossils are known from the Upper Permian. They are related to the extant Nannochoristidae. The oldest representatives of Pistillifera go back to the Triassic. Fossils of Meropeidae were found in Asian Jurassic deposits and the oldest records for the southern family Choristidae are from the Lower Cretaceous. Species of Panorpidae, Panorpodidae and Bittacidae are preserved in Baltic amber. Representatives of the extinct Pseudopolycentropodidae were characterized by strongly elongated mouthparts apparently specialized on the uptake of liquid food. The postcephalic body is similar to that of Nannochoristidae (Grimaldi et al. 2005).

Economic importance. Mecoptera are economically irrelevant.

[Grell (1939); Bierbrodt (1943); Hepburn (1969, 1970); Mickoleit (1975, 1976, 1978, 2008); Kaltenbach (1978); Russell (1979); Byers (1987, 1991); Willmann (1989; 2005b); Whiting (2002); Grimaldi et al. (2005); Beutel & Baum (2008); Beutel et al. (2009, 2011); Wiegmann et al. (2009); Friedrich & Beutel (2010a); Friedrich et al. (2013)]

6.34 Siphonaptera (Greek *siphon* = tube, *aptera* = without wings, English common name: fleas)

Diversity and distribution. With approximately 2,000 described species and subspecies (ca. 15% of the names are invalid according to Lewis [1998]) fleas belong to the medium sized groups of insects. The highly specialized ectoparasites occur on all continents including Antarctica. The diversity appears to be highest in mountainous areas of the moderate subtropical regions and most species are described from Eurasia (Kinzelbach 2005). Hosts are primarily mammals, especially rodents and insectivores, and secondarily birds (ca. 5% of the species).

Autapomorphies
- Ectoparasitic habits
- Body laterally compressed
- Complete loss of wings (Figs 6.34.1, 6.34.2)
- Laciniae and labrum (or epipharynx) form sucking stylets
- Regular rows of flattened setae (ctenidia) (not present in all species) (Figs 6.34.1, 6.34.2)
- Hindlegs modified as jumping legs, with basal pads of resilin (Figs 6.34.1, 6.34.2)
- Compound eyes strongly reduced (Fig. 6.34.2)
- Ocelli absent
- Antennae short (Figs 6.34.1, 6.34.2)
- Antennal groove (Fig. 6.34.1)

Fig. 6.34.1: *Ctenocephalides felis* (Pulicidae), lateral view, SEM micrograph. Abbr.: ant – antenna, cten – ctenidia, lac – lacinia, ley – latera eye, pmx – palpus maxillaris, sVIII – abdominal sternum VIII. Scale bar: 250 µm. Courtesy H. Pohl.

- Large male copulatory organ functioning as clasping apparatus (Fig. 6.34.2)
- Larvae legless (alternatively a possible synapomorphy with Diptera) (Fig. 6.34.4)
- Larval eyes absent (Fig. 6.34.5)

Taxonomy. Siphonaptera are sometimes divided into three suborders Pulicomorpha, Pygiopsyllomorpha and Hystrichopsyllomorpha (see e.g., Kinzelbach 2005), but this traditional classification is not based on a robust phylogeny (see below). Only families, subfamilies and tribes were used as intraordinal categories by Lewis (1998), who listed 15 families (and 29 subfamilies) in alphabetical order. In total they comprise 44 tribes, 220 genera and 1,957 species. The aberrant monogeneric family Ancistropsyllidae (three species) contains parasites of artiodactyls in the Oriental region (Lewis 1998). Ceratophyllidae (384 spp.) are almost exclusively found in the Holarctic region, but few species occur in the Southern Hemisphere, including one restricted to Antarctica. Ischnopsyllidae (122 spp.) are widely distributed and mainly parasitize bats (Lewis 1998). Leptopsyllidae (230 spp.) comprise two subfamilies in Lewis' (1998) classification. The small families Chimaeropsyllidae (26 spp.) and Xiphiopsyllidae (8 spp.) occur in sub-Saharan Africa and parasitize rodents and insectivores. The monogeneric Coptopsyllidae (19 spp.) are distributed in the Mediterranean region and specialized on gerbils. Ctenophtalmidae (548 spp.) is a large group divided into nine subfamilies and 42 genera (Lewis 1998). Hystrichopsyllidae (36 spp.) are subdivided into two subfamilies. Malacopsyllidae (two monotypic genera) are specialized

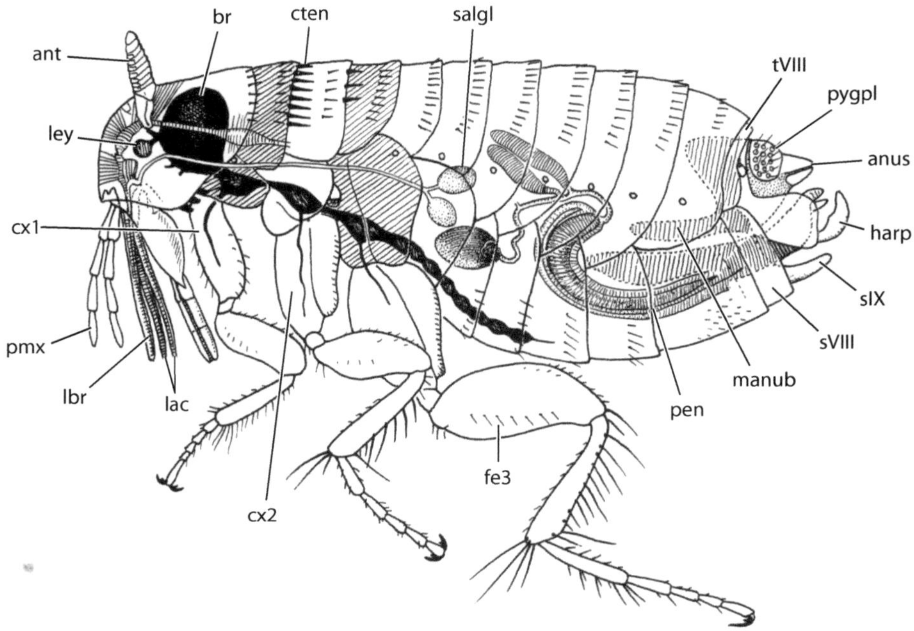

Fig. 6.34.2: Male, lateral view, schematized. Abbr.: ant – antenna, br – brain, cten – ctenidia, cx1/2 – pro-/mesocoxa, fe3 – metathoracic femur, harp – harpago, lac – lacinia, lbr – labrum, ley – lateral eye, manub – manubrium, pen – penis, pmx – palpus maxillaris, pygpl – pygidial plate, sVIII/IX – abdominal sternum VIII/IX, salgl – salivary gland, tVIII – tergum VIII. Redrawn from Kinzelbach (2005), after Weber & Weidner (1974).

on armadillos in the Neotropical region (Lewis 1998). Pygiopsyllidae (163 spp.) were subdivided into three subfamilies by Lewis (1998), and Rhopalopsyllidae and Stephanocircidae both in two. Vermipsyllidae (39 spp.) are restricted to the Northern Hemisphere and mainly specialized on carnivores. The best known siphonapteran species belong to Pulicidae (182 spp. in Lewis' concept), which mainly occur in Africa. This family includes the human flea *Pulex irritans*, the rat flea *Xenopsylla cheopis* (vector of the Plague), the dog flea *Ctenocephalides canis*, and the cat flea *C. felis* (Fig. 6.34.1). Tunginae (five genera and 23 spp.) or "jiggers" (or "chigoe fleas") were treated as a pulicid subfamily by Lewis (1998).

A comprehensive analysis of Siphonaptera was presented by Whiting et al (2008). Sequence data of four genes (18S and 28S rDNA, Cytochrome Oxidase II, Elongation Factor 1-alpha) were analyzed using POY (direct optimization). Sixteen families (family concept differing from that of Lewis [1998]) and 26 subfamilies were included. The results suggest a basal position of Tungidae and monophyletic lineages Pygiopsyllomorpha (Lycopsyllidae + [Pygopsyllidae + Stivaliidae (=Stivaliinae)]) and Ceratophyllomorpha (Ischnopsyllidae, Ceratopsyllidae and Leptopsyllidae). Macro-psyllidae were placed as sister group of Coptopsyllidae, and Pulicidae as sister group of Chimaeropsyllidae. Hystrichopsyllidae and Leptopsyllidae turned out as paraphyletic and Ctenophtalmidae as “grossly non-monophyletic”.

The species level taxonomy of Siphonaptera is treated in a remarkable multivolume monograph (Hopkins & Rothschild 1953–1971) and in other contributions based on the famous Rothschild collection at The Natural History Museum in London (see Grimaldi & Engel 2005).

Diagnosis. Small, completely wingless, with piercing sucking mouthparts and short antennae. Body laterally compressed. Hindlegs modified as jumping legs (Figs 6.34.1, 6.34.2). Larvae slender, with prognathous head and completely reduced legs (Fig. 6.34.4).

Morphology, adults. Fleas are small or very small insects ranging in size between 0.5 and 8 mm. The largest European species, the mole flea *Hystrichopsylla talpae*, is 6 mm long. The body is completely wingless and laterally compressed. The cuticle is smooth and strongly sclerotized. A characteristic feature is the presence of rows of strongly developed flattened setae, the ctenidia.

The helmet-shaped head is usually orthognathous (Fig. 6.34.2). Its posterior margin usually overlaps with the anterior margin of the prothorax. The cervical membrane is not exposed. The foramen occipitale is partly closed by a postgenal bridge. Head sutures are absent. Ctenidia are often present along the genal margin. The compound eyes (sometimes erroneously referred to as ocelli) are largely or completely reduced; separate ommatidia are never present; true ocelli are missing. The labrum forms the unpaired stylet of the piercing sucking apparatus (Michelsen 1997). The antenna is short and club-shaped and inserted in a groove (scrobe) at rest; the antennal grooves are connected with each other by a ridge or an interantennal groove; the scapus is club-shaped and curved; the pedicellus is short and bears several very long setae; the flagellum is unusually compact, with short flagellomeres with groups of shorter setae. The mandibles are reduced. The structure formerly referred to as “lacinia lever” was identified as the stipes by Michelsen (1997) based on the musculature; the laciniae and labrum (Michelsen 1997) are transformed into stylets (traditionally the unpaired stylet is interpreted as epipharynx); together they enclose the food canal; narrow closed salivary channels are present on the mesal side of the laciniae (Fig. 6.34.3); their distal part is serrated; the galea is absent; a sheath is formed by the large, leaf-shaped first maxillary palpomere (Michelsen 1997), the distal maxillary palpomeres (maxillary palp usually 5-segmented), and 5-segmented, mesally concave labial palps (Fig. 6.34.2). The structure traditionally interpreted as postmentum is the prementum according to Michelsen (1997). The structure formerly interpreted as a short hypopharynx is referred to as lingua by Michelsen (1997); its apex releases the saliva. A well-developed cibarial pumping apparatus is present and also a postcerebral pumping chamber with strongly developed dilators (similar to that of Nanno-

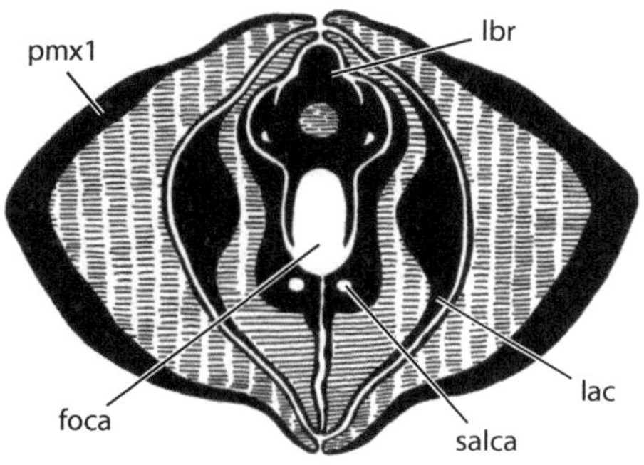

Fig. 6.34.3: Mouthparts, cross section.
Abbr.: foca – food canal, lac – lacinia, lbr – labrum, pmx1 – maxillary palpomere 1, salca – salivary canal.
Redrawn from Weber & Weidner (1974).

choristidae [Mecoptera] and dipterans). The brain and suboesophageal ganglion form a compact structure around the anterior foregut (autapomorphy of Antliophora).

The laterally compressed thorax is mainly characterized by the complete absence of the wings and correlated simplifications (Figs 6.34.1, 6.34.2). The prothorax is often partly overlapped by the posterior margin of the head (see above). The propleurosternum supports the head from below and is immovably connected with it. Posteriorly the prothorax is movably connected with the compact wingless pterothorax. The terga of all three segments are simple plates and about equally sized; they overlap with each other; at their posterior margin they often bear ctenidia. The size of the pleura, coxae and femora increases from the prothorax to the metathorax; the legs bear a fairly dense vestiture of setae; pads with resilin are present at the leg bases; the metacoxae and metafemora are distinctly enlarged and equipped with strongly developed muscles; the tarsi are always 5-segmented; the claws are strongly curved; small smooth pulvilli are present; an arolium is not developed.

The abdomen is broadly connected with the metathorax. Eight segments are distinctly recognizable. Like in the thorax the plate-like tergites overlap with the tergites of the following segment. The terminal segments are more or less strongly retracted, modified or reduced. Tergite X is transformed into a pygidial plate, which is densely covered with microtrichia; posterior to it an anal cone is present (Fig. 6.34.2); it bears the anus and in females anal stylets, which possibly represent vestigial cerci. Sternite VII of females overlaps with the corresponding tergite. The female genital opening lies within a genital chamber which is enclosed by a subgenital plate formed by sternite VIII and a ventrally extended tergite VIII; posterior to it lies the ostium bursae, the opening of the bursa copulatrix, which receives two ductus receptaculi; one of them is often distally closed or absent, whereas the other bears the strongly sclerotized species specific receptaculum seminis; the vagina is continuous with the paired oviducts. The male copulatory apparatus (Fig. 6.34.2) is unusually large and complicated. It is retracted in its resting position.

The postcephalic digestive tract is characterized by a narrow oesophagus, a proventriculus with spiny acanthae, and a straight, inflated midgut. Four free Malpighian tubules are present and six rectal papillae. The heart lies in the posterior body region. Only two ostia and wing muscles are present. The ovaries comprise 4–8 polytrophic ovarioles. The testes are pear-shaped and unifollicular.

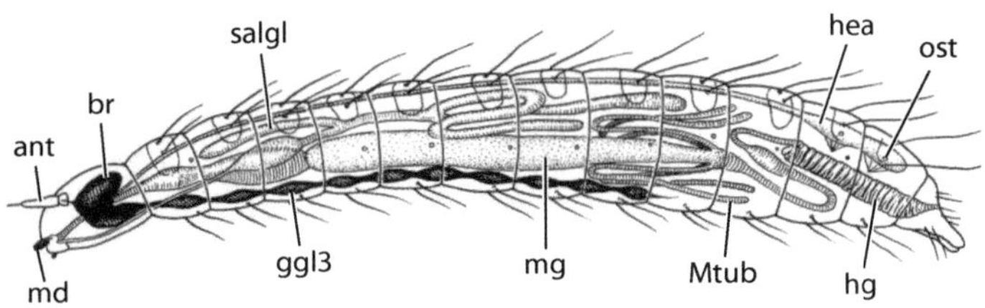

Fig. 6.34.4: Larva, lateral view. Abbr.: ant – antenna, br – brain, ggl3 – metathoracic ganglion, hea – heart, hg – hindgut, md – mandible, mg – midgut, Mtub – Malpighian tubules, ost – ostium, salgl – salivary gland. Redrawn from Weber & Weidner (1974).

Morphology, larva. The mature larvae (Figs 6.34.4) are usually 4–5 mm long. They are slender, more or less cylindrical, and appear curved in lateral view. Legs are completely absent. The color is whitish. The cuticle is smooth but bears a vestiture of long posteriorly directed setae.

The head is sclerotized and prognathous (Fig. 6.34.5). Eyes are absent. The labrum is free and bears three characteristic fields of papillae which probably serve as attachment devices. The antennae are relatively long. The mouthparts are well-developed. The labial glands of the larvae are very large and reach the middle region of the abdomen posteriorly.

The thorax is completely legless (Fig. 6.34.4). The meso- and metathorax are slightly larger than the prothorax and bear a pair of slightly convex setiferous pleurites on both sides. A pair of spiracles is present on the lateral side of the prothorax.

The abdomen is 10-segmented and slightly widening posteriorly (Fig. 6.34.4). Setiferous pleurites are present on segments I–IX. A pair of anal struts is present on the terminal segment X and a dense comb of fine setae above them. Seven pairs of spiracles are present on segments I–VII.

Morphology, pupae. The pupa adectica exarata libera has a whitish coloration first but is getting darker during the pupal stage. Lateral processes on the mesothorax (present in Ceratophyllidae, absent in Pulicidae) were interpreted as wing buds, but form basal parts of the mesepimeron after the metamorphosis (Poenicke 1969).

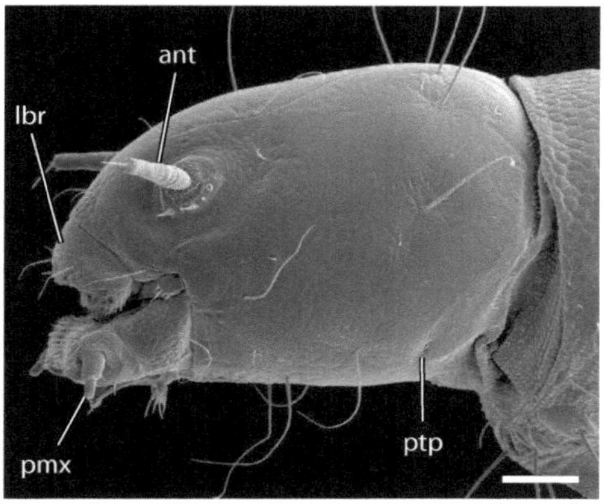

Fig. 6.34.5: *Ceratophyllus* sp. (Ceratophyllidae), larval head, lateral view, SEM micrograph. Abbr.: ant – antenna, lbr – labrum, pmx – palpus maxillaris, ptp – posterior tentorial pit. Scale bar: 50 µm. Courtesy H. Pohl.

Biology. Fleas are highly modified in correlation with their ectoparasitic habits. However, they depend less strongly on their hosts than anoplurans and are often quite flexible and able to feed on different species. The host spectrum of fleas is narrower. In contrast to lice they do not occur on Dermoptera and elephants, and primarily they do not infest primates (humans are only surrogate hosts). However, one group of fleas occurs on bats (Ischnopsyllidae) which do not belong to the anopluran host spectrum. The most important group of siphonapteran hosts are the rodents (ca. 75% of all flea species). Only about 1% of all species occur on monotremes (*Bradiopsylla*, Lycopsyllidae), Edendata, Pholidota, Hyracoidea, and Artiodactyla, 8% on insectivores, 5% on marsupials, 5% on bats, 3% on carnivores, and 3% on lagomorphs. Parasitism of birds has evolved secondarily and is the exception (ca. 5% of all species). Some bird parasites can be very flexible (like some ectoparasites of mammals). *Ceratophyllus gallinae*, the chicken flea, was found on more than hundred bird species.

Numerous adaptations are related with the highly specialized ectoparasitic life style, among them the piercing sucking mouthparts (labrum and lacinia), the reduced light sense organs, the loss of wings, the ctenidia, and the remarkable jumping capacity (up to 200 times the body length). The musculature of the hindlegs is strongly developed and pads of highly elastic resilin at the leg base contribute to the leap.

The hosts are detected using body temperature, olfactory signals or mechanically. The skin is perforated by alternating movements of the apically serrated laciniae. Saliva is released in the wound. Blood is sucked back using the strongly developed cibarial and postcerebral pumping chambers. The fine acanthae of the proventriculus likely prevent the backflow of blood, which can be stored in the voluminous midgut.

Fleas are mainly controlled by their hosts. Birds use baths in dust and anthills to control the populations of the parasites. Parasites of fleas include viruses, tape-worms (cysts), and nematodes. Larvae are also attacked by mites and beetles in the host's nest.

The larvae usually live in the host's den or nest. In human habitations they prefer cracks in the floor, carpets or similar dark and protected places. An exception to the rule is *Uropsylla tasmanica* which parasitizes on Quolls (dasyurid marsupials). The females attach the eggs to hair and the larvae burrow into the host's skin and stay there permanently. Only the pupation takes place on the ground in this species (Grimaldi & Engel 2005).

Reproduction and development. Female fleas lay eggs on the host few days after the first blood meal, which is required for successful reproduction. The eggs are deposited in batches of 6–10. The total number is ca. 400–500 in the human flea and up to 1,000 in the cat flea. The eggs usually roll off the host and fall in the host's den or nest, which is the primary habitat of the larvae. The embryonic development usually takes between two days and two weeks. The hatching eyeless primary larvae pierce the chorion with a cephalic egg burster. They avoid exposure to light and start feeding on any organic material including granules of dried blood, dead insects, feces or detritus accumulating in the nest. Some larvae are apparently carnivorous or even cannibalistic. There are three larval instars. After 2–3 weeks, shortly after the third molt, the pupation takes place. The 3rd instar larvae releases feces, bends its body and produces a cocoon using secretions of the large labial glands. The cocoon is camouflaged with minute dirt particles (Kinzelbach 2005).

The reproductive cycle of the rabbit flea (*Spilopsyllus cuniculi*) is regulated by the corticosteroid level of the host. Its increase prior to delivery induces increased feeding activities of the female fleas, storage of yolk substances in the oocytes, and an increased production of feces. The fleas leave the adult host shortly before, during or after birth. The copulation of the fleas is triggered by kairomones in the urine of the newborn rabbits. Their return to the adult host is induced by a decreasing level of the corticosteroids.

Fossil record. The fossil record of siphonapterans is limited and does certainly not reflect the true age of the group. *Tarwinia australis* from the Lower Cretaceous sediments of Australia (Koonwarra, Victoria) was disputed but probably belongs to the stem group of Siphonaptera (Grimaldi & Engel 2005). Other fossil fleas are known from the Tertiary. The very few fleas embedded in Baltic amber are very similar to extant species of the genera they were assigned to. A "giant flea" from the Cretaceous was described recently (Huang et al. 2012) but its systematic placement needs further confirmation.

Economic importance. Fleas are primarily a nuisance of humans and livestock. They are still rather frequent parasites in human habitations, but *Pulex irritans* is clearly on the decline in regions with improved hygiene conditions. Many species are potential vectors of viruses (e.g., myxomatosis), bacteria (e.g., Murine typhus) and protozoans (*Trypanosoma*). In contrast to some statements in the literature fleas do not transmit the Hyper Tape-worm (*Echinococcus granulosus*).

Historically the rat flea (*Xenopsylla* spp.) had a tremendous impact as the vector of the bacterium *Yersinia pestis*. Entire regions in Asia and Europe were depopulated by the plague ("Black Death"), with the death toll reaching ca. 100 millions in Europe in the 14th century.

[Snodgrass (1946); Seguy (1951); Hopkins & Rothschild (1953–1971); Wenk (1953); Poenicke (1969); Weber & Weidner (1974); Medvedev 1994); Michelsen (1997); Lewis (1998); Whiting et al. (2008); Grimaldi & Engel (2005); Kinzelbach (2005); Huang et al. (2012)]

6.35 Diptera (Greek *di* = two, *pteron* = wing, refers to the presence of only one pair of wings, English common name: true flies) (Katharina Schneeberg & Rolf G. Beutel)

Diversity and distribution. Diptera are one of the megadiverse orders with presently ca. 154,000 described species. The actual number may be twice as high. They occur on all continents including coastal regions of Antarctica, in all zoogeographic regions, and in a very broad spectrum of habitats. The number of individuals is often extremely high.

Autapomorphies

- Labrum, hypopharynx, laciniae and mandibles modified as piercing/slicing stylets (groundplan) (Figs 6.35.1, 6.35.2)
- Haustellum (proboscis) formed by the prementum (Fig. 6.35.4)
- Labrum forms the dorsal cover of the piercing apparatus (Figs 6.35.1, 6.35.2)
- Labial palps transformed into a labellum (Figs 6.35.4, 6.35.5)
- Hindwings modified as halteres (Fig. 6.35.7)
- Large mesothorax; metathorax and prothorax reduced in size
- Males with a maximum number of seven abdominal spiracles
- Reduced number of chromosomes (maximum n = 10, n = 3–6 in most groups)
- Larvae (Figs 6.35.9, 6.35.10) without true thoracic legs (many called maggots) (convergent in Siphonaptera?)

Taxonomy. The group is traditionally subdivided into the two suborders "Nematocera" and Brachycera. The former (now often referred to as "lower Diptera") is a paraphyletic grade containing groups mostly characterized by plesiomorphies, especially in the larval stages, whereas Brachycera are definitely a clade. Most nematoceran adults are more or less slender, whereas brachycerans are usually charac-

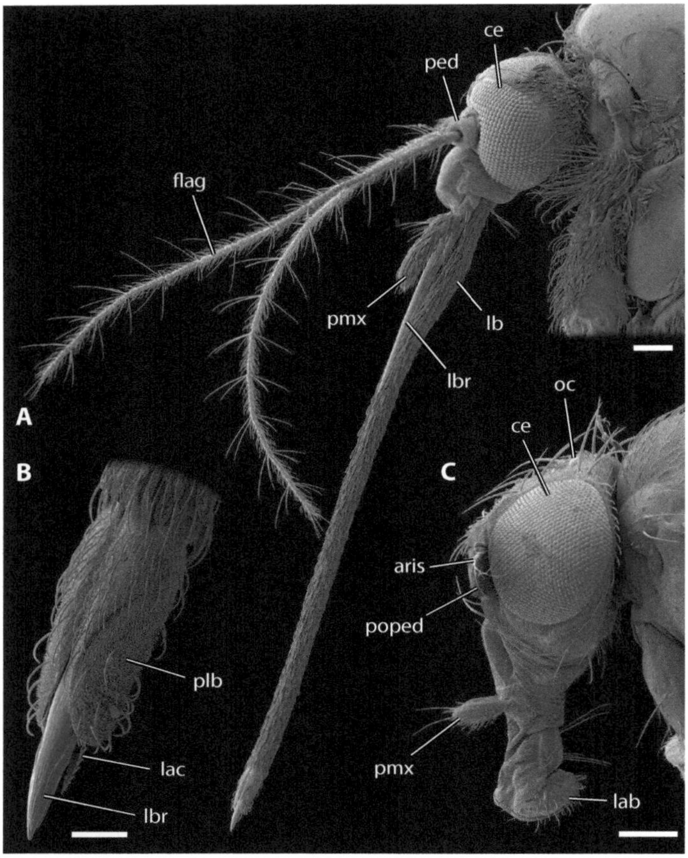

Fig. 6.35.1: Diptera, adult heads, lateral view, SEM micrographs. A, *Aedes punctor* ("Nematocera", Culicidae), female, B, distal part of stinging apparatus (magnified from A), C, *Drosophila melanogaster* (Brachycera, Drosophilidae). Abbr.: aris – arista, ce – compound eye, flag – flagellum, lab – labellum, lac – lacinia, lb – labium, lbr – labrum, oc – ocellus, ped – pedicellus, plb – palpus labialis, pmx – palpus maxillaris, poped – postpedicellus. Scale bars: A, C: 200 μm, B: 50 μm.

terized by a more robust body. Nematoceran lineages are Tipulomorpha (crane flies, e.g., Pediciidae, Tipulidae, Limoniidae), Psychodomorpha (sand flies, e.g., Tanyderidae, Psychodidae), Culicomorpha (e.g., mosquitoes, blackflies, e.g., Culicidae, Simuliidae, Ceratopogonidae, Chironomidae), and Bibionomorpha (e.g., Bibionidae, Axymyiidae, Cecidomyiidae, Sciaridae, Mycetophilidae, Anisopodidae) (Wiegmann et al. 2011). The composition and relationships of the infraorders/suborders is still under debate. With ca. 9,000 described species Limoniidae is the nematoceran family with the highest diversity. Recent molecular studies suggest that the highly special-

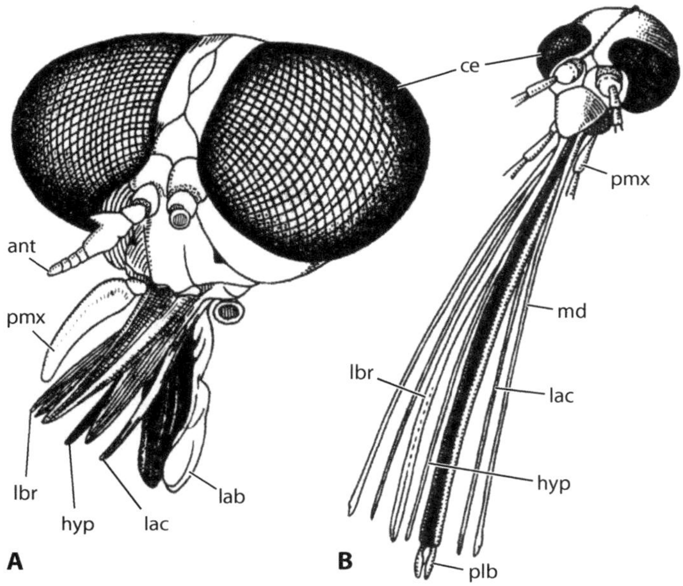

Fig. 6.35.2: Heads with piercing sucking mouthparts. A, Tabanidae; B, Culicidae. Abbr.: ant – antenna, ce – compound eye, hyp – hypopharynx, lab – labellae, lac – lacinia, lbr – labrum, md – mandibular stylet, plb – palpus labialis, pmx – palpus maxillaris. Redrawn from Ziegler (2005), after Mc Alpine (1981).

ized Deuterophlebiidae and Nymphomyiidae may represent the most basal lineages of Diptera, and that Bibionomorpha are the sistergroup of Brachycera (Wiegmann et al. 2011). Brachycera are characterized as a clade by antennae with a maximum number of ten segments, a maxillary palp with two segments or less, and larvae with vertically oriented hook-shaped mandibles and a partly or almost completely reduced head capsule. Within Brachycera, the "Orthorrhapha" are a paraphyletic grade. They include Tabanomorpha (horse flies), Stratiomyomorpha (soldier flies), Asiloidea (robber flies and relatives), and Empidoidea (dance flies). The latter probably form a clade Eremoneura with the monophyletic Cyclorrhapha (Wiegmann et al. 2011), which are characterized by the almost completely reduced larval head capsule, a larval cephaloskeleton (=cephalopharyngeal skeleton), and pupation within a puparium formed of the final larval integument. Important and successful cyclorrhaphan families are the Syrphidae (ca. 6,000 spp.), Tephritidae (ca. 4,500 spp.), Drosophilidae (ca. 3,500 spp.), Muscidae (ca. 4,000 spp.) and Tachinidae (ca. 8,000 spp.). The Schizophora ("Acalypratae" + Calyptratae) are a monophyletic cyclorrhaphan subgroup, which comprises ca. 80 families. It is characterized by the ptilinum, a membranous head sac which can be extended with haemolymph pressure and helps the teneral adult to emerge from the puparium.

Diagnosis. Very small to medium sized insects with strongly modified mouthparts and hindwings transformed into halteres. The larvae lack true thoracic legs and many are referred to as maggots, especially in the Cyclorrhapha (Figs 6.35.9, 6.35.10).

Morphology, adults. Dipterans range in size between 0.5 mm (e.g., Cecidomyiidae, Ceratopogonidae) and 60 mm (*Mydas heros*, Mydidae). The cuticle is relatively thin, sometimes fragile, and often bears a vestiture of long setae (macrotrichia), which are usually sensorial hairs or associated with glands. Yellow, brown and black are frequent colors but more conspicuous pigmentations occur and also metallic colorations (e.g., Calliphoridae). Sometimes the surface is covered with extremely small microtrichia or waxy secretions.

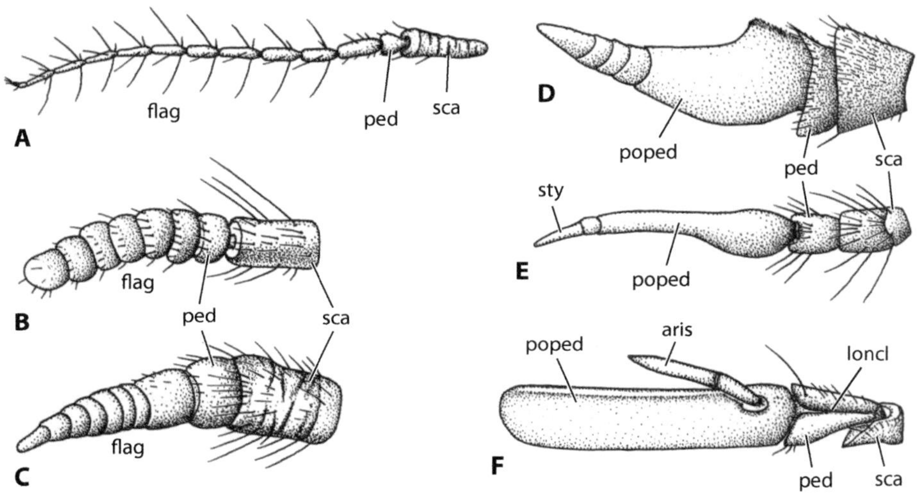

Fig. 6.35.3: Antennae. A, *Tipula maxima* (Tipulidae); B, *Bibio* sp. (Bibionidae); C, *Coenomyia ferruginea* (Brachycera, Xylophagidae); D, *Hybomitra montana* (Brachycera, Tabanidae); E, *Rhamphomyia anthracina* (Brachycera, Empididae); F, *Onychogonia suggesta* (Brachycera, Tachinidae). Abbr.: aris – arista, flag – flagellum, loncl – longitudinal cleft, ped – pedicellus, poped – postpedicellus, sca – scapus, sty – stylus. Redrawn from Ziegler (2005).

The head (Figs 6.35.1, 6.35.2) is almost always orthognathous but exceptions occur. A prognathous condition occurs in Tipulomorpha, Culicidae, Bibionidae and some other groups. The posterior cephalic region is fully exposed. The foramen occipitale is primarily strengthened by a postoccipital ridge and narrowed by a hypostomal bridge; the narrow cervical region results in an increased movability of the head. The head capsule is often globular or compressed in antero-posterior direction, with a concave posterior side. A rostrum is characteristic for many Tipulidae (crane flies) and their relatives. In schizophoran flies the frontal region is subdivided into a median

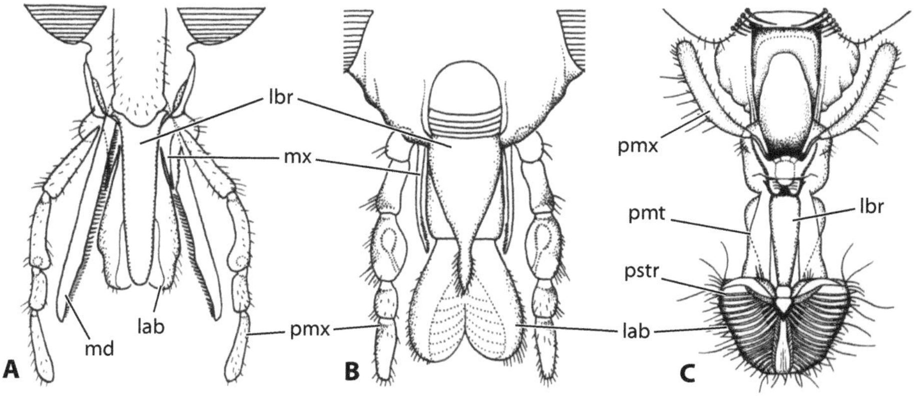

Fig. 6.35.4: Ventral part of heads with mouthparts. A, Blephariceridae; B, Anisopodidae; C, Calliphoridae (Brachycera). Abbr.: lab – labellae, lbr – labrum, md – mandible, mx – maxilla, pmt – prementum, pmx – palpus maxillaries, pstr – pseudotracheae. Redrawn from Ziegler (2005).

"vitta" and the two lateral orbital and frontal plates (parafrontalia); the extrusible frontal ptilinum, which facilitates hatching from the puparium, is characteristic for this lineage. The frontal region below the antennal insertions is referred to as the epistoma or praefrons ("face"). The frontoclypeal strengthening ridge is present in the groundplan. The compound eyes are well-developed and strongly enlarged in some groups, or even connected with each other on the dorsal (some Sciaroidea) or ventral side (some Nymphomyiidae); in some groups the ommatidia are distinctly larger on the dorsal side, and regions with different cuticular lenses sometimes form seemingly separate eyes; in most Diopsidae, and some members of other groups, the compound eyes are stretched or located at the ends of stalks; three ocelli are usually present on the dorsal side of the head but reductions occur; some for instance Mydidae have only one median ocellus, while Tipulidae, among others, lack ocelli altogether. The articulated labrum is often more or less strongly elongated and dorsally covers the sucking apparatus formed by the modified mouthparts (Figs 6.35.1, 6.35.2, 6.35.4, 6.35.5); in basal groups with a sucking-piercing apparatus (e.g., Culicidae) it forms one of the stylets (Fig. 6.35.1B), in addition to the stylet-like mandibles, laciniae, and hypopharynx; in some groups the labrum is entirely membranous, but it is entirely sclerotized dorsally and ventrally in most cyclorrhaphans. The antennae (Fig. 6.35.3) are inserted anterad the compound eyes and often more or less closely adjacent medially; they are often sexually dimorphic; the scapus and pedicellus are distinctly developed but sometimes strongly modified in shape; a dorsolateral longitudinal cleft of the pedicellus is a presumptive autapomorphy of Calyptratae; the Johnston's organ can provide information on the flight speed and the orientation of the body and also serve as acoustical receptor; the flagellum is usually filiform (e.g., Tipulidae, Blepharicer-

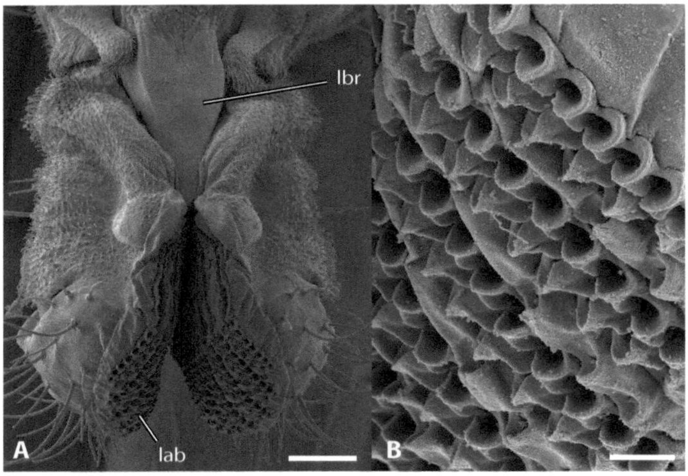

Fig. 6.35.5: Diptera, mouthparts of *Drosophila melanogaster* (Brachycera, Drosophilidae), SEM micrographs. A, anterior view, B, pseudotracheae on labellar surface. Abbr.: lab – labellum, lbr – labrum.
Scale bars: A, 50 μm, B: 10 μm.

idae), plumose (e.g., Culicidae, Chironomidae) or moniliform (e.g., Bibionidae) in the basal lineages, usually with a vestiture of long and short setae; only eight segments of the flagellum are present in the groundplan of Brachycera; the basal segment fuses with several primary flagellomeres to form an enlarged postpedicellus with chemoreceptive functions; the two terminal segments form a narrowed mechanoreceptive stylus in orthorrhaphan flies; the intermediate segments tend to fuse with the postpedicellus; only scapus, pedicellus and postpedicellus are present in Cyclorrhapha, the latter usually distinctly enlarged, but sometimes also retracted into the pedicellus (e.g., Hippoboscoidea); the terminal arista of Cyclorrhapha, usually a bristle-like 3-segmented appendage (Fig. 6.35.3), is a mechanoreceptor. The mandibles are primarily present as stylets (see above), but are very often reduced (Fig. 6.35.4B, C), more often in males than in females. In the groundplan the maxilla is composed of cardo, stipes, 5-segmented palp and the lacinia, which is primarily a stylet but reduced in most groups; the number of palpomeres is also frequently reduced; a deep sensorial groove is present on palpomere 3 in several nematoceran groups (also in the mecopteran family Nannochoristidae). The labium is composed of postmentum and prementum; the postmentum is often reduced to different degrees, correlated with a widened hypostomal bridge; the prementum forms the haustellum (proboscis) (Figs 6.35.4, 6.35.5); ventrally, it is strengthened by the theca, a trough or boat-shaped sclerotized plate; the primarily 2-segmented palps are modified as labellae; on their surface they usually bear pseudotracheae for an optimized uptake of liquid food (capillary forces) (Fig. 6.35.5), but these specialized structures are missing in some basal

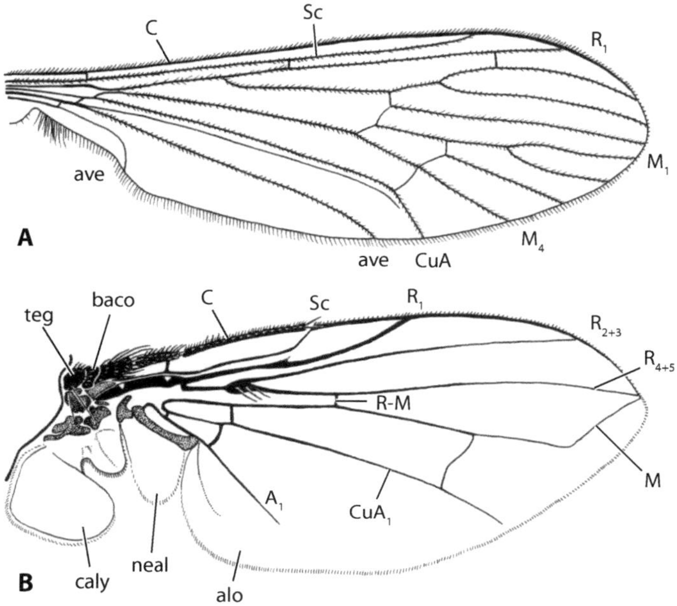

Fig. 6.35.6: Forewings. A, *Nothotrichocera* (Trichoceridae); B, *Eurysthaea* (Brachycera, Tachinidae). Abbr.: A_1 – first anal vein, alo – anal lobe, ave – anal veins, baco – basicosta, C – costa, caly – calyptra, CuA_1 – cubitus anterior, $M_{1,4}$ – medial veins, neal – neala, R_{1-5} – radial veins, Sc – subcosta, teg – tegula. Redrawn from Ziegler (2005).

lineages (e.g., Axymyiidae); in Cyclorrhapha the entire prementum is separated from the remaining parts of the head by a broad membranous zone. The tentorium is primarily composed of posterior, dorsal and anterior arms and a tentorial bridge, but the dorsal arm and the bridge are vestigial or absent in most groups; the anterior arm is probably wide and hollow in the groundplan. Three pairs of cervical sclerites are present in the groundplan, but only the laterocervicalia are well-developed and articulate with a process of the foramen occipitale. The hypopharynx is primarily a stylet-like element of the piercing-sucking apparatus; its suspensoria are part of the plate-like floor of the cibarium. The cephalic part of the digestive tract is characterized by a precerebral cibarial pump and a postcerebral pharyngeal pumping chamber. The head musculature is distinctly reduced compared to other groups of Holometabola (e.g., Neuroptera); an extreme degree of reduction (eight paired muscles) occurs in Deuterophlebiidae. The brain and suboesophageal ganglion form a compact complex around the pharynx (autapomorphy of Antliophora, similar to the condition in Hemiptera). The tube-like and sometimes branched salivary (labial-) glands open on the proboscis but there main part lies within the thorax or even the abdomen.

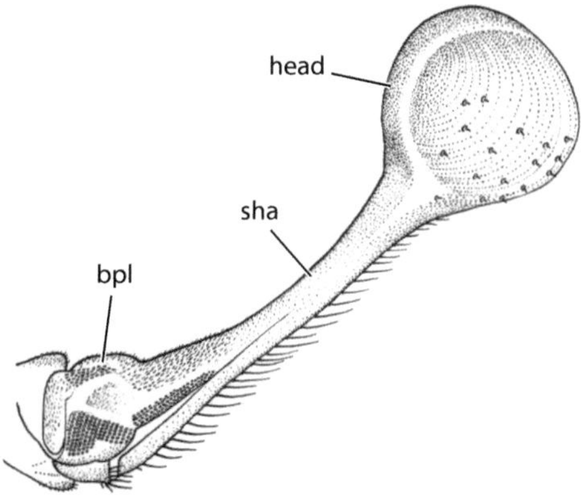

Fig. 6.35.7: Haltere, *Bibio* sp. (Bibionidae). Abbr.: bpl – basal plate, sha – shaft. Redrawn from Ziegler (2005).

The thorax is mainly characterized by the partial reduction of the pro- and metathorax and the large mesothorax. The small prothorax is immovably attached to the mesothorax. The pronotum is firmly connected with the propleura. A precoxal bridge occurs in some groups. The mesonotum is subdivided into a prescutum, scutum and scutellum in the usual manner; the mesoscutum is subdivided by a transverse scutal suture (V-shaped in Tipulomorpha); the mesoscutellum is reduced in many flightless dipterans; a convex bulge, the postscutellum, occurs in some groups; the anterior phragma is small but the posterior phragma (mesopostnotum) is strongly developed. The pleural parts of the thoracic segments vary strongly in different taxa; the mesopleural elements comprise the anepisternum, katepisternum, anepimeron and katepimeron; in Brachycera, especially in Cyclorrhapha, the pleural suture is S-shaped, but it is straight in the groundplan; an important trend is the increasing degree of fusion of the mesocoxal meron with the katepimeron. The metathorax is strongly reduced. Episternum and epimeron are present as separate elements in the groundplan; in Cyclorrhapha, both pleural elements form a structural unit with the mesothoracic katepimeron and mesocoxal meron, the hypopleura. The metanotum is a narrow strip connecting the insertion areas of the halteres between the mesopostnotum and the abdominal tergite I. Two pairs of thoracic spiracles are present, one on the mesothorax (on or near the prothoracic-mesothoracic border) and one on the metathorax. The forewings (Fig. 6.35.6) vary considerably in different dipteran lineages; they are membranous and their base is distinctly narrowed; specific structures occurring at the wing base are the alula, neala and calyptra (squama); the 3rd anal vein is generally missing; there is a distinct trend to strengthen the longitudinal veins ("costalisation") and reduce the rigidity of the posterior part; a reticulate pattern is a rare exception (as in some Nemestrinidae); the hindwings are transformed into halteres (Fig. 6.35.7) composed of basal plate, shaft and head; they serve as gyroscopic

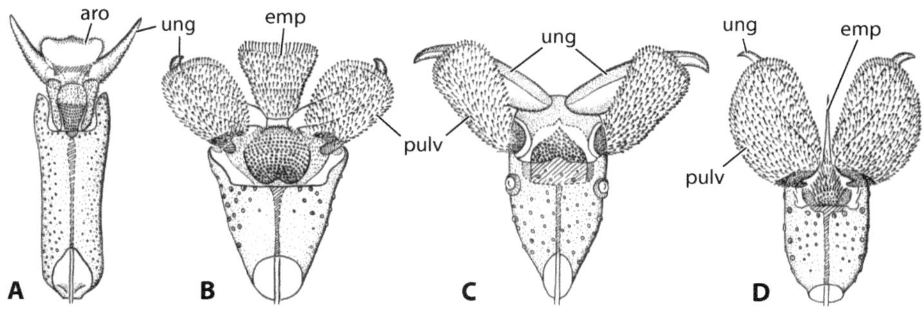

Fig. 6.35.8: Apical part of leg with pretarsal attachment structures. A, *Tipula* (Tipulidae); B, *Stratiomys* (Brachycera, Stratiomyidae); C, *Leptomydas* (Brachycera, Mydidae), female; D, *Musca* (Brachycera, Muscidae). Abbr.: aro – arolium, emp – empodium, pulv – pulvilli, ung – ungues (claws). Redrawn from Ziegler (2005), A, B, D after Mc Alpine (1981).

sensorial organs and their high speed rotation in opposite direction to the wings stabilize the flight. The flight musculature in the mesothorax is strongly developed. The more or less slender legs (very thin in most nematoceran lineages) are composed of the usual elements; specialized cleaning spurs, sometimes combined with notches, occur frequently; the tarsi are 5-segmented in almost all species; an arolium is arguably present in the dipteran groundplan (Fig. 6.35.8A) but replaced by an empodium and paired hairy pulvilli in all groups except for Tipulomorpha (Fig. 6.35.8A); an expanded empodium densely covered with adhesive microtrichia is present in many groups.

The abdomen is primarily composed of ten segments (segment XI is fused to X) and possibly a telson. It can be divided into a preabdomen and a postabdomen. The number of preabdominal (pregenital) segments can vary between two and nine. Dipterans are characterized by a trend to shorten the entire abdomen. Segment V (or even IV in some males) is the posteriormost exposed segment in some derived forms. The relatively weakly sclerotized tergites and sternites are usually connected by pleural membranes, which also contain the abdominal spiracles. Few female dipterans have eight abdominal spiracles, whereas the maximum number is seven in males. Broadening of tergites and sometimes also of sternites can lead to a strongly armored preabdomen (Calyptratae). Tergites I and II are fused in Cyclorrhapha. The posterior segments (postabdomen) are more or less strongly inclined and bent anteriorly; they are often retracted or reduced. The male postabdomen is characterized by the extremely variable and often very complicated copulatory apparatus; the homologisation of its elements is often problematic; in the groundplan modifications are restricted to segments IX and X, which together form the hypopygium; a temporary or permanent rotation of this unit through 180° (hypopygium inversum) is a characteristic feature of some nematoceran and orthorrhaphan groups. Tergite IX (epandrium) often bears articulated processes (surstyli); sternite IX (hypandrium) bears the coxites

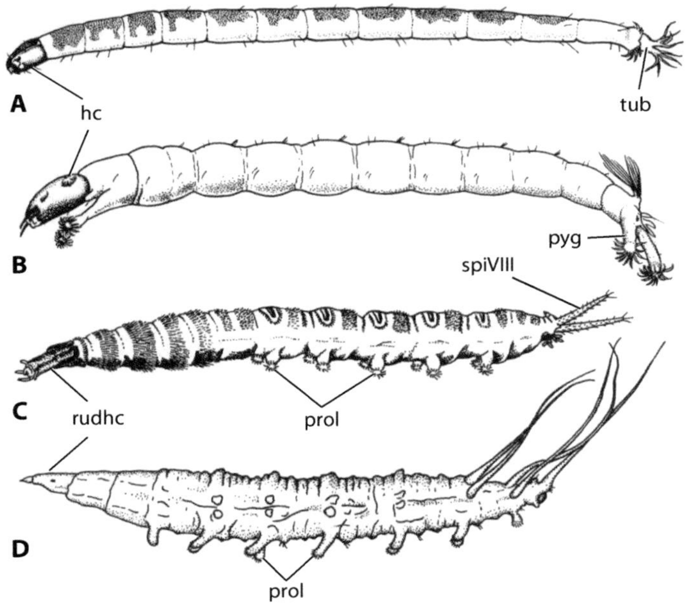

Fig. 6.35.9: Larvae, final instars, lateral view. A–C, nematoceran groups, D, Brachycera. A, *Dasyhelea* sp. (Ceratopogonidae); B, *Ablabesmyia* sp. (Chironomidae); C, *Dicranota* sp. (Pediciidae); D, *Atrichops* sp. (Brachycera, Atheceridae). Abbr.: hc – head capsule, prol – abdominal prolegs, pyg – pygopod, rudhc – rudimentary head capsule, spiVIII – spiracle VIII, tub – tubuli. Redrawn from Ziegler (2005), after Mc Alpine (1981).

and the aedeagus, the median copulatory organ; the divided coxites bear apical styli and sometimes other lobes or appendages can be present; in "higher brachycerans" mesal valves ("cerci") are present in addition to the lateral valves (surstyli); they can be fused to an unpaired structure; the aedeagus is often divided into a basiphallus and distiphallus; at its base it bears two pairs of lobe-, plate- or hook-shaped appendages, the pregonites and postgonites. In Cyclorrhapha segment sternite V is functionally a part of the postabdomen; during copulation it forms functional forceps together with the cerci and surstyli. The female genital opening lies between sternites VIII and IX; a secondary ovipositor formed by telescoping terminal segments can be present; it is equipped with sensilla and cerci (valves) are present in the groundplan; specialized secondary ovipositors with strongly sclerotized elements occur in parasitic groups (e.g., Tachinidae).

The largest part of the digestive tract lies within the abdomen; the voluminous ingluvies, a storage organ, is sometimes subdivided into three parts; a specialized proventriculus is not developed; the mid-gut is simple and tube-like in nematocerans but modified to different degrees in Brachycera; the balloon-shaped rectum is equipped with 2–6 papillae; extensions containing endosymbionts occur in some groups. Four Malpighian tubules are present in most cases, but the number varies between two and five. The testes are coiled, tube-like structures; a sperm pump is usually present, but is missing in Culicomorpha and some other groups. The ovaries comprise both polytrophic and meroistic ovarioles; the number varies between one and several hundred; a specialized part of the genital chamber ("uterus") and an associated gland ("milk gland") occurs in Glossinidae and Hippoboscidae in the context of vivipary.

Morphology, larvae. The aquatic, semiaquatic or terrestrial larvae are usually elongated and the thorax is always legless (Figs 6.35.9, 6.35.10).

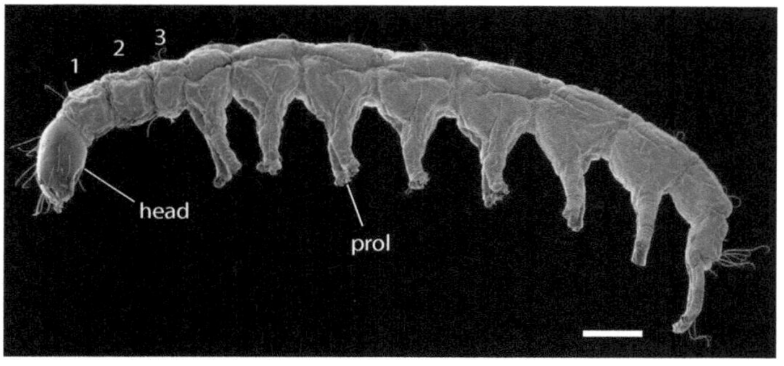

Fig. 6.35.10: *Nymphomyia dolichopeza* (Nymphomyiidae), larva, lateral view, SEM micrograph. Abbr.: 1–3 – 1st–3rd thoracic segment, prol – abdominal proleg. Scale bar: 100 µm. From Schneeberg et al. (2012), modified.

The head is almost always prognathous (Fig. 6.35.9A, B). The head capsule is well-developed in the groundplan (e.g., Bibionomorpha, Culicomorpha, Ptychopteromorpha) but more or less strongly reduced in Tipuloidea, Cecidomyiidae, Canthyloscelidae and especially in the brachyceran lineages (Fig. 6.35.9D); it is strongly retracted in Tipuloidea and some other groups; in Cyclorrhapha the external skeleton of the head is almost completely reduced, but a cephaloskeleton is developed (Fig. 6.35.11). In the groundplan the ecdysial frontal and coronal sutures are present (eucephalic head: e.g., Tanyderidae, Culicidae), as well as the transverse frontoclypeal strengthening ridge. The frons is primarily V-shaped like in most other insect lineages, but U-shaped in Culicidae and related groups. The labrum is primarily well-developed and articulated with the anterior clypeal margin (e.g., Tanyderidae); a characteristic feature of some nematoceran lineages (e.g., Tanyderidae, most Culicomorpha) are dense labral brushes; premandibles or messores, more or less movable labral structures involved in food acquisition, are present in several basal dipteran groups. The antennae are often 1-segmented and strongly shortened and simplified, but antennae with more than one antennomere occur in several lineages (e.g., Psychodidae, Tipulidae, Nymphomyiidae). The mandibles are articulated in the typical dicondylic manner with a horizontal axis of movement in few nematoceran groups, but usually they move in an oblique plane or even vertically. The ventral mouthparts are strongly modified. The endite lobes of the maxilla are indistinctly separated in few groups but more or less completely fused in most cases. The palp is 1-segmented. The labium is reduced to varying degrees.

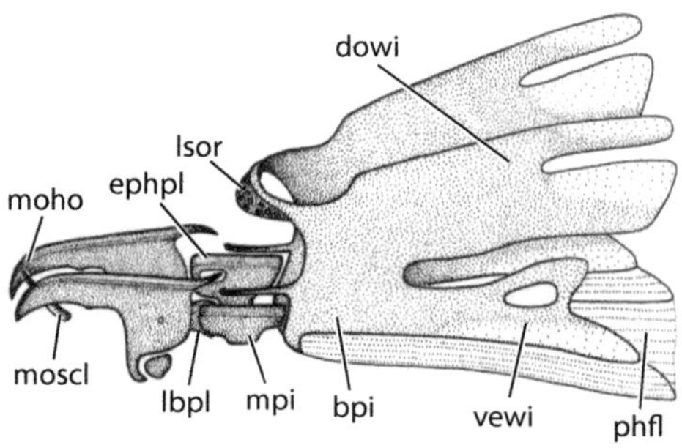

Fig. 6.35.11: Cephaloskeleton of a final instar larva of Oestridae (Brachycera). Abbr.: bpi – basal piece, dowi – dorsal wing with incision, ephpl – epipharyngeal plate, lbpl – labial plate, lsor – light sense organ, moho – mouthhooks, moscl – mouth sclerite, mpi – median piece, phfl – pharyngeal floor with filter apparatus, vewi – ventral wing with window. Redrawn from Ziegler (2005), after Ziegler (1998).

The thorax is always legless and usually more or less cylindrical (Fig. 6.35.9). Prothoracic prolegs occur in Simuliidae, Chironomidae and some Ceratopogonidae; different substitutes of legs, usually bearing hooks or bristles, occur sometimes on the abdomen.

The abdomen is 10-segmented. Abdominal prolegs (Fig. 6.35.10: prol), usually bearing hooks or bristles, occur in few presumably basal groups (Nymphomyiidae, Deuterophlebiidae, Blephariceridae, Pediciidae), but also in some higher dipteran lineages (e.g., Athericidae, some Ephydridae, Sciomyzidae and Muscidae).

A holopneustic tracheal system is present in few cases. The larvae of most groups are amphipneustic or metapneustic, but peri-, pro- or apneustic tracheal systems also occur; the posterior spiracles are often enlarged or otherwise modified, for instance forming snorkels in aquatic larvae; reductions of the tracheal system occur in aquatic (Chaoboridae) and endoparasitic groups (Oestridae). Haemolymph pigments resembling haemoglobin are an adaptation to low environmental oxygen levels in some species of Chironomidae and Oestridae.

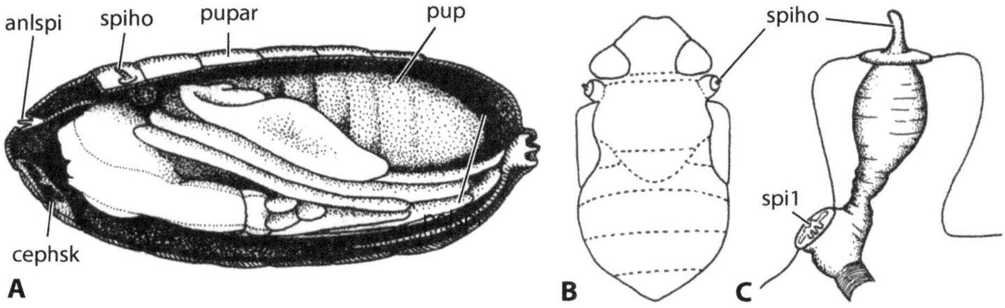

Fig. 6.35.12: Pupa and puparium of cyclorrhaphan flies. A, lateroventrally opened puparium; B, dorsal view of pupa; C, prothoracic spiracle with spiracular horn, lateral view (Brachycera, Tachinidae). Abbr.: anlspi – anterior larval spiracle, cephsk – cephaloskeleton, polspi – posterior larval spiracle, pup – pupa, pupar – puparium, spiho – spiracular horn, spi1 – prothoracic spiracle. Redrawn from Ziegler (2005), after Ziegler (1998).

Morphology, pupae. Different types of pupae occur in Diptera (e.g., pupa obtecta, pupa exarata), but the mandibles are always unsclerotized and immobilized (pupa adectica). The pupa coarctata with a puparium formed by the hardened last larval exuvia is a specific type occurring in Cyclorrhapha (Fig. 6.35.12) and probably an autapomorphy of this lineage. The name Cyclorrhapha refers to the round lid of the puparium, which is shed when the adults emerge. However, Stratiomyidae and Xylomyidae (non-cyclorrhaphan sister taxa) also pupate in the last larval skin.

Biology. The adults are usually free-living with the exception of some specialized ectoparasites (e.g., Streblidae). They occur in a very broad variety of habitats, such as deserts, tropical rain forests, high alpine zones, and even Antarctic coastal regions. The larvae usually develop in moist environments (e.g., mud) but also in many different types of aquatic habitats including fast flowing streams (e.g., Deuterophlebiidae, Blephariceridae and many others) and hot springs, and some specialists are even found in marine habitats (as deep as 30 m) or in petroleum puddles. Pupation usually takes place in a less moist environment compared to that of the larvae. Species of Diptera are normally characterized by excellent flying abilities, but reductions occur in different lineages, sometimes related to ectoparasitism (e.g., Hippoboscidae). Flies can occur in extreme numbers of individuals, and especially small species can drift as aerial plankton over very long distances. Some endoparasitic larvae of Oestridae can migrate in the body of their host and reach different organs, the spinal canal or even embryos.

In few cases the adults are short-lived, with almost completely reduced mouthparts not suitable for food uptake and an extremely reduced head muscle system (e.g., Nymphomyiidae, Deuterophlebiidae, Oestridae). Generally, the adult mouthparts are designed to consume liquid food. Aside from water this can include excretions or secretions of animals (e.g., honeydew) and plants (e.g., nectar), feces, liquids developing in decaying organisms, blood or haemolymph, and also more solid substances dissolved or diluted by saliva (e.g., sugar, pollen). Predaceous habits occur in lineages with well-developed sucking-piercing mouthparts (e.g., Asilidae), but also in other groups (e.g., Dolichopodidae, Scathophagidae). A few adult flies are associated with ants, termites (Phoridae), bees or various large predaceous arthropods (Milichiidae). Like the habitats, the food spectrum is very broad in larvae, including detritus, various body fluids (e.g., Oestridae), feces, cadavers (e.g., Calliphoridae.), or living plants (e.g., Tephritidae). Predaceous or ecto- or endoparasitic larvae occur in different lineages (e.g., Tachinidae, Oestridae).

The more or less exposed flies have a very broad spectrum of predators. This includes birds, bats, reptiles, amphibians (especially frogs), dragonflies, and spiders. Larvae are used by sphecid wasps as food substrate for their offspring. Predators of the larvae include fish, birds, odonatan larvae, adult and immature beetles, and mites. Larvae and adults are also attacked by predaceous dipterans such as for instance species of Empididae, Asilidae, and Dolichopodidae. Predaceous maggots of Muscidae can also play a certain role in decimating other dipteran larvae. Certain fish species, especially *Gambusia*, are used to control biting fly larvae (especially *Anopheles*) and microorganisms are also applied in pest management (e.g., *Bacillus thuringiensis*). Aquatic dipteran larvae are also affected by epibiontic algae, fungi and protozoans (e.g., *Vorticella*). Further parasites are viruses, bacteria, endoparasitic fungi, pathogenic flagellates (e.g., Microsporidia), and species of Nematoda and Nematomorpha. Parasitic Hymenoptera usually use larvae and pupae as hosts, but some species deposit eggs in adults. Hyperparasitic hymenopteran species (Trigonalidae, Perilampidae) are known from Tachinidae. Parasitic dipterans parasitizing other flies occur in Tachinidae, Pipunculidae and Bombyliidae.

Reproduction and development. Dipterans are usually bisexual. Parthenogenesis occurs as paedogenesis in Cecidomyiidae and Sciaridae, and as true parthenogenesis in Sciaridae, Chironomidae, Psychodidae and others. Strategies for finding partners include the formation of swarms, different forms of territorial behavior (e.g., Lek polygyny), hill topping, the use of pheromones, acoustical and optic signals (e.g., color patterns on wings), and specialized mating behavior (e.g., Sepsidae). Nuptial gifts are offered by males of Empididae to demonstrate fitness and to avoid being killed and consumed by the females. The copulation is often initiated during flight. Males often use specific clasping devices for immobilizing the females. The position of males and females during the copulation depends on the torsion of the male genital segments (hypopygium).

The uptake of protein-rich food ("blood meal") by the female is often necessary for the normal development of the eggs (e.g., Culicidae), which is often accomplished within a few days under suitable conditions. The fertilized eggs are deposited at substrates suitable for the development of larvae (e.g., water bodies, feces, cadavers, plant tissue). The larvae usually hatch from the egg after a short period, usually using preformed zones of weakness and sometimes also egg teeth. Ovovivipary and larvipary as a specific form of vivipary have evolved several times. Females of Hippoboscidae and Glossinidae nourish the developing larvae with gland secretions ("milk glands") and the entire development takes place in a specific organ of the females ("uterus"). The offspring is deposited as final larval stage ready for pupation (adenotrophic vivipary). The number of larval stages varies. Three instars is likely a groundplan feature of Cyclorrhapha, but up to eight can occur in other brachyceran lineages. The usual number of instars is four in lower Diptera.

Fossil record. The oldest fossil flies are known from Early Triassic deposits in France (ca. 230 Ma). The most diverse and best preserved early fauna is that of the Late Carnian (Late Triassic) of Virginia (USA) (Blagoderov et al. 2007). Many families appear in the Jurassic and Cretaceous. Parallel diversification of brachyceran groups and the angiosperms is documented in the fossil record. Numerous amber fossils are known from the Tertiary.

Economic importance. In their economic and, especially, medical impact Diptera clearly surpass all other groups of insects. Even the seemingly harmless housefly (*Musca domestica*) is a potential vector of typhoid fever, dysentery, cholera, anthrax and amoebiasis. Blood sucking species can transmit malaria (*Anopheles*), filariasis (Simulidae: e.g., river blindness, elephantiasis), leishmaniasis, and trypanosomiasis (e.g., sleeping sickness), and arboviruses (**ar**thropod-**bo**rne viruses) causing for instance yellow fever (*Aedes aegypti*), dengue fever (*Aedes* spp.), and different forms of encephalitis (e.g., *Culex* spp.). Diseases of livestock transmitted by dipterans are avian malaria, surra (horses), the bluetongue disease (mainly sheep and goats, buffalo, dromedaries etc.), and others.

Dipterans, especially species of Tephritidae and Anthomyiidae, can also cause enormous damage in cultivated plants. Larvae can mine in fruit, leaves and stems. A certain positive role is that of pollinating plants. Some species are highly specialized on specific species of orchids (Tephritidae) and ceratopogonids are pollinating the cocoa plant. An important function of larvae is the processing of feces and cadavers, and immature stages of Calliphoridae and some other groups (e.g., Phoridae) are used in forensic entomology to estimate the post mortem interval (PMI). Some dipterans are used to control weeds and others were successful in Australia as vectors of the *Myxoma* virus, thus helping to decimate populations of rabbits. Larvae of Tachinidae, Bombyliidae, Nemestrinidae and other groups parasitize potentially harmful insect species. Sterile maggots are successfully applied in microsurgery to clean poorly healing wounds, for instance in cases of diabetes.

[Hennig (1973); Oosterbroek & Courtney (1995); Michelsen (1996); Ziegler (1998, 2005); Yeates & Wiegmann (2005); Blagoderov et al. (2007); Yeates et al. (2007); Wiegmann et al. (2011); Schneeberg et al. (2012)]

7 Literature

7.1 Textbooks and comprehensive works

Aubert, J. 1959. Plecoptera. Insecta Helvetica. (Fauna), 139 pp.

Beier, M. 1968a. Mantodea (Fangschrecken). Handbuch der Zoologie, IV, Arthropoda: Insecta. Inst. 4. Walter de Gruyter, Berlin.

Beier, M. 1968b. Phasmida (Stab- oder Gespenstheuschrecken). Handbuch der Zoologie, IV, Arthropoda: Insecta. Inst. 6. Walter de Gruyter, Berlin.

Beier, M. 1972. Saltatoria (Grillen und Heuschrecken). Handbuch der Zoologie, IV, Arthropoda: Insecta. Inst. 17. Walter de Gruyter, Berlin.

Beier, M. 1974. Blattariae (Schaben). Handbuch der Zoologie, IV, Arthropoda: Insecta. Inst. 22. Walter de Gruyter, Berlin.

Beutel, R.G. & Leschen, R.A.B. (vol. eds.) 2005. Coleoptera, Vol. I. Morphology and Systematics (Archostemata, Adephaga, Myxophaga, Polyphaga partim). Handbook of Zoology Vol. IV, Arthropoda: Insecta. De Gruyter, Berlin, New York.

Chapman, R.F. 1998. The Insects. Structure and Function. 4th ed. Cambridge University Press, Cambridge, New York, Melbourne.

Chapman, R.F. & Joern, A. (eds). 1990. Biology of Grasshoppers. Wiley-Interscience, New York.

Crowson, R.A. 1981. The Biology of Coleoptera. John Murray, London, UK.

CSIRO Division of Entomology (ed.) 1991. The Insects of Australia: a Textbook for Students and Research Workers. 2nd ed. Cornell University Press, Ithaca, N.Y.

Dathe, H. (ed.) 2005. Lehrbuch der Speziellen Zoologie. Begründet von A. Kaestner. 2. Aufl. Bd. I: Wirblellose Tiere. 5. Teil: Insecta. Spektrum, Gustav Fischer, Heidelberg, Berlin.

Dettner, K. & Peters, W. (eds.) 2003. Lehrbuch der Entomologie. 2nd ed. Spektrum Akademischer Verlag, Berlin, Heidelberg.

Eisenbeis, G. & Wichard, W. 1985. Atlas zur Biologie der Bodenarthropoden. G. Fischer, Stuttgart, New York. 434 pp.

Gilbert, S.F. 2003. Developmental Biology. Seventh Ed. Sinauer Associates, Sunderland, Massachusetts.

Grassé, P.P. 1949, 1951, 1973, 1976, 1977, 1979. Traité de Zoologie. Insectes. Tome VIII, Fasc. 1 (1973) Insectes. Tête, Aile, Fasc. 2 (1979) Insectes. Thorax, Abdomen, Fasc. 4 (1976), 5A (1977) Insectes. Gamétogenèses, fécondation, métamorphoses. Fasc. 5B (1977) Insectes. Embryologie, Cécidogenèse, Inscects venimeux. Tome IX (1949) Insectes : paléontologie, géonémie, Aptérygotes, Ephéméroptères, Odonatoptères, Blattoptéroïdes, Orthoptéroïdes, Dermaptéroïdes, Coléoptères. Tome X, Fasc. 1 (1951) Nevropteroides, Mecopteroides, Hymenopteroides (Symphytes, Apocrites Terebranta). Fasc. 2. Hymenopteroides (Apocrites Aculeata), Psocopteroides, Hemipteroides, Thysanopteroides. Masson, Paris.

Grimaldi, D.A. & Engel, M.S. 2005. Evolution of the Insects. Cambridge University Press, Cambridge, New York, Melbourne etc.

Gullan, P.J. & Cranston, P.S. 1994. The Insects: an Outline of Entomology. Chapman & Hall, London, Glasgow, Weinheim, New York, Tokyo, Melbourne, Madras.

Günther, K. & Herter, K. 1974. Dermaptera (Ohrwürmer). Handbuch der Zoologie, IV, Arthropoda: Insecta. Inst. 23. Walter de Gruyter, Berlin.

Handschin, E. 1929. Urinsekten oder Apterygota. Die Tierwelt Deutschlands 16: 1 – 150.

Harrison, F.W. & Locke, M. (eds.) 1991, 1998. Microscopic Anatomy of Invertebrates. Vol. 11: Insecta. John Wiley & Sons, Hoboken, N.J.

Heming, B. 2003. Insect Development and Evolution. Cornell University Press, Ithaca, London.

Hennig, W. 1950. Grundzüge einer Theorie der phylogenetischen Systematik. Deutscher Zentralverlag, Berlin.
Hennig, W. 1966. Phylogenetic Systematics. University of Illinois Press, Urbana, Illinois.
Hennig, W. 1969. Die Stammesgeschichte der Insekten. Kramer, Frankfurt (Main), Germany.
Hennig, W. 1973. Diptera (Zweiflügler). Pp. 1–200 in Helmcke, J.G., Starck, D. & Wermuth, H. (eds) Handbuch der Zoologie, IV. Band: Arthropoda – 2. Hälfte: Insecta, 2. Teil.. Walter de Gruyter, Berlin, New York.
Janetschek, H. 1970. 3. Protura (Beintastler). Handbuch der Zoologie, IV, Arthropoda: Insecta. Inst. 14. Walter de Gruyter, Berlin.
Kaltenbach, A. 1968. Embioptera (Spinnfüßer). Handbuch der Zoologie, IV, Arthropoda: Insecta. Inst. 1. Walter de Gruyter, Berlin.
Kaltenbach, A. 1978. Mecoptera (Schnabelhafte, Schnabelfliegen). Pp. 1–111 in Handbuch der Zoologie IV. Insecta. Inst. 25. Walter de Gruyter, Berlin.
v. Kéler, S. 1963. Entomologisches Wörterbuch. Akademie Verlag, Berlin.
v. Kéler, S. 1969. Mallophaga (Federlinge und Haarlinge). In: Handbuch der Zoologie, gegr. von W. Kükenthal, 2. Aufl., 4. Bd., 2. Hälfte, 2. Teil, Lief. 10. Walter de Gruyter & Co., Berlin.
Kristensen, N.P. (ed.) 1999b. Lepidoptera, Moths and Butterflies. Vol. 1: Evolution, Sytematics and Biogeography. Handbook of Zoology. Vol. IV Arthropoda: Insecta. Part 35. Walter de Gruyter, Berlin, New York.
Kristensen, N.P. (ed.) 2003. Lepidoptera, Moths and Butterflies. Vol. 2: Morphology, Physiology and Development. Handbook of Zoology. Vol. IV Arthropoda: Insecta. Part 36. Walter de Gruyter, Berlin, New York.
Leschen, R.A.B., Beutel, R.G. & Lawrence, J.F. (vol. eds.) 2010. Coleoptera, Vol. II. Morphology and Systematics (Elateroidea, Bostrichiformia, Cucujiformia partim). Handbook of Zoology Vol. IV, Arthropoda: Insecta. De Gruyter, Berlin, New York.
Matsuda, R. 1965. Morphology and evolution of the insect head. Memoirs of the American Entomological Institute 4: 1–334.
Matsuda, R. 1970. Morphology and evolution of the insect thorax. Memoirs of the Entmological Society of Canada 76: 1–431.
Matsuda, R. 1976. Morphology and evolution of the insect abdomen, with special reference to developmental patterns and their bearings upon systematics. Pergamon Press Ltd., Oxford.
Minelli, A. 1993. Biological Systematics: The Stare of the Art. Chapman & Hall, London.
Mitchell, T. B. 1960. Bees of the Eastern United States. Volume 1, Tech Bull No 141., 538 pp.
Moritz, G. 2006. Die Thripse. Pflanzensaftsaugende Insekten Bd. 1. 1. Auflage. Westarp Hohenwarsleben. 384 S.
Naumann, I.D., Carne, P.B., Lawrence, J.F., Nielsen, E.S., Spradberry, J.P., Taylor, R.W., Whitten, M.J. & Littlejohn, M.J. (eds) 1991. Insects of Australia: A Textbook for Students and Research Workers. Volume I. Second Edition. Carlton, Victoria, Melbourne University Press.
New, T.R. 1989. Planipennia, Lacewings. Handbuch der Zoologie, Vol. IV, Arthropoda: Insecta. De Gruyter, Berlin, New York.
Priesner, H. 1968. Thysanoptera (Physopoda, Blasenfüßer). Handbuch der Zoologie, IV, Arthropoda: Insecta. Inst. 5. Walter de Gruyter, Berlin.
Rasnitsyn, A.P., Quicke, D.L.J. (eds.) 2002. History of Insects. Kluwer Academic Publishers, Dordrecht.
Schaller, F. 1970. 1. Collembola (Springschwänze). Handbuch der Zoologie, IV, Arthropoda: Insecta. Inst. 12. Walter de Gruyter, Berlin.
Schedl, W. 1991. Hymenoptera, Unterordnung Symphyta. Handbuch der Zoologie, IV, Arthropoda: Insecta. Inst. 31. Walter de Gruyter, Berlin.
Seifert, G. 1995. Entomologisches Praktikum. 3. Aufl. Thieme Verlag, Stuttgart, New York.

Snodgrass, R.E. 1935. The Principles of Insect Morphology. McGraw-Hill Book Company, New York, USA.
St. Quentin, D. & Beier, M. 1968. Odeonata (Libellen). Handbuch der Zoologie, IV, Arthropoda: Insecta. Inst. 3. Walter de Gruyter, Berlin.
Stehr, F. (ed.) 1987, 1991. Immature Insects. Volumes I, II. Kendall & Hunt, Dubuque, Iowa.
Strausfeld, N.J. 1976. Atlas of an Insect Brain. Springer, Heidelberg.
Strümpel, H. 1983. Homoptera, Pflanzensauger. Pp. I–IV + 1–22 in Fischer, M. (ed.) Handbuch der Zoologie, IV, Arthropoda: Insecta. Teilband/part 28. Walter de Gruyter, Berlin.
Sturm, H. & Machida, R. (eds.) 2001. Archaeognatha. Handbook of Zoology. Vol. IV Arthropoda: Insecta. Part 37. Walter de Gruyter, Berlin, New York.
Sudhaus, W. & Rehfeld, K. 1992. Einführung in die Phylogenetik und Systematik. Gustav Fischer Verlag, Stuttgart.
Swofford, D.L., Olsen, G.J., Waddell, P.J., & Hillis, D.M. 1990. Phylogeny reconstruction. Molecular Systematics 3: 407–514.
Wägele, W. 2001. Grundlagen der der Phylogenetischen Systematik. Verlag Dr. Friedrich Pfeil, München.
Weber, H. 1933. Lehrbuch der Entomologie. Gustav Fischer, Jena.
Weber, H. 1938. Grundriß der Insektenkunde. Gustav Fischer, Jena.
Weber, H. & Weidner, H. 1974. Grundriss der Insektenkunde. Gustav Fischer Verlag, Stuttgart.
Weidner, H. 1972. Copeognatha (Psocodea). Handbuch der Zoologie, IV, Arthropoda: Insecta. Inst. 18. Walter de Gruyter, Berlin.
Zwick, P. 1980. Plecoptera (Steinfliegen). Handbuch der Zoologie, IV, Arthropoda: Insecta. Inst. 26. Walter de Gruyter, Berlin.

7.2 Review articles

Beutel, R.G. & Pohl, H. 2006. Endopterygote systematics – where do we stand and what is the goal (Hexapoda, Arthropoda). Systematic Entomology 31: 202–219.
Beutel, R.G., Leschen, R.A.B. & Friedrich, F. 2009. Darwin, Beetles and Phylogenetics. Naturwissenschaften 96: 1293–1312.
Bookstein, F.L. 1998. A hundred years of morphometrics. Acta Zoologica Academiae Scientiarum Hungaricae 44: 7–59.
Buschbeck, E.K. & Friedrich, M. 2008. Evolution of insect eyes: tales of ancient heritage, deconstruction, reconstruction, remodeling, and recycling. Evolution, Education and Outreach 1: 448–462.
Klass, K.-D. 2009. A critical review of current data and hypotheses on hexapod phylogeny. Proceedings of the Arthropodan Embryological Society of Japan 43: 3–22.
Kristensen, N.P. 1981. Phylogeny of insect orders. Annual Review of Entomology 26, 135–157.
Kristensen, N.P. 1991. Phylogeny of extant hexapods. Pp. 125–140 in CSIRO (ed.) The Insects of Australia, 2nd ed., 2 Vols., Carnell University Press.
Kristensen, N.P. 1995. Forty years' insect phylogenetic systematics. Zoologische Beiträge N.F. 36(1): 83–124.
Kristensen, N. P. 1997. The groundplan and basal diversification of the hexapods. Fortey, R. A., Thomas, R. H. (eds.). Arthropod Relationships. London: Chapman & Hall. Systematics Association Special Volume Series 55: 281–293.
Kristensen, N. P. 1999. Phylogeny of endopterygote insects, the most successful lineage of living organisms. European Journal of Entomology 96: 237–253.

Pass, G. 2000. Accessory pulsatile organs: evolutionary innovations in insects. Annual Review of Entomology 45: 495–518.
Trautwein, M.D., Wiegmann, B.M., Beutel, R.G., Kjer, K.M., Yeates, D.K. 2012. Advances in insect phylogeny at the dawn of the postgenomic era. Annual Review of Entomology 57: 449–468.
Weirauch, C. & Schuh, R.T. 2011. Systematics and Evolution of Heteroptera: 25 Years of Progress. Annual Review of Entomology 56, 487–510.
Willmann, R. 2003. Die phylogenetischen Beziehungen der Insecta: offene Fragen und Probleme. Verhandlungen Westdeutscher Entomologentag 2001: 1–64.

7.3 Cladistic software and related studies

Goloboff, P.A., 1995. NONA Version 1.5. Tucuman: Fundacion e Instituto Miguel Lillo.
Goloboff, P.A., 2005. Estimating character weights during tree search. Cladistics 9 (1): 83–91.
Goloboff, P.A., Farris, J.S., Nixon, K.C. 2003. TNT: tree analysis using new technology, program and documentation, available from the authors, and at www.zmuc.dk/public/phylogeny.
Goloboff, P.A., Farris, J.S. & Nixon, K.C. 2008. TNT, a free program for phylogenetic analysis. Cladistics 24 (5): 774–786.
Maddison, W.P. & Maddison, D.R. 2003. Mesquite: a modular system for evolutionary analysis. Ver. 1.0. http://mesquitepoject.org.
Maddison, W.P. & Maddion, D.R. 2005. Mesquite: a modular system for evolutionary analysis. Biological Sciences, Miscellaneous Papers 62 (5): 11–03–1118.
Nixon, K.C. 1999. Winclada (BETA). Version 1.00.08. Ithaca, NY: Nixon, K.C.
Rohlf, F.J. 2006. tps-DIG, Digitize Landmarks and Outlines, Version 2.05. [Software and Manual]. Department of Ecology and Evolution, State University of New York at Stony Brook, New-York.
Swofford, D.L. 1998. PAUP*. Phylogenetic Analysis using Parsimony (* and other methods). Sinauer Associates, Sunderland.

7.4 Complete references

Achtelig, M. 1967. Über die Anatomie des Kopfes von *Raphidia flavipes* Stein und die Verwandtschaftsbeziehungen der Raphidiidae zu dem Megaloptera. Zoologische Jahrbücher für Anatomie 84: 249–312.
Achtelig, M. 1975. Die Abdomenbasis der Neuropteroidea (Insecta, Holometabola). Zoomorphologie 82: 201–242.
Adams, D.C. & Rohlf, F.J. 2000. Ecological character displacement in Plethodon: biomechanical differences found from a geometric morphometric study. Proceedings of the National Academy of Sciences U.S.A. 97: 4106–4111.
Adams, D.C., Rohlf, F.J. & Slice D.E., 2004. Geometric morphometrics: ten years of progress following the 'revolution'. Italian Journal of Zoology 71: 5–16.
Albouy, V. & Caussanel, C., 1990. Dermaptères ou Perce-Oreilles. Faune de France, 75: 1–245.
Andersen, N.M. 1998. Water striders from the Paleogene of Denmark with a review of the fossil record and evolution of semiaquatic bugs (Hemiptera: Gerromorpha). Det Kongelige Danske Videnskabernes Selskabs Biologiske Skrifter 50: 1–152.
Anderson, D.T. 1972a. The development of hemimetabolous insects. Pp. 95–165 in Counce, S.J. & Waddington, C.H. (eds.) Developmental Systems: Insects. Vol. I. Academic Press, London.

Anderson, D.T. 1972b. The development of holometabolous insects. Pp. 165–242 in Counce, S.J. & Waddington, C.H. (eds.) Developmental Systems: Insects. Vol. I. Academic Press, London.

Ando, H. (ed.) 1982. Biology of the Notoptera. Kashiyo-Insatu Co. Ltd., Nagano.

Ansorge, J. 2001. *Dobbertinia reticulata* Handlirsch 1920 from the Lower Jurassic of Dobbertin (Mecklenburg/Germany) – the oldest representative of Sialidae (Megaloptera). Neues Jahrbuch fur Geologie und Paläontologie Monatshefte 9: 553–564.

Arikawa, K., Eguchi, E., Yoshida, A. & Aoki, K. 1980. Multiple extraocular photoreceptive areas on genitalia of butterfly *Papilio xuthus*. Nature 288: 700–702.

Armstrong, J.D., Kaiser, K., Müller, A., Fischbach, K.F., Merchant, N., & Strausfeld, N.J. 1995. Flybrain, an on-line atlas and database of the *Drosophila* nervous system. Neuron 15(1): 17.

Aspöck, H. & Aspöck, U. 1971. Raphidioptera (Kamelhalsfliegen). Pp. 1–48 in Handbuch der Zoologie, IV, Arthropoda: Insecta. Inst. 15. Walter de Gruyter, Berlin.

Aspöck, H. & Aspöck, U., Hölzel, H. 1980. Die Neuropteren Europas. Eine zusammenfassende Darstellung der Systematik, Ökologie und Chorologie der Neuropteroidea (Megaloptera, Raphidioptera, Planipennia) Europas. 2 Bde: 495/355 pp. – Goecke und Evers, Krefeld.

Aspöck, H., Aspöck, U. & Rausch, H. & 1991. Die Raphidiopteren der Erde. 2 Volumes. Goecke & Evers, Krefeld.

Aspöck, U. 2002. Phylogeny of the Neuropterida (Insecta: Holometabola). Zoologica Scripta 31: 51–55.

Aspöck, U. & Aspöck, H. 2005. Neuropterida (Neuropteroidea, Neuroptera sensu lato). Pp. 542–584 in Dathe, H.H. (ed.) Lehrbuch der Speziellen Zoologie, begründet von A. Kaestner, 2. Aufl. Bd. I: Wirbellose Tiere. 5. Teil: Insecta. Spektrum, Gustav Fischer, Heidelberg, Berlin.

Aspöck, U. & Aspöck, H. 2007. Verbliebene Vielfalt vergangener Blüte. Zur Evolution, Phylogenie und Biodiversität der Neuropterida (Insecta: Endopterygota). Denisia 20: 451–516.

Aspöck, U. & Aspöck, H. 2008. Phylogenetic relevance of the genital sclerites of Neuropterida (Insecta: Holometabola). Systematic Entomology 33: 97–127.

Aspöck, U., Plant, J.D. & Nemeschkal, H.L. 2001. Cladistic analysis of Neuroptera and their systematic position within Neuropterida (Insecta: Holometabola: Neuropterida: Neuroptera). Systematic Entomology 26: 73–86.

Averof, M. &. Cohen, S.M. 1997. Evolutionary origin of insect wings from ancestral gills. Nature 385: 627–30.

Badonnel, A. 1934. Recherche sur l'anatomie des Psoques. Bulletin scientifique de la France et de la Belgique 18(Suppl): 11–241.

Bai, M., Beutel, R.G., Song, K.Q., Liu, W.G., Malqin, H., Li, S., Hu, X.Y., Yang, X.K. 2012. Evolutionary patterns of hind wing morphology in dung beetles (Coleoptera: Scarabaeinae). Arthropod Structure & Development 41: 505–513.

Bai, M., Jarvis, K., Wang, S.Y., Song, K.Q., Wang, Y.P., Wang, Z.L., Li, W.Z., Wang, W., Yang, X.K. 2010. A second new species of ice crawlers from China (Insecta: Grylloblattodea), with thorax evolution and the prediction of potential distribution. PLoS ONE 5: e12850

Bai, M., McCullough, E., Song, K.Q., Liu, W.G., Yang, X.K. 2011. Evolutionary constraints in hind wing shape in Chinese dung beetles (Coleoptera: Scarabaeinae). PLoS ONE 6: e21600

Barber-James, H.M., Gattolliat, J.L., Sartori, M. & Hubbard, M.D. 2008. Global diversity of mayflies (Ephemeroptera, Insecta) in freshwater. Hydrobiologie 595: 339–350.

Bauchhenss, E. 1979. Die Pulvillen von *Calliphora erythrocephala* Meig. (Diptera, Brachycera) als Adhäsionsorgane. Zoomorphologie 93: 99–123.

Bauernfeind, E. 2005. 6. Ordnung Ephemeroptera (Eintagsfliegen). Pp. 108–120 in Dathe, H.H. (ed.) Lehrbuch der Speziellen Zoologie, begründet von A. Kaestner, 2. Aufl. Bd. I: Wirbellose Tiere. 5. Teil: Insecta. Spektrum, Gustav Fischer, Heidelberg, Berlin.

Baum, E., Dressler, C. & Beutel, R.G. 2007. Head structures of *Karoophasma* sp. (Mantophasmatodea) with phylogenetic implications. Journal of Zoological Systematics and Evolutionary Research 45: 104–119.

Bechly, G. 1996. Fossil Odonates in Tertiary amber. Petalura 2: 1–15.

Bechly, G., Brauckmann, C., Zessin, W. & Gröning, E. 2001. New results concerning the morphology of the most ancient dragonflies (Insecta: Odonatoptera) from the Namurian of Hagen-Vorhalle (Germany). Journal of Zoological Systematics and Evolutionary Research 39: 209–226.

Beier, M. 1968a. Mantodea (Fangschrecken). Handbuch der Zoologie, IV, Arthropoda: Insecta. Inst. 4. Walter de Gruyter, Berlin.

Beier, M. 1968b. Phasmida (Stab- oder Gespenstheuschrecken). Handbuch der Zoologie, IV, Arthropoda: Insecta. Inst. 6. Walter de Gruyter, Berlin.

Beier, M. 1972. Saltatoria (Grillen und Heuschrecken). Handbuch der Zoologie, IV, Arthropoda: Insecta. Inst. 17. Walter de Gruyter, Berlin.

Beier, M. 1974. Blattariae (Schaben). Handbuch der Zoologie, IV, Arthropoda: Insecta. Inst. 22. Walter de Gruyter, Berlin.

Bennett, M.F. 1979. Extraocular light receptors and circadian rhythms. Pp. 641–663 in Handbook of Sensory Physiology, Vol. VII/6A, Autrum, H. (ed.). Springer, Berlin.

Bergsten, J. 2005. A review of long-branch attraction. Cladistics 21: 163–193.

Berlese, A. 1909. Redia, Firenze VI, 1: 1–182.

Béthoux, O. 2009. The earliest beetle identified. Journal of Paleontology 83: 931–937.

Béthoux, O. & Nel, A. (2002). Venation pattern and revision of Orthoptera sensu nov. and sister groups. Phylogeny of Palaeozoic and Mesozoic Orthoptera sensu nov. Zootaxa, 96: 1–88.

Béthoux, O., Nel, A., Gand, G., Lapeyrie, J. & Galtier, J. (2002). Discovery of the genus *Iasvia* Zalessky, 1934 in the Upper Permian of France (Lodève Basin) (Orthoptera, Ensifera, Oedischiidae). Geobios 35: 293–302.

Beutel, R.G. 1997. Über Phylogenese und Evolution der Coleoptera (Insecta), insbesondere der Adephaga. Verhandlungen des Naturwissenschaftlichen Vereins in Hamburg NF 31: 1–164.

Beutel R.G. & Hörnschemeyer, T. 2002. Description of the larva of *Rhipsideigma raffrayi* (Coleoptera: Archostemata), with phylogenetic and functional implications. European Journal of Entomology 99: 53–66.

Beutel, R.G., Ge, S.-Q. & Hörnschemeyer, T. 2008a. On the head morphology *Tetraphalerus*, the phylogeny of Archostemata and the basal branching events in Coleoptera. Cladistics 24: 270–298.

Beutel, R.G. & Leschen, R.A.B (vol. eds.) 2005 Handbook of Zoology, Vol. IV Arthropoda: Insecta. Part 38. Coleoptera, Vol. 1: Morphology and Systematics (Archostemata, Adephaga, Myxophaga, Polyphaga (partim). Walter De Gruyter, Berlin, New York.

Beutel, R.G. & Baum, E. 2008. A longstanding entomological problem finally solved? Head morphology of *Nannochorista* (Mecoptera, Insecta) and possible phylogenetic implications. Journal of Zoological Systematics and Evolutionary Research 46: 346–367.

Beutel, R.G. & Friedrich, F. 2008. Comparative study of larval head structures of Megaloptera (Hexapoda). European Journal of Entomology 105: 917–938.

Beutel, R.G., Friedrich, F. & Aspöck, U. 2010. The larval head of Nevrorthidae and the phylogeny of Neuroptera (Insecta). Zoological Journal of the Linnean Society 158: 533–562.

Beutel, R. G., Friedrich, F., Hörnschemeyer, T., Pohl, H., Hünefeld, F., Beckmann, F., Meier, R., Misof, B., Whiting, M. F. & Vilhemsen, L. 2011. Morphological and molecular evidence converging upon a robust phylogeny of the megadiverse Holometabola. Cladistics 26: 1–15.

Beutel, R.G. & Ge, S.-Q. 2008. The larval head of *Raphidia* (Raphidioptera, Insecta) and its phylogenetic significance. Zoology 11: 89–113.

Beutel, R.G. & Gorb, S. 2001. Ultrastructure of attachment specializations of hexapods (Arthropoda): evolutionary patterns inferred from a revised ordinal phylogeny. Journal of Zoological Systematics and Evolutionary Research 39: 177–207.

Beutel, R.G. & Gorb, S. 2006. A revised interpretation of the evolution of attachment structures in Hexapoda (Arthropoda), with special emphasis on Mantophasmatodea. Arthropod Systematics & Phylogeny 64: 3–25.

Beutel, R.G. & Gorb, S. 2008. Evolutionary scenarios for unusual attachment devices of Phasmatodea and Mantophasmatodea (Insecta). Systematic Entomology 33: 501–510.

Beutel, R.G., Kristensen, N.P. & Pohl, H. 2009. Resolving insect phylogeny: The significance of cephalic structures of the Nannomecoptera in understanding endopterygote relationships. Arthropod Structure & Development 38: 427–460.

Beutel, R.G., Krogmann, L. & Vilhelmsen, L.T. 2008. The larval head morphology of *Xyela* sp. (Xyelidae, Hymenoptera) and its phylogenetic implications. Journal of Zoological Systematics and Evolutionary Research 46: 118–132.

Beutel, R.G., Leschen, R.A.B. & Friedrich, F. 2009. Darwin, Beetles and Phylogenetics. Naturwissenschaften 96: 1293–1312.

Beutel, R.G. & Pohl, H. 2005. Head structures of males of Strepsiptera (Hexapoda) with emphasis on basal splitting events within the Order. Journal of Morphology 244: 1–14.

Beutel, R.G., Pohl, H. & Hünefeld, F. 2005. Strepsipteran brains and effects of miniaturisation. Arthropod Structure and Development 34: 301–313.

Beutel, R.G. & Vilhelmsen, L.B. 2007. Head anatomy of Xyelidae (Hexapoda: Hymenoptera) and phylogenetic implications. Organisms, Diversity and Evolution 7: 207–230.

Beutel, R.G. & Weide, D. 2005. Cephalic anatomy of *Zorotypus hubbardi* (Hexapoda: Zoraptera): new evidence for a relationship with Acercaria. Zoomorphology 124: 121–136.

Bierbrodt, E. 1943. Der Larvenkopf von *Panorpa communis* L. und seine Verwandlung, mit besonderer Berücksichtigung des Gehirns und der Augen. Zoologische Jahrbücher (Abteilung für Anatomie und Ontogenie der Tiere) 68: 49–136.

Bininda-Emonds, O.R.P., Bryant, H.N. & Russell, A.P., 1998. Supraspecific taxa as terminals in cladistic analysis: implicit assumptions of monophyly and a comparison of methods. Biological Journal of the Linnean Society 64: 101–133.

Bitsch, C. & Bitsch, J. 2000. The phylogenetic interrelationships of the higher taxa of apterygote hexapods. Zoologica Scripta 29: 131–156.

Bitsch, C. & Bitsch, J. 2004. Phylogenetic relationships of basal hexapods among the mandibulate arthropods: a cladistic analysis based on comparative morphological characters. Zoologica Scripta 33: 511–550.

Blagoderov, V., Grimaldi, D.A. & Fraser, N.C. 2007. How time flies for flies: diverse Diptera from the triassic of Virginia and the early radiation of the order. American Museum Novitates 3572: 1–39.

Blanke, A., Wipfler, B., Letsch, H., Koch, M., Beutel, R. & Misof, B. 2012. Revival of Palaeoptera – head characters support a monophyletic origin of Odonata and Ephemeroptera (Insecta). Cladistics 28: 560–581.

Blanke, A., Greve, C., Wipfler, B., Beutel, R., Holland, B. & Misof, B. 2013. The identification of concerted convergence in insect heads corroborates Palaeoptera. Systematic Biology 62 : 250–263.

Bohn, H. 2005. 14. Ordnung Blattoptera, Schaben. 15. Ordnung Isoptera, Termiten. Pp. 197–250 in Dathe, H.H. (ed.) Lehrbuch der Speziellen Zoologie, begründet von A. Kaestner, 2. Aufl. Bd. I: Wirbellose Tiere. 5. Teil: Insecta. Spektrum, Gustav Fischer, Heidelberg, Berlin.

Bookstein, F.L. 1998. A hundred years of morphometrics. Acta Zoologica Academiae Scientiarum Hungaricae 44: 7–59.

Bookstein, F. L., Schäfer, K., Prossinger, H., Seidler, H., Fieder, M., Stringer, C., Weber, G. W., Arsuaga, J.-L., Slice, D.E., Rohlf, F.J., Recheis, W., Mariam, A.J. & Marcus, L.F. 1999. Comparing frontal cranial profiles in archaic and modern *Homo* by morphometric analysis. Anatomical Record (New Anatomist) 257: 217–224.
Bradler, S. 2003. Phasmatodea, Gespenstschrecken. Pp. 251–261 in Dathe, H.H. (ed.) Lehrbuch der Speziellen Zoologie, begründet von A. Kaestner, 2. Aufl. Bd. I: Wirbellose Tiere. 5. Teil: Insecta. Spektrum, Gustav Fischer, Heidelberg, Berlin.
Bradler, S. 2009. Die Phylogenie der Stab- und Gespenstschrecken (Insecta: Phasmatodea) (2.1.). Species, Phylogeny and Evolution 2: 3–139.
Bremer, K., 1994. Branch support and tree stability. Cladistics 10, 295–304.
Brittain, J.E. & Sartori, M. 2003. Ephemeroptera. Pp. 373–380 in Resh, V.H. & Cardé, R. (eds) Encyclopedia of Insects. Academic Press, New York.
Brodsky, A.K. 1994. The Evolution of Insect Flight. Oxford University Press, London.
Brower, A.V.Z & Schawaroch, V. 1996. Three steps of homology assessment. Cladistics 12: 265–272.
Brown, B.V. 1993. A further chemical alternative to critical- point-drying for preparing small (or large) flies. Fly Times 11: 10.
Boudreaux, B.H. 1979. Arthropod Phylogeny with Special Emphasis to Insects. John Wiley & Sons Inc, New York, Chichester, Brisbane, Toronto.
Burckhardt, D. 2009. Taxonomy and phylogeny of the Gondwanan moss bugs or Peloridiidae (Hemiptera, Coleorrhyncha). Deutsche Entomologische Zeitschrift 56: 173–235.
Burckhardt, D. 2010. Mooswanzen – Peloridiidae (Hemiptera, Coleorrhyncha), eine enigmatische Insektengruppe. Entomologica Austriaca 17: 9–22.
Buschbeck, E.K. & Friedrich, M. 2008. Evolution of insect eyes: tales of ancient heritage, deconstruction, reconstruction, remodeling, and recycling. Evolution, Education and Outreach 1: 448–462.
Bybee, S.M., Ogden, T.H., Branham, M.A. & Whiting, M.F. 2008. Molecules, morphology and fossils: a comprehensive approach to odonate phylogeny and the evolution of the odonate wing. Cladistics 24: 477–514.
Byers, G.W. 1987. Order Mecoptera. Pp. 246–252 in Stehr F.W. (Ed.) Immature Insects, Vol. I. Kendall/Hunt Publishing Company, Dubuque, Iowa.
Byers, G.W. 1991. Mecoptera (Scorpion-flies, hanging flies). Pp. 696–704 in CSIRO (Ed.), The Insects of Australia, Vol. 2. Melbourne University Press, Carlton, Victoria.
Caldecutt, W.C. & Adams D.C. 1998. Morphometrics of trophic osteology in the threespine stickleback, *Gasterosteus aculeatus*. Copeia 1998: 827–838.
Carlson, S.D. & Chi, C. 1979. The functional morphology of the insect photoreceptor. Annual Review of Entomology 24: 379–416.
Chapman, R.F. 1998. The Insects. Structure and Function. 4th ed. Cambridge University Press, Cambridge, New York, Melbourne.
Chapman, R.F. & Joern, A. (eds). 1990. Biology of Grasshoppers. Wiley-Interscience, New York.
Chaudonneret, J. 1948. Le labium des Thysanoures (Insectes Aptérygotes). Annales des Sciences Naturelles – Zoologie et Biology Animale 10 : 1–27.
Chaudonneret, J. 1950. La morphologie céphalique de *Thermobia domestica* (Packard) (Insecte Aptérygote Thysannoure). Annales des Sciences Naturelles, Zoologie 12: 145–302.
Childers, C.C. 1997. Feeding and oviposition injuries to plants. Pp. 505–37 in Lewis, T. (ed.) Thrips as Crop Pests. CABI, Wallingford, UK.
China, W.E. 1962. South American Peloridiidae (Hemiptera-Homoptera: Coleorrhyncha). Transactions of the Royal Entomological Society of London 114: 131–161.

Cobben, R.H. 1968. Evolutionary Trends in Heteroptera: Part I Eggs, Architecture of the Shell, Gross Embryology and Eclosion. Wageningen, The Netherlands: Centre for Agricultural Publishing and Documentation, Wageningen, Nederland.

Cobben, R.H. 1978. Evolutionary Trends in Heteroptera. Part 2: Mouthpart Structures and Feeding Strategies. H. Veeman and Zonen B.V., Wageningen, Nederland.

Corbet, P.S. 1999. Dragonflies: behavior and ecology of Odonata. Comstock Publishing Associates, Ithaca, New York.

Crichton, M.I. 1957. The structure and function of the mouth parts of adult caddis flies (Trichoptera). Philosophical Transactions of the Royal Society (B) 241: 45–91.

Crowson, R.A. 1981. The Biology of Coleoptera. John Murray, London, UK.

Crowson, R.A. 1995. Some interesting evolutionary parallels. Pp. 63–85 in Pakaluk, J. & Ślipiński, S.A. (eds) 1995. Biology, Phylogeny, and Classification of Coleoptera: Papers Celebrating the 80th Birthday of Roy Crowson. Museum I Instytut Zoologii PAN, Warszawa.

Cryan, J.R. & Urban, J.M. 2012: Higher level phylogeny of the insect order Hemiptera: is Auchenorrhyncha really paraphyletic. Systematic Entomology 37: 7–31.

Damgaard, J., Klass, K.D., Picker, M.D. & Buder, G. 2008. Phylogeny of the Heelwalkers (Insecta: Mantophasmatodea) based on mtDNA sequences, with evidence for additional taxa in South Africa. Molecular Phylogenetics and Evolution 47: 443–462.

Dallai, R., Gottardo, M., Mercati, D., Machida, R., Mashimo, Y., Matsumura, Y. & Beutel, R.G. 2013. Divergent mating patterns and a unique mode of external sperm transfer in Zoraptera: an enigmatic group of pterygote insects. Naturwissenschaften 100: 581–594.

Dallai, R., Mercati, D., Gottardo, M., Machida, R., Mashimo, Y. & Beutel, R.G. 2011. The male reproductive system of *Zorotypus caudelli* Karny (Zoraptera): sperm structure and spermiogenesis. Arthropod Structure and Development 40: 531–47.

Dallai, R., Mercati, D., Gottardo, M., Machida, R., Mashimo, Y. & Beutel, R.G. 2012a. The fine structure of the female reproductive system of *Zorotypus caudelli* Karny (Zoraptera). Arthropod Structure and Development 41: 51–63.

Dallai, R., Mercati, D., Gottardo, M, Dossey, A.T., Machida, R, Mashimo, Y. & Beutel, R.G. 2012b. The male and female reproductive systems of *Zorotypus hubbardi* Caudell, 1918 (Zoraptera). Arthropod Structure and Development 41: 337–359.

Dashman, T. 1953: Terminology of the pretarsus. Annals of the Entomological Society of America 46: 56–62.

Deckert, J. & Göllner-Scheiding, U. 2005. 24. Ordnung Heteroptera, Wanzen. Pp. 402–425 in Dathe, H.H. (ed.) Lehrbuch der Speziellen Zoologie, begründet von A. Kaestner, 2. Aufl. Bd. I: Wirbellose Tiere. 5. Teil: Insecta. Spektrum, Gustav Fischer, Heidelberg, Berlin.

Dean, J.C. & D.I. Cartwright, D.I. 1987. Trichoptera of a Victorian forest stream: species composition and life histories. Marine and Freshwater Research 38: 845–860.

Deans, AR., Mikó, I., Wipfler, B. & Friedrich, F. 2012. Evolutionary phenomics and the emerging enlightenment of arthropod systematics. Invertebrate Systematics 26: 323–330.

DeBrye, R.W. 2001. Improving interpretation of the decay index for DNA sequence data. Systematic Biology 50, 742–752.

Denholm, B., Sudarsan, V., Pasalodos-Sanchez, S., Artero, R., Lawrence, P., Maddrell, S., Baylies, M. & Skaer, H. 2003. Dual origin of the renal tubules in Drosophila: mesodermal cells integrate and polarize to establish secretory function. Current Biology 13:1052–1057.

Denis, J.R. & Bitsch, J. 1973. Morphologie de la tête des insectes. Pp. 1–593 in Grassé, P.P. (ed.) Traité de Zoologie 8 (1). Masson, Pairis.

Dettner, K. & Peters, W. (eds.) 2003. Lehrbuch der Entomologie. 2nd ed. Spektrum Akademischer Verlag, Berlin, Heidelberg.

Di Giulio, A., Maurizi, E., Stacconi, M.V.R. & Romani, R. 2012. Functional structure of antennal sensilla in the myrmecophilous beetle *Paussus favieri* (Coleoptera, Carabidae, Paussini). Micron 43: 705–719.

Drilling, K. & Klass, K.-D. 2010. Surface structures of the antenna of Mantophasmatodea (Insecta). Zoologischer Anzeiger 249: 121–137.

Dunger, W. 2005. 1. Ordnung Protura, Beintaster. 2. Ordnung Collembola, Springschwänze. 3. Ordnung Diplura, Doppelschwänze. 4. Ordnung Archaeognatha (Microcoryphia), Felsenspringer. 5. Ordnung Zygentoma, Silberfischchen. Pp. 66–107 in Dathe, H. (ed.) Lehrbuch der Speziellen Zoologie. Begründet von A. Kaestner. 2. Aufl. Bd. I: Wirbellose Tiere. 5. Teil: Insecta. Spektrum, Gustav Fischer, Heidelberg, Berlin.

Eberhard, M.J.B, Pass, G., Picker, M.D., Beutel, R., Predel, R. & Gorb, S. N. 2009. Structure and function of the arolium of Mantophasmatodea (Insecta). Journal of Morphology 270: 1247–1261.

Eldredge, N. & Crawcraft, J. 1980. Phylogenetic Patterns and the Evolutionary Process. Columbia University Press, New York.

Engel, M.S. & Grimaldi, D.A. 2002. The first Mesozoic Zoraptera (Insecta). American Museum Novitates 3362: 1–20.

Engel, M.S. & Grimaldi, D.A. 2004. New light shed on the oldest insect. Nature 427: 627–630.

Engel, M.S. & Grimaldi, D.A. 2006. The earliest webspinners (Insecta: Embiodea). American Museum Novitates 3514: 1–15.

Essig, E.O. 1926. Insects of Western North America. The Macmillan Company, New York.

Evans, J.W. 1937. The morphology of the head of Homoptera. Papers and Proceedings of the Royal Society of Tasmania 1937: 1–20.

Evans, J.W. 1981. A review of present knowledge of the family Peloridiidae and new genera and new species from New Zealand and New Caledonia (Hemiptera: Insecta). Records of the Australian Museum 34: 381–406.

Fanenbruck, M., Harzsch, S. & Wägele, J.W. 2004. The brain of the Remipedia (Crustacea) and an alternative hypothesis on their phylogenetic relationships. Proceedings of the National Academy of Sciences of the United States of America 101: 3868–3873.

Farris, J.S. 1988. Hennig86 version 1.5 manual. Software and MSDOS program.

Farris, J.S. 1989. The retention index and the rescaled consisteny index. Cladistics 5: 417–419.

Farris, J.S., Albert, V.A., Källersjö, M., Lipscomb, D. & Kluge, A.G. 1996. Parsimony jacknifing outperforms neighbour-joining. Cladistics 12: 99–124.

Federle, W., Maschwitz, U., Fiala, B., Riederer, M., Hölldobler, B. 1997. Slippery ant-plants and skilful climbers: selection and protection of specific ant partners by epicuticular wax blooms in *Macaranga* (Euphorbiaceae). Oecologia 112: 217–224.

Felsenstein, J. 1983. Statistical inference of phylogenies. Journal of The Royal Statistical Society, A, 164: 246–272.

Felsenstein, J. 2003. Inferring Phylogenies. Sinauer Associates Inc.

Fick, W. 1981. Über den Darmtrakt der Fulgeromorpha. Unpublished PhD-Thesis. University Hamburg.

Flint, O.S. 1973. The Megaloptera of Chile (Neuroptera). Revista Chilean de Entomología 7: 32–45.

Forey, P.L., Humphries, C.J., Kitching, I.L., Scotland, R.W., Siebert, D.J. & Williams, D.M. 1992. Cladistics. A Practical Course in Systematics. The Systematics Association Publication No. 10. Clarendon Press, Oxford.

François, J. 1970. Squelette et musculature céphalique de Campodea chardardi Condé (Diplura: Campodeidae). Zoologische Jahrbücher / Abteilung Anatomie und Ontogenie 87: 331–376.

François, J., Dallai, R. & Yin, W.Y. 1992. Cephalic anatomy of *Sinentomon erythranum* Yin (Protura: Sinentomidae). International Journal of Insect Morphology and Embryology 21: 199–213.

Frania, H.E. & Wiggins, G.B. 1997. Analysis of morphological and behavioral evidence for the phylogeny and higher classication of Trichoptera (Insecta). Royal Ontatio Museum for Life Sciences Contributions 160: 1–67.

Friedemann, K., Wipfler, B., Bradler, S., & Beutel, R. G. 2012. On the head morphology of *Phyllium* and the phylogenetic relationships of Phasmatodea (Insecta). Acta Zoologica, 93: 184–199.

Friedrich, F., Beckmann, F. & Beutel, R.G. 2013. The head of *Merope tuber* (Meropeidae) and the phylogeny of Mecoptera (Hexapoda). Arthropod Structure and Development 42: 69–88.

Friedrich, F. & Beutel, R.G. 2008. The thorax of *Zorotypus* (Hexapoda, Zoraptera) and a new nomenclature for the musculature of Neoptera. Arthropod Structure and Development 37: 29–54.

Friedrich, F. & Beutel, R.G. 2010a. The thoracic morphology of *Nannochorista* (Nannochoristidae) and its implications for the phylogeny of Mecoptera and Antliophora. Journal of Zoological Systematics and Evolutionary Research 48: 50–74.

Friedrich, F. & Beutel, R.G. 2010b. Goodbye Halteria? The thoracic morphology of Endopterygota (Insecta) and its phylogenetic implications. Cladistics 26: 579–612.

Friedrich, F., Farell, B.D. & Beutel, R.G. 2009. The thoracic morphology of Archostemata and the relationships of the extant suborders of Coleoptera (Hexapoda). Cladistics 25: 1–37.

Friedrich, F., Pohl, H., Hünefeld, F., Beckmann, F., Herzen, J. & Beutel, R.G. 2008. SRμCT-based study of external and internal structures of adults and larvae of Endopterygota (Hexapoda). Hasylab Annual Report 2007: 1527–1528.

Friedrich, M. & Tautz, D. 2001. Arthropod rDNA phylogeny revisited: A consistency analysis using Monte Carlo simulation. Annales de la Societe Entomologique de France 37: 21–40.

Gäde, G. & Šimek, P. 2010. A novel member of the adipokinetic peptide family in a "living fossil", the ice crawler *Galloisiana yuasai*, is the first identified neuropeptide from the order Grylloblattodea. Peptides 31: 372–376.

Gaedike, R. & Häuser, C.L. 2005. 33. Ordnung Lepidoptera, Schmetterlinge. pp. 672–733 in in Dathe, H. (ed.) Lehrbuch der Speziellen Zoologie. Begründet von A. Kaestner. 2. Aufl. Bd. I: Wirbellose Tiere. 5. Teil: Insecta. Spektrum, Gustav Fischer, Heidelberg, Berlin.

Gangwere S.K., Muralirangan M.C. & Muralirangan M. 1997. The Binomics of Grasshoppers, Katydids and Their Kin. CAB International, Wallingford.

Garrouste, R., Clemente, G., Nel, P., Engel, M.S., Grandcolas, P., D'Haese, C., Lagebro, L., Denayer, J., Gueriau, P., Lafaite, P., Olive, S., Prestianni, C. & Nel, A. 2012. A complete insect from the Late Devonian period. Nature 488: 82–85.

Gharaibeh, W.S., Rohlf, F.J., Slice, D.E., DeLisi, L.E. 2000. A geometric morphometric assessment of change in midline brain structural shape following a first episode of schizophrenia. Biological Psychiatry 48: 398–405.

Gilbert, S.F. 2003. Developmental Biology. Seventh Ed. Sinauer Associates, Sunderland, Massachusetts.

Giribet, G., Edgecombe, G.D. & Wheeler, W.C. 2001. Arthropod phylogeny based on eight molecular loci and morphology. Nature 413: 157–161.

Glorioso, M.J. 1981. Systematics of the dobsonfly subfamily Corydalinae (Megaloptera: Corydalidae). Systematic Entomology 6: 253–290.

Gokan, N. & Meyer-Rochow, V.B. 2000. Morphological comparisons of compound eyes in Scarabaeoidea (Coleoptera) related to the beetles' daily activity maxima and phylogenetic positions. Journal of Agricultural Science – Tokyo University of Agriculture 45: 15–61.

Goloboff, P.A., 1995. NONA Version 1.5. Tucuman: Fundacion e Instituto Miguel Lillo.

Goloboff, P.A., 2005. Estimating character weights during tree search. Cladistics 9: 83–91.

Goloboff, P.A., Farris, J.S., Nixon, K.C. 2003. TNT: tree analysis using new technology, program and documentation, available from the authors, and at www.zmuc.dk/public/phylogeny.

Goloboff, P.A., Farris, J.S. & Nixon, K.C. 2008. TNT, a free program for phylogenetic analysis. Cladistics 24: 774–786.

Gorb, S. 2012. Adhesion and Friction in Biological Systems. Springer, Dordrecht, Heidelberg, London, New York.

Grabarek, V. 2008. Die Kopfmorphologie von *Neodiprion sertifer* (Geoffroy, 1785) (Diprionidae) und ihre phylogenetische Bedeutung. Unpublished Diploma thesis, FSU Jena, Germany.

Gray, E.G. 1960. The fine structure of the insect ear. Philosophical Transactions of the Royal Society B 243: 75–94.

Grell, K.G. 1939. Der Darmtraktus von *Panorpa communis* L. und seine Anhänge bei Larve und Imago. Zoologische Jahrbücher (Abteilung für Anatomie und Ontogenie der Tiere) 64: 1–86.

Grimaldi, D.A. 2010. 400 million years on six legs: on the origin and early evolution of Hexapoda. Arthropod Structure & Development 39: 191–203.

Grimaldi, D.A. & Engel, M.S. 2005. Evolution of the Insects. Cambridge University Press, Cambridge, New York, Melbourne etc.

Groll, E.K. & Günther, K.K. 2005. 17. Ordnung Saltatoria (Orthoptera), Heuschrecken, Springschrecken. Pp. 261–290 in Dathe, H.H. (ed.) Lehrbuch der Speziellen Zoologie, begründet von A. Kaestner, 2. Aufl. Bd. I: Wirbellose Tiere. 5. Teil: Insecta. Spektrum, Gustav Fischer, Heidelberg, Berlin.

Günther, K.K. 2005. 11. Ordnung Embioptera, Tarsenspinner, Spinnfüßer, Embien, 19. Ordnung Psocoptera (Copeognatha, Corrodentia), Staubläuse und Flechtlinge. Pp. 167–172, 296–308 in Dathe, H.H. (ed.) Lehrbuch der Speziellen Zoologie, begründet von A. Kaestner, 2. Aufl. Bd. I: Wirbellose Tiere. 5. Teil: Insecta. Spektrum, Gustav Fischer, Heidelberg, Berlin.

Günther, K. & Herter, K. 1974. Dermaptera (Ohrwürmer). Handbuch der Zoologie, IV, Arthropoda: Insecta. Inst. 23. Walter de Gruyter, Berlin.

Günther, K.K. 1974. Staubläuse, Psocoptera. Pp. 1–314 in Dahl (ed.) Die Tierwelt Deutschlands, 61. Teil, Gustav Fischer, Jena.

Gwynne, D.T. 2001. Katydids and bush-crickets: reproductive behavior and evolution of the Tettigoniidae. Cornell University Press Ithaca, London.

Haas, M.S., Brown, S.J. Beeman, R.W. 2001. Pondering the procephalon: the segmental origin of the labrum. Development, Genes and Evolution 211: 81–95.

Haas, F. & Kukalová-Peck, J. 2001. Dermapteran hindwing structure and folding: New evidence for familial, ordinal and superordinal relationships within Neoptera (Insecta). European Journal of Entomology 98: 445–509.

Haas, F., Waloszek, D. & Hartenberger, R. 2003. *Devonohexapodus bocksbergensis*, a new marine hexapod from the Lower Devonian Hunsrück Slates, and the origin of Atelocerata and Hexapoda. Organisms Diversity & Evolution 3: 39–54.

Haas, F. 2005. 12. Ordnung Dermaptera, Ohrwürmer. Pp. 173–180 in Dathe, H. (ed.) Lehrbuch der Speziellen Zoologie. Begründet von A. Kaestner. 2. Aufl. Bd. I: Wirbellose Tiere. 5. Teil: Insecta. Spektrum, Gustav Fischer, Heidelberg, Berlin.

Hamilton, K. G. A. 1971. The insect wing. Part I. Origin and development of wings from notal lobes. Journal of the Kansas Entomological Society 44: 421–33.

Hannemann, H. J. 1956. Die Kopfmuskulatur von *Micropteryx calthella* (L.) (Lep.). Zoologische Jahrbücher Abteilung für Anatomie und Ontogenie der Tiere 75: 177–206.

Hasenfuss, I. 2002. A possible evolutionary pathway to insect flight starting from lepismatid organization. Journal of Zoological Systematics and Evolutionary Research 40: 65–81

Hayat, M.A. 1981. Fixation for Electron Microscopy. Academic Press, New York.

Heddergott, H. 1939. Kopf und Vorderdarm von Panorpa communis L. Zoologische Jahrbücher / Abteilung Anatomie und Ontogenie 65: 229–294.

Heming, B. 2003. Insect Development and Evolution. Cornell University Press, Ithaca, London.

Heming, B.S. 1970. Postembryonic development of the female reproductive system in *Frankliniella fusca* (Thripidae) and *Haplothrips verbasci* (Phlaeothripidae) (Thysanoptera). Miscellaneous Publications of the American Entomological Society 7: 197–234.
Heming, B.S. 1975. Antennal structure and metamorphosis in *Frankliniella fusca* (Hinds) (Thripidae) and *Haplothrips verbasci* (Osborn) (Phlaeothripidae) (Thysanoptera). Quaestiones Entomologicae 11: 25–68.
Heming, B.S. 1991. Thysanoptera. Pp. 1–21 in Immature Insects, Vol. 2. Kendall/Hunt Publishing Co., Dubuque, Iowa.
Heming, B.S. 1993. Structure, function, ontogeny, and evolution of feeding in thrips (Thysanoptera). Pp. 3–41 in Schaefer, C.S. & Leschen, R.A.B. (eds) Functional Morphology of insect feeding. Entomological Society of America, Thomas Say Publications in Entomology Lanham, MD.
Heming, B.S. 1995. History of the germ line in male and female thrips (Thysanoptera). Pp. 505–535 in Proceedings of the 1993 International Conference on Thysanoptera, Towards Understanding Thrips Management, Burlington, Vermont.
Hennig, W. 1950. Grundzüge einer Theorie der phylogenetischen Systematik. Deutscher Zentralverlag, Berlin.
Hennig, W. 1966. Phylogenetic Systematics. University of Illinois Press, Urbana, Illinois.
Hennig, W. 1969. Die Stammesgeschichte der Insekten. Kramer, Frankfurt (Main), Germany.
Hennig, W. 1973. Diptera (Zweiflügler). Pp. 1–200 in Helmcke, J.G., Starck, D. & Wermuth, H. (eds) Handbuch der Zoologie, IV. Band: Arthropoda – 2. Hälfte: Insecta, 2. Teil. Walter de Gruyter, Berlin, New York.
Hepburn, H.R. 1969. The skeleto-muscular system of Mecoptera: the head. University of Kansas Science Bulletin 48: 721–765.
Hepburn, H.R. 1970. The skeleto-muscular system of Mecoptera: the thorax. University of Kansas Science Bulletin 48: 801–844.
Hertel, W. & Pass, G. 2002. An evolutionary treatment of the morphology and physiology of circulatory organs in insects. Comparative Biochemistry and Physiology Part A: Molecular & Integrative Physiology 133: 555–575.
Hinton, H.E. 1958. The phylogeny of the panorpoid orders. Annual Review of Entomology 3: 181–206.
Hoch, H., Deckert, J. & Wessel, A. 2006. Vibrational signalling in a Gondwanan relict insect (Hemiptera: Coleorrhyncha: Peloridiidae). Biology Letters 2: 222–224.
Hodkinson, I.D. 1971. A new species of Strophingia Enderlein (Homoptera: Psylloidea) from Britain. Journal of Entomology B 40: 1–5.
Holland, B. R., H. G. Spencer, T. H. Worthy, and M. Kennedy. 2010. Identifying cliques of convergent characters: concerted evolution in the cormorants and shags. Systematic Biology 59:433–445.
Hopkin, S.P. 1997. Biology of the Springtails:(Insecta: Collembola). Oxford University Press on Demand, London.
Hopkins, G.H.E. & Rothschild, M. (eds.) 1953–1987. An Illustrated Catalogue of the Rothschild Collection of Fleas (Siphonaptera) in the British Museum (Natural History) with Keys and Short Descriptions for the Identification of Families, Genera, Species and Subspecies. Vols. I–VII. Oxford University Press, London.
Hörnschemeyer, T. 1998. Morphologie und Evolution des Flügelgelenks der Coleoptera und Neuropteridae. Bonner Zoologische Monographien 43: 1–127.
Hörnschemeyer, T., Beutel, R.G., Pasop, F. 2002. Head structures of *Priacma serrata* LeConte (Coleoptera, Archostemata) inferred from x-ray tomography. Journal of Morphology 252: 298–314.
Hörnschemeyer, T., Fischer, C., Klug, R., Ries, R. & Willmann, R. 2012. Schnitte am laufenden Nanometer: Moderne Ultrastrukturforschung mit Serial Block-Face Scanning Elektronenmikroskopie. DgaaE Nachrichten 26: 16–20.

Hörnschemeyer, T., Haug, J., Bethoux, O., Beutel, R.G., Charbonier, S. Thomas A. Hegna, T. A., Koch, M., Rust, S., Wedmann, S., Bradler, S., Willmann, R. 2013. Is *Strudiella* a Devonian insect? Nature 488: 82–85.

Huang, D.Y., Nel, A., Zompro, O. & Waller, A. 2008. Mantophasmatodea now in the Jurassic. Naturwissenschaften 95: 947–952.

Huang, D., Engel, M.S., Cai, C., Wu, H. & Nel, A. 2012. Diverse transitional giant fleas from the Mesozoic era of China. Nature 483(7388): 201–204.

Hünefeld, F. 2007. The genital morphology of *Zorotypus hubbardi* Caudell, 1918 (Insecta: Zoraptera: Zorotypidae). Zoomorphology 126: 135–151.

Hünefeld, F. & Beutel, R.G. 2005. The sperm pumps of Strepsiptera and Antliophora (Hexapoda). Journal of Zoological Systematics and Evolutionary Research 43: 297–306.

Hünefeld, F., Kristensen, N.P., 2010. The female postabdomen and internal genitalia of the basal moth genus *Agathiphaga* (Insecta: Lepidoptera: Agathiphagidae): morphology and phylogenetic implications. Zoological Journal of the Linnean Society 159: 905–920.

Hünefeld, F. & Beutel, R.G. 2012. The female postabdomen of the enigmatic Nannochoristidae (Insecta: Mecopterida) and its phylogenetic significance. Acta Zoologica 93: 231–238.

Imms, A.D. 1936. The ancestry of insect. Transactions of the Society for British Entomology 3: 1–2.

Illies, J., Helmcke, J.-G. & Kükenthal, W. 1968. Ephemeroptera (Eintagsfliegen). Handbuch der Zoologie, IV, Arthropoda: Insecta. Inst. 6. Walter de Gruyter, Berlin.

Ishiwata, K., Sasaki, G., Ogawa, J., Miyata, T. & Su, Z.H. 2011. Phylogenetic relationships among insect orders based on three nuclear protein-coding gene sequences. Molecular Phylogenetics and Evolution 58: 169–180.

Jałoszyński, P. & Beutel, R.G. 2012. Functional morphology and evolution of specialized mouthparts of Cephenniini (Scydmaeninae, Staphylinidae). Arthropod Structure and Development 41: 593–607.

Jamieson, B.G.M. 1987. The Ultrastructure and Phylogeny of Insect Spermatozoa. Cambridge University Press, Cambridge, UK.

Janetschek, H. 1970. 3. Protura (Beintastler). Handbuch der Zoologie, IV, Arthropoda: Insecta. Inst. 14. Walter de Gruyter, Berlin.

Jarvis, K.J., Haas, F. & Whiting, M.F. 2005. Phylogeny of earwigs (Insecta: Dermaptera) based on molecular and morphological evidence: reconsidering the classification of Dermaptera. Systematic Entomology 30: 442–453.

Jordan, K.H. 1950. Wasserwanzen. Die neue Brehm-Bücherei, Heft 23, Akademische Verlagsgesellschaft Geest & Portig K.-G, Leipzig.

Jordan, K.H. 1952. Wasserläufer. Die Neue Brehm-Bücherei, Heft 52, Akademische Verlagsgesellschaft Geest & Portig (1952)

Jordan, K.H.C. 1972. Heteroptera (Wanzen). Handbuch der Zoologie, IV, Arthropoda: Insecta. Inst. 16. Walter de Gruyter, Berlin.

Kaltenbach, A. 1968. Embioptera (Spinnfüßer). Handbuch der Zoologie, IV, Arthropoda: Insecta. Inst. 1. Walter de Gruyter, Berlin.

Kaltenbach, A. 1978. Mecoptera (Schnabelhafte, Schnabelfliegen). Pp. 1–111 in Handbuch der Zoologie IV. Insecta. Inst. 25. Walter de Gruyter, Berlin.

v. Kéler, S. 1963. Entomologisches Wörterbuch. Akademie Verlag, Berlin.

v. Kéler, S. 1969. Mallophaga (Federlinge und Haarlinge). Handbuch der Zoologie, IV, Arthropoda: Insecta. Inst. 18. Walter de Gruyter, Berlin.

Kelsey, L. P. 1954. The skeleto-motor mechanism of the dobson fly, *Corydalus cornutus*. Part I. Head and prothorax. Cornell University Memoirs 334: 1–51.

Kelsey, L.P. 1957. The skeleto-motor mechanism of the dobson fly, *Corydalus cornutus*. Part II. Pterothorax. Cornell University Memoirs 346: 1–42.

Kendall, D.G. 1977. The diffusion of shape. Advances in Applied Probability 9: 428–430.
Kendall, D.G. 1984. Shape-manifolds, Procrustean metrics and complex projective spaces. Bulletin of the London Mathematical Society 16: 81–121.
Kinzelbach, R. 2005. Ordnung Siphonaptera (Aphaniptera), Flöhe. Pp. 734–745 in Dathe, H. (ed.) Lehrbuch der Speziellen Zoologie. Begründet von A. Kaestner. 2. Aufl. Bd. I: Wirbellose Tiere. 5. Teil: Insecta. Spektrum, Gustav Fischer, Heidelberg, Berlin.
Kinzelbach R.K. 1971a. Morphologische Befunde an Fächerflüglern und ihre phylogenetische Bedeutung (Insecta: Strepsiptera). Zoologia 41:1–256.
Kinzelbach, R. 1971b. Strepsiptera (Fächerflügler). Pp. 1–68 in Handbuch der Zoologie, IV, Insecta. Inst. 24. Walter de Gruyter, Berlin.
Kirk, W.D.J. 1997. Feeding. Pp. 217–57 in Lewis, T. (ed.) Thrips as Crop Pests. CABI, London.
Kjer, K. 2004. Aligned 18S and insect phylogeny. Systematic Biology 53: 506–514.
Kjer, K.M., Blahnik, R.J. & Holzenthal, R.W. 2002. Phylogeny of caddisflies (Insecta, Trichoptera). Zoologica Scripta 31: 83–91.
Kjer, K.M., Carle, F.L., Litman, J. & Ware, J. 2006. A molecular phylogeny of Hexapoda. Arthropod Systematics & Phylogeny 64: 35–44.
Klass, K. 1995. Die Phylogenie der Dictyoptera. Dissertation, Universität München. Cuvier, Göttingen.
Klass, K.-D. 2005. 9. Ordnung Notoptera (Grylloblattodea). 10. Ordnung Mantophasmatodea. Pp. 155–166 in Dathe, H.H. (ed.) Lehrbuch der Speziellen Zoologie, begründet von A. Kaestner, 2. Aufl. Bd. I: Wirbellose Tiere. 5. Teil: Insecta. Spektrum, Gustav Fischer, Heidelberg, Berlin.
Klass, K.-D. 2009. A critical review of current data and hypotheses on hexapod phylogeny. Proceedings of the Arthropodan Embryological Society of Japan 43: 3–22.
Klass, K.D. & Ehrmann, R. 2005. 13. Ordnung Mantodea, Fangschrecken, Gottesanbeterinnen. Pp. 155–161 in Dathe, H.H. (ed.) Lehrbuch der Speziellen Zoologie, begründet von A. Kaestner, 2. Aufl. Bd. I: Wirbellose Tiere. 5. Teil: Insecta. Spektrum, Gustav Fischer, Heidelberg, Berlin.
Klass K.-D. & Matushkina N.A. 2012. The exoskeleton of the female genitalic region in *Petrobiellus tokunagae* (Insecta: Archaeognatha): insect-wide terminology, homologies, and functional interpretations. Arthropod Structure & Development 41: 575–591.
Klass, K.D. & Meier, R. 2006. A phylogenetic analysis of Dictyoptera (Insecta) based on morphological characters. Entomologische Abhandlungen 63: 3–50.
Klass, K.D., Picker, M.D., Damgaard, J., Van Noort, S. & Tojo, K. 2003. The taxonomy, genitalic morphology, and phylogenetic relationships of southern African Mantophasmatodea (Insecta). Entomologische Abhandlungen 61: 3–67.
Klass, K.D., Zompro, O., Kristensen, N.P. & Adis, J. 2002. Mantophasmatodea: a new insect order with extant members in the Afrotropics. Science, 296: 1456–1459.
Klass, K.D., Zompro, O. & Adis, J.U., 2005. 10. Ordnung Dermaptera, Ohrwürmer. Pp. 161–166 in Dathe, H. (ed.) Lehrbuch der Speziellen Zoologie. Begründet von A. Kaestner. 2. Aufl. Bd. I: Wirbellose Tiere. 5. Teil: Insecta. Spektrum, Gustav Fischer, Heidelberg, Berlin.
Klaus, AV. & Schawaroch, V. 2006. Novel methodology utilizing confocal laser scanning microscopy for systematic analysis in arthropods (Insecta). Integrative & Comparative Biology 46: 207–214.
Klingenberg, C.P. & Leamy, L.J. 2001. Quantitative genetics of geometric shape in the mouse mandible. Evolution 55: 2342–2352.
Koeth, M., Friedrich, F., Pohl, H. & Beutel, R.G. (2012) The thoracic skeleto-muscular system of *Mengenilla* (Strepsiptera: Mengenillidae) and its phylogenetic implications. Arthropod Structure and Development 41: 323–335.
Koch, M. 2001. Mandibular mechanisms and the evolution of hexapods. Annales de la Société entomologique de France 37: 129–174.

Komarek, A. & Beutel, R.G. 2007. Problems in taxonomy and suggestions for a standardized description of new insect taxa. Entomological Problems 36: 55–70.

Kranz, B.D., Schwarz, M.P., Morris, D.C. & Crespi, B.J. 2002. Life history of *Kladothrips ellobus* and *Oncothrips rodwayi*: insight into the origin and loss of soldiers in gall-inducing thrips. Ecological Entomology *27*: 49–57.

Krenn, H. W., & Aspöck, H. 2012. Form, function and evolution of the mouthparts of blood-feeding Arthropoda. Arthropod Structure & Development 41: 101–118.

Kristensen, N.P. 1975. The phylogeny of hexapod "orders". A critical review of recent accounts. Journal of Zoological Systematics and Evolutionary Research 13: 1–44.

Kristensen, N.P. 1981. Phylogeny of insect orders. Annual Review of Entomology 26: 135–157.

Kristensen, N.P. 1991. Phylogeny of extant hexapods. Pp. 125–140 in CSIRO (ed.) The Insects of Australia 2nd ed., 2 Vols, Cornell University Press.

Kristensen, N.P. 1995. Forty years' insect phylogenetic systematics. Zoologische Beiträge N.F. 36: 83–124.

Kristensen, N.P. 1997. The groundplan and basal diversification of the hexapods. Fortey, R. A., Thomas, R. H. (eds.). Arthropod Relationships. London: Chapman & Hall. Systematics Association Special Volume Series 55: 281–293.

Kristensen, N.P. 1999a. Phylogeny of endopterygote insects, the most successful lineage of living organisms. European Journal of Entomology 96: 237–253.

Kristensen, N.P. (ed.) 1999b. Lepidoptera, Moths and Butterflies. Vol. 1: Evolution, Sytematics and Biogeography. Handbook of Zoology. Vol. IV Arthropoda: Insecta. Part 35. Walter de Gruyter, Berlin, New York.

Kristensen, N.P. (ed.) 2003. Lepidoptera, Moths and Butterflies. Vol. 2: Morphology, Physiology and Development. Handbook of Zoology. Vol. IV Arthropoda: Insecta. Part 36. Walter de Gruyter, Berlin, New York.

Kristensen, N.P., Scoble, M.J. & Karsholt, O. 2007. Lepidoptera phylogeny and systematics: the state of inventorying moth and butterfly diversity. Zootaxa 1668: 699–747.

Kühl, G. & Rust, J. 2009. *Devonohexapodus bocksbergensis* is a synonym of *Wingertshellicus backesi* (Euarthropoda) – no evidence for marine hexapods living in the Devonian Hunsrück Sea. Organisms Diversity & Evolution 9: 215–231.

Kukalová-Peck, J. 1983. Origin of the insect wing and wing articulation from the arthropodan leg. Canadian Journal of Zoology 61: 1618–1669.

Kukalová-Peck, J. 1987. New Carboniferous Diplura, Monura, and Thysanura, the hexapod ground plan, and the role of thoracic side lobes in the origin of wings (Insecta). Canadian Journal of Zoology 65: 2327–2345.

Kukalová-Peck, J. 1991. Fossil history and the evolution of hexapod structures. Pp. 141–179 in CSIRO (ed.) Insects of Australia, Vol. 1. Cornell University Press, Ithaca, New York.

Kukalová-Peck, J. 2008. Phylogeny of higher taxa in Insecta: finding synapomorphies in the extant fauna and separating them from homoplasies. Evolutionary Biology 35: 4–51.

Kukalová-Peck, J., Lawrence, J.F. 1993. Evolution of the hind wing in Coleoptera. Canadian Entomologist 125: 181–258.

Kukalová-Peck, J. & Lawrence, J.F. 2004. Use of hind wing characters in assessing relationships among Coleopteran Suborders and major endoneopteran lineages. European Journal of Entomology 101: 95–144.

Kukalová-Peck, J. & Beutel, R.G. 2012. Is the Carboniferous †*Adiphlebia lacoana* really the "oldest beetle"? Critical reassessment and description of a new Permian beetle family. European Journal of Entomology 109 633–645.

Land, M.F. & Nilsson, D.-E. 2002. Animal Eyes. Oxford University Press, London.

Larivière, M.-C., Burckhardt, D. & Larochelle, A. 2011. Peloridiidae (Insecta: Hemiptera: Coleorrhyncha). Fauna of New Zealand, Number 67. Manaaki Whenua Press, Lincoln, Canterbury, New Zealand.

Lawrence, J.F. & Newton, A.F. 1982. Evolution and classification of beetles. Annual Review of Ecology and Systematics 13: 261–290

Lawrence, J.F., Ślipiński, A., Seago, A.E., Thayer, M.K., Newton, A.F., & Marvaldi, A.E. 2011. Phylogeny of the Coleoptera based on morphological characters of adults and larvae. Annales Zoologici 61: 1–217.

Lehane, M.J. 1998. Chapter 30. The midgut. Pp. 725–746 in Harrison, F.W. & Locke, M. (eds.) Microscopic Anatomy of Invertebrates. Vol. 11B: Insecta. John Wiley & Sons, Hoboken, N.J.

Leschen, R.A.B., Beutel, R.G. & Lawrence, J.F. (vol. eds.) 2010. Coleoptera, Vol. II. Morphology and Systematics (Elateroidea, Bostrichiformia, Cucujiformia partim). Handbook of Zoology Vol. IV, Arthropoda: Insecta. De Gruyter, Berlin, New York.

Lewis, R.E. 1998. Résumé of the Siphonaptera (Insecta) of the world. Journal of Medical Entomology 35: 377–389.

Lewis, T. 1973. Thrips: Their Biology, Ecology and Economic Importance. Academic Press, London, New York.

Lewis, T. (ed.) 1997. Thrips as Crop Pests. CABI, Wallingford, UK.

Lienhard, C. & Smithers, C.N. 2002 Psocoptera (Insecta): World Catalogue and Bibliography. Museum d'histoire Naturelle, Geneva.

Lin, C.P., Chen, M.Y. & Huang, J.P. 2010. The complete mitochondrial genome and phylogenomics of a damselfly, *Euphaea formosa* support a basal Odonata within the Pterygota. Gene 468: 20–29.

Liu, X., Aspöck, H. & Aspöck, U. 2012a. *Sinoneurorthus yunnanicus* n. gen. et n. sp.–a spectacular new species and genus of Nevrorthidae (Insecta: Neuroptera) from China, with phylogenetic and biogeographical implications. Aquatic Insects 34: 131–141.

Liu, X., Hayashi, F. & Yang, D. 2012b. A New Species of Alderfly (Megaloptera: Sialidae) from Yunnan, China. Entomological News 122: 265–269.

Lo, N., Tokuda, G., Watanabe, H., Rose, H., Slaytor, M., Maekawa, K., Bandi, C. & Noda, H. 2000. Evidence from multiple gene sequences indicates that termites evolved from wood-feeding cockroaches. Current Biology 10: 801–804.

Lohmann, H. 1996. Das phylogenetische System der Anisoptera (Odonata). Entomologische Zeitschrift 106: 209–266.

Lyal, C. 1985. Phylogeny and classification of the Psocodea, with particular reference to the lice (Psocodea: Phthiraptera). Systematic Entomology 10: 145–165.

Machida, R., Tojo, K., Tsutsumi, T., Uchifune, T., Klass, K.D., Picker, M.D., & Pretorius, L. 2004. Embryonic development of heel-walkers: reference to some prerevolutionary stages (Insecta: Mantophasmatodea). Proceedings of the Arthropodan Embryological Society of Japan 39: 31–39.

Machida, R., Nagashima, T., Ando, H. 1990. The early embryonic development of the jumping bristletail *Pedetontus unimaculatus* Machida (Hexapoda: Microcoryphia, Machilidae). Journal of Morphology 206: 181–195.

Maddison, W.P., Donoghue, M.J. & Maddison, D.R. 1984. Outgroup analysis and parsimony. Systematic Biology 33: 83–103.

Maddison, W.P. & Maddison, D.R. 2003. Mesquite: a modular system for evolutionary analysis. Ver. 1.0. http://mesquitepoject.org.

Maddison, W.P. & Maddion, D.R. 2005. Mesquite: a modular system for evolutionary analysis. Biological Sciences, Miscellaneous Papers 62: 11–03–1118.

Mahner, M. 1993. Systema Cryptoceratorum Phylogeneticum (Insecta, Heteroptera). Zoologica 48: 1–302.

Maki, T. 1936. Studies on the skeletal structure musculature and nervous system of the Alder Fly Chauliodes formosanus Petersen. Memoirs of the Faculty of Science and Agriculture Taihoku Imperial University 16: 117–243.

Malicky, H. 1973. Trichoptera (Köcherfliegen). Pp. 1–114 in Handbuch der Zoologie IV, Insecta. Inst. 29. Walter de Gruyter, Berlin.

Marden, J. H. 2003. The surface-skimming hypothesis for the evolution of insect flight. Acta Zoologica Cracoviensia 46 (Suppl.): 73–84.

Martoja, R. 1977. Organes génitaux femelles: Voies génitales et glandes annexes. Pp. 1–123 in Grassé, P.P. (ed.) Traité de Zoologie, Anatomie, Systématique, Biologie: Insectes: Gamétogeneses, Fécondation, Métamorphoses. Masson, Paris.

Mashimo, Y., Machida, R., Dallai, R. Gottardo, M. & Beutel, R.G. 2011. Egg structure of *Zorotypus caudelli* Karny (Insecta, Zoraptera, Zorotypidae). Tissue & Cell 43: 230–237.

Matsuda, R. 1956. Musculature of the head of *Agulla adnixa* (Hagen) (Neuroptera, Raphidiidae). Journal of the Kansas Entomological Society 29: 146–155.

Matsuda, R. 1965. Morphology and evolution of the insect head. Memoirs of the American Entomological Institute 4: 1–334.

Matsuda, R. 1970. Morphology and evolution of the insect thorax. Memoirs of the Entmological Society of Canada 76: 1–431.

Matsuda, R. 1976. Morphology and evolution of the insect abdomen, with special reference to developmental patterns and their bearings upon systematics. Pergamon Press Ltd., Oxford.

Mauersberger, G. & Mey, E. 1993. Mallophagen und Vogelsystem – Beitrag zur Diskussion der Parasitophyletik. Mitteilungen aus dem Zoologischen Museum, Berlin. 69 (Suppl.) 17: 3–30.

Mayr, E. 1975. Grundlagen der Zoologischen Systermatik. Verlag Paul Parey, Hamburg, Berlin.

McCafferty, W.P. 1991. Toward a phylogenetic classification of the Ephemeroptera (Insecta): a commentary on systematics. Annals of the Entomological Society of America 84: 343–360.

McKittrick, F.A. 1964. Evolutionary studies of cockroaches. Cornell University Agricultural Experimental Station Memoirs 389: 1–197.

Medvedev, S.G. 1994. Morphological basis of the classification of fleas (Siphonaptera). Entomological Review 73: 30–51.

Meier, R. 1994. On the inappropriateness of presence/absence recoding for nonadditive multistate characters in computerized cladistic analyses. Zoologischer Anzeiger 232: 201–212.

Meier, R. 1997. A test and review of the empirical performance of the ontogenetic criterion. Systematic Biology 46: 699–721.

Mendes, L.F. 2002. Taxonomy of Zygentoma and Microcoryphia: historical overview, present status and goals for the new millenium. Pedobiologia 46: 225–233.

Metscher, B.D. 2009. MicroCT for comparative morphology: simple staining methods allow high-contrast 3D imaging of diverse non-mineralized animal tissues. BMC Physiology 9: 11.

Meusemann, K., von Reumont, B.M., Simon, S., Roeding, F., Strauss, S., Kück, P., Ebersberger, I., Walzl, M., Pass, G., Breuers, S., Achter, V., von Haeseler, A., Burmester, T., Hadrys, H., Wägele, J.W. & Misof, B. 2010. A phylogenomic approach to resolve the arthropod tree of life. Molecular Biology and Evolution 27: 2451–2464.

Mey, E. 1998. Zur Taxonomie, Lebensweise und parasitophyletischen Evidenz der Federlingsgattung Struthiolipeurus sensu lato (Insecta, Phthiraptera, Ischnocera). Mitteilungen aus dem Museum für Naturkunde Berlin, Zoologische Reihe. 74: 65–93.

Mey, E. 2005. 20. Ordnung Phthiraptera, Tierläuse, Lauskerfe. Pp. 308–330 in Dathe, H.H. (ed.) Lehrbuch der Speziellen Zoologie, begründet von A. Kaestner, 2. Aufl. Bd. I: Wirbellose Tiere. 5. Teil: Insecta. Spektrum, Gustav Fischer, Heidelberg, Berlin.

Mey, W. 2005. 32. Ordnung Trichoptera, Schmetterlinge. pp. 652–672 in in Dathe, H. (ed.) Lehrbuch der Speziellen Zoologie. Begründet von A. Kaestner. 2. Aufl. Bd. I: Wirbellose Tiere. 5. Teil: Insecta. Spektrum, Gustav Fischer, Heidelberg, Berlin.

Meyer-Rochow, V.B. 1999. Compound eye: circadian rhythmicity, illumination, and obscurity. Pp. 97–124 in Eguchi, E. & Tominaga, Y. (eds.) Atlas of Arthropod Sensory Receptors. Springer. Tokyo, Berlin, New York.

Miazaki, M. 1987. Morphology of aphids. Pp. 1–25 in Minks, A.K., Harrewijn, P. (eds.) Aphids, their Biology, Natural Enemies and Control. Vol. A. Elsevier, Amsterdam.

Michels, J. 2007. Confocal laser scanning microscopy: using cuticular autofluorescence for high resolution morphological imaging in small crustaceans. Journal of Microscopy 227: 1–7.

Michels, J. & Büntzow, M. 2010. Assessment of Congo red as a fluorescence marker for the exoskeleton of small crustaceans and the cuticle of polychaetes. Journal of Microscopy 238: 95–101.

Michels, J. & Gorb S.N. 2012. Detailed three-dimensional visualization of resilin in the exoskeleton of arthropods using confocal laser scanning microscopy. Journal of Microscopy 245: 1–16.

Michelsen, V. 1996. Neodiptera: New insights into the adult morphology and higher level phylogeny of Diptera (Insecta). Zoological Journal of the Linnean Society 117: 71–102.

Michelsen, V. 1997. A revised interpretation of the mouthparts in adult fleas (Insecta Siphonaptera). Zoologischer Anzeiger 235: 217–223.

Mickoleit, G. 1961 Zur Thoraxmorphologie der Thysanoptera. Zoologische Jahrbücher Abteilung Anatomie und Ontogenie der Tiere 79: 1–92.

Mickoleit, E. 1963: Untersuchungen zur Kopfmorphologie der Thysanopteren. Zoologische Jahrbücher Abteilung Anatomie und Ontogenie der Tiere 81: 101–150.

Mickoleit, G. 1973. Über den Ovipositor der Neuropteroidea und Coleoptera und seine phylogenetische Bedeutung (Insecta, Holometabola). Zeitschrift für Morphologie der Tiere 74: 37–64.

Mickoleit, G. 1975. Die Genital- und Postgenitalsegmente der Mecoptera-Weibchen (Insecta, Holometabola) I Das Exoskelet. Zeitschrift für Morphologie der Tiere 80: 97–135.

Mickoleit, G. 1976. Die Genital- und Postgenitalsegmente der Mecoptera-Weibchen (Insecta, Holometabola) II Das Dach der Genitalkammer. Zeitschrift für Morphologie der Tiere 88: 133–156.

Mickoleit, G. 1978. Die phylogenetischen Beziehungen der Schnabelfliegen-Familien aufgrund morphologischer Ausprägungen der weiblichen Genital- und Postgenitalsegmente (Mecoptera). Entomologica Germanica 4: 258–271.

Mickoleit, G. 2008. Die Sperma-Auspreßvorrichtung der Nannochoristidae (Insecta: Mecoptera). Entomologica Generalis 31: 193–226.

Mickoleit, G. 1966. Zur Kenntnis einer neuen Spezialhomologie (Synapomorphie) der Panorpoidea. Zoologische Jahrbücher / Abteilung für Anatomie und Ontogenie der Tiere 83: 483–496.

Mickoleit, G. 1969. Vergleichend-anatomische Untersuchungen an der pterothorakalen Pleurotergalmuskulatur der Neuropteria und Mecopteria (Insecta, Holometabola). Zeitschrift für Morphologie der Tiere 64: 151–178.

Miller, K.B., Hayashi, C., Whiting, M.F., Svenson, G.J. & Edgerly, J.S. 2012. The phylogeny and classification of Embioptera (Insecta). Systematic Entomology 37: 550–570.

Miller, K.B., Hayashi, C., Whiting, M.F., Svenson, G.J. & Edgerly, J.S. 2012. The phylogeny and classification of Embioptera (Insecta). Systematic Entomology 37: 550–570.

Minelli, A. 1993. Biological Systematics. Chapman & Hall, London.

Mitchener, C.D. & Sokal, R.R. 1957. A quantitive approach to a problem in classification. Evolution 11: 130–162.

Miyako, Y., Arikawa, K. & Eguchi, E. 1993. Ultrastructure of the extraocular photoreceptor in the genitalia of a butterfly *Papilio xuthus*. Journal of Comparative Neurology 327: 458–468.
Moritz, G. 1991. Phylogenie der Thysanoptera (Insecta) aus ontogenetischer Sicht. Mitteilungen aus dem Museum für Naturkunde in Berlin. Zoologisches Museum und Institut für Spezielle Zoologie (Berlin) 67: 141–155.
Morse, J.C. 1997. Checklist of world Trichoptera. Pp. 339–342 in: Holzenthal, R.W., Flint, O.S. (Eds) Proceedings of the 8th International Symposium on Trichoptera. Ohio Biological Survey, Columbus.
Mound, L.A. & Morris, D.C. 2003. The morphological background to Thysanoptera phylogeny. Entomologische Abhandlungen 61: 151–53.
Moulds, M.S. 2005. An appraisal of the higher classification of cicadas (Hemiptera: Cicadoiea) with special reference to the Australian fauna. Records of the Australian Museum 57: 375–446.
Müller, C., Rosenberg, J., Richter, S., Meyer-Rochow, V.B. 2003. The compound eye of *Scutigera coleoptrata* (Linnaeus, 1758) (Chilopoda: Notostigmophora): an ultrastructural re-investigation that adds support to the Mandibulata-concept. Zoomorphology 122: 191–209.
Nardi, F., Spinsanti, G., Boore, J.L., Carapelli, A., Dallai, R., & Frati, F. 2003. Hexapod origins: monophyletic or paraphyletic? Science 299: 1887–1889.
New, T.R. 1987. Biology of the Psocoptera. Oriental Insects 21: 1–109
New, T.R. 1989. Planipennia, Lacewings. Handbuch der Zoologie, Vol. IV, Arthropoda: Insecta. De Gruyter, Berlin, New York.
New, T.R. & Theischinger, G. 1993. Megaloptera (Alderflies, Dobsonflies). Handbuch der Zoologie, Vol. IV, Arthropoda: Insecta. De Gruyter, Berlin, New York.
Niehuis, O., Hartig, G., Grath, S., Pohl, H., Lehmann, J., Tafer, H., Donath, A., Krauss, V., Eisenhardt, C., Hertel, J., Petersen, M., Mayer, C., Meusemann, K., Peters, R.S., Stadler, P.F., Beutel, R.G., Bornberg-Bauer, E., McKenna, D.D., Misof, B. 2012. Genomic and morphological evidence converge to resolve the enigma of Strepsiptera. Current Biology 22: 1309–1313.
Nielsen, A. 1980. A comparative study of the genital segments and the genital chamber in female Trichoptera. Kongelige Danske Videnskabernes Selskab Biologiske Skrifter 23: 1–200.
Nieukerken, E.J. van, Kaila, L., Kitching, I.J., Kristensen, N. P., Lees, D.C., Minet, J., Mitter, C., Mutanen, M., Regier, J.C., Simonsen, T.J., Wahlberg, N., Yen, S.-H., Zahiri, R., Adamski, D., Baixeras, J., Bartsch, D., Bengtsson, B.A., Brown, J.W., Bucheli, S.R., Davis, D.R., De Prins, J., De Prins, W., Epstein, M.E., Gentili-Poole, P., Gielis, C., Hattenschwiler, P., Hausmann, A., Holloway, J.D., Kallies, A., Karsholt, O., Kawahara, A., Koster, J.C., Kozlov, M., Lafontaine, J.D., Lamas, G., Landry, J.-F., Lee, S., Nuss, M., Park, K.-T., Penz, C., Rota, J., Schmidt, B.C., Schintlmeister, A., Sohn, J.-C., Solis, M.A., Tarmann, G.M., Warren, A.D., Weller, S., Yakovlev, R.V., Zolotuhin V.V. & Zwick, A. 2011. Order Lepidoptera Linnaeus, 1758. In Zhang, Z.-Q. (ed.) Animal biodiversity: An outline of higher-level classification and survey of taxonomic richness. Zootaxa 3148: 212–221.
Nilsson, D.-E. 1989. Optics and evolution of the compound eye. Pp. 30–73 in Stavenga, D.G. & Hardie, R.C. (eds.) Facets of Vision. Springer, Berlin, Heidelberg.
Nixon, K.C & Carpenter, J.M. 1993. On outgroups. Cladistics 9: 413–426.
Noirot, C. & Quennedey, A. 1974. Fine structure of insect epidermal glands. Annual Review of Entomology 19: 61–80.
Novokschonov, V. & Novokschonova, E. 1997. *Okolpania favorabilis* n. sp. (Planipennia; Neuroptera: Permithonidae) aus dem unteren Perm des Urals. Paläontologische Zeitschrift 71: 89–90.
Oeser, R. 1961. Vergleichend-morphologische Untersuchungen über den Ovipositor der Hymenopteren. Mitteilungen aus dem Zoologischen Museum in Berlin 37: 3–119.

Ogden, T.H., Gattolliat, J.L., Sartori, M., Staniczek, A.H., Soldán, T. & Whiting, M.F. 2009. Towards a new paradigm in mayfly phylogeny (Ephemeroptera): combined analysis of morphological and molecular data. Systematic Entomology 34: 616–634.

Ogden, T.H. & Whiting, M.F. 2003. The problem with the Paleoptera problem: sense and sensitivity. Cladistics 19: 432–442.

Ogden, T.H. & Whiting, M.F. 2005. Phylogeny of Ephemeroptera (mayflies) based on molecular evidence. Molecular Phylogenetics and Evolution 37: 625–643.

Oosterbroek, P. & Courtney, G. 1995. Phylogeny of the nematocerous families of Diptera (Insecta). Zoological Journal of the Linnean Society 115: 267–311.

Ouvrard, D., Burckhardt, D., Azar, D. & Grimaldi, D. 2010 Non-jumping plant-lice in Cretaceous amber (Hemiptera: Sternorrhyncha: Psylloidea). Systematic Entomology 135: 172–180.

Pakaluk, J. & Ślipiński, S.A. (eds) 1995. Biology, Phylogeny, and Classification of Coleoptera: Papers Celebrating the 80th Birthday of Roy Crowson. Museum I Instytut Zoologii PAN, Warszawa.

Parrish, W.B. 1967. The origin, morphology, and innervation of aphid stylets (Homoptera). Annals of the Entomological Society of America 60: 273–276.

Pass, G. 2000. Accessory pulsatile organs: evolutionary innovations in insects. Annual Review of Entomology 45: 495–518.

Pass, G., Gereben-Krenn, B.A., Merl, M., Plant, J., Szucsich, N.U., & Tögel, M. 2006. Phylogenetic relationships of the orders of Hexapoda: contributions from the circulatory organs for a morphological data matrix. Arthropod Systematics & Phylogeny 64: 165–203.

Pass, G. & Szucsich, N.U. 2011. 100 years of research on the Protura: many secrets still retained. Soilorganisms 83: 309–334.

Paulus, H.F. 1979. Eye structure and the monophyly of the Arthropoda. Pp. 299–384 in Gupta, A.P. (ed.) Arthropod Phylogeny. Van Nostrand Reinhold Co., New York, London.

Paulus, H.F. 1986. Comparative morphology of the larval eyes of Neuropteroidea. Pp. 157–164 in Gepp J., Aspöck, H. & Hölzel, H. (eds.) Recent Research in Neuropterology. Proceedings of the 2nd Symposium on Neuropterology, Hamburg.

Pendergrast, J.G. 1962. The internal anatomy of the Peloridiidae (Homoptera: Coleorrhyncha). Transactions of the Royal Entomological Society of London 114: 49–65.

Pfau, H.K. 2002. Tandem grip mechanics and tandem linkage shifting in Odonata – reconstruction of evolution and phylogenetic significance. International Journal of Odonatology 5: 129–179.

de Pinna, C.C., 1991. Concepts and tests of homology in the cladistic paradigm. Cladistics 7: 367–394.

Poenicke, H.-W. 1969. Über die postlarvale Entwicklung von Flöhen (Insecta, Siphonaptera), unter besonderer Berücksichtigung der sogenannten "Flügelanlagen". Zeitschrift für Morphologie der Tiere 65: 143–186.

Pohl, H., 2000. Die Primärlarven der Fächerflügler – evolutive Trends (Insecta, Strepsiptera). Kaupia 10: 1–144.

Pohl, H. 2009. The oldest fossil strepsipteran larva (Insecta: Strepsiptera) from the Geisel Valley, Germany (Eocene). Insect Systematics & Evolution 40: 333–347.

Pohl, H. 2010. A scanning electron microscopy specimen holder for viewing different angles of a single specimen. Microscopy Research and Technique 73: 1073–1076.

Pohl, H. & Beutel, R.G. 2005. The phylogeny of Strepsiptera (Hexapoda). Cladistics 21: 1–47.

Pohl, H. & Beutel, R.G. 2008. The evolution of Strepsiptera. Zoology 111: 318–338.

Pohl, H., Beutel, R.G. & Kinzelbach, R. 2005. Protoxenidae fam. n. (Insecta, Strepsiptera) from Baltic amber – a 'missing link' in strepsipteran phylogeny. Zoologica Scripta 31: 123–134.

Pohl, H., Wipfler, B., Grimaldi, D., Beckmann, F. & Beutel, R.G. 2010. Reconstructing the anatomy of the 42 million-year-old fossil †*Mengea tertiara* (Insecta, Strepsiptera). Naturwissenschaften 97: 855–859.

Ponomarenko, A.G. 1969. The historical development of archostematan beetles. Trudy Paleontologicheskogo Instituta Akademiya Nauk SSSR 125: 1–240 (in Russian).

Ponomarenko, A. G. 1995. The geological history of beetles. Pp. 155–171 in Pakaluk, J. & Ślipiński, S.A. (eds) Biology, Phylogeny, and Classification of Coleoptera: Papers Celebrating the 80th Birthday of Roy Crowson. Museum I Instytut Zoologii PAN, Warszawa.

Popov, Y.A. & Shcherbakov, D.E. 1991. Mesozoic Peloridioidea and their ancestors (Insecta: Hemiptera, Coleorrhyncha). Geologica et Palaeontologica 30: 215–235.

Predel, R., Neupert, S., Huetteroth, W., Kahnt, J., Waidelich, D. & Roth, S. 2012. Peptidomics-based phylogeny and biogeography of Mantophasmatodea (Hexapoda). Systematic Biology 61: 609–629.

Prendini 2001. Supraspecific taxa as terminals in cladistic analysis. Groundplans versus exemplars revisited. Systematic Biology 50: 290–300.

Priesner, H. 1968. Thysanoptera (Physopoda, Blasenfüßer). Handbuch der Zoologie, IV, Arthropoda: Insecta. Inst. 5. Walter de Gruyter, Berlin.

Rähle, W. 1970. Untersuchungen an Kopf und Prothorax von *Embia ramburi* Rimsky-Korsakow 1906 (Embioptera, Embiidae). Zoologische Jahrbücher, Abteilung für Anatomie und Ontogenie der Tiere 87: 248–330.

Rasnitsyn, A.P. 1999. Taxonomy and morphology of *Dasyleptus* Brogniart, 1885, with description of a new species (Insecta: Machilidae: Dasyleptidae). Russian Entomological Journal 8: 145–154.

Rasnitsyn, A.P. 2003. On the skimming hypothesis of the origin of insect flight. Acta Zoological Cracoviensia 46 (Suppl.): 85–88.

Rasnitsyn, A.P., Quicke, D.L.J. (eds.) 2002. History of Insects. Kluwer Academic Publishers, Dordrecht.

Regier, J.C., Shultz, J.W., Zwick, A., Hussey, A., Ball, B., Wetzer, R., Martin, J.W. & Cunningham, C.W. 2010. Arthropod relationships revealed by phylogenomic analysis of nuclear protein-coding sequences. Nature 463: 1079–1083.

Rehn, A.C. 2003. Phylogenetic analysis of higher-level relationships of Odonata. Systematic Entomology 28: 181–240.

v. Reumont, B.M., Jenner, R.A., Wills, M. A.. Dell´Ampio, E, Pass, G., Ebersberger, I. Meusemann, K., Meyer, B., Koenemann, S., Iliffe, T. M., Stamatakis, A., Niehuis, O. Misof, B. 2012. Pancrustacean phylogeny in the light of new phylogenomic data: support for Remipedia as the possible sister group of Hexapoda. Molecular Biology and Evolution 29: 1031–1045.

Reynolds, E.S. 1963. The use of lead citrate at high pH as electronopaque stain in electron microscopy. Journal of Cell Biology 17: 208–212.

Richter, S. 2002. The Tetraconata concept: hexapod-crustacean relationships and the phylogeny of Crustacea. Organisms Diversity and Evolution 2: 217–237.

Richter, S. & Meier, R. 1994. The development of phylogenetic concepts in Hennig's early publications. Systematic Biology 43: 212–221.

Röber, H. 1942. Morphologie des Kopfes und des Vorderdarmes der Larve und Imago von *Sialis flavilatera*. Zoologische Jahrbücher für Anatomie 67: 61–118.

Rohlf, F.J. 1999. Shape statistics: Procrustes superimpositions and tangent spaces. Journal of Classification 16: 197–223.

Rohlf, F.J. & Bookstein, F.L. 2003. Computing the uniform component of shape variation. Systematic Biology 53: 66–69.

Rohlf, F.J. 2006. Virtual reconstruction: a primer in computer-assisted paleontology and biomedicine. Quarterly Review of Biology 81: 93.

Ross, E.S. 1966. The Embioptera of Europe and the Mediterranean region. Bulletin of the British Museum (Natural History), Entomology Series 17: 273–326.

Ross, E.S. 1970. Biosystematics of the Embioptera. Annual Review of Entomology 15: 157–172.

Ross, E.S. 2000. Embia: Contributions to the biosystematics of the insect order Embiidina. Part 1: Origin, relationships and integumental anatomy of the insect order Embiidina. Part 2: A review of the biology of Embiidina. Occasional papers of the California Academy of Sciences 149: 1–53.

Ross, E.S. 2007. The Embiidina of Eastern Asia, Part I. Proceedings of the California Academy of Sciences 58: 575–600.

Roth, L. M. 1991. A new cave-dwelling cockroach from Western Australia (Blattaria: Nocticolidae). Records of the Western Australian Museum 15: 17–21.

Roth, L.M. & Alsop, D.W. 1978. Toxins in Blattaria. Pp. 465–484 in Bettini, S. (ed.) Arthropod Venoms. Springer-Verlag, New York.

Rüber, L. & Adams, D.C. 2001. Evolutionary convergence of body shape and trophic morphology in cichlids from Lake Tanganyika. Journal of Evolutionary Biology 14: 325–332.

Russell, L.K. 1979. A study of the armoured boreid *Caurinus dectes* (Mecoptera). Unpublished PhD thesis, Oregon State University.

Schaller, F. 1970. 1. Collembola (Springschwänze). Handbuch der Zoologie, IV, Arthropoda: Insecta. Inst. 12. Walter de Gruyter, Berlin.

Schawaroch, V. & Li, S.C. 2007. Testing mounting media to eliminate background noise in confocal microscope 3-D images of insect genitalia. Scanning 29: 177–184.

Schedl, W. 1991. Hymenoptera, Unterordnung Symphyta. Handbuch der Zoologie, IV, Arthropoda: Insecta. Inst. 31. Walter de Gruyter, Berlin,.

Schepotieff, A. 1909. Studien über niedere Insecten. I. *Protapteron indicum* n. g., n. sp. Zoologische Jahrbücher, Abteilung für Systematik 28: 121–138 + 5 pls.

Schmidt, F. 1982. Revision des trichoptères Canadiens II. Les Glossosomatidae et Philopotamidae (Annulipalpia). Mémoires de la Societe Entomologique du Canada 122: 1–76.

Schmidt, F. 1983. Revision des trichoptères Canadiens III. Les Hyalopsychidae, Psychomyiidae, Goeridae, Brachycentridae, Sericostomatidae, Helicopsychidae, Beraeidae, Odontoceridae, Calamoceratidae et Molannidae. Mémoires de la Societe Entomologique du Canada 125: 1–109.

Schneeberg, K., Friedrich, F., Courtney, G.W., Wipfler, B. & Beutel, R.G. 2012. The larvae of Nymphomyiidae (Diptera, Insecta) – Ancestral and highly derived? Arthropod Structure & Development 41: 293–301.

Schneider K. & Klass K.-D. 2013. The female genitalic region in Eudermaptera (Insecta: Dermaptera). Zoologischer Anzeiger 252: 183–203.

Schuh, R.T. 1979. Evolutionary trends of Heteroptera. Part II. Mouthpart-structures and feeding strategies. Systematic Zoology 28: 653–656.

Schuh, R.T. & Slater, J.A. 1995. True Bugs of the World (Hemiptera: Heteroptera) – Classification and Natural History. Cornell University Press, New York.

Schuh, R.T. & Brower, A.V.Z. 2009. Biological Systematics: Principles and Applications. Cornell University Press, Ithaca, NY.

Schuh, R.T. Weirauch, C. & Wheeler, W.C. 2009. Phylogenetic relationships within the Cimicomorpha (Hemiptera: Heteroptera): a total-evidence analysis. Systematic Entomology 34: 15–48.

Seeger, W. 1979. Spezialmerkmale an Eihüllen und Embryonen von Psocoptera im Vergleich zu anderen Paraneoptera (Insecta): Psocoptera als monophyletische Gruppe. Stuttgarter Beiträge zur Naturkunde (A) 329: 1–57.

Seguy, E. 1951. Ordre des Siphonaptères. Pp. 745–769 in Grassé, P.P. (ed.) Traité de Zoologie. Insectes. Tome X, Fasc. 1. Nevropteroides, Mecopteroides, Hymenopteroides (Symphytes, Apocrites Terebranta). Mason et Cie., Paris.

Seifert, G. 1995. Entomologisches Praktikum. 3. Aufl. Thieme Verlag, Stuttgart, New York.

Simon, S., Strauss, S., von Haeseler, A. & Hadrys, H. 2009. A phylogenomic approach to resolve the basal pterygote divergence. Molecular Biology and Evolution 26: 2719–2730.

Slice, D.E. 2001. Landmark coordinates aligned by Procrustes analysis do not lie in Kendall's shape space. Systematic Biology 50: 141–149.

Smith, D.R. & Middlekauff, W.W., 1987. Suborder Symphyta. In: Stehr F. W. (Ed.), Immature Insects, Vol. I. Kendall/Hunt Publishing Company, Dubuque, Iowa, pp. 618–648.

Sneath, P.H.A. & Sokal, R.R. 1973. Numerical Taxonomy: The Principles and Practice of Numerical Classification. Freeman, San Francisco.

Snodgrass, R.E. 1935. The Principles of Insect Morphology. McGraw-Hill Book Company, New York, USA.

Snodgrass, R.E. (1946) The skeletal anatomy of fleas (Siphonaptera). Smithsonian Miscellaneous Collections 104: 1–89.

Soldán, T. 2003. Ephemeroptera phylogeny and higher classification: present status and conflicting hypotheses. Entomologische Anhandlungen 61: 125–126.

Song, N., Liang, A.P. & Bu, C.P. 2012. A molecular phylogeny of Hemiptera inferred from mitochondrial genome sequences. PloS one 7: e48778.

Spangenberg, R., Wipfler, B., Friedemann, K., Pohl, H., Weirauch, C., Hartung, V. & Beutel, R.G. (2013) The cephalic morphology of the Gondwanan key taxon *Hackeriella* (Coleorrhyncha, Hemiptera). Arthropod Structure and Development 42: 315–337.

Staniczek, A.H., 2000: The mandible of silverfish (Insecta: Zygentoma) and Mayflies (Ephemeroptera): its morphology and phylogenetic significance. Zoologischer Anzeiger 239: 147–178.

Staniczek, A.H. 2001. Der Larvenkopf von *Oniscigaster wakefieldi* McLachlan, 1873 (Insecta: Ephemeroptera: Oniscigastridae). Ein Beitrag zur vergleichenden Anatomie und Phylogenie der Eintagsfliegen. PhD thesis, Eberhard-Karls-Universität Tübingen.

Stehr, F. (ed.) 1987. Immature Insects. Vol. I. Kendall & Hunt, Dubuque, Iowa.

Stehr, F. (ed.) 1991. Immature Insects. Vol. II. Kendall & Hunt, Dubuque, Iowa.

Stirling, J.W. & Woods, A.E. 2002. Transmission electron microscopy: diagnostic applications. Pp. 701–728 in Bancroft, J.D. & Gamble, M. (eds) Theory and Practice of Histological Techniques. 5th edition. Churchill Livingstone, Edinburgh, London, New York etc.

Storozhenko, S. Y. (1997) Classification of order Grylloblattida (Insecta), with description of new taxa. Far Eastern Entomologist 42: 1–20.

St. Quentin, D. & Beier, M. 1968. Odeonata (Libellen). Handbuch der Zoologie, IV, Arthropoda: Insecta. Inst. 3. Walter de Gruyter, Berlin.

zur Strassen, R. & Göllner-Scheiding, U. 2005. 21. Ordnung Thysanoptera (Physopoda), Fransenflügler, Thripse, „Blasenfüße". Pp. 331–342 in Dathe, H.H. (ed.) Lehrbuch der Speziellen Zoologie, begründet von A. Kaestner, 2. Aufl. Bd. I: Wirbellose Tiere. 5. Teil: Insecta. Spektrum, Gustav Fischer, Heidelberg, Berlin.

Strausfeld, N.J. 1976. Atlas of an Insect Brain. Springer, Heidelberg.

Strausfeld, N.J., Hansen, L., Li, Y., Gomez, R.S. & Ito, K. 1998. Evolution, discovery, and interpretations of arthropod mushroom bodies. Learning & Memory 5: 11–37.

Strausfeld, N.J. 2009. Brain organization and the origin of insects: an assessment. Proceedings of the Royal Society B 276: 1929–1937.

Strümpel, H. 1983. Homoptera, Pflanzensauger. Handbuch der Zoologie/Handbook of Zoology, IV, Arthropoda: Insecta. Inst. 28. Walter de Gruyter, Berlin.

Strümpel, H. 2005. Hemiptera (Rhynchota), Schnabelkerfe). Pp. 343–425 in Dathe, H.H. (ed.) Lehrbuch der Speziellen Zoologie, begründet von A. Kaestner, 2. Aufl. Bd. I: Wirbellose Tiere. 5. Teil: Insecta. Spektrum, Gustav Fischer, Heidelberg, Berlin.

Strümpel, H. 2010. Die Zikaden – Auchenorrhyncha. Die Neue Brehm-Bücherei Bd. 668. Pflanzensaugende Insekten Band 6. Westarp Wissenschaften-Verlagsgesellschaft mbH, Hohenwarsleben

Stüben, M. & Linsenmair, E. 2008. Advances in insect preparation: bleaching, clearing and relaxing ants (Hymenoptera: Formicidae). Myrmecological News 12: 15–21.

Sturm, H. & Machida, R. (eds.) 2001. Archaeognatha. Handbook of Zoology. Vol. IV Arthropoda: Insecta. Part 37. Walter de Gruyter, Berlin, New York.

Swofford, D.L., Olsen, G.J., Waddell, P.J., & Hillis, D.M. 1990. Phylogeny reconstruction. Molecular systematics 3: 407–514.

Swofford, D.L. 1998. PAUP*. Phylogenetic Analysis using Parsimony (* and other methods). Sinauer Associates, Sunderland.

Szumik, C., Edgerly, J.S., & Hayashi, C.Y. 2008. Phylogeny of embiopterans (Insecta). Cladistics 24: 993–1005.

Tafforeau, P., Boistel, R., Boller, E., Bravin, A., Brunet, M., Chaimanee, Y., Cloetens, P., Feist, M., Hoszowska, J., Jaeger, J.-J., Kay, R. F., Lazzari, V., Marivaux, L., Nel, A., Nemoz, C., Thibault, X., Vignaud, P. & Zabler, S. 2006. Applications of X-ray synchrotron microtomography for non-destructive 3D studies of paleontological specimens. Applied Physics Series A 83: 195–202.

Tan, J. & Ren, D. 2009. Mesozoic Archostematan Fauna from China. Science Press, Beijing.

Terry, M. D. & Whiting, M.F. 2005. Mantophasmatodea and phylogeny of the lower neopterous insects. Cladistics 21: 240–257.

Theischinger, G. 1991. Plecoptera. Pp. 311–319 in CSIRO (ed.) The Insects of Australia: a Textbook for Students and Research Workers. 2nd ed. Cornell University Press, Ithaca, N.Y.

Tröster, G. 1990: Der Kopf von *Hybophthirus notophallus* (Neumann) (Phthiraptera: Anoplura). Eine funktionsmorphologishce und konsequent-phylogenetische Analyse. Stuttgarter Beiträge zur Naturkunde (A) 442: 1–89.

Tjeder, B. 1966. Neuroptera-Planipennia. The Lace-wings of Southern Africa. 5. Family Chrysopidae. Pp. 228–534 in: Hanström, B., Brinck, P. & Rudebeck, G. (eds) South African Animal Life 12. Statens Naturvetenskapliga Forskningsråd, Stockholm.

Tuxen, S.L. 1931. Monographie der Proturen. I. Morphologie. Nebst Bemerkungen über Systematik und Ökologie. Zeitschrift für Morphologie und Ökologie der Tiere 22: 671–720.

Tuxen, S. 1970. The systematic position of entognathous Apterygotes. Anales de la Escuela Nacional de Ciencias Biologicas 17: 65–79.

Uvarov, B.P. 1966, 1977. Dynamics of Populations: Outbreaks and Plagues. Grasshoppers and Locusts: a Handbook of General Acridology. Volumes 1, 2. University Press, Cambridge.

Vilhelmsen, L. 2001. Phylogeny and classification of the extant basal lineages of the Hymenoptera (Insecta). Zoological Journal of the Linnean Society 131: 393–442.

Wägele, W. 2001. Grundlagen der der Phylogenetischen Systematik. Verlag Dr. Friedrich Pfeil, München.

Walker, E.M. 1919. On the male and immature state of Grylloblatta campodeiformis Walker. Canadian Entomologist 51: 131–139.

Walker, E.M. 1914. A new species of Orthoptera, forming a new genus and family. Canadian Entomologist 46: 93–99.

Walker, E.M. 1931. On the anatomy of *Grylloblatta campodeiformis* Walker 1. Exoskeleton and musculature of the head. Annals of the Entomological Society of America 24: 519–536.

Walker, E.M. 1933. On the anatomy of *Grylloblatta campodeiformis* Walker 2. Comparisons of head with those of other orthopteroid insects. Annals of the Entomological Society of America 26: 309–344.

Walker, E.M. 1938. On the anatomy of Grylloblatta campodeiformis Walker 3. Exoskeleton and musculature of the thorax. Annals of the Entomological Society of America 31: 588–640.

Wasserthal, L.T. 2003. Circulation and thermoregulation. Pp. 205–228 in Kristensen, N.P. (ed.) Lepidoptera, Moths and Butterflies. Vol. 2: Morphology, Physiology and Development. Handbook of Zoology. Vol. IV Arthropoda: Insecta. Part 36. Walter de Gruyter, Berlin, New York.

Watrous, L.E. & Wheeler, Q.D. 1981. The out-group comparison method of character analysis. Systematic Biology 30: 1–11.

Watson J.A.L. & Gay F. J. (1991) Isoptera (termites). Pp. 330–347 in Division of Entomology, CSIRO (ed.) The Insects of Australia (ed.). Melbourne University Press, Melbourne.

Weber, H. 1930. Biologie der Hemipteren. Eine Naturgeschichte der Schnabelkerfe. I–VII. Biologische Studienbücher XI. Julius Springer, Berlin.

Weber, H. 1931. Lebensweise und Umweltbeziehungen von Traleurodes vaporariorum (Westw.). Zeitschrift für Morphologie und Ökologie der Tiere 23: 575–753.

Weber, H. 1933. Lehrbuch der Entomologie. Gustav Fischer, Jena.

Weber, H. 1936. Copeagnatha, Flechtlinge. In Schulze, P. (ed.) Biologie der Tiere Deutschlands, Lief. 39, Teil 27.

Weber, H. 1938. Grundriß der Insektenkunde. Gustav Fischer, Jena.

Weber, N.A. 1949. Late summer invertebrates, mostly insect, of the Alascan Arctic Slope. Entomology News 5: 118–128.

Weesner, F.M. 1969. External anatomy. Pp. 161–192 in Krishna, K., Weesner, F.M. (eds.) Biology of termites. Academic Press, New York.

Weidner, H. 1972. Copeognatha (Psocodea). Handbuch der Zoologie, IV, Arthropoda: Insecta. Inst. 18. Walter de Gruyter, Berlin.

Weirauch, C. 2007. Hairy attachment structures in Reduviidae (Cimicomorpha, Heteroptera), with observations on the fossula spongiosa in some other Cimicomorpha. Zoologischer Anzeiger 246: 155–175.

Weirauch, C. & Cassis, G. 2009. Frena and druckknopf: a synopsis of two fore wing-to-body coupling mechanisms in Heteropterodea (Hemiptera). Insect Systematics & Evolution 40: 229–252.

Weirauch, C. & Schuh, R.T. 2011. Systematics and evolution of Heteroptera: 25 years of progress. Annual Review of Entomology 56: 487–510.

Wenk, P. 1953. Der Kopf von *Ctenocephalus canis* (Curt.) (Aphaniptera). Zoologische Jahrbücher (Abteilung für Anatomie und Ontogenie der Tiere) 73: 103–164.

Wheat, C.W. & Wahlberg, N. 2013. Phylogenomic insights into the Cambrian Explosion, the colonization of land and the evolution of flight in Arthropoda. Systematic Biology 62: 93–109.

Wheeler , W.C., Whiting, M.F., Carpenter, J.C. & Wheeler, Q.D. 2001. The phylogeny of the insect orders. Cladistics 12: 1–57.

Wheeler, W.C., Schuh, R.T. & Bang, R. 1993. Cladistic relationships among higher groups of Heteroptera: congruence between morphological and molecular data sets. Entomologica Scandinavica 24: 121–138.

Whiting, M.F. 2002. Mecoptera is paraphyletic: multiple genes and phylogeny of Mecoptera and Siphonaptera. Zoologica Scripta 31: 93–104.

Whiting, M.F., Whiting, A.S., Hastriter, M.W., & Dittmar, K. 2008. A molecular phylogeny of fleas (Insecta: Siphonaptera): origins and host associations. Cladistics, 24: 677–707.

Wiegmann, B.M., Trautwein, M.D., Kim, J.-W., Cassel, B.K., Bertone, M.A., Winterton, S.L., Yeates, D.K. 2009. Single-copy nuclear genes resolve the phylogeny of the holometabolous insects. BMC Biology 7.

Wiegmann, B.M., Trautwein, M.D., Winkler, I.S., Barr, N.B., Kim, J., Lambkin, C., Bertone, M.A., Cassel, B.K., Bayless, K.M., Heimberg, A.M., Wheeler, B.M., Peterson, K.J., Pape, T., Sinclair, B.J., Skevington, J.S., Blagoderov, V., Caravas, J., Kutty, S.N., Schmidt-Ott, U., Kampmeier, G.E., Thompson, F.C., Grimaldi, D.A., Beckenbach, A.T., Courtney, G.W., Friedrich, M., Meier,

R. & Yeates, D.K. 2011. Episodic radiations in the fly tree of life. Proceedings of the National Academy of Sciences, USA. 108: 5690–5695.

Wiggins, G.B. 1984. Trichoptera, some concepts and questions. Pp. 1–12 in Proceedings of the 4th International Symposium on Trichoptera. Series Entomologica Vol. 30, Dr. W. Junk Publishers, The Hague.

Wiggins, G.B. 1987. Order Trichoptera. Pp. 253–287 in: Stehr F.W. (ed.) Immature Insects, Vol. I. Kendall/Hunt Publishing Company, Dubuque, Iowa.

Wigglesworth, V.B. 1973. Evolution of Insect Wings and Flight. Nature 246: 127–129.

Willkommen, J., Hörnschemeyer, T. 2007. The homology of wing base sclerites and flight muscles in Ephemeroptera and Neoptera and the morphology of the pterothorax of *Habroleptoides confusa* (Insecta: Ephemeroptera: Leptophlebiidae). Arthropod Structure & Development 36: 253–269.

Willmann, R. 1989. Evolution und phylogenetisches System der Mecoptera (Insecta: Holometabola). Abhandlungen der Senckenbergischen Naturforschenden Gesellschaft 544: 1–153.

Willmann, R. 2003. Die phylogenetischen Beziehungen der Insecta: offene Fragen und Probleme. Verhandlungen Westdeutscher Entomologentag 2001: 1–64.

Willmann, R. 2005a. Phylogenese und Systemk der Insecta. Pp. 1–65 in Dathe, H.H. (ed.) Lehrbuch der Speziellen Zoologie, begründet von A. Kaestner, 2. Aufl. Bd. I: Wirbellose Tiere. 5. Teil: Insecta. Spektrum, Gustav Fischer, Heidelberg, Berlin.

Willmann, R. 2005b. 35. Ordnung Mecoptera, Schnabelfliegen. Pp. 746–755 in Dathe, H. (ed.) Lehrbuch der Speziellen Zoologie. Begründet von A. Kaestner. 2. Aufl. Bd. I: Wirbellose Tiere. 5. Teil: Insecta. Spektrum, Gustav Fischer, Heidelberg, Berlin.

Wipfler, B., Machida, R., Müller, B. & Beutel, R.G. 2011. On the head morphology of Grylloblattodea (Insecta) and the systematic position of the order, with a new nomenclature for the head muscles of Dicondylia. Systematic Entomology 36: 241–266.

Wipfler, B., Pohl, H. & Predel, R. 2012. Two new genera and two new species of Mantophasmatodea (Insecta, Polyneoptera) from Namibia. ZooKeys 166: 75–98.

Woods, A.E. & Stirling, J.W. 2002. Electron microscopy: the preparative techniques. Pp. 679–700 in Bancroft, J.D. & Gamble, M. (eds) Theory and Practice of Histological Techniques. 5th edition. Churchill Livingstone, Edinburgh, London, New York etc.

Woodworth, C.W. 1906. The wing veins of insects. University of California Publications, Technical Bulletins, Entomology 1: 1–152.

Wygodzinsky, P. 1961. On a surviving representatitive of the Lepidotrichidae (Thysanura). Annals of the Entomological Society of America. 54: 621–627.

Xué, L., Dallai, R., & Yin, W.Y. 1993. Tracheal ultrastructure of Protura (Insecta: Apterygota). International Journal of Insect Morphology and Embryology 22: 13–23.

Xylander, W.E.R. & Günther, K.K. 2005. 7. Ordnung, Odonata Libellen. Pp. 121–142 in Dathe, H.H. (ed.) Lehrbuch der Speziellen Zoologie, begründet von A. Kaestner, 2. Aufl. Bd. I: Wirbellose Tiere. 5. Teil: Insecta. Spektrum, Gustav Fischer, Heidelberg, Berlin.

Yang, D. & Liu, X.Y. 2010. Megaloptera. Fauna Sinica Insecta, Vol. 51. Science Press, Beijing.

Yeates, D.K. & Wiegmann, B.M. 2005. The Evolutionary Biology of Flies. Columbia University Press, New York.

Yeates, D.K., Wiegmann, B.M., Courtney, G.W., Meier, R., Lambkin, C. & Pape, T. 2007. Phylogeny and systematics of Diptera: two decades of progress and prospects. Zootaxa 1668: 565–590.

Yin, W.Y. 1965. Studies on Chinese Protura II. A new family of the suborder Eosentomoidea. Acta Entomologica Sinica 14: 186195.

Yoshizawa, K. 2011. Monophyletic Polyneoptera recovered by wing base structure. Systematic Entomology 36: 377–394.

Yoshizawa, K. & Johnson, K.P. 2003 Phylogenetic position of Phthiraptera (Insecta: Paraneoptera) and elevated rate of evolution in mitochondrial 12S and 16S rDNA. Molecular Phylogenetics and Evolution 29: 102–114.

Yoshizawa, K. & Johnson, K.P. 2005. Aligned 18S for Zoraptera (Insecta): phylogenetic postion and molecular evolution. Molecular Phylogeny and Evolution 37: 572–580.

Yoshizawa, K. & Johnson, K.P. 2006. Morphology of male genitalia in lice and their relatives and phylogenetic implications. Systematic Entomology 31: 350–361.

Yoshizawa, K. & Johnson, K.P. 2010. How stable is the "Polyphyly of Lice" hypothesis (Insecta: Psocodea)? A comparision of phylogenetic signal in multiple genes. Molecular Phylogeny and Evolution 55: 939–951.

Yoshizawa, K. & Saigusa,T. 2001. Phylogenetic analysis of paraneopteran orders (Insecta: Neoptera) based on forewing base structure, with comments on monophyly of Auchenorrhyncha (Hemiptera). Systematic Entomology 26: 1–13.

Yoshizawa, K., Lienhard, C. & Johnson, K.P. 2006. Molecular systematics of the suborder Trogiomorpha (Insecta: Psocodea: "Psocoptera"). Zoological Journal of the Linnean Society 146: 287–99.

Yuasa, H. 1920. The anatomy of the head and mouth-parts of Orthoptera and Euplexoptera. Journal of Morphology 33: 251–307.

Zankel A., Kraus, B., Poelt, P., Schaffer, M. & Ingolic, E. 2009. Ultramicrotomy in the ESEM, a versatile method for materials and life sciences. Journal of Microscopy 233: 140–148.

Zhang, Y.-Y., Xuan, W.-J., Zhao, J.-L., Zhu, C.-D. & Jiang, G.-F. 2010. The complete mitochondrial genome of the cockroach *Eupolyphaga sinensis* (Blattaria: Polyphagidae) and the phylogenetic relationships within the Dictyoptera. Molecular Biology Reports 37: 3509–3516.

Zhao, J., Zhao, Y., Shih, C., Ren, D. & Wang, Y. 2010. Transitional fossil earwigs – a missing link in Dermaptera evolution. BMC Evolutionary Biology 10: 344.

Ziegler, J. 1998. Die Morphologie der Puparien und der larvalen Cephalopharyngealskelette der Raupenfliegen (Diptera: Tachinidae) und ihre phylogenetische Bewertung. Studia Dipterologica 3 (Suppl.): 1–244.

Ziegler, J. 2005. 36. Ordnung Diptera, Zweiflügler (Fliegen und Mücken). Pp. 756–860 in Dathe, H. (ed.) Lehrbuch der Speziellen Zoologie. Begründet von A. Kaestner. 2. Aufl. Bd. I: Wirbellose Tiere. 5. Teil: Insecta. Spektrum, Gustav Fischer, Heidelberg, Berlin.

Zrzavý, J. 1990. Evolution of Antennal Sclerites in Heteroptera (Insecta). Acta Universitatis Carolinae – Biologica 34: 189–227.

Zucker, RM. 2006. Whole insect and mammalian embryo imaging with confocal Microscopy: Morphology and apoptosis. Cytometry 69A: 1143–1152.

Zwick, P. 1980. Plecoptera (Steinfliegen). Handbuch der Zoologie, IV, Arthropoda: Insecta. Inst. 26. Walter de Gruyter, Berlin.

Zwick, P. 2000. Phylogenetic system and zoogeography of the Plecoptera. Annual Review of Entomology 45: 709–746.

Zwick, P. 2005. 8. Ordnung Plecoptera, Steinfliegen, Frühlingsfliegen, Uferbolde. Pp. 144–154 in Dathe, H.H. (ed.) Lehrbuch der Speziellen Zoologie, begründet von A. Kaestner, 2. Aufl. Bd. I: Wirbellose Tiere. 5. Teil: Insecta. Spektrum, Gustav Fischer, Heidelberg, Berlin.

Taxonomic Index

R

Bei Fragen zur Produktsicherheit wenden Sie sich bitte an:
If you have any questions regarding product safety,
please contact:

Walter de Gruyter GmbH
Genthiner Straße 13
10785 Berlin
productsafety@degruyterbrill.com